David M. Egglestew

# VOLTS TO HERTZ . . . the rise of electricity

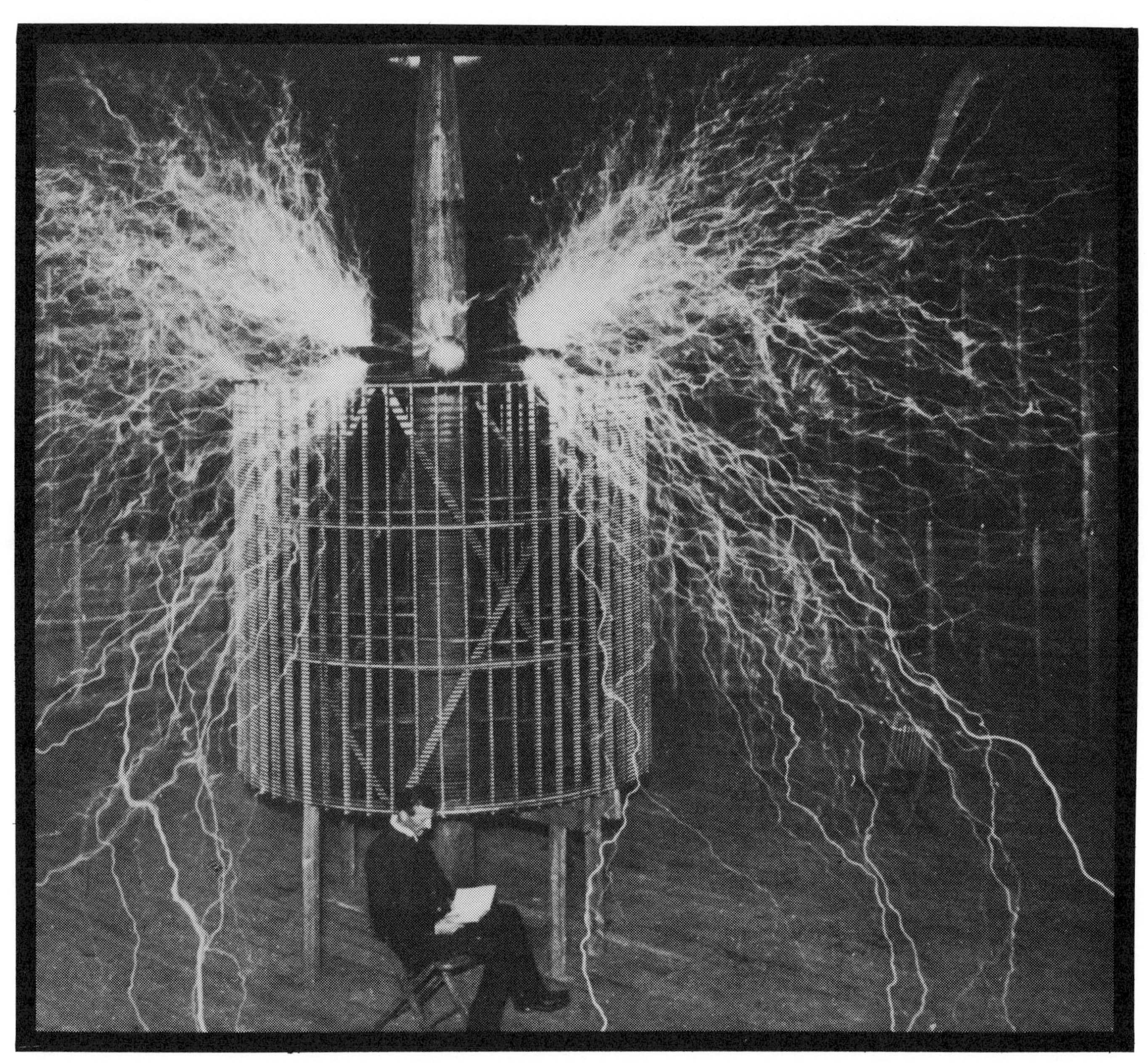

THERE IS NO SUBJECT MORE CAPTIVATING, MORE WORTH OF STUDY, THAN NATURE. TO UNDERSTAND THIS GREAT MECHANISM, TO DISCOVER THE FORCES WHICH ARE ACTIVE, AND THE LAWS WHICH GOVERN THEM, IS THE HIGHEST AIM FOR THE INTELLECT OF MAN - Tesla

# VOLTS TO HERTZ . . .
# the rise of electricity

FROM THE COMPASS TO THE RADIO
through the works of
SIXTEEN GREAT MEN OF SCIENCE
whose names are used
in Measuring Electricity and Magnetism

Sanford P. Bordeau
SENIOR MEMBER, IEEE

ISBN 0-8087-4908-0
Printed in the United States of America.

Burgess Publishing Company
7108 Ohms Lane
Minneapolis, Minnesota 55435



J I H G F E D C B A

Reproduced by photo-offset directly from the author's camera-ready manuscript.

**Bordeau, Sanford P.**
Volts to Hertz—the rise of electricity.

Bibliography: p.
Includes index.
1. Electricity—History. 2. Magnetism—History.
3. Physicists—Biography. I. Title.
QC507.B73 1982 537'.09 82-17702
ISBN 0-8087-4908-0

**Additional copies of this book may be ordered from the publisher.**

# FOREWORD

THE WONDERS OF ELECTRICITY never cease . . . .

. . . in nature it makes the flash of lightning and the flickering beauty of the Auroral Lights. In the hands of man electricity is harnessed to light and power the civilized world . . . it is a willing worker in the home, the office and factory . . . it speeds information and weaves a web of communication around the globe . . . it can easily bounce a signal off the moon or leap the vastness of outer space. Electricity is our most versatile servant. What it will do tomorrow we can only guess . . . the future seems limitless.

The evolution of electricity as a created energy is one of the greatest feats in the history of science - and it has come to pass in a relatively short span of time. The science of electricity and magnetism began to be lifted out of the mists of antiquity about three and a half centuries ago. In the last two centuries it has been propelled with increasing speed into the reality serving mankind today. The history of this advance is one of the best documented and interesting of the sciences, and consists in large part in the stories of the gifted men who were the founding fathers.

This presentation is an episodic time-strip highlighting the rise of electricity and magnetism through the works of sixteen pioneers upon whose names posterity has put a special stamp. These sixteen men put together the formative pieces that have led to the modern world of electromagnetic technology. They were privileged to produce a succession of epoch-making discoveries that determined the course and destiny of the electrical science.

In the early days of the work of these men in electricity and magnetism, there was no formal system of units and names for the quantities they measured. The determinations they made were expressed in arbitrary figures dependent on the methods and apparatus employed; thus their results might not be generally usable by others. With the advance of electrical technology this situation became intolerable, and action for standardization was taken up by leaders in the electrical community. In the late 19th century, in an international effort unparalleled in other branches of science, a common system of units and names was proposed and promulgated. This has finally given us the electrical and magnetic terminology in universal use today.

By necessity, the numerous units of measurement that resulted from the continuous advance in the electrical science, required new, simple terms for identification. In a novel and inventive use of men's names, a grateful scientific posterity has given the founding discoverers a form of immortality. Their names were chosen for fundamental units of electricity and magnetism. Thus the fame and names of sixteen pioneers lives after them, by common assent and usage, in the literature and language of their discoveries. No other science can show such a succession of luminous names.

And by a kind of subliming chronology, as shown by the following chart, the span of lives of these sixteen men, from William Gilbert in 17th century England, to Nikola Tesla, who is still remembered personally by some in our century, gives an overlapping sequence of the mainstream of electrical history up into the 20th century. The chapters in this book follow that chronology in terms of the names of electrical and magnetic units, and the lives and works of the men behind them.

The founding fathers of electrical science reviewed herein were lone workers operating in perhaps less complicated and less specialized times. They proceeded with broad strokes in fields that nowadays have been taken over by corporate effort and by collective specialization. The electrical founders had a high degree of imagination and a passion for exact knowledge. Some had "luck" in their findings, and this has sometimes been called providential. It was more likely the result of prepared and inspired minds.

Some of the early scientists had handicaps of poverty and lack of recognition, but all participated in the labor and love of seeking the secrets of nature and finding new truths. Exploring the darkness of the unknown they found luminous reality. Each, building on the work of their predecessors and the experiments of others, added the essential steps in the movement upward from the mysterious attractions of the amber and the lodestone of antiquity to the all-pervading electrotechnology that lights, powers and communicates the world today. It is not possible to present fully the elemental complex of their superior intellects that led to discovery, but the documented stories of their lives and accomplishments can be told for what inspiration it will be to all who are interested in electricity and magnetism.

References at the end of this volume indicate sources from which much background for these histories has been derived. Also I have used thankfully the resources of the Burndy Library, the James Jerome Hill Reference Library, the Smithsonian Institution, the Royal Institution, the Science Museum, the Nikola Tesla Museum, the Deutsches Museum, the Institution of Electrical Engineers, the Institute of Electrical and Electronics Engineers, the National Bureau of Standards and the International Electrotechnical Commission, for facts and illustrations.

I am thankful to Dr. Bern Dibner of the Burndy Library for the encouragement given me in the early stages of this publication, and also for the comments of Professors John H. Kuhlmann, Paul A. Cartwright, E. Bruce Lee and Vernon D. Albertson of the University of Minnesota, and of Dr. Richard B. Whittington of the University of Liverpool.

Honolulu, Hawaii

S.P.Bordeau

## THE MEN BEHIND THE NAMES OF THE ELECTRIC AND MAGNETIC UNITS

1750 1800 1850 1900

|  |  | ELECTRICAL UNIT DESIGNATION |
|---|---|---|
| GILBERT | 1544–1603 | Magnetomotive Force* |
| COULOMB | 1736–1806 | Electric Charge |
| WATT | 1738–1819 | Power |
| VOLTA | 1745–1827 | Electric Potential |
| OERSTED | 1777–1851 | Magnetic Field Intensity* |
| AMPÈRE | 1775–1836 | Current |
| OHM | 1788–1854 | Resistance |
| FARADAY | 1791–1867 | Capacitance |
| HENRY | 1797–1878 | Inductance |
| GAUSS | 1777–1855 | Magnetic Field Strength* |
| WEBER | 1804–1891 | Magnetic Flux |
| MAXWELL | 1831–1879 | Magnetic Flux* |
| SIEMENS | 1816–1892 | Conductance |
| HERTZ | 1857–1894 | Frequency |
| TESLA | 1856–1943 | Magnetic Flux Density |

*CGS System

# CONTENTS

# VOLTS TO HERTZ . . . the rise of electricity

# GILBERT... the earth is a giant magnet!

## a founding father of electrical and magnetic science, William Gilbert pioneered the experimental method of inquiry

**THE ROYAL PHYSICIAN,** William Gilbert, was in attendance to the Queen, as recorded in the scene below - but not because of any indisposition on the part of Her Majesty. Elizabeth I was in good health and at the peak of her rule. She was holding an unusual royal court. Dr. Gilbert, who was not only a practicing physician, but also a keen experimenter, was demonstrating to her the strange powers of electrical and magnetic attractions. The Queen's interest in science was in response to the forward thrust of the times and the broader outlook for her nation. England had just conquered Spain's "Invincible Armada," and had become the world's greatest naval power. English ships were now circling the globe, and navigation was open for commerce and for acquiring colonies.

Good navigation was a prime necessity for ocean supremacy. In this respect Her Majesty's Navy was making a call on science to improve nautical skills. This involved the ability to chart accurately the location and course of a ship. Dr. Gilbert had just published his monumental treatise *De Magnete,* the earliest work on magnetism. By treating the earth for the first time as a huge magnet, Gilbert had given a rational basis for the response of the compass and the dip needle to the earth's north-south magnetic poles, an understanding of the location and variation of which was essential in order to improve their use for navigation.

William Gilbert's *De Magnete* was not only a record of his investigations, but also a first giant stride in establishing a scientific methodology - and

FOR THE QUEEN'S NAVY....

WILLIAM GILBERT shows the strange attractions of electrics and magnetics to Elizabeth I at his London residence, Wingfield House. With the Queen in the background are Sir Walter Raleigh and Sir Francis Drake, two of England's greatest explorers and navigators. They were vitally interested in Gilbert's work with magnetism and the compass to improve reliability and speed of England's ships which were circling the world for goods and treasure. From a painting in the Town Hall of Colchester, the birthplace of Gilbert, and where he did some of his experiments. Gilbert is a modification of Gilberd, the original name of the family.

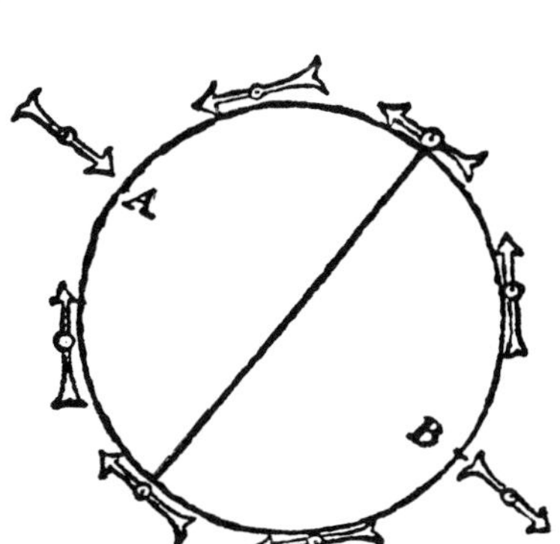


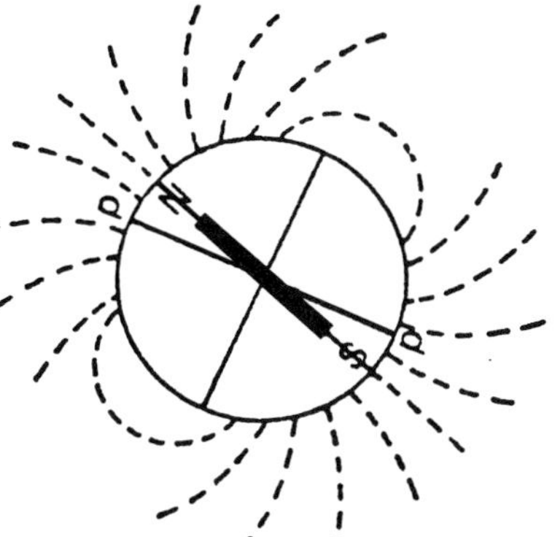

in demonstrating the use of experiment to separate fact from fancy. Motivated by practical needs of navigation, he took magnetism out of the speculation and mysticism of antiquity and placed it on the firm footing of verifiable factual knowledge. He blazed the path of experimental electrical science along which later footsteps were to lead to the world of electrical technology, as outlined in these historical sketches.

**THE RISE OF MODERN SCIENCE** must rank as one of the greatest events in the history of man. It gave birth to the technology we see all about us today, that has so revolutionized our civilization. Modern science has relieved human drudgery and enhanced standards of living around the world. How did it evolve? Modern science came about in large part by the use of inspired imagination, observation and experiment to discover new truths about nature for the enlightenment of humanity.

The new method arose in Europe four and a half centuries ago when inquiring and gifted minds started to make the sciences independent branches of learning and began to adopt a rational approach in examining the structure and natural forces of the world. Its beginnings were also a revolt against the scientific dictates, often erroneous and confused, handed down in writings from antiquity. Up to about the year 1500, ancient knowledge as revealed by religious and philosophical scholars, was often accepted as final authority on matters of nature and the universe.

Modern science started by breaking with the traditional lore in favor of experimental scientific inquiry designed to produce truths that could be exploited for the benefit of mankind. As Francis Bacon (1561-1626), the great English essayist champion of a rebirth in scientific thinking, wrote of the future: *"The sure and lawful goal of science is none other than this: that human life be endowed with new discoveries and power."*

The beginnings of the scientific method and the bold revolt against ancient thinking are dramatized by two names of the sixteenth century. They shine like beacons in signaling the dawn of a new age. One is Galileo Galilei (1564-1642) the Italian whose scientific pioneering was mainly in mechanics and astronomy, and who was the creator of the science of dynamics; the other name is that of William Gilbert (1544-1603) an English physician, a founder of magnetic science and the subject of this historical outline.

**GALILEO and GILBERT** were men of the Renaissance, the era of the great European rebirth in art and literature, and a new enthusiasm for all forms of learning. The Renaissance began with a great revival of interest in the wisdom of the ancient world. The vast writings of the Near East and Greece, first translated into Arabic, were transmitted westward through the spread of the Islamic religion and culture, and by the Crusades. Translated into Latin by monastic scholars, these writings provided Europeans with a stimulating new field of knowledge based on the achievements of classical antiquity.

**THE RENAISSANCE** was a period of increasing mercantile wealth, of the invention of the printing press, of venturous navigation to discover the outward world, and of new and wider investigations seeking the workings of nature. Universities were promoted for the propagation of knowledge. The Renaissance shed new light on the dignity of man and awakened him to the wonders of this world, which in the Middle Ages, had been regarded merely as a brief and troubled preparation for the next. And it produced minds like those of Galileo and Gilbert who pioneered science to reveal the secrets of the natural world.

Before the dawn of the Renaissance, scientific investigation proceeded haltingly. Scholastics derived their science mainly from the works of the ancient authors, and there was little use of their own senses in testing the information. Foremost among the ancient authorities was Aristotle, the master mind of ancient Greece, whose contributions to all forms of learning, handed down through two thousand years, were highly influential and had been accepted by scholastics and by the Church almost as Holy Writ.

To the old question: "which falls faster from a given height, a light body or a heavy body," the 16th century student, lacking recourse to an objective experiment, would likely seek an answer in "Let's see what is said by ancient authorities." Consulting Aristotle's "Physics," the student would find among other statements on falling bodies: "that which is heavier than another in equal bulk moves downward the quicker."

Aristotle's physics had been accepted as adequate for many centuries, but in the course of time many of his teachings based on philosophical speculation and without basis of fact, were questioned, challenged and found incorrect. Change was at hand!

One of those who rebelled against ancient authority was Galileo. Among his many investigations he took up the study of motion, including that of falling bodies. He pioneered the methodology of experimental science by devising appropriate tests under controlled conditions and with careful measurement. In 1590, in memorable experiments made at the University of Pisa, by rolling balls down inclined planes, and timing with the swing of pendulums, he removed any doubt that, falling freely, all bodies accelerated at the same rate.

The legacy from ancient writings is great. The Greeks laid the foundations of philosophy and ethics, and established the canons of Western beauty. But in the physical sciences, which are the basis for technology, many of their speculations were flawed by lack of purposeful experimental support. One early shortcoming of Greek science was an almost total reduction of mechanics to a servile level, and a feeling that manual effort was unsuited to scientific investigation.

An introduction to objective experiment was made by the Islamic Arabic scholars, an improvement over the mysticism and speculation of the Greeks. But in observing natural phenomena and accumulating facts, the Arabs lacked elaboration of a system of projecting hypotheses for obtaining the truly scientific conclusions.

The rise of Christianity and its struggles with the ancient religions tended to hamper and pervert the forward movement from Arabic scholarship. Revolutionary teachings that implied doubt of ecclesiastical authority might be branded as heresy. Experimental science was sometimes held suspect as leading to atheism. Galileo espoused the sun-centered planetary theory of Copernicus, which was universally acknowledged later. But in his time this teaching was seen as a menace to faith, and brought an interdict from the Inquisition that threatened Galileo's life. Science that upset tradition and religious acceptance was held dangerous.

**THE PRIME EFFORT** of the new scientific inquiry was a creative daring to establish unquestioned fact leading to unerring sequence, a pursuit of the unexpected, and the imagination and will to propose new doctrine. Observations using the tools of experimental demonstration were to provide the needed data. These data might, by inductive reasoning then yield hypotheses that could lead to laws revealing the structure and the forces of nature, laws that could stand the tests of repeated verification.

The importance of Galileo and Gilbert lies partly in their discoveries and writings. But of more significance for posterity was their priority in advocating controlled experiment as the proper procedure of scientific investigation. The unknown was to be accounted for by the known. Experience, measurement and analysis, rather than the uncritical acceptance of existing scholastic tradition was to be the method. Both men laid ground rules for new and pragmatic procedures that initiated a revolution in scientific thinking. Both were exemplars of the shift from the Philosophical to the Scientific Age. Both braved anathema in defying ancient dogma.

**NEW SCIENTIFIC BRAVERY**, such as that of Galileo, in the more favorable atmosphere of the Renaissance, began the movement to sweep away obstacles to scientific progress. This daring was found in those outstanding individuals to whom the natural delight in study and investigation of nature outweighed the penalty of disfavor or persecution. And this new daring in science tended to speed up in those countries, such as England, where religious disfavor did not stifle initiative.

One of those who by word and deed was an outspoken champion of science by experiment and who countered vendors of pseudo-science with the logic of demonstrated proof, was William Gilbert. He was a product of an era in English history - the reign of Elizabeth I - when the frontiers of the world and society were expanding as never before. In the field of physical science he took up for new inquiry the strange attractive powers of amber and lodestone that had puzzled men throughout the ages. He established experimental facts and gave nomenclature to what we now know as static electricity and magnetism. All is contained in his great book *De Magnete.*

In prefacing the book Gilbert expressed an atmosphere of the era in whch he worked for experimental science. Explaining a delay in publishing his book, he showers contempt on the dogma of the scholars:

. . . "why should I, in so vast an ocean of books whereby the minds of the studious are befuddled and vexed; of books of the more stupid sort whereby the common herd and fellows without a spark of talent are made intoxicate, crazy, puffed up, are led to write numerous books and to profess themselves philosopher, physicians, mathematicians, and astrologers, the while ignoring and condeming men of learning; why, I say, should I submit this noble and inadmissible philosophy to the judgement of men who have taken the oath to follow the opinion of others . . ."

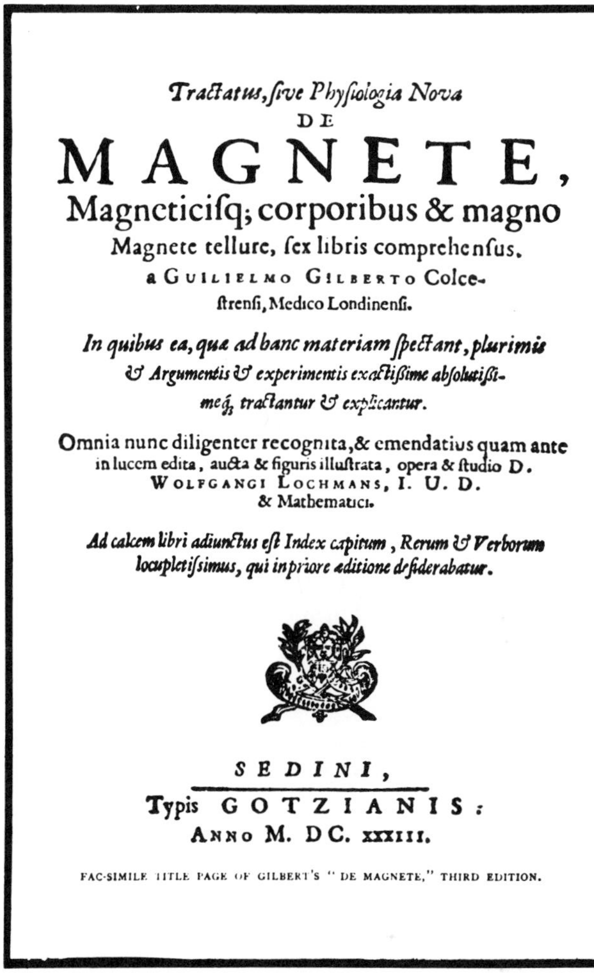

And Gilbert appeals to the open mind:

. . . "To you alone, true philosophers, and ingenuous minds, who not only in books but in things themselves look for knowledge, have I dedicated these foundations of magnetic science - a new style of philosophising."

*Tractatus, sive Physiologia Nova*
DE
MAGNETE,
Magneticisq; corporibus & magno
Magnete tellure, sex libris comprehensus.
a GUILIELMO GILBERTO Colce-
strensi, Medico Londinensi.

*In quibus ea, quæ ad hanc materiam spectant, plurimis & Argumentis & experimentis exactissime absolutissimeq; tractantur & explicantur.*

Omnia nunc diligenter recognita, & emendatius quam ante in lucem edita, aucta & figuris illustrata, opera & studio D. WOLFGANGI LOCHMANS, I. U. D. & Mathematici.

*Ad calcem libri adiunctus est Index capitum, Rerum & Verborum locupletissimus, qui in priore editione desiderabatur.*

*SEDINI,*
Typis GOTZIANIS:
ANNO M. DC. XXXIII.

FAC-SIMILE TITLE PAGE OF GILBERT'S "DE MAGNETE," THIRD EDITION.

Fig. 1.1- GILBERT'S BOOK WAS PUBLISHED IN LATIN, the universal scientific language of the time. Title page from 1633 German edition. From a translation of De Magnete by P. Fleury Mottelay, Dover Publications Inc., N.Y.

ALLEGORICAL REPRESENTATION . . .

Fig. 1.2 - Early science was often presented in an aura of mysticism, and it relied on ancient symbolizing and authority where there was lack of exact knowledge. Gilbert proposed "a new style of philosophizing" in which the truth was arrived at by the procedure of experiment to find the facts. Illustration courtesy the Burndy Library

**WILLIAM GILBERT** was born in 1544 at Colchester, about 50 miles northeast of London, the eldest of a family of eleven. His father, Jerome Gilbert, rose to be a recorder of the city, and was held in high professional esteem. After grammar school, William, in 1558, entered St. John's College Cambridge, to study mathematics, and in 1564 he became an examiner in that subject for the Royal College of Surgeons. His interest then turned to medicine and he graduated as a doctor in 1569.

As was a custom of the educated English of his day having the means, Gilbert visited the continent spending time in Italy which was considered to have the best medical schools. Back in London he went into medical practice and after some years established a reputation as an expert on maritime diseases. He was elected to a Fellow of the Royal College of Physicians. Medical men, then as now, were specially regarded and Dr. Gilbert's skills soon caught the eye of the reigning monarch, Elizabeth, who appointed him her "physician in ordianry."

Endowed with a restless and inquiring mind, Gilbert soon ranged beyond private practice into the field of scientific investigation. He centered first on chemistry, in which he showed an exactness of experimental method.

Before long, Gilbert's attention turned to the strange phenomena of electric and magnetic attraction. These mysterious forces, commented on by the Greek philosophers Thales and Theophrastus two thousand years ago, had excited wonder through the ages.

To carry on his Royal practice and private investigations, Gilbert established his residence near St. Paul's Cathedral. Here he worked out the famous experiments that were to occupy the rest of his life and make him one of the founders of the electric and magnetic sciences. His quarters became a meeting place for some of the active philosophers of the day, an assemblage that was instrumental in the promotion of science as a fruitful adventure, one that should be recognized as an independent branch of learning, and worthy of being pursued with vigor.

**GILBERT'S REPUTATION** grew, both as a physician and experimenter, resulting in his selection in turn as Censor, Treasurer, Councilarious and finally as President of the Royal College of Physicians. Capping his career was appointment in 1600 as Chief Physician of the Court of Queen Elizabeth. His service there was of short duration, however, as the Queen died in March 1603, and Gilbert's death followed in November of the same year. He was buried in the city of his birth, Colchester.

The year 1600 of his appointment to the Court also saw the publication of Gilbert's great work

*De Magnete,* with the full title: "On the Magnet, Magnetick Bodies also, and on the great magnet the Earth; a new Philosophy demonstrated by many arguments and experiments." The book was published in Latin and was reissued in several editions in the same language in the early sixteen hundreds. *De Magnete* was translated into English in 1893 by P. Fleury Mottelay. It is from this work that the quoted excerpts in this chapter are taken.

*De Magnete* brought to fruition Gilbert's years of patient and exhausting inquiry to create a great work of experimental investigation. It was the first treatise to establish electricity and magnetism as sciences. The book created a profound impression. It inspired men of learning everywhere, and earned the admiration of Galileo. Ever since, *De Magnete* has generally been considered a watershed in the flow of science from the ancient to the modern.

*De Magnete* is also a fascinating historical study on the difficulties and deviations that beset the transition from one thought process to another. Gilbert, on the one hand, forcefully sheds the magic and mysticism of the ancients with his rational procedure of experimental demonstration. On the other hand his explanations on the unknown, when he strays from his field, often revert to an occult speculation that borders on natural magic. It remained for the experimenters in the centuries after Gilbert to drop the appeal to a spiritual motivation of the natural forces, and then finally to codify science by means of the rigors of mathematics.

*De Magnete* has three general divisions. The first covers his experimental work on lodestone and the properties of magnetism. This is Gilbert's most valuable and memorable contribution to science. The second part covers navigation, and reveals Gilbert's association with mathematicians and the instrument makers. Solving the problems of navigation was one of Gilbert's great motivations for his book, although some of his conclusions proved erroneous and impractical. The last division of *De Magnete* deals with astronomy. Gilbert was a strong supporter of the new Copernican heliocentric planetary system. His arguments on its merits were a stimulus to succeeding observers who then carried on the establish the framework of the modern concept of the solar system.

Fig. 1.3- GILBERT'S BIRTHPLACE, Colchester, England. Part of the house remains today. Gilbert died in London in the great plague of 1603, and is buried in Colchester.

**THE LODESTONE** - The earliest Greek writings referred to certain stones that had the property of attracting iron. A massive body of this stone near the ancient city of Magnesia in Asia Minor was reported to pull at the iron tips of sheperd's staffs and at the nails in their boots. From this location the terms "magnet" and "magnetism" were derived. The name "magnetite" has now been given to the mineral oxide of iron that exhibits natural magnetism. It was early found that the magnetism of the stone could be transferred by contact to a steel needle, to the ends of which was imparted a polarity difference, or directional quality. From this came the word "lodestone," also spelled "loadstone," for the stone which could give direction to a needle.

The strange force of the lodestone had long puzzled man. It had also served him through the ages because of the discovery that a rod of lodestone, free to turn, would orient itself in a north-south direction. This led to invention of the "compass." which used a pivoted steel needle that had been magnetized by contact with lodestone. Invention of the compass, and its first use for navigation, is

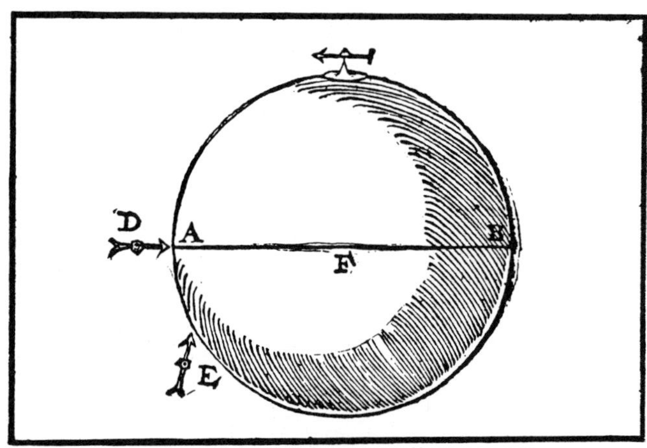


GILBERT USED A LODESTONE "LITTLE EARTH" . . .

Fig. 1.4- "Terrella's" magnetic field simulated the field of the earth so a compass needle (D) stood vertically at the north-south poles (A-B), and horizontally at the equator.

credited to the Chinese. The compass, in addition to being a direction finder, was the first scientific electrical indicating instrument, and it was to play a critical part in future magnetic discoveries.

**USE OF THE COMPASS** for navigation was old by Gilbert's time. It is possible that the Vikings, in their voyaging, used its guidance. Columbus kept compass records, and had noted that the compass direction differed from the true north, a deviation already known to navigators, and called the "declination."

London was a center of navigation, and of instrument making. Robert Norman, a manufacturer of compasses, found, in 1756, that a compass needle pivoted vertically, dipped downward at an angle of about 70 degrees. He suggested this was due to a magnetic body inside the earth. Gilbert was familiar with instrument makers such as these, and with navigators in the City. He was also well versed in the historical background of this subject.

Gilbert opens *De Magnete* with a review of ancient opinions and delusions on the lodestone. He cites legends of amazing powers . . ."that in the Indian seas have certain sharp-pointed rocks that abound in lodestone which will draw every nail out of ships that land alongside them . . ." and stories of travelers having nails in their boots, being held fast when walking on lodestone mountains. On these Gilbert remarks: "Men are deplorably ignorant with respect to natural things, and modern philosophers, as though dreaming in the darkness, must be aroused and taught the uses of things, the dealing with things; they must be made to quit the sort of learning that comes only from books, and that rests only on vain arguments from probability and upon conjectures." With that Gilbert begins the account of his true method of getting facts.

**GILBERT'S INVESTIGATIONS** of lodestone are ingenious and exhaustive. He is not only content with observation, but frequently leaps into creative proposals. This is shown by his first great conclusion that the Earth itself is a giant magnet, with magnetic poles in the neighborhood of the geographic north and south poles. He was led to this by experiments with his favorite device, the "Terrella," or "little earth," a piece of lodestone shaped into a globe-like magnetic ball.

The Terrella was envisioned magnetically as a miniature world. It had north and south poles, a magnetic alignment along meridians and an equator which was not merely a mathematical circle but a natural demarcation between the poles. This he proved by tracing a pivoted needle around the Terrella, shown in Fig. 1.4. Orientation of the needle ranged from horizontal at the equator to vertical at the poles in conformity with the magnetic field. Gilbert projected this to be true of the earth. When free to move, the Terrella itself also revolved into alignment with the poles of the earth. The Terrella was one of the first uses of a small-scale model to test an hypothesis.

Gilbert proved by many experiments that "magnetic bodies are governed and regulated by the earth . . ." Consequently he reasoned that lodestone was an iron ore directionally magnetized when it lay along a meridian, and thus was affected by the earth's magnetism. Gilbert also argued that magnets are subject to the earth in all their movements. These movements he described as "coition" or attraction, "direction," meaning toward the earth's poles, "variation," or deflection from the meridian, "declination," or dip, and circular movement or "revolution."

The earth as a magnet easily explained the operation of the compass, and Gilbert rejects all previous opinions that ascribe direction of the compass needle to a power in the heavens, such as the "second star in the tail of the Great Bear selected by one philosopher." And Gilbert also made a distinction arising from the fact that unlike magnetic poles attract, and like poles repel. He insisted that poles of the compass be called "north-seeking" and "south-seeking," rather than north and south, since the north pole of the earth actually attracts the south pole of the compass needle, and vice versa.

**GILBERT'S EXPERIMENTS** with iron in various shapes highlighted its property of intensifying the magnetic field. He capped the poles of a lodestone, as in Fig. 1.5, and found that lifting power greatly increased. He showed that long and slender magnets had more definite poles. He found that splitting a magnet repeatedly resulted in smaller magnets, each with its north and south poles, the first demonstration that poles cannot be isolated. He measured magnetic attraction and found the pull was not linear; the intensity increased rapidly with reduction in the gap. He showed that there was no insulator for magnetism; nothing that he interposed had an effect on a compass. Magnetism from a lodestone was not even shut off by iron, but only diverted.

*De Magnete* expounds Gilbert's findings on the compass, which was the only practical use for magnetism at the time. For centuries it had been known that the compass needle deviated from the astronomical north, depending on geographical location. Gilbert ascribed this to the fact that the earth is not a smooth surface, but has low spots at the oceans and high spots at the continents, as in Fig. 1.6. Thus in the northern hemisphere the compass would be attracted to the continents, hence to the east of the true north on the European side of the Atlantic, and to the west on the American side of the Atlantic. In the early 19th century it was found that the compass points west of north on both sides of the Atlantic at a location in the extreme north of Canada.

Fig. 1.5- IRON CAPS on ends of the lodestone intensified its magnetism, increasing its attraction for the iron bars.

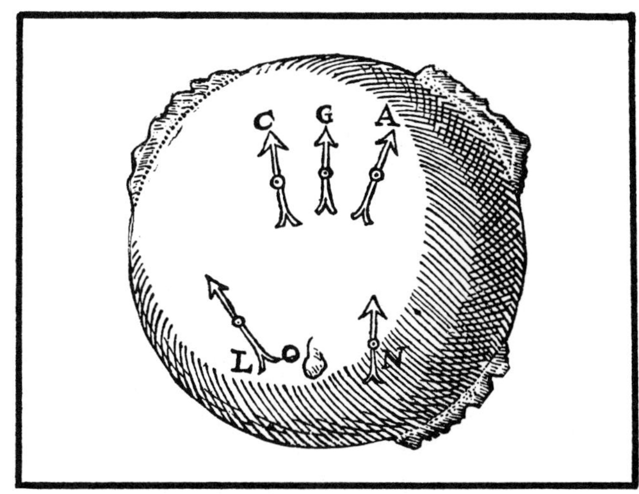


THE COMPASS DEVIATES FROM THE TRUE NORTH

Fig. 1.6 - Declination of the compass needle (deviation from true north-south) was due, said Gilbert, to attraction by large masses of lodestone in the continents of the earth.

Gilbert proposed to catalog the declination and dip for various locations as an aid to navigation, based on a belief that the earth's magnetic poles remained invariable, and hence that the declination and dip at any one place would be constant. Hopes for this scheme faded, however, with the discovery later that, while the oceans and the continents are fixed, the magnetic poles drifted unpredictably from year to year so that the charts would have to be revised continually.

In astronomy Gilbert speculated on the operations of the universe, though this was outside the realm of experimental proof. He is full of praise for Copernicus and his sun-centered planetary system. Gilbert examined the earth's diurnal rotation and the rotation of the planets in magnetic terms. He attributes to "magnetical energy" the orbital rotation of the earth, the moon and the other planets. While this theory was not the correct one it did attempt the first proposal for causes that previously were ascribed to undefinable cosmology, and he set the groundwork for Newton's future theory of gravitation.

**AMBER** - *De Magnete* discusses another material highly prized in antiquity for its supposed magical properties. *Amber,* an ancient fossilized tree resin found in the Baltic area, and named by the Greeks

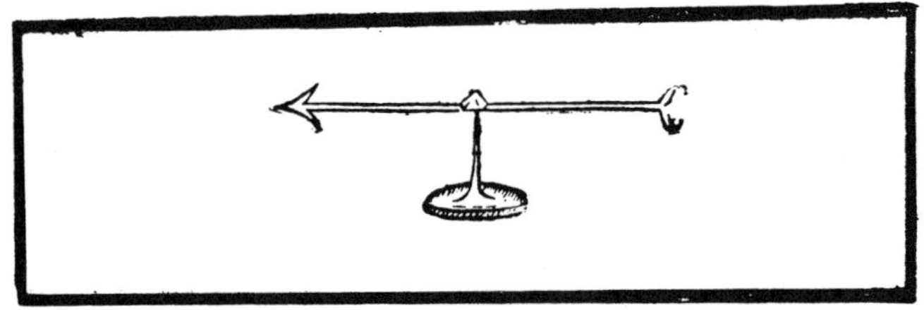

Fig. 1.7 - GILBERT'S "VERSARIUM," an electroscope, used a pivoted light gilt needle to test electrostatic attraction due to rubbing amber, sealing wax and other materials.

"elektron," had a very special quality. When rubbed with a cloth it could make chaff, bits of thread and other light particles jump and stick to it. In this attractive ability amber resembled the powers of lodestone, and the unexplained action of these two materials had long intrigued the curious.

Amber and its attractive property, when rubbed, came under Gilbert's scrutiny and he proceeded to explore the effects of rubbing action on other materials. Using a pivoted metallic needle, Fig. 1.7, which he called a "versarium," actually a form of "electroscope," which is an instrument responding to the presence of an electric charge, Gilbert found that amber was not the only material that could attract. He showed, for the first time, that many substances, such as glass, rock crystal, sulfur, sealing wax and some minerals, when rubbed, affected the needle. He called these "electrics," and those materials that were inactive when rubbed he called "anelectrics," such as wood, bone and the metals.

Gilbert considered that the attraction was caused by an "electric effluvia," which was brought into action by the rubbing. The power of this effluvia was affected by many external conditions and was particularly sensitive to moisture. Condensation on the surface, for example, instantly checked the effluvium, so that it was difficult to perform the experiments in humid weather. Liquids such as alcohol prevented attraction, but olive oil did not affect the action. Excessive heat of any kind drove out the effluvium, destroying the action.

The ancients had considered the forces of lodestone and amber as being due to similar causes, both being endowed with a kind of "self-power." Gilbert separated the powers, making a distinction between the magnetics and the electrics - thus:

"In all bodies everywhere are presented two causes or principles whereby the bodies are produced, to wit, matter *(materia)* and form *(forma).* Electrical movements come from the *materia* , but magnetic from the prima *forma;* and these two differ widely from each other and become unlike, the one (magnetics) ennobled by many virtues, and prepotent; the other (electrics) lowly, of less potency, wherefore its nature has to be awakened by friction til the substance attains a moderate heat, and gives out an effluvium, and its surface is made to shine. Moist air blown upon it from the mouth or a current of humid air chokes its powers, and if a sheet of paper or linen cloth be interposed there is no movement. But lodestone, neither rubbed nor heated, and even though it be drenched with liquid, and whether in air or water, attracts magnetic bodies. A lodestone attracts only magnetic bodies, electrics attract everything. A lodestone lifts great weights; a strong one weighing two ounces lifts half an ounce or an ounce, Electrics attract only light weights: e.g., a piece of amber three ounces in weight lifts only one-fourth of a barleycorn weight."

Fig. 1.8- HAMMERING IN THE MAGNETISM - A red-hot iron bar is held along the earth's meridian, and is hammered as it cools down. The bar then becomes a linear magnet in being reworked in the influence of the earth's magnetism.

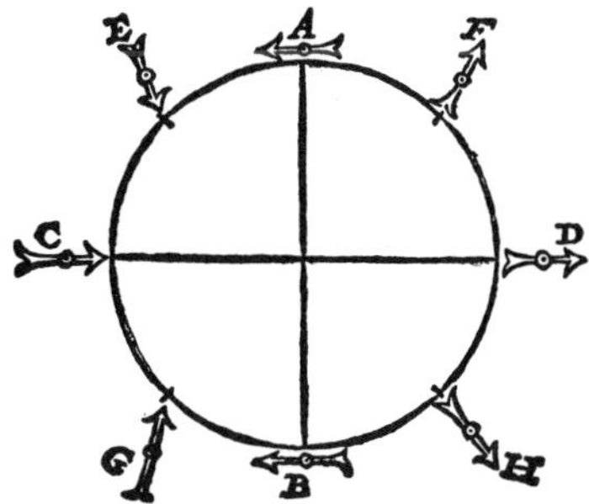


GILBERT EXPLAINED THE DIP NEEDLE. . .

Fig. 1.9- The COMPASS indicates only the horizontal component of the earth's magnetic field. Since that field usually also has a vertical component, the non-horizontal component is indicated by a DIP NEEDLE, a compass needle suspended vertically. Gilbert attributed its operation as shown above. Dip at the equator, A-B, he pictured to be nil, because there the earth's magnetism was horizontal, and at the poles C-D, to be maximum, pointing perpendicularly to the earth's north-south magnetic poles. Variations in the dip, Gilbert hypothesized, were due to unusual local concentrations of lodestone in the interior of the earth.

**MAGNETIC ATTRACTION**, which Gilbert preferred to call magnetic "coition," was the subject of several chapters in *De Magnete.* In one chapter he comments on the reports of the learned through the ages as they tried to light up the mystery of the lodestone. Their explanations involved such agencies as vapors, emanations and invisible rods, all of which lacked experimental verification. Gilbert rejected them as convoluted and full of doubt and uncertainty. In subsequent chapters Gilbert tries to bring within reason an explanation of the action of the losestone.

A lodestone was energized, Gilbert proposed, because it partook of some of the magnetism of the earth. But what was the cause of the earth's magnetism? To Gilbert the magnetic forces were a primal energy from an animate mother earth. They were part of a grand nature of the earth, innate in all parts and diffused throughout. These answers were the best that could be had at the time, and they did link magnetism with the reality of its earthly origin.

Gilbert is at his best in the remainder of his book in which he reports on a great series of painstaking experiments and demonstrations on the force characteristics of the lodestone and the exertion of magnetism under various conditions. The concluding chapters are concerned with the compass and the dip needle and their relation to the earth's magnetic field. The prime problem was accounting for and determining the variations of the compass and dip needles with location, in an effort to make the readings most effective for navigation.

A better answer than Gilbert's for the cause of magnetism had to wait more than 250 years in coming. In 1830 the French scientist, Ampère proposed an explanation based on his speculations in electromagnetism, the magnetism produced by an electric current, as discovered by Oersted in 1820. Ampère theorized that the molecules making up an iron bar had currents forever flowing around them in paths of zero resistance. The circulating currents made each molecule a tiny magnet. The poles of these magnets were normally randomly oriented, but under the influence of an external field of magnetism, they would line up to give the bar definite north and south poles, Fig. 1.8. Some materials such as steel, could maintain their molecular polar alignment after removal from the magnetic field that excited them, to become permanent magnets.

Ampère's explanation of magnetic materials anticipated by over seventy years the modern atomic theory of magnetism. The electrons of the

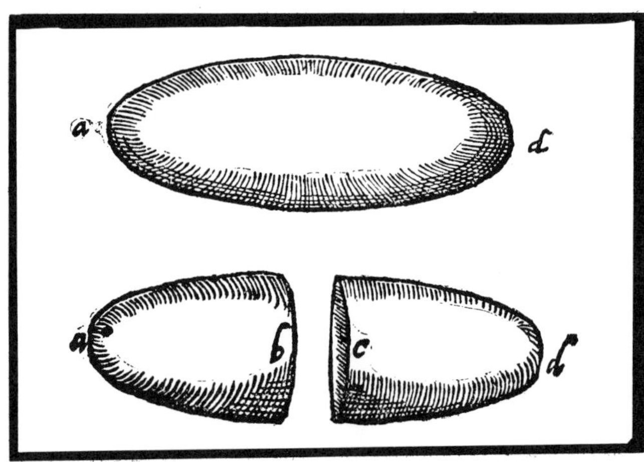


Fig. 1.10 - THERE ARE NO ISOLATED MAGNET POLES
Splitting up the lodestone repeatedly simply made smaller magnets; each piece still had north-south magnetic poles.

atoms in the modern theory produce magnetism in ways which are equivalent to the amperian circulating currents. This is done by the orbiting of the electrons around the nucleus, and the body spin of the electron around its own axis. The circulating currents create tiny magnetic dipoles, which when directionally aligned, produce external magnetic fields. Since magnetism is inherent in the structure of the atoms, all substances are magnetic, some very slightly, others, like ferromagnetics, very strongly magnetic.

Gilbert ascribed the earth's magnetism to a large body of lodestone oriented permanently in the earth's interior. But the evidence today is that the earth's core is a hot and partly molten mass of iron composition. These "fluid metals" circulate as a hydrodynamic dynamo because of the earth's rotation, generating electric currents that produce the north-south field of magnetism encircling the earth. Variations of the internal electric currents cause unpredictable changes in the strength and direction of the magnetic field, with consequent shifting of the earth's magnetic poles. Readings of the compass are affected by these shifts, hence for navigational purposes it was essential that the field be constantly charted. This, begun in the 18th century, initiated the science of "terrestrial magnetism," dealing with the direction, intensity and variations of the earth's field. Until the 19th century this was the main thrust of magnetic science.

**MAGNETIC SCIENCE** changed abruptly in the 19th century with the discovery of electromagnetism, the magnetism produced by electric current. This brought on the development of machinery like electric generators and motors operating by the interaction of magnetic fields. It became necessary then to deal, not with the lodestone type of permanent magnetism, or terrestrial magnetism, but with the magnetism produced by electric current in coils of wire.

The rapid evolution of electrical machinery in the 19th century made it imperative that this type of magnetism be investigated. Its laws of behavior in magnetic circuits needed formulation to achieve proper design. It was essential to be able to calculate the amount of magnetism in the magnetic circuit of an electrical machine resulting from a given current in the coils.

Some of the earliest investigations of the electromagnetic circuit were carried out by physicists who initiated the first electrical engineering departments in the various universities. One of the pioneering physicists to study and formulate the properties of the magnetic circuit was the eminent American scientist, Henry A. Rowland (1848-1901). He was a graduate of the Rensselaer Polytechnic Institute, did post-graduate work in Germany, and carried out his research, together with teaching, at Johns Hopkins University.

Using the "lines of force" concept developed by Michael Faraday in England, Rowland proceeded to show that the magnetic lines of force resulting from an electric current admitted of definite calculation using absolute units. Lines of magnetism were envisioned as traversing a circuit, driven by a "force," and meeting an "opposition," to establish a resultant " magnetic flux."

In the electric circuit, a work potential is provided by an electromotive force which, when applied to a circuit, produces a current through a resistance, as expressed by Ohm's law:

$$\text{Current} = \frac{\text{Electromotive Force}}{\text{Resistance}}$$

In analogy to the electric circuit, Rowland envisioned a work potential for a magnetic circuit as being provided by a "magnetomotive force." This force, applied to a magnetic circuit having a magnetic resistance, called "reluctance," produced a "magnetic flux." Thus, as a kind of Ohm's law for the magnetic circuit:

$$\text{Flux} = \frac{\text{Magnetomotive Force}}{\text{Reluctance}}$$

**MAGNETOMOTIVE FORCE** thus represented the work potential of a magnetic circuit, similar to electromotive force representing the work potential of an electric circuit.

Rowland expressed magnetomotive force in terms of the work required to carry a "unit magnet pole" once around the magnetism encircling a one-turn, air-core coil in which flowed a unit current. In the centimeter-gram-second (CGS) system a unit magnet pole is defined as a theoretical point pole which repels an equal pole at a distance of one centimeter, in a vacuum, with a force of one dyne. Unit current is the abampere, equal to ten amperes.

A unit pole, having an area of $4\pi r^2$, emanates $4\pi$ lines of magnetic flux. These, interlinking the flux produced by the current, $I$, in the coil, require $4\pi\ I$ ergs of work for passing once around the conductor of the coil, as shown by Fig. 1.11. This magnetic work function he called the "magnetomotive" force.

**The "GILBERT"** - In the rise of science there also arose the need of a terminology for physical units to designate the attributes and laws of newly discovered phenomena. In the standardization of the units it has been traditional to identify them by use of the names of the founding geniuses in the various fields of science.

In honor of Gilbert, the American Institute of Electrical Engineers, in 1894, proposed his name for the unit of magnetomotive force in the CGS system of electromagnetic units, and the name was approved in 1930 by the International Electrotechnical Commission.

The CGS unit of magnetomotive force, the gilbert, is $4\pi\ I$ abamperes, equal to $0.4\pi\ I$ amperes.

The term gilbert was generally used in text books in the United States in the early 20th century. But it was not accepted in Europe. And in engineering circles it was generally considered simpler to use "ampere-turns" in magnetomotive force calculations rather than the gilbert unit involving the factor $0.4\pi$.

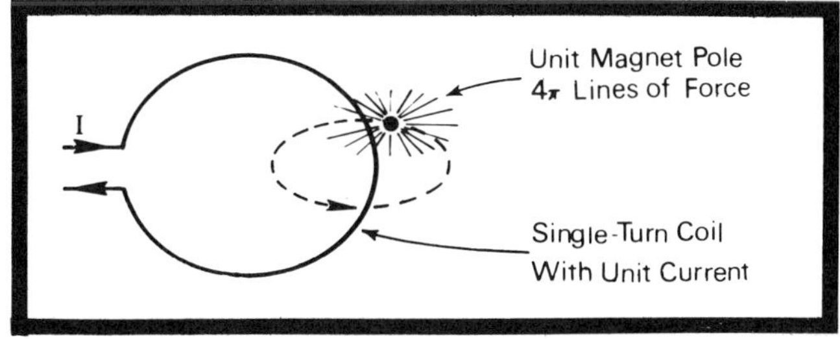


MONOPOLE CIRCLES THE CURRENT

Fig. 1.11- A monopole, or unit magnet pole, if such were found to exist, would generate $4\pi I$ ergs of work, or one *gilbert,* in encircling a one-turn coil carrying an electric current of unit value, *I.* This is *magnetomotive force.*

In 1960 the General Conference on Weights and Measures set up the International System of Units (SI). The MKSA (meter-kilogram-second-ampere) system of electrical units, first adopted in 1935, is a part of the SI system, and is now in general use. Magnetomotive force in the MKSA system is measured in *ampere-turns,* equal to $0.4\pi$ gilberts.

Although the gilbert has been superseded as a unit, we continue to give honor to the name, as one of the founders of the electrical science. Until the work of Coulomb in the 18th century, and the discovery of electromagnetism by Oersted in 1820, Gilbert's book *De Magnete* had about everything there was to say about magnetism. It was a prelude to the future when the forces of magnetism would run the world. Gilbert's personality is given in the book *William Gilberd,* a biography by E.B. Gilberd and Lord Penney.

*This lively bachelor - he remained unmarried all his life - appears to have made quite an impression on London. He was tall, with a spade beard typical of his time, and with a cheerful complexion; an happiness not ordinary in so hard a student and retired person. He had the clearness of Venice glass, without the brittleness thereof; soon ripe, and long-lasting in his perfections . . . One saith of him "that he was stoical, but not cynical, reserved but not morose."*

*A sensible man . . . he seems to have been something of a talker, perhaps a charmer in spite of his reserve.*

*He was a man, moreover, of wealth and wide interests, who devoted much of his time and money to science.*

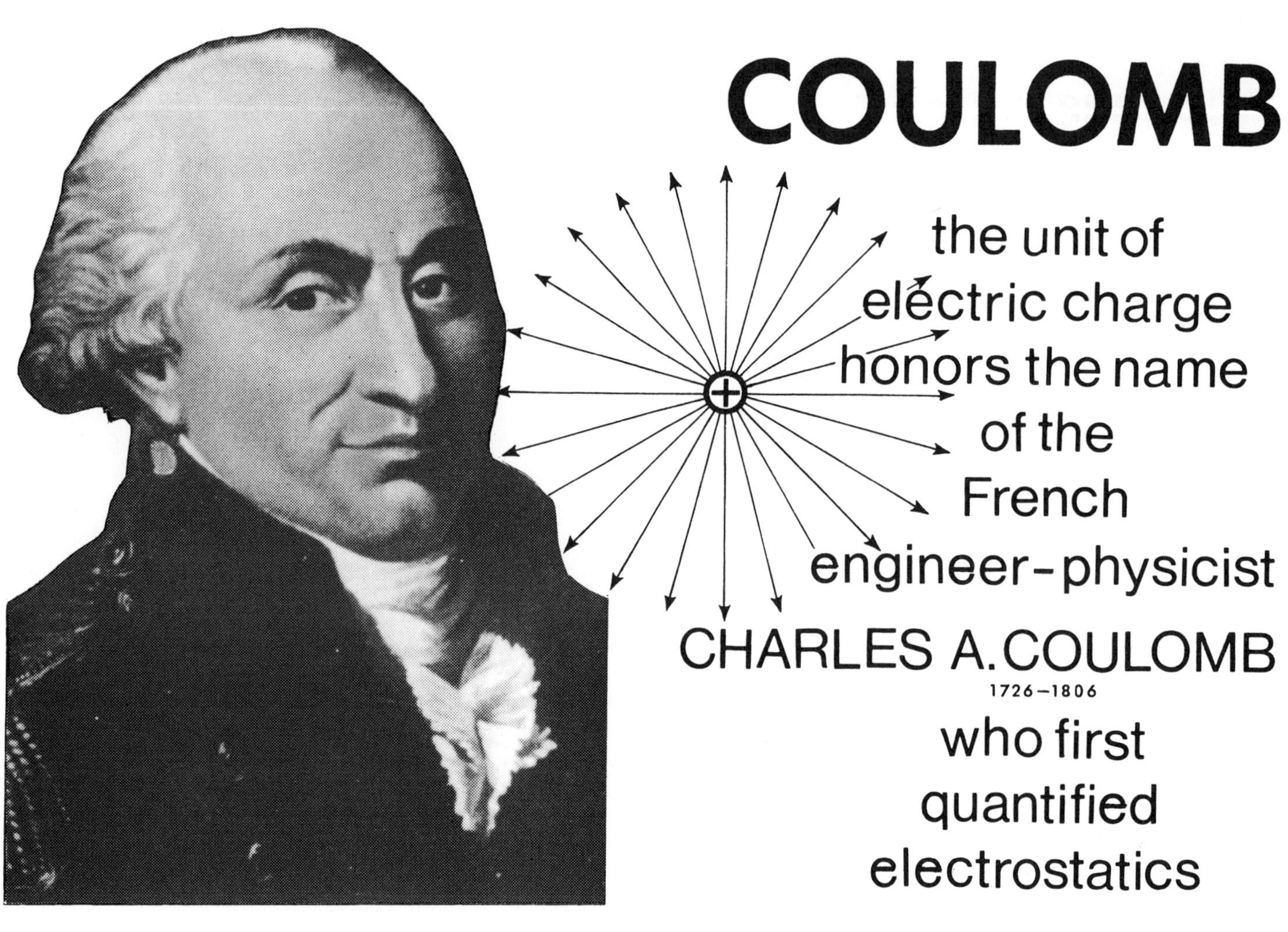


**The "SEA OF DARKNESS"** was never ending. Week after week Columbus had sailed his three small ships on an unknown path toward the setting sun seeking a westward route to the fabled spice islands and the Indies. Day followed day with increasing worry - there was no land in sight.

A storm came up . . . a blackened sky . . . the wind howled, the sea raged, the air crackled. Columbus was faced with a mutinous crew. Then came a wondrous sight - from the tips of the masts, shooting heavenward, a glow of purple streamers filled the air. To his awe-struck crew, Columbus shouted that this was St. Elmos's "holy fire" sent as a sign of blessing for the voyage. The storm abated and the crew was assured by a seeming miracle, and Columbus sailed on to discover America.

"St. Elmo's Fire," shooting from the masts, had been known to the ancients. But no one in antiquity had attributed these discharges to an electrical origin. Then, at the end of the 17th century, picking up where Dr. Gilbert had left off a hundred years before, curious investigators, producing electricity by friction in large quantities, began to observe sparks and glow in the electrical discharges, and the connection between light and electricity became apparent, though not understood. A new era had been opened - that of "electrostatics."

**A BLINDING FLASH** of lightning piercing the sky of a summer storm, the flickering beauty of the auroral lights, and the crackle when walking on a rug on a dry winter's day, are all evidences of electrostatics. It is the oldest, and until development of the electric battery in 1800, was the only known kind of electricity and the only branch of electrical science. Electrostatic history dates from antiquity and reaches one of its high points in the

# . . . he measured the force in "St. Elmo's Fire"

investigations of Charles A. Coulomb (1726-1806), who first established the fundamental laws of static electricity.

To the ancient Greeks we owe the earliest record of the effect now known as "triboelectrification," or the production of static electricity by frictional contact of certain materials. Thales of Miletus reported in 600 B.C. that a piece of amber, when vigorously rubbed with a cloth, responded with an attractive power. Light particles such as chaff, bits of papyrus and thread, jumped to the amber from a distance and were held to it. Amber, a fossil evergreen tree resin, called "elektron," was thought to possess a "soul." There was no other way of accounting for a strange effect like this and it was wrapped in an aura of myth and magic down through the ages.

More than two thousand years elapsed between the Greek experiences with amber and the English physician who picked up where Thales left off. William Gilbert decided to re-examine in a scientific way the lore handed down from antiquity on the rubbed amber, and on the magnetic lodestone which the ancient Chinese had used to make compasses. In doing so he rediscovered and enlarged on the properties of electrification and and magnetism, recorded in his great book *De Magnete,* discussed in the previous chapter.

Gilbert showed that not only amber, but also

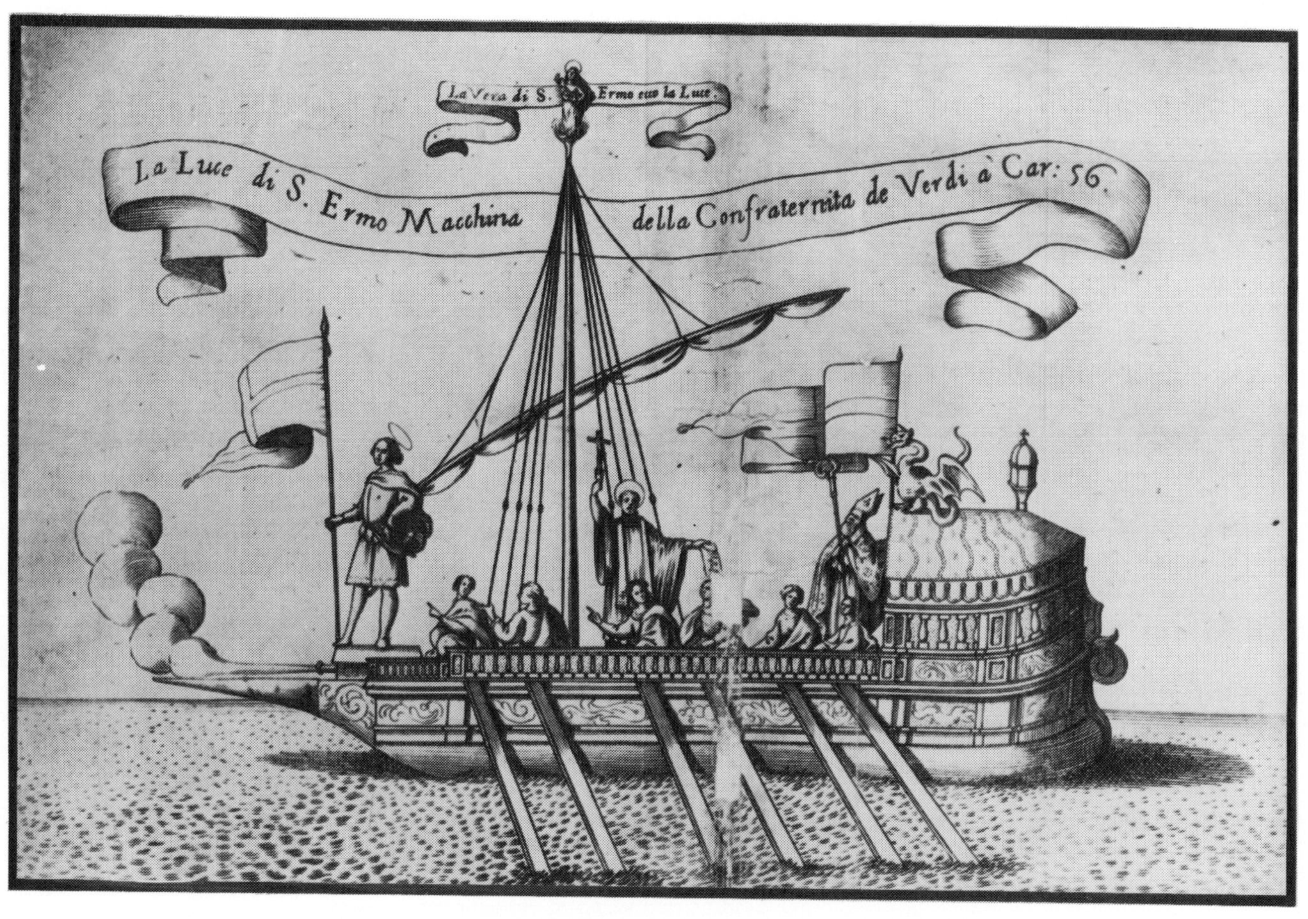

SPARKING AT SOCIAL OCCASIONS . . .

Fig. 2.1-Laying hands on the rotating glass ball of this 18th century hand-cranked electrostatic machine provided sparks and shocks for entertainment. People also submitted to shock treatments forsupposed health benefits.

"ELECTRIC FIRE"

Fig. 2.2- Electrostatic machine, shown on the opposite page, ignites candles and spirits of wine. More serious experiments were performed on animals and plants to determine the effect on growth habits. Illustration Burndy Library.

glass, agate, diamond, sapphire and many other materials had, when rubbed, the attractive power. He noted the attraction was in a straight line, that moisture hindered the electrification. He classified substances as electrics and non-electrics, and noted that electric attraction was much weaker than magnetic. Gilbert had trouble explaining attraction at a distance. He ascribed it to "arms of effluvia" which were ejected from the amber to pull in the particles. Gilbert did not report that particles could also be repelled. Not until a century later was electrostatic repulsion noted by DuFay in France.

Next step in the progress of electrification was an improvement of the friction process. Rotating tubbing machines were developed to give continuous and large-scale production of electrostatic charges. Otto von Guericke (1602-1685), in Germany, the well-known inventor of the vacuum pump, mounted a sulfur ball on a shaft so that it could be turned by a hand crank. When rubbed with the dry hands the sulfur ball generated electrostatic charges strong enough to show strange new effects. Guericke reported sparks and crackling sounds.

**FRANCIS HAUKSBEE** (?-1713), experimentalist to the Royal Society in London, developed a similar apparatus using blown glass spheres and cylinders. Hauksbee improved on the vacuum pump invented by Guericke, and experimented with many electrostatic effects in evacuated vessels to produce novel effects. He observed weird and beautiful glow when laying hands on a rotating evacuated glass sphere. With a nine inch diameter globe he generated enough light to read by. Hauksbee suspected that electrification by friction, and production of light by friction, were somehow related. He called the effect "phosphorescence."

Hauksbee was the first to show that "barometric light," produced when the mercury in a barometer tube was shaken, was due to an electrification friction between the mercury and the glass. The phenomena Hauksbee reported raised many questions that were taken up for investigation by later experimenters.

**IN THE 18th CENTURY** electrification became a popular science and experimenters discovered many new attributes of electrical behavior. Stephen Gray (1666-1736), in England, proved that electrification could flow hundreds of feet through ordinary twine when suspended by silk threads. Thus he theorized that electrification was a "fluid." Substituting metal wires for the support threads he found the charges would quickly dissipate, thus the difference in materials that either conduct or insulate began to take shape. The "electrics," like silk, glass and resin, held charge. The "non-electrics," like metals and water, conducted charges. Gray also found that electrification could be transferred by proximity of one charged body to another without direct contact. This was evidence of electrification by induction, a principle that was to be made use of in later machines for producing electrostatic charges.

In France, aroused by Gray's experiments, Charles F. DuFay (1698-1739), a member of the French Academy of Science, showed by extensive

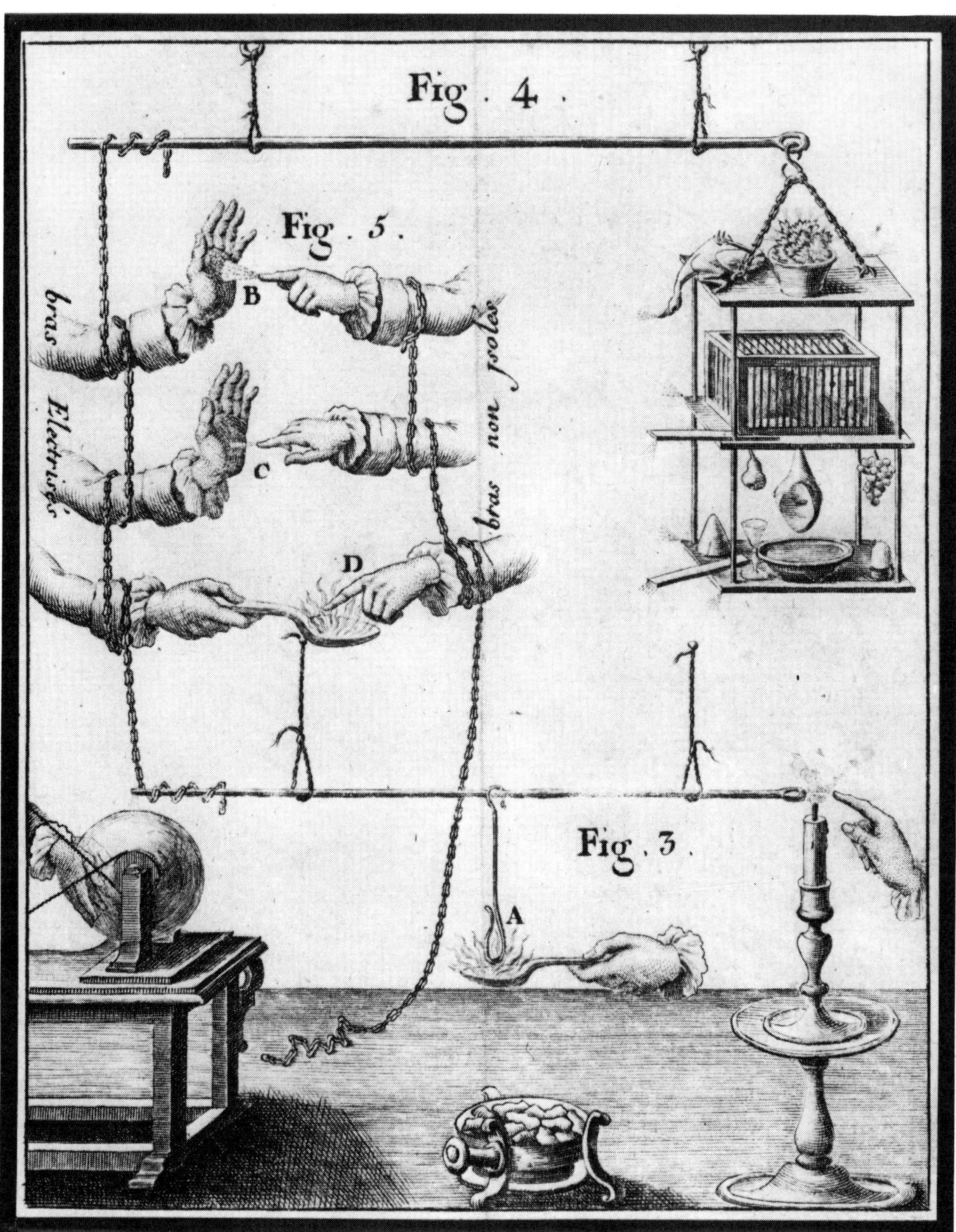
Fig. 4.
Fig. 5.
B
C
D
bras Electrisés
bras non isolés
Fig. 3.
A

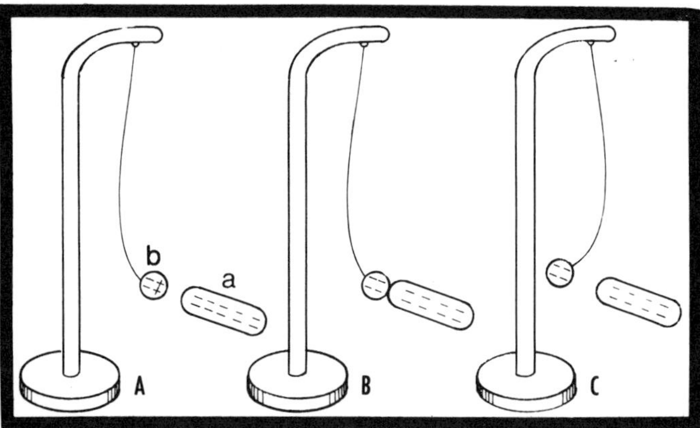


Fig. 2.3- HISTORY of the sequence in the early discoveries of electrostatic action are shown by the pith ball experiments above. A- The rubbed amber rod, a, is left negatively charged. Brought near the pith ball, b, it induces a nearby positive charge, causing the ball to be attracted. B- On contact of the rod and ball the charges on rod and ball are shared and they assume similar charges. C- Like charges then repel the ball. This ATTRACTION, then COMMUNICATION, then REPULSION, indicated two electricities.

tests that practically all materials with the exception of metals and those too soft or fluid to be rubbed, could be electrified. Later, however, he discovered that if metals were insulated they could hold the largest electric charge of all. DuFay found that rubbed glass would repel a piece of gold leaf whereas rubbed amber, gum or wax attracted it. He concluded there were two different kinds of electric "fluids," which he labelled "vitreous" and "resinous." He found that while unlike charges attracted each other, like charges repelled. Fig.2.3 shows the sequence of *attraction,* then, *communication,* then, *repulsion,* that indicated there were two kinds of electricity.

Experimenters began to enlarge and perfect their frictional electric machines. Instead of hands held against the rotating glass, leather pads, pressed to the glass were substituted. With increase in size of the apparatus, powerful electric sparks could be created to give spectacular affects. Public demonstrations became popular, and many experimented for their amusement and instruction. The mystery and glamor of electricity began to be exploited with promises and implied hopes that it could cure any ill. Actually, at that time, electricity had no practical use other than that for medicine.

**THE LEYDEN JAR** -E.G. von Kleist (died 1748) and Pieter van Musschenbroek (1692-1761), a physicist in Leyden, Holland, were the first, independently, to discover, by a painful experience, that electric charge could be stored to high strength, then instantly released. Musschenbroek was trying to electrify water in a bottle by dipping a wire from the electric machine into it. Holding the bottle in one hand he, after a few moments, proceeded to remove the wire with the other hand. An astounding shock convulsed his arms and breast. With this dramatic incident there was revealed to the electric science a new phenomenon - that an electric charge could be accumulated.

The Leyden Jar was the first means of storing static electricity. It was simply a glass jar coated inside and outside with metal foil. Further experiments showed that the only requirement for storage was two conductive coatings separated by an insulator such as glass. This is the basic principle of the "capacitor," a means of holding charge.

Leyden Jars permitted the accumulation of large charges which could produce enormous sparks. They opened exciting new opportunities for electrical investigation and for spectacular effects. Abbé Nollet, a French physicist and popularizer of electric experiments, arranged a demonstration for the King in Paris in 1750. A group of Carthusian monks, holding hands to make a 900-foot circle, were made to spring suddenly into the air, en masse, on being shocked by the discharge of a Leyden Jar.

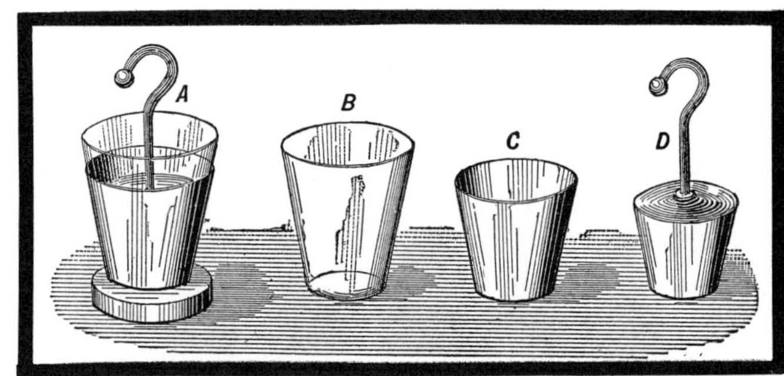


WHERE THE ELECTRICITY IS STORED . . .

Fig. 2.4- Benjamin Franklin prepared a Leyden Jar, A, with removable coatings, C and D. He strongly charged the jar and removed the coatings which then showed little charge. But when reassembled, the original strong charge was restored. Franklin thus proved that the electrification was stored in the glass, B, and not in the metal coatings.

ELECTRICITY FROM THE SKY ON A KITE STRING

Fig. 2.5 - That lightning is a giant electric spark was proved by Franklin and his son, June 1752. By charging a Leyden jar during a thunderstorm they demonstrated the identity of lightning and electrification from electrostatic machines.

**THE SCENE NOW SHIFTS** across the Atlantic to Benjamin Franklin (1706-1790), who was America's first electrical scientist. In addition to a career as statesman, diplomat, publisher, signer of the Declaration of Independence and the Constitution, Franklin was an avid experimenter and inventor. In 1743, at the age of thirty seven, Franklin witnessed, with excited interest, a demonstration of static electricity in Boston, and resolved to pursue the strange effects with investigations of his own.

Purchasing and devising various apparatus, Franklin became a practically full-time electrical enthusiast. He launched into many years of experiments with electrostatic effects. He corresponded on his results with scientists engaged in similar work in England. A gift of an electrical rubbing tube sent to Franklin for the Philadelphia Library Company in 1746 by Peter Collison of London, was put to use immediately by Franklin and other interested members. The results stimulated Franklin into original thinking on causes and effects of electrification. Carrying out his experiments far removed from the background of European conceptions, enabled Franklin to evaluate his observations with new and independent insight. He made discoveries surpassing the level of his European counterparts.

**FRANKLIN** as an electrician is most popularly known for his risky kite experiment during a thunderstorm, in June 1752, at Philadelphia. Although various European investigators had surmised the identity of electricity and lightning, Franklin was the first to prove by an experimental procedure and demonstration that lightning was a giant electrical spark. Having previously noted the advantages of sharp metal points for drawing "electrical fire," Franklin put them to use as "lightning rods." Mounted vertically on roof tops they would dissipate the thundercloud charge gradually and harmlessly to the ground. This was a first practical application in electrostatics.

The importance of Franklin's work led to reading of his papers before the Royal Society in London. Correspondence flourished between Franklin and European savants working with electrostatics. In 1751, Franklin's letters to Peter Collison were published in London in a book: *Experiments and Observations on Electricity made at Philadelphia in America by Mr. Benjamin Franklin.* Translated into French, German and Italian, the book made Franklin famous overseas.

The origin of such noteworthy output from remote and colonial America made Franklin especially marked. The lightning rod was a subject of intense interest and application, with much controversy as to its propriety and protection. With proper installation it was found very protective. In his many trips to Europe as statesman and experimenter, Franklin was lionized in social circles, and eminently regarded by scientists. He was awarded the Copely Medal by the Royal Society in 1753.

Franklin's investigations led to his proposal of a "one-fluid" theory of electricity as against DuFay's two-fluid description. He argued that "electrical fire" is a common element in all bodies and is normally in a balanced or neutral state. Excess or deficiency of charge such as produced by the friction between materials, created an imbalance.

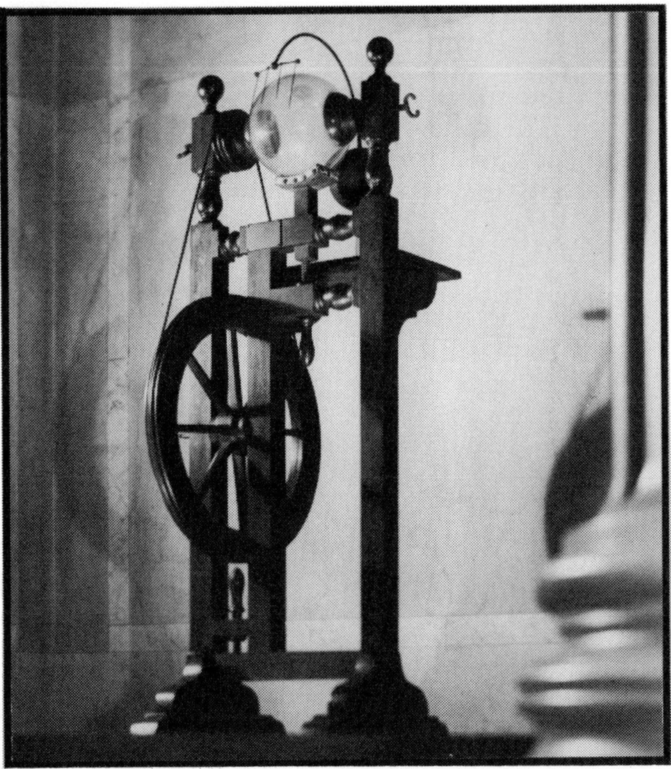

Fig. 2.6- ELECTRIC MACHINE thought to have been used by Benjamin Franklin. The glass globe is rubbed by a friction pad. Courtesy the Franklin Institute, Philadelphia

Electrification by friction was thus a process of *separation* rather than creation of charge. By balancing a charge gain with an equal charge loss Franklin had implied a law, namely, that *the quantity of the electric charge is conserved.*

Franklin guessed that when glass was rubbed the excess charge appeared on the glass, and he called that *positive* electricity. He thus established the direction of conventional current from positive to negative. It is now known that the electrons producing a current move in the opposite direction.

**IN ALL THE SCIENCES** the ability to formulate relationships is essential to orderly interpretation. Measurements are necessary for mathematical resolution from the unknown to the known. Up to the end of the 18th century the knowledge of electrostatics was mainly *qualitative.* There were means for detecting but not for measurement, and relationships between the charges had not been formulated. The next needed step was to *quantify* the phenomena of electrostatic charge forces.

For this determination the scientific scene shifts back to France, which was developing a tradition of fine instrumentation, and a mathematical and analytical proficiency for formulating scientific laws. A pioneer is establishing the basic relationships of electrostatics was the engineer - turned - physicist, Charles A. Coulomb, a product of France's excellence in technical education.

**CHARLES AUGUSTIN COULOMB** was born 14 June 1736 into a well-to-do family in the city of Angouleme, in southwestern France. During his youth the family moved to Paris where Coulomb attended the Collége de France. A breakup between the parents resulted in the return of the father, Henry, to his ancestral home in Montpellier in 1746, accompanied by his son. Here young Coulomb continued his schooling at the Montpelier Academy. He showed an early predilection and capability in science and his talent was recognized by invitations to read papers on astronomy and mathematics before a local learned society.

With his ambition set on the best in an engineering education, Coulomb returned to Paris in 1758, and applied for admission to the famous Ècole du Génie, operated by the military at Mèziéres. After some tutoring he passed the examinations and was enrolled in 1760. He graduated in civil engineering with honors a year later, and with a rank of first lieutenant in the Corps of Engineers.

Coulomb's first post was at the naval works of Brest, but in 1764 he was suddenly transferred to the West Indies French colony of Martinique, and put in charge of rebuilding Fort Bourbon which had been destroyed by the British. This was a major asignment involving direction of hundreds of laborers in all phases of military consruction, and a tough proving ground for a young engineer. Coulomb carried through competently, and used the opportunity to gain a good foundation in the application of mechanics to materials and structures. But several bouts with fever overtaxed Coulomb's health, and forced his return to France in 1772.

Coulomb early exhibited an inquiring mind for scientific fundamentals, and while in Martinique had expanded from military engineering into physics. On return to France he began a lifetime of broad-ranging investigations in mechanics, and in electricity and magnetism that would eventually place him in the forefront of the science of his time.

France was in an era of political and social unrest with movements toward liberalism and democracy, culminating in the period of the Revolution. As part of this movement there was a renaissance in scientific inquiry. Investigations into natural phenomena became popular, and it was in this atmosphere that Coulomb carried on with his work.

As a member of the Corps of Engineers, Coulomb was involved with a great array of projects, both in military and public service. Included in the civil engineering were the handling of structures, machines, canals, and the harbors and waterworks. All involved the applications of mechanics, so that Coulomb turned his attention and talents to fundamental investigations on the strength of materials as related to the forces imposed on them. This led to his paper on the influence of friction and cohesion on some problems in statics, presented to the French Academy of Sciences in 1773. Coulomb's work in this field was a major contribution in the application of flexure mechanics to the stresses and strains of masonry structures which were the important building elements at the time.

Coulomb's interest in friction and cohesion led to a study of the problem of shear, as represented by bodies under torsional stress. As part of his investigations he experimented with the twisting characteristics of various fibers in tension. He found that for moderate angles of twist the force of torsion was proportional to the twist. He was struck with the delicacy of turning movement provided by a balance arm hung perpendicularly to, and suspended by, a long, light fiber. The arrangement was practically friction-free, thus very sensitive to a delicate torque force at the end of the balance arm. Also the proportionality of fiber twist to movement of the torque arm made it an instrument with useful characteristics for measuring the force of actions at a distance.

**THE FRENCH ACADEMY of SCIENCE** had, for a number of years, offered a prize for an improvement in the construction of the magnetic compass. The purpose of the award was a design that would provide a higher order of stability and sensitivity for scientific observations of the earth's magnetic field. The usual maritime compass, with its pivoted needle, was unsuitable for precision work because of the erratic behavior introduced by the friction of the bearing.

A high-grade compass was required for recording and computing the continual shift in the pattern and strength of the magnetism of the earth. The magnetic variations included those of gradual long range, those of sudden large irregularity due to magnetic storms, and small daily changes, all attributed to changes in activity of the sun. Readings of diurnal swings were often unsatisfactory and confused because bearing friction exceeded the magnetic force on the needle. Coulomb's torsion studies of fibers led him to a solution of the problem.

Coulomb found in his torsional balance a superior principle for a compass required to indicate small deviations. The fiber suspension permitted use of a heavier needle of greater magnetic strength and hence more responsive to small magnetic changes. By shielding the instrument against air currents, and using a long pointer and magnifier glass, readings of great precision were obtained.

In 1776 Coulomb made a model of the torsion arm compass and put it through tests in which he recorded the diurnal variations in the magnetic field. Submitting a paper on *Investigations of the Best Method of Making Magnetic Needles,* he was awarded the prize by the Academy in 1777. The sensitivity of the Coulomb torsion arm suspension method changed the whole future of precise measurement of magnetic fields, and his compass was adopted by the Paris Observatory in 1780.

Coulomb continued his work on friction and shear, and in 1781 won another prize from the Academy for his friction study titled: *The Theory of Simple Machines.* This was a pioneering work establishing the science of friction. It brought him wide acclaim and the highly prized election to the Academy. With his admission to that prestigious body, and having secured a permanent post in Paris Cooulomb was now able to turn his main efforts from applied engineering to research in physics, and this marks the beginning of his celebrated investigations in electricity, magnetism and the properties of matter.

In 1784 Coulomb reported, in an Academy paper on his further torsional experiments dealing with the elasticity of metal filaments. Coulomb found that when twisting a metal thread the torsional force depended on the metal, and for a given metal

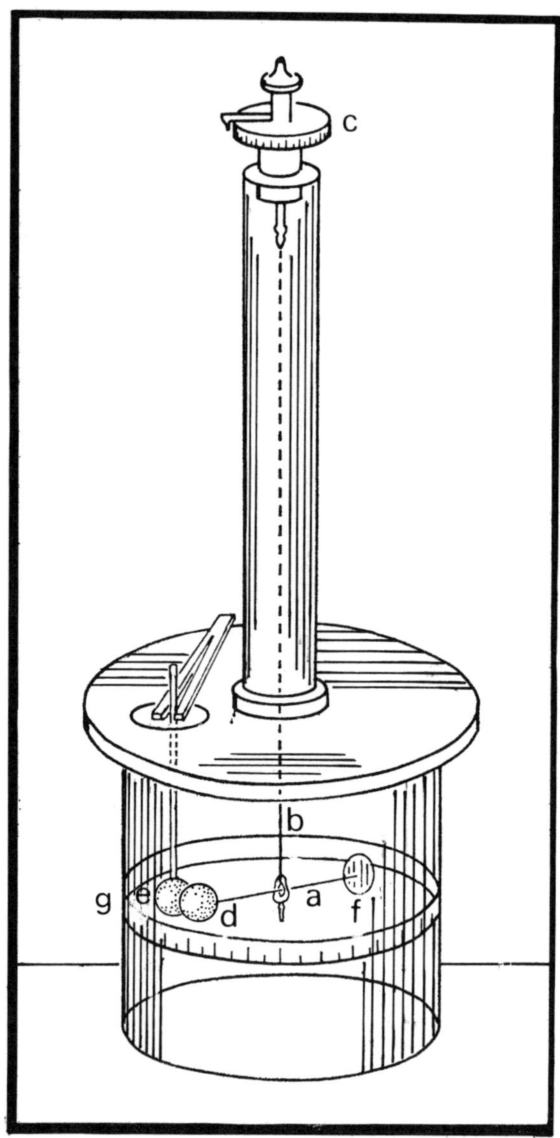


THE FIRST PRECISION ELECTRICAL INSTRUMENT

Fig. 2.7- Coulomb invented the torsion balance, above, to measure the mutual forces of electrostatic charges. Stationary pith ball, e, and movable pith ball, d. are energized by an electrostatic charge inserted from the outside. The identical charges then repel each other and their separation is read on circular scale, g. The graduated head, c, is then turned in a direction to force the separation of the balls to one-half the distance, and the scale reading is taken. Coulomb found that the electrostatic force varied as the inverse square of the separation. Sensitivity of the balance required a high degree of manipulative skill.

was proportional to the angle of twist, to the fourth power of the diameter of the filament and inversely proportional to its length.

Using fine metal threads with definite torsional characteristics to support a torque arm provided a new instrument of greater sensitivity. It now gave the opportunity for measurement of forces that were very small and at a distance from each other.

One of the pressing problems of Coulomb's time was precise determination of the relationship of forces exerted by one electrically charged body on another. That these forces existed and were both attractive and repulsive had long been known qualitatively. But how did they relate quantitatively? Coulomb, a follower of Newtonian physics, had suspected that the force between two electrostatically charged bodies might, like the force of gravity between bodies, vary inversely as the square of their separation.

Since the electrostatic forces between small charged objects are very small, how could they be measured? The possible answer to this problem was doubtless in Coulomb's mind when he found that that the fine wire of his torque arm torsion balance could be twisted by a very minute force. Now he was in a position to adapt the balance to electrostatic operation, in the effort to test the inverse square law, and to put electrostatics on a quantitative basis. Coulomb's work on this determination in electrostatics and his other investigations in magnetism, are the subject of seven memoirs presented to the Academy from 1785 to 1791.

**COULOMB'S TORSION BALANCE,** Fig. 2.7, used a light straw torque arm, a, suspended by the torsion wire, b, the twist of which was controlled by the graduated head, c. On one end of the torque arm, a, was a small pith ball, d, covered with gold foil and conterbalanced by the vane, f, which served also to dampen oscillations. The pith ball,d, was adjusted by means of knob, c, to zero on the circular scale, g, of the container cylinder. A similar pith ball, e, was inserted through the hole in the cover and adjusted so that it just touched pith ball, d. Both balls were uncharged. Coulomb now inserted an insulated pin, which had been charged from a Leyden Jar, and touched both pith balls. The same charge was thus distributed evenly over the two balls.

The two pith balls immediately repelled each other, and the torque arm with ball, d, turned to an angle such that the twist in the wire balanced the force of repulsion. The number of degrees was then read on scale, g. Now the twist of the wire was increased by turning the graduated head, c, to return ball, d, against the force of repulsion, toward ball, e, by one-half of the original angle of twist, and a reading was made on the graduated head.

In one of his many trials Coulomb found that after the balls were charged the swing of the torque arm was such as to twist the wire 36 degrees, read on the circular scale. Coulomb then turned the graduated head to increase the twist, forcing the separation back to one-half, or 18 degrees. He found the head had been turned 126 degrees, so that the total repulsive force was equivalent to 144 degrees. That is, at half the angle the repulsion was 4 times 36, or 4 times as great. Separation was again reduced by one-half giving a total torsion of 576 degrees, or 4 times 144, again 4 times the previous trial, indicating an inverse square relationship, as summarized in Table 2.1.

Coulomb realized that the force between the two balls was not as the arc in degrees but as the chord of the arc, and this altered the effective lever arm. However, for small angles the effects nearly cancelled, so that he concluded that the angles could be used for calculation.

Coulomb made several tests in confirming the inverse square results, then presented his findings in a First Memoir to the French Academy. He stated a basic law of electrics:

*The force of repulsion between two small spheres which are electrified with the same kind of electricity is inversely proportional to the square of the distance between the centers of the two spheres.*

**COULOMB'S** First Memoir was, in a sense, provisional. It gave evidence only for repulsive charges, and established the principles of his torsion balance method.

Final proof required similar tests on the attractive forces of charged spheres. For this testing the torsion balance, as previously used, proved unstable because whereas the attractive forces varied inversely as the square of the distance, the opposing twist force of the suspension wire varied only directly. Close approach simply brought the balls in contact, nullifying the experiment.

Coulomb solved the problem by a dynamic oscillation method. A small gilded disk was attached to one end of the torsion balance arm. The arm was lined up with, and placed a few inches from, a large charged metal globe. The disk was then grounded by being touched lightly, to give it an induced charge opposite to that on the globe so that it would feel a force of attraction. Deflecting the arm slightly caused the disk at its end to oscillate in the electrostatic field of force between the two charged bodies. A count of vibration frequency was made for various distances between the centers of the disk and globe. For an inverse square relationship the frequency of oscillation should be proportional to the distance. Experimental data obtained by Coulomb, typified in Table 2.2, showed the law to be true. His results were reported in a Second Memoir to the Academy.

In the Second Memoir Coulomb reported also on the extension of his investigations to the attractive and repulsive forces of magnetism. These forces, he reasoned, should act the same over a distance as those of electricity. Thus he sought to show whether the same inverse square laws would apply. Coulomb's experiments with magnetism used the same static torque arm method and the dynamic oscillation method as for electric charges.

In contrast to the electrostatic tests, a magnetic point source could not be realized because magnet poles could not be isolated. Coulomb found that,

*Summary of Coulomb's Tests for Repulsion and Attraction*

Table 2.1-Force Measurements for Repulsive Charges

| Measurement Number | Separation Angle | Micrometer Reading | Total Twist |
|---|---|---|---|
| 1 | 36° | 0° | 36° |
| 2 | 18 | 126 | 144 |
| 3 | 8.5 | 567 | 576 |

Table 2.2- Force Measurements for Attractive Charges

| Measurement Number | Separation Inches | Seconds for 15 Oscillations Actual | Predicted* |
|---|---|---|---|
| 1 | 9 | 20 | 20 |
| 2 | 18 | 41 | 40 |
| 3 | 24 | 60 | 54 |

*For 15 Oscillations by the Inverse Square Law

for his purposes, point poles of magnetism could be approximated by the use of long, slender needles for the magnets. For magnetic investigation, Coulomb also had to contend with, and make use of, the earth's magnetic field. In one of his experiments a short suspended magnetic needle was lined up with the earth's meridian, and its period of oscillation timed. A much longer suspended needle was then lined up perpendicular with the axis of the small needle, and the oscillations of the small needle recorded for various separations of the two needles.

Coulomb hypothesized that for the inverse square law to hold the force between the magnets should be as the square of the number of oscillations of the small needle. For small distances, corrected for the earth's effect, the experimental results confirmed the hypothesis. For longer separations the effect of the far pole of the long magnet altered the results, and Coulomb took this into consideration in his calculations. Coulomb repeated the experiment with needles of various lengths and concluded that the action of what he called *magnetic fluid,* whether it was repulsive or attractive, was always as the inverse square of the distance.

Summarizing in his Second Memoir, the results of his studies, Coulomb stated his law:

For Electrics: *The electric forces of two electrified spheres, being either repulsive or attractive, and therefore of two electrified molecules, is in the ratio compounded of the densities of the electric fluid of the two electrified molecules and inversely as the square of the distances.*

For Magnetics: *The magnetic fluid acts by attraction and repulsion according to the ratio compounded directly to the density of the fluid and inversely to the square of the distances between the molecules.*

Thus for both electrostatics and magnetics Coulomb had proved that the relationship of repulsive or attractive force, $F$, to distance of separation, $r$, of charges or poles, is:

$$F \text{ varies as } 1/r^2$$

In addition to the inverse square relationship indicated in his law above, Coulomb implied that the quantity of the "densities" was involved, i.e., that the forces were also proportional to the product of the strength of the electric charges or the magnetic pole strengths. Initially Coulomb offered no proof for this.

In a later experiment, to determine the relationship for electric charges, Coulomb devised an experiment in which he used a ball suspended in front of a charged surface. He successively reduced the charge on the ball by one-half by touching it with an identical uncharged ball. The resulting separation of distance between the ball and the plate indicated the force was directly proportional to the amount of charge. This, in addition to the inverse square relationship showed that the force was also a product of the two charges $Q_1$ and $Q_2$.

Combining the two relationships:

$$F = K\frac{Q_1 Q_2}{r^2}$$

and for magnetic poles, $M_1$ and $M_2$.

$$F = K\frac{M_1 M_2}{r^2}$$

Coulomb had no way of knowing the value of $K$, a proportionality factor with dimension depending on the system of units used. But he noted the similarity of his formulas to that of Newton's law of gravitation, involving also an inverse square action at a distance.

**COULOMB'S FORCE EQUATIONS** apply to point charges and poles, which are much smaller than their separation. This is represented on the smallest scale by the Coulomb force relationship between electrons and protons in the atom. The equations apply only to static electric charges and stationary magnetism. Moving charges and magnetic fields bring into action new electric and magnetic forces.

Accuracy of Coulomb's results was limited to a few percent, and he worked with distances of about 10 centimeters. But later tests have all confirmed that the exponent of $r$ must be 2, and the inverse square law must hold for all distances.

After Coulomb verified his inverse square law, he moved on to examine the configuration of the electrostatic strength around the surface of various charged objects. To do this he designed a probe consisting of a small gold leaf disk attached to a thin insulating handle. Touched to a point on a charged body, the disk absorbed a charge proportional to the electrostatic density at that point. Charge on the disk was then measured on the torsion balance. By repeated tests, Coulomb could picture the complicated charge distribution on any particular charged object. This was a significant step in the understanding of the nature of the electrostatic forces.

Coulomb continually sought explanations for phenomena encountered in his studies. In electrostatics, bodies tended to lose their charge by some kind of leakage. How and why did this occur? How was charge carried, and how did materials act in this respect? On subjects like these Coulomb presented his theories in his Third to Sixth Memoirs.

Coulomb determined that the rate of leakage, for a given atmospheric condition, was directly as the quantity of the original charge and inversely as the square of length of the dielectric support. Leakage was pictured by Coulomb as a continuous process, on a molecular level, in which there was a constant charge-sharing with the air or contiguous bodies. A time element was involved in which the resistance interval between the molecules had to be overcome, a process which harkened back to his concept of cohesion as a limiting factor in the structural strength of bodies.

Coulomb considered all matter to consist of either perfect conductors or what we now call dielectrics or insulators - imperfect conductors. Conductors permitted free flow of electricity over the surface. Dielectrics permitted the charge to reside on the surface or within, in a static condition. But no dielectric was immune to conduction because imperfections provided conducting molecules and these would allow flow of electricity if the electric intensity was high enough to overcome the cohesion of the molecules.

**MAGNETICS** had a particular fascination for Coulomb and was the subject of the seventh and last of his celebrated Memoirs. Coulomb examined the structure of magnets in terms of momenta, and the magnetic intensity in terms of the magnet dimensions. He attempted to resolve the ultimate processes accounting for magnetic performance.

Various investigators in the 18th century ascribed magnetism to a magnetic fluid, an excess

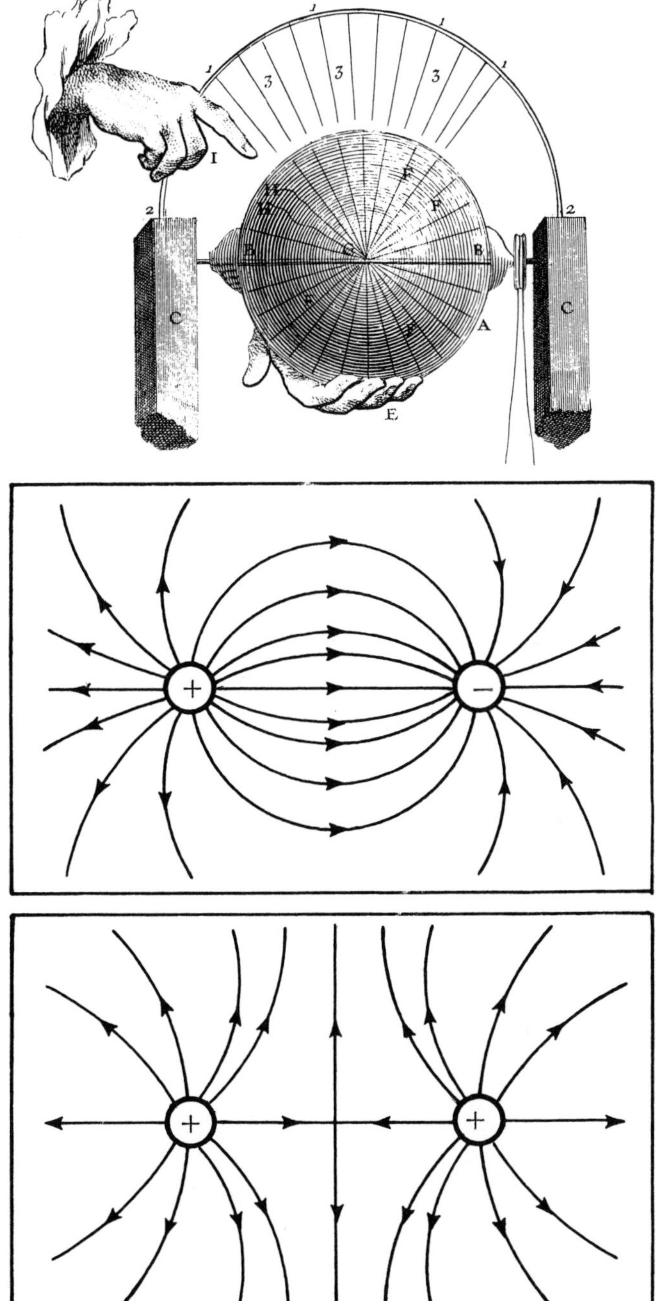


PICTURING THE ELECTRIFICATION . . .

Fig. 2.8- Shown above is an 18th century demonstration of electric force. The threads on an arc support were stiffened by attraction to the charged sphere. In the 19th century, experimenter Michael Faraday envisioned the electrification as a strain in the space around charges which he pictured by "lines of force," above, for the attraction of opposite charges, and repulsion of like electrical charges.

or deficiency of which produced the polar properties. Polarity was also thought to represent positive and negative magnetic fluids. Physicists were divided into one-fluid and two-fluid proponents of the magnetic makeup.

Coulomb originally subscribed to the two-fluid theory: the flow of fluids determined the magnetic gradient, and the transference of fluid to the ends of a magnet established the poles. The fluid theory was at a loss, however, to explain the fact that magnets could be indefinitely broken up and still maintain polarity. Also, magnets made up of small wires showed greater strength than the equivalent solid magnet. This led to a revision in Coulomb's thinking.

Coulomb eventually decided to discard the macroscopic fluid concept, and he made a prescient stride in hypothesizing that the ultimate magnetic particle was in fact none other than a polarized molecule. The magnetic fluid within the molecule was free to move to provide it with two poles, but the fluid could not pass from molecule to molecule. A thin magnet could be envisioned as a chain of magnetic molecules lined up, pole to pole, with evidence of the magnetism apparent at the ends of the chain. A long, thin magnet, then, however minutely cut up, would still retain the strength and polar properties of the original magnet. As an illustration, Coulomb pointed out that a thin tube filled with steel filings acted as a magnet even though its components were all separate.

Coulomb's molecular theory received a general approval. It did not destroy the fluid concept, but limited it to within the molecule. Coulomb's molecular magnet set the stage for Ampère, some decades later, to hypothesize molecular polarization, not by a fluid, but by an electric current flowing around the molecule, normal to its axis.

**THE FRENCH REVOLUTION,** beginning in 1789, brought drastic changes in government, and the overturning of many of the older institutions. The Corps du Génie was reorganized, and Coulomb retired from it in 1791. The Academy was abolished in 1793. Public works departments were also purged. Coulomb found it so dangerous in Paris that he exiled himself to a small estate he owned at Blois, where he continued his investigations. When Napoleon became the emperor of France, Coulomb came back to Paris, where he was able to resume his research. He was elected a member of of the new Institut de France, as an experimenter in physics. Coulomb was made inspector-general for public instruction in 1802, and served with distinction in organizing the French secondary school system. Coulomb's health, which had long ago been weakened by the fevers in Martinique, began to fail in the early summer of 1806. He died on the 23rd of August of that year.

As an engineer and physicist Coulomb was a true exponent of the scientific method. His work was based on use of rational analysis, supported by experimental proof. He also was a pioneer in bringing into physics the mathematization of experiment. He raised investigative science to a new high in France. And as a lifetime public servant he gave freely of his time for civil and military projects requiring high skill and insight.

**COULOMB -** In recognition of Coulomb's contribution to electrical science, the International Electrical Congress, at its meeting in Paris in 1881, named the practical unit of electrical charge the "coulomb."

In the CGS (centimeter-gram-second) electrostatic system of units, Coulomb's equation is measured in "statcoulombs." The unit charge, or one statcoulomb, is that quantity of point charge, when placed one centimeter from an equal charge in free space, repels it with a force of one dyne. The constant, *K,* relating force and charge in the CGS Coulomb equation, is equal to unity.

The unit of charge in the MKSA (meter-kilogram-second-ampere) system of units is the "coulomb," equal to $3 \times 10^9$ statcoulombs. In the MKSA system, force, *F,* is measured in newtons and the separation, *r,* in meters. Then, to make the two sides of Coulomb's equation balance, *K* is given a dimensional factor equal to $1/(4\pi\varepsilon_o)$, where $\varepsilon_o$ is the "permittivity constant," or capacity to accomodate electric charge in free space, with a value of about $8.9 \times 10^{-12}$ coulomb/newton-meters.

In the MKSA system Coulomb's law then becomes:

$$F = \frac{1}{4\pi\varepsilon_o} \frac{Q_1 Q_2}{r^2}$$

**COULOMB RELATIONSHIPS** - The coulomb, as a unit in the MKSA system, has a one-to-one relationship with other units. One *volt* is defined as a potential difference between points in an electric field such that one *joule* of work must be expended to move a charge of one *coulomb* against the field between the points, Thus:

*One volt = one joule per COULOMB*

The coulomb also has a relationship to the *ampere.* Electric current is measured as the rate of flow of charge per unit time, and

*One ampere = one COULOMB per second*

The coulomb also has a relationship to the *farad* The capacitor stores electric charge in proportion to the voltage across it, and

*One farad = one COULOMB per volt*

The COULOMB is equal in charge to that of $6.24 \times 10^{18}$ electrons. Two coulombs of like charge, one meter apart, would repel each other with a force of $9 \times 10^9$ newtons - 1 million tons.

**ELECTROSTATIC FORCES** - Coulomb considered the forces developed by charges as an "electrical action." This action was localized in the charges, and evidenced by the attraction and repulsion with other charges; the action occurred instantly over the intervening space. This was in keeping with Newtonian physics in which gravitation was an instant action-at-a-distance between bodies, with no role assigned to intervening space.

Michael Faraday, in the next century, considered that electrostatic action was something more than merely attraction and repulsion at a distance. He considered electrostatic action in terms of a *field of force,* and he visualized the field as made up of *lines of force.* The lines of force between electric charges were under strains of tension and compression, and their density was proportional to the strength of the electric field of force.

The lines of electrostatic force represent an electric field, **E.** A charged particle situated in the field will experience an electric force, **F**, on it. The amount of the force will be the product of the field strength and the strength of the charge, or, $\mathbf{F} = \mathbf{E}Q$. In the MKSA system of units the force, **F**, is measured in newtons, the electric field, **E**, in volts per meter, and the charge, $Q$, in coulombs.

Since an electrostatic charge produces an electric

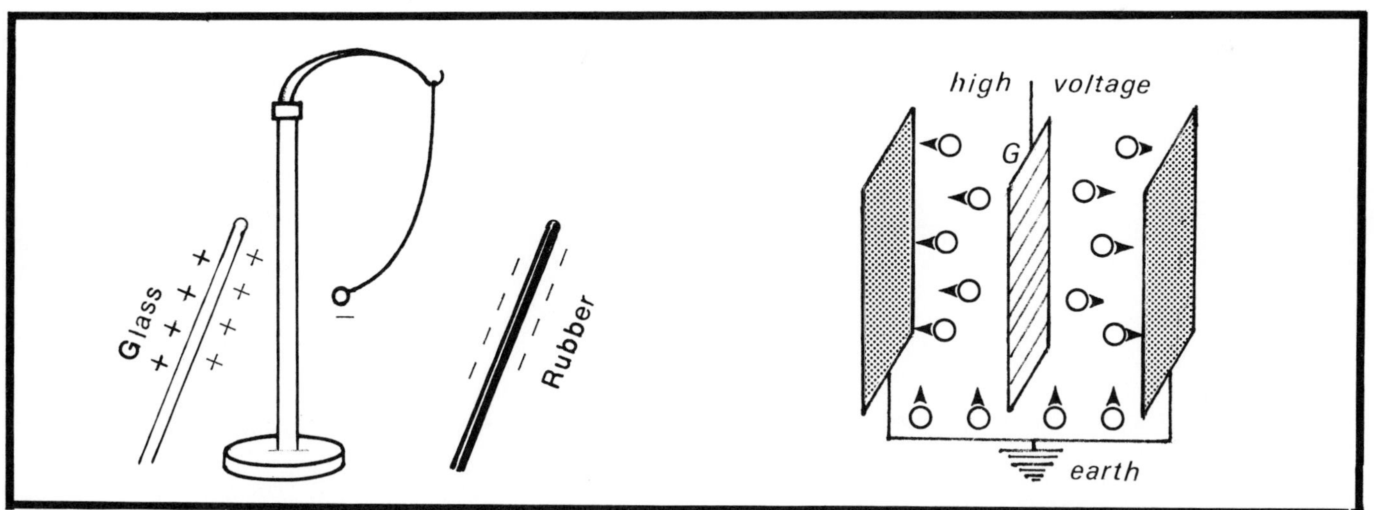


ELECTROSTATIC CHARGE EXERTS A FORCE

Fig. 2.9 - That electrostatic charge is accompanied by a field of force is shown by the power of attraction (unlike charges) and repulsion (like charges) on the pith ball.

THE CHARGE FORCE CAN MOVE PARTICLES

Fig. 2.10 - Smoke particles moving up a chimney are charged by high-voltage grid, G, from which they are then repelled to grounded metal plates for mechanical removal.

field of force emerging from the charge, it has the potential for doing work. If another charge is moved from one point to another in the field, Fig. 2.9, so as to utilize the force, work is done. Work done per unit charge is the "potential difference" between the points, and in the practical system of units is measured in volts.

**NATURE OF ELECTRIC CHARGE-** In his writings Coulomb discussed his concepts of the nature of charges in conductors and non-conductors in terms of the two-fluid theory of electrics. He did not attempt a definition of the nature of electrification itself; he was interested in how charges behaved. Not until a century after Coulomb, in the 1890s, did discovery of the electron provide one answer to "what is a unit electric charge?"

The electron charge is the smallest individual quantity of electric charge. It is the least negative charge and equal to $1.6 \times 10^{-19}$ coulombs.

The least positive charge is associated with the proton, and is exactly equal to the unit electronic charge. It is identified as the charge on the nucleus of the hydrogen atom.

The atom is normally electrically neutral, the positive and negative charges exactly balance. By various means, however, electrons may be moved to alter the balance. One means, the most ancient, is friction contact on certain classes of materials, a behavior called "triboelectrification." In this mode friction changes the electron-proton balance. Triboelectrification is considered fundamentally as a contact phenomenon between the different materials. The charges produced by contact are augmented by the greater number of contacts achieved by the rubbing or friction process.

When glass is rubbed with silk some electrons are transferred. By convention the glass is then called *positively* charged, leaving the silk *negatively* charged. There is now an attractive force between the positive glass and the negative silk, that is, the unlike charges attract each other.

Two glass rods rubbed with silk repel other, indicating that positive charges repel.

When amber, resin and hard rubber are rubbed with silk, the two attract each other. But the charged amber now attracts the previously charged glass rod. Thus we assume there must be a negative sign to the charge on the amber, and now a positive charge to the silk.

Two amber rods rubbed with silk repel each other, and so we conclude that lke negative charges also repel each other.

The attraction of unlike charges and repulsion of like charges are facts of experimental observation. No net charge, that is no excess + or - charge, is created by friction contact, only electron transfer. Regardless of the action or redistribution of charge, none is gained or lost. The total charge is conserved.

**ELECTROSTATICS SINCE COULOMB** - Up to the middle of the 19th century, electrification generators were friction type, using glass spheres or cylinders rubbed usually by silk covered leather pads. But frictional electricity was hard to handle and control. About 1860 new machines were developed, eliminating friction by using the induction principle. Typical of these inductors was the Holtz Machine, developed in 1864, a giant example of which is shown in Fig. 2.11.

In this generator circular glass plates, having radial sectors of tinfoil, are revolved in opposite directions, one in front of the other. Initial charges on the sectors of one rotating plate continuously induce opposite charges on sectors of the other rotating plates. These could be collected in Leyden Jars so that powerful high-voltage electrostatic effects could be obtained.

In the 18th and 19th century heydays of electrostatics, powerful induction machines were built for experimental and therapeutic purposes. Because it lacked industrial uses, however, electrostatics languished as a neglected science, having little more than academic interest.

**ELECTROSTATICS REVIVED** in the 20th century with new industrial developments that afforded the opportunity of putting the forces of charged fields to practical use. New technologies focused new attention to the possibilities of electrostatic applications. The first of these uses was for electrostatic precipitation.

Sir Oliver Lodge (1851-1940), in England, did the first experimenting on a method of removing dust from the air electrically. Pioneering in this

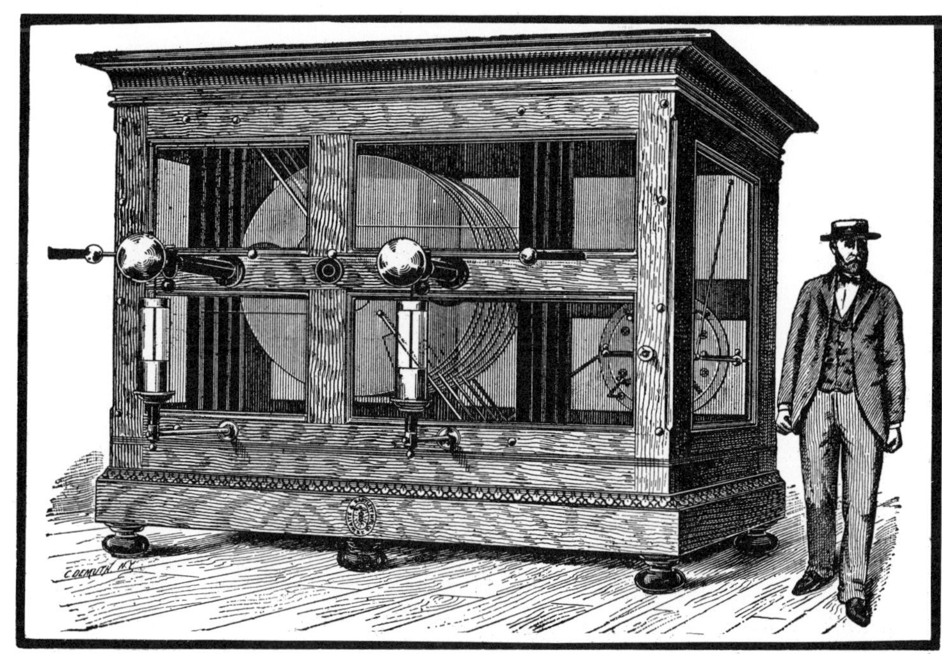

ELECTROSTATIC CURE

Fig. 2.11 - Early electrostatics had no other uses than for medical treatment. With the discovery and application of X-rays, electrostatic machines like this were used to excite the tubes.

This Holst electrostatic generator, claimed to be the largest in the world, was at the Battle Creek Sanitorium, Battle Creek, Michigan. Eight stationary and eight rotating disks, each 60 inches in diameter, produced spark discharges up to 30 inches. From *The Electrical Experimenter,* New York, 1915.

field was then carried on by Frederick G. Cottrell (1877-1950), an American engineer.

Electrostatic precipitation is done by passing air or gas through a screen of metal wires or bars which is separated a short distance from a metal plate connected electrically to the ground. The screen and plate are electrified to set up a strong directional field between them. Particles in the air or gas, passing through the screen, become electrically charged. They are then propelled from the screen to the plate by the force of the electric field. The collected particles are removed mechanically for disposal.

Cottrell installed the first industrial electrostatic precipitator in 1907 in a plant near San Francisco, California, where it successfully separated sulfuric acid mist from a gas stream. In 1923 he pioneered the use of electrostatic precipitation for fly-ash control at the coal-fired electric power plants of the Detroit Edison Company. Today tens of thousands of precipitators are in use for dust, fly-ash and other particulate removal in cement, steel, pulp and paper and other types of mills. The general public concern with air polution has brought new activity in the further possibilities of electrostatics for polutant control.

Electrostatic forces are also highly effective for manipulating small particles of solids or liquids. Paint is atomized to form a cloud of fine particles or droplets, which are charged by a high voltage. The object to be painted, electrically grounded, is enveloped by the charged paint cloud and rapidly covered with a uniform film. The automobile industry is a major user of this technique. Charged ink droplets can be guided by electrostatic fields to perform printing operations. Xerography, or dry printing, uses the light sensitivity of selenium elements, electrostatically charged, to transfer images to paper. Electrostatic precipitation is also used in mineral processing.

Electrostatic forces are an essential element in the mechanics of the atom, and modern physics makes reference to them. Electronic apparatus relies on electrostatic force. The industrial uses have given rise to "electrostatic engineering" as a branch of electrical engineering.

**ELECTROSTATIC MOTORS** - The earliest conversion of electrical energy to continuous mechanical motion is credited to a Scot, Andrew Gordon, a Benedictine monk, and a professor of philosophy at Erfert, Germany. In the early 1740s he devised the "electric bells" shown in Fig. 2.12. Benjamin Franklin's "electric spin wheel," described in 1748, was apparently the first machine for a conversion of electrical energy into rotating mechanical energy, as shown in Fig. 2.13.

Gordon's "electric bells" was an example of a "pendulum type" motor, using a metal clapper

Fig. 2.12- "ELECTRIC BELLS," invented about 1742 by Andrew Gordon at Erfurt, Germany. Probably the first device for converting electrical energy into continuous motion. The suspended metal clapper alternately contacts a bell, receives a charge, is then repelled by the like charges to swing to the other bell. From Oleg E. Jefimenko, "Electrostatic Motors," Electret Scientific Co., Star City, W. Va.

suspended by a silk thread. From an initial contact with one of the bells the clapper took on a charge of the same polarity as the bell. It was then repelled by that bell and attracted to the other. On touching the second bell the same action took place and the clapper swung back. In the Franklin motor the movement was effected by the alternate attraction and repulsion of the charges present on the rotating arms and those on the terminals of the two Leyden Jars.

Andrew Gordon invented another type of motor called the "electric pinwheel." The rotary element used radial wire arms with sharp-pointed ends bent at right angles to the arm. When connected to an electrostatic generator, the air next to the sharp ends was electrified with the same polarity as the points, thus repelling them to spin the wheel.

Subsequent to these early devices many variations were made. all using the basic principle of the attraction and repulsion of electrostatic charges.

In 1867, Holtz, the inventor of the electrostatic generator shown in Fig. 2.11. found that when one generator was electrically connected to and powered by a smiliar machine, it would run as a motor with a considerable mechanical output.

Another type of electrostatic motor used a rotor with fan-like blades which turned between pairs of similarly shaped stationary sectors. A commutator on the rotor switched the polarity of the charge on the rotor and stationary blades to that the rotor blades repelled on leaving and attracted on entering one sector for the next.

The newest type of electrostatic motor uses "electrets" for the driving elements. Electrets are dielectric materials which can be permanently electrified. They are the electrostatic analogs of permanent magnets, and exhibit definite polarity.

While electrostatic forces are not a match for electromagnetic forces where high powers are concerned, in the low wattage range they afford possibilities for small drives that are advantageous

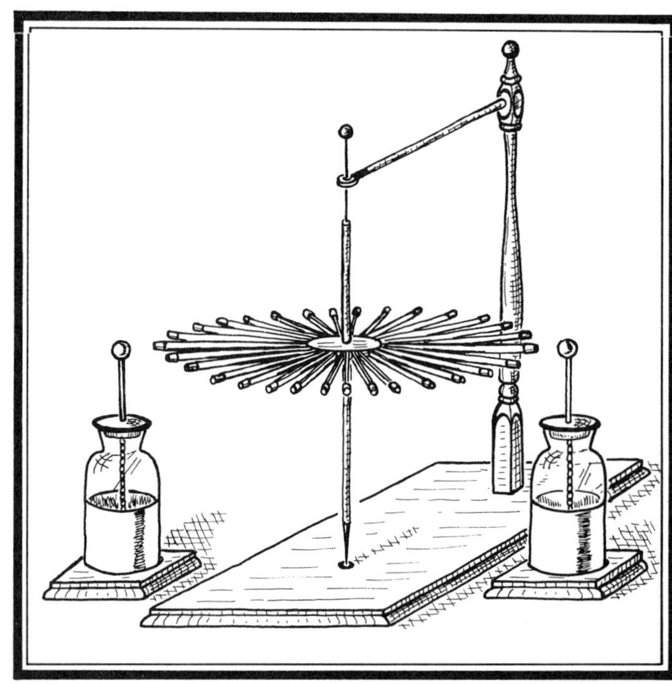

Fig. 2.13- "ELECTRIC WHEEL," invented by Benjamin Franklin, was one of the earliest means of producing rotary motion by electricity. As the thimbles moved past the Leyden Jars they were alternately attracted and repelled by the change of polarity. From Olef E. Jefimenko, "Eledtrostatic Motors," Electret Scientific Co., Star City, W. Va.

for some applications. With a suitable antenna, small electrostatic motors can be operated from atmospheric electricity.

**ELECTROSTATIC GENERATORS** - In 1931 R.J. Van de Graaff developed a new method for generating electrical potentials of several million volts. It employs the principle of charging a large outer surface, usually a spherical shell, by the continuous feeding of charge to the inside. A traveling belt moves charges imposed on one end, to a collector connected to the shell at the other end. The belt, Fig. 2.14, is positively charged at a. These charges are then moved up the the belt and exchanged at, b, for negative charges on the interior of the sphere. The negative charges are grounded by the returning belt. The accumulated + charges develop a high voltage difference, V, between the sphere and ground.

High-speed rotary inductor variable capacitance type electrostatic generators, at rated speeds up to 30,000 rpm, have been envisioned or experimentally developed. They are operated under pressurized gas, or in a vacuum, in order to increase the voltage output.

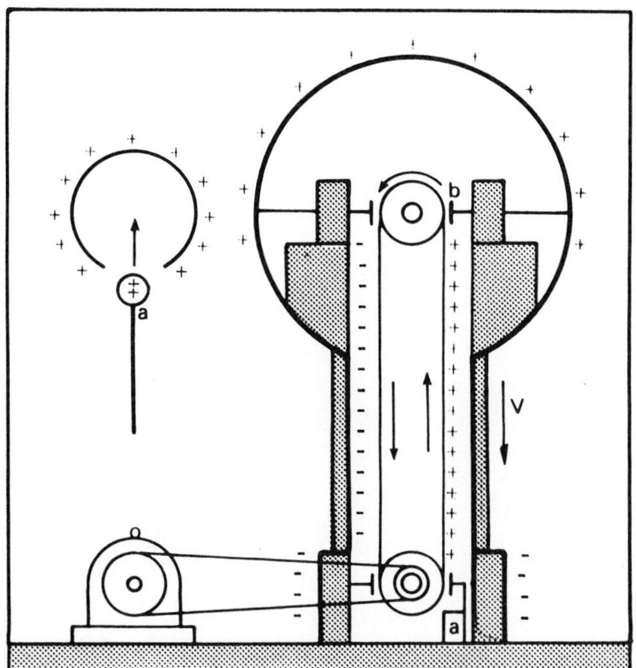


Fig. 2.14- VAN de GRAAFF electrostatic generator, put into use in 1931, develops potentials of several million volts. Principle is shown in the inset. Positive charges sprayed on vertical belt at, a, are collected at, b, to charge the sphere. Negative charges are returned to earth.

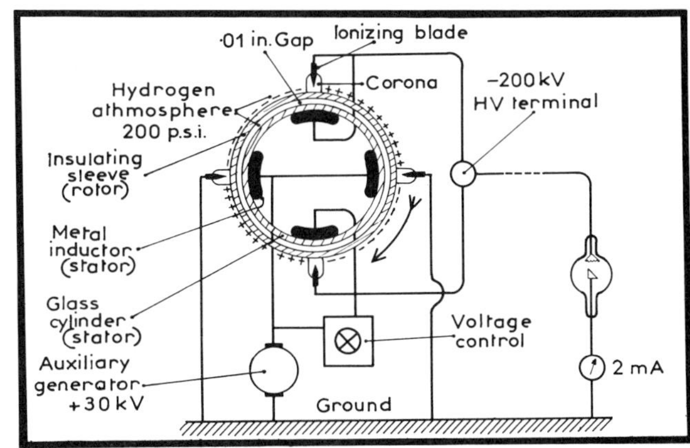


Fig. 2.15- FELICI CYLINDRICAL TYPE electrostatic generator. Metal inductors on stationary glass cylinder transmit charge, by induction, to rotating insulating sleeve, from which collection is made by the ionizing blades. Illustration courtesy Prof.N.J.Felici, National Center for Scientific Research, Grenoble, France.

The cylindrical rotating type high-voltage generator, Fig. 2.15, uses a rotating insulator sleeve that functions like the traveling belt of a Van de Graaff machine. The rotating sleeve collects charges from the inductors on the inner stator glass cylinder. Intended for industrial use, these generators operate under pressurized hydrogen gas. Speed is about 3,000 rpm. The rated output ranges up to about one million volts, and 10 kilowatts.

*THE FORCES QUANTIFIED by Coulomb have thus now been elevated to an important field of electrical science and technology. The delicate torsion balance and the method used by Coulomb opened a new era of detection and measurement, paving the way for better and more accurate observation of the electrostatic forces. With Coulomb, the first phase of early electrical exploration , begun by Gilbert, and covering two centuries, came to an end. The next century would open a spectacular new age of electricity and magnetism, and initiate the era of electric power.*

# WATT. . . he launched the Industrial Revolution

## that marked the rise of the factory and decline of the farm as the central focus of work in Western society

"**I SELL HERE** what all the world desires to have . . . **POWER!**" Thus proclaimed Matthew Boulton, James Watt's business partner, who was showing an important visitor the famous Soho Works of Boulton & Watt, near Birmingham, England. The Soho Works was producing the world's first line of steam engines.

After overcoming initial technical and economic difficulties, the Watt steam engines were winning their way into industry, not only in England, but in other countries in Europe. The engines were beginning to revolutionize the old state of trade and labor, and they would profoundly change the world's civilization. With an ample and cheap source of steam power, the old domestic system of manufacture was giving way to the new factory system. Instead of the small master, working in his own house, and with a few local apprentices and journeymen, the rich capitalist employer, with his army of factory hands, was taking over.

The Soho Works was the first large-scale product of the great talents of a scientist-inventor and the business acumen of an entrepreneur and financier, a type of partnership that was an indication of things to come. Taking the steam engine from its crude beginnings, inventor James Watt, with the manufacturing abilities of Matthew Boulton, successively provided the engine with a separate condenser, the flywheel, the throttle, the centrifugal governor and other basics that would supply the world's first universal industrial power.

Soho Works, Birmingham 1830 — Ronan Picture Library

# the unit of power is named for JAMES WATT who developed the steam engine

**SOME MEN STAND PRE-EMINENT** as great innovators - men whose lives are decisive turning points in world history. One of them is James Watt the celebrated symbol of the rise of the Industrial Revolution. Development of the steam engine by Watt in the eighteenth century marks the great divide between old and modern history.

With Watt's inventions, mankind would no longer be dependent on the efforts of animals, or the uncertainties of wind and water, for power. The steam engine would provide the world with an unlimited new source of energy. Steam power would bring unimagined new growth of industrial production and it would open an era of unparalleled social revolution.

The ability of producing a spinning motion by use of steam had been known for a long time. In the 3rd century A.D., an ingenious mechanic, Heron of Alexandria, built a toy steam turbine that whirled by reaction like a rotary lawn sprinkler. In the 17th century, Giovanni Branca, an Italian inventor, reported on a small turbine he devised in which a bladed impulse wheel was spun by a steam jet. Other acute minds had envisioned the possibilities of steam for power, but the spur for development did not exist at the time. An urgent need, and the right time and place, were the incentives for practical invention.

**ENGLAND,** in the eighteenth century, was a country with a growing interest in science, and with expectations for social improvement by its use. Francis Bacon, the philosopher-essayist, in

Fig. 3.1- JAMES WATT (1736-1819), and a drawing from his famous patent for a steam engine which used a separate condenser. The Science Museum, London

## TAKING MANKIND OUT OF THE TREAD MILL ....

WATT'S steam engines were the world's first technology to make possible the use of energy on a gigantic scale . . . they brought relief from such wearying burdens as pictured in this mine pumping scene of 1500 A.D. Two men tread the power wheel, while a man (and a dog) rest awhile. Tread mill, B, turns cogwheel, D. A chain pulls balls, G, up through the cylinder to raise the water. Illustration from Agricola's "De Re Metallica" (1556). originally printed in Latin in Germany, and translated by Herbert C. and L.H. Hoover.

his books *Novum Organum* and *The New Atlantis,* had proposed a total reconstruction of knowledge to extend man's real domination of the universe. In London, a group of scholars formed the Royal Society in 1660 to promote development of the new science that had been proposed by Bacon, both for philosophical and practical purposes. The early members of the Society were such outstanding talents as the first experimenter on the properties of air, Robert Boyle; the man who made the first great discoveries with the microscope, Robert Hooke, and Sir Isaac Newton, famous for his work on the laws of motion, gravitation, optics, and the development of the calculus.

Rise of the merchant class in England and the great expansion of navigation, brought an increase in recognition of the value of scientific and mechanical progress that would improve manufactures for the import and export trade. The great land owners of the country were coming out of their ancient way of feudal life and going into the pursuit of their land holdings as a source of financial gain, both from agriculture through better cultivation, and from exploitation of the minerals beneath their land.

Recovering more land for agriculture by drainage required surveying, excavating and pumping. Mining provided the need and opportunity for new machines to dewater the underground operations. Textile mills and other factories needed better sources of machine power. The stage was set for the aggressive use of science, and for development of new mechanisms to further material progress.

**ENGLAND** had the right resources for the birth of the most momentous economic revolution the world has ever known. She had coal for fuel. Most important, she had an iron industry capable of producing the parts needed for the building of machines such as steam engines. England was foremost in sea trade, with ample risk capital devoted to the mercantile enterprises. The government was stable, a body of labor was available, and the middle class was favorable to commercial activity. The value of science and productive mechanisms to the market place began to be recognized.

England and Scotland had a supply of scientific talent. Inventors and mechanics were turning to investigations of natural laws and of the country's resources for their practical uses. But early industrial progress had been limited by lack of adequate power. To overcome this the tremendous capabilities of steam began to be commandeered and put to work. At first the engines were crude and inefficient, but with succeeding developments and improvements they finally provided the country and mankind with an unlimited source of power that was superior to anything available before.

**PUMPING** was the first big incentive for steam power. Through the ages underground mining operations had been hampered and endangered by flooding. The depth and capacity of mining depended on the primitive pumps, operated by humans, by animals and by waterwheels. The underground mines of Cornwall, in southwestern England, that had been worked and deepened since antiquity, provided one of the first areas for

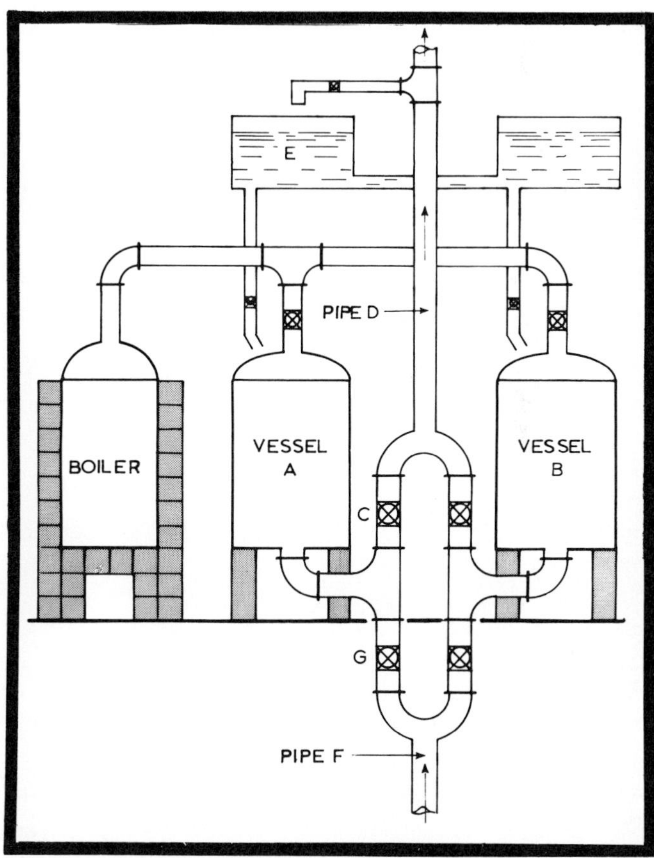

Fig. 3.2- SAVERY'S PUMP of 1628. Steam from boiler is first injected into vessel, A, driving water through check-valve, C, to discharge, D. Water from, E, then splashes over A, condensing steam, creating a vacuum and sucking up water through check-valve, G from suction pipe, F, to refill A. The same process then occurs in the alternate vessel, B.

use of steam-powered pumps.

A workable apparatus for lifting water was first developed by Thomas Savery (1650-1715), an Englishman who started life as a miner, but later became a well-known London scientist and inventor. Savery patented a "water-raising-engine" in 1608, based on the principle that "nature abhors a vacuum," a principle which, applied to pumping, was a radical idea at the time.

As shown by Fig. 3.2, water in vessel, A, is pushed upward by steam pressure from the boiler. The steam is then condensed by cooling water from, E, and the resulting vacuum refills the vessel by suction through pipe, F. Meanwhile the process is repeated in vessel, B.

Savery's pump was used in some mines, but the steam pressures required by high lifts were beyond the safety limits of existing boilers. Also the efficiency was extremely low because of the need of reheating the steam vessels after each cooling.

In 1679 a French scientist, Denys Papin, inventor of the pressure cooker, proposed that a piston could be raised by heating a little water in a cylinder to produce steam pressure. Condensing the steam by water sprayed into the cylinder would then create a vacuum to return the piston. The scheme was impractical because of using the cylinder both for boiler and for piston chamber.

But, in England, another inventor saw that he could adopt both Savery's and Papin's ideas and modify them to produce the world's first practical engine.

**THOMAS NEWCOMMEN** (1663-1729), a tool merchant in the mining country of southwest England, had no scientific training, but was a skilled mechanic, a logical thinker, and aware of the pumping problems in his area. Perceiving that Savery's and Papin's principles could be combined, he set about to construct the first successful engine using steam. Newcommen made Papin's principle effective by separating the boiler and the cylinder. He then used Savery's principle of a vacuum to utilize the force of atmospheric pressure.

Newcommen, in 1705, gave his ideas the form and working arrangement shown in Fig. 3.3. Steam was valved into the cylinder as the piston rose due to the counteracting weight of the pump rod on the rocker arm. When the piston reached the top, a chilling jet of water was sprayed into the cylinder to condense the steam. Thereupon atmospheric pressure pushed the piston down, thus raising the mine pump rod to lift the water.

Newcommen's engine was, for its time, a clever construction. The steam pressure required was low, only that to fill the cylinder. The work was done by the weight of the atmosphere. To get more power you made the cylinder bigger. The engine was first introduced in 1717, and was immediately successful for pumping in mines, giving a big push to mine development. Though ponderous by modern standards, the Newcommen engines were widely used and were rugged. They were used to operate both mine pumps and blowers, and some were in service for more than a hundred years.

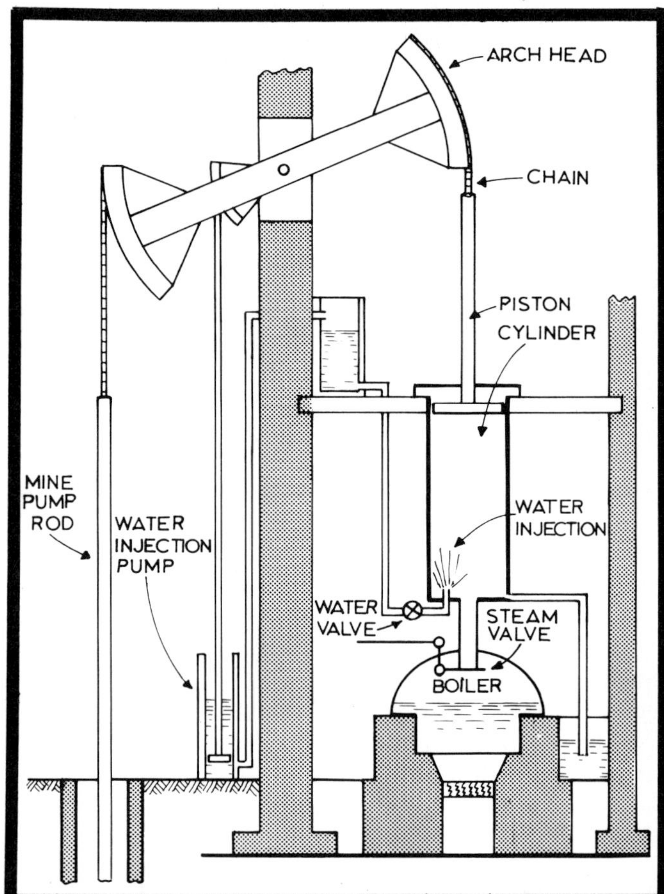


ATMOSPHERIC PRESSURE FURNISHED THE POWER

Fig. 3.3- Newcommen rocker engine of 1705. Steam is admitted to the cylinder from the boiler as the piston is raised by the weight of the pump rod on the rocker arm. Injection of water then condenses the steam, thus creating a vacuum. Atmospheric pressure then forces the piston down, lifting the pump rod to operate the pump in the mine shaft.

THE MODEL THAT LED WATT TO INVENTION . . .

Fig. 3.4- This little Newcommen engine at the University of Glasgow was used by Watt in work leading to invention of the separate condenser. The Science Museum, London.

Despite their usefulness, Newcommen engines were very inefficient. The high fuel consumption in alternately heating and cooling the cylinder limited application of the engines mainly to coal mines where they could use the waste coal, and in the richer metal mines where the expense could be afforded. The way was open for new ideas. For the most revolutionary of these the scene now shifts to Scotland and a gifted young man names James Watt.

**AT GREENOCH,** near Glasgow, James Watt was born in 1746, being descended from a family of practical mathematicians and navigators. It was intended that young Watt would follow in his father's business as a supplier of navigation instruments, and eventually as the operator of a family-owned ship. Hence he was not apprenticed to a craft, nor sent to a university. However, the family fortune was ruined by loss of the ship in a storm, and James had to seek a new career. Having a bent for mechanical pursuits, he decided at the age of nineteen to become an instrument maker.

Watt was then too old for acceptance by the guild at Glasgow, so was apprenticed to a London instrument concern. After a year, ill health forced his return to Glasgow, where he tried to set up an instrument business. The guild, however, would not recognize him because he had not served a full term of apprenticeship. But the University of Glasgow was not under that restriction, and recognizing Watt's talents and usefulness, allowed him to open an instrument repair shop in the University laboratory precincts in 1757.

Glasgow was one of the intellectual and creative, busy cities of the world, and the University was a center of scientific talent. In this atmosphere Watt, although relatively unschooled, quickly absorbed knowledge and was inspired by such eminent lecturers as James Black, discoverer of the latent heat of steam. Watt had the opportunity for frequent discussions with Black and others on the characteristics of heat and its relation to steam. This was a lively subject because of the growth in the use of steam engines. In these discussions there were speculations on the future possibilities of steam power. Newcommen engines were in wide use at the time.

**WATT'S** keen interest led him into experiments with the force of steam. By a fortunate coincidence a model of Newcommen's engine for classroom demonstration, Fig. 3.4, did not operate properly, and was turned over to Watt for repair. Trying various alterations he got the machine going and was quickly struck with its extravagant use of steam. The boiler was not able to keep the engine running.

Watt began a systematic study to find out why. He was led to a detailed examination of the engine and the properties of steam. He was particularly impressed with the *latent heat* of steam, that is the amount of heat hidden in the steam that required so much cooling to condense it. Effective disposal of the heat latent in steam appeared to be a factor that would improve the efficiency of the Newcommen engine.

Every stroke of the engine required heating and cooling of the cylinder. Steam was first admitted to raise the piston, and them cooled and condensed

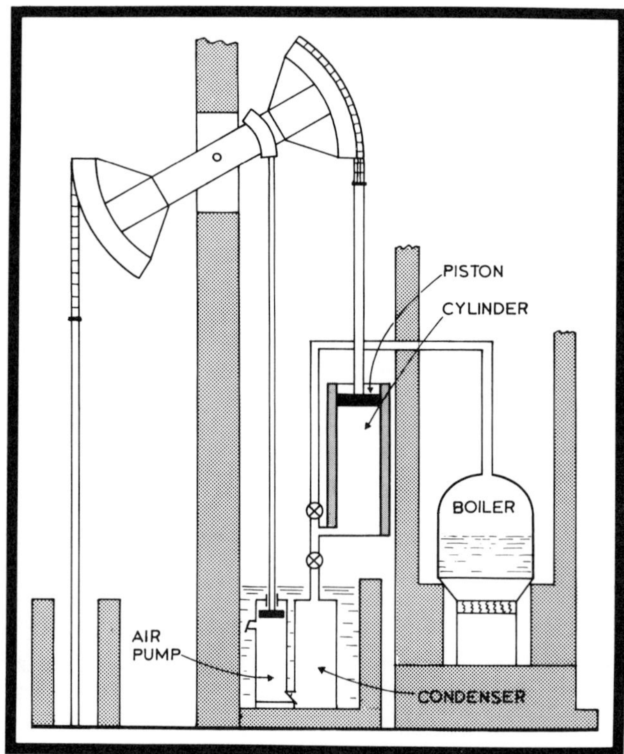


Fig. 3.5- WATT'S FIRST ENGINE, patented in 1768. The drawing shows the essential features of an insulated steam cylinder exhausting into a separate water-cooled condenser. Air pump removes distillate from the condenser cylinder.

by the jet of water. A lot more steam was then required just to reheat the cylinder walls. This was wasted. Furthermore, jet cooling did not remove all the latent heat, hence did not provide a good vacuum. Watt concluded that some new method of condensing the steam was needed to improve the performance of the engine.

For several years Watt pondered on the gain in achieving condensation without cooling the cylinder, and how to accomplish it. Then, in a flash of inspiration came an idea. He would simply exhaust and condense the steam in a separate vessel, or "condenser." Thereupon a high vacuum could be obtained, and the cylinder kept hot. Put to test on the Newcommen model, the separate condenser did what he anticipated. The engine became much more efficient; it could operate with about one-fourth the steam it took previously.

Watt then worked out other features for improving operation of the engine - a little pump to remove the condenser distillate, and a steam jacket for keeping the cylinder hot. These features were included in Watt's first patent of 1768, one of the most famous industrial patents ever issued. Watt had little resources of his own to demonstrate his engine principles on a large scale. He needed financial backing, and eventually found the support that was to help make his invention a commercial success.

To meet the cost of his experiments Watt first got help from his friend, Dr. Black, at the University. Later, John Roebuck, a coal mine operator, bought a two-thirds interest in Watt's patent. But in 1773 Roebuck's business failed and Matthew Boulton, one of Roebuck's creditors. took over his shares of Watt's patent. Boulton met Watt and was impressed with him and with the patent. He proposed a partnership to build the improved engine. Watt thus became associated with Boulton, an outstanding manufacturer of metal goods in Birmingham.

**MATTHEW BOULTON,** a friend of Benjamin Franklin, was a founder of the Lunar Society, a group which met to discuss technical innovation. He realized the potential for engines that would supplant waterpower in his factories for greater and steadier production. Boulton recognized Watt's genius and saw a future in Watt's engine patent. He founded a separate firm, Boulton and Watt, to manufacture the steam engine. With this partnership was launched a new era of power. The first factory was located at Soho, near Birmingham. With the help of John Wilkinson, an expert machinist who supplied the means for precision boring of the cylinders, the new factory began turning out their first engines.

Watt's first engines used the Newcommen rocker-arm arrangement, with the new separate condenser. While more economical and faster, they worked only single-acting, up and down, using atmospheric pressure. Boulton, a forward thinker, realized that the engine would be greatly superior for his mills if it could produce a revolving motion rather than the rocking-arm reciprocating type. He pressed Watt for the development of a rotary engine with a flywheel, more suitable for general use.

Watt responded with an engine in which expanding steam pressure rather than the atmospheric pressure furnished the power. By connecting the piston, through a crank, to a flywheel, he emancipated the engine from its limited up and down role for operating pumps and blowers, into a universal type drive for all kinds of mills and machines. In 1782 Watt received two patents covering improvements in the efficiency of engine operation.

**WATT'S PATENTS** of 1782 brought the engine up to its final state of most effective performance. One patent related to the use of double action, with steam pressure applied to each side of the piston, thus giving full utilization of the cylinder. The second patent, relating to steam economy, was to use the expansive property of steam itself. Instead of admitting steam for the full capacity of the cylinder, it was cut off when the piston had completed only part of the stroke, allowing the full expansive power of the steam to be used. These major modifications opened the way for use of higher steam pressures and higher speeds. The new rotary engines began to power a great and revolutionary expansion in industry.

THE ROCKING BEAM PUMP IS IN USE TODAY . . .

Fig. 3.6- Present-day oil-field pump. Except for the modern method of motor or engine drive, the pumping rig uses the same rocking beam as Newcommen or early Watt engines.

Further contributions by Watt to engine operational improvement were a throttle valve for regulating steam flow, and a centrifugal governor to automatically actuate the throttle valve to maintain the desired speed and to prevent runaway. Watt's governor was the first example of a feedback control, which today is an essential element in technology. Watt also devised a steam inducator, which traced a curve showing the relation of steam pressure to movement of the piston. This was used for analysis of the engine performance.

**"HORSEPOWER"** - The customary charge for the Boulton and Watt engines was based on the savings effected in fuel consumption over the older engines. But after the development and wider application of the new rotary engines, Watt and Boulton were faced with the question of proper charge for them. More than a hundred years earlier Savery had compared the work of his engines to the horses required to lift an equivalent amount of water. Others had subsequently made the same comparisons but no unit of how power should be measured had been established.

When the rotative engine was introduced it frequently replaced animal power, so that the number of horses replaced seemed an obvious way to measure and charge for performance. Watt was also concerned with systematizing his line of engines, and he proceeded with calculations to define the magnitude of horsepower. Data given to him indicated that a certain mill horse walked in a path 24 feet in diameter, 2½ turns per minute. Watt assumed the horse exerted a pull of 180 pounds, and arrived at a figure of 32,400 foot-pounds per minute for one horsepower. Watt later rounded off the figure to 33,000 foot-pounds.

Another version of the origin of determining a horsepower is that Watt measured the rate of work exerted by a horse drawing rubbish up an old mine shaft and found it amounted to about 22,000 ft-lbs per minute. He added a margin of 50%, arrving at 33,000 ft-lbs. In 1783 he stated that: "Each horse equals 33,000 lb., one foot high, per minute." The firm then began to rate their engines in so many horses. Watt had set up a unit for defining the rate of doing work that would come into universal use.

The partnership of Boulton and Watt built and sold engines with wide success, including use on the first primitive steam locomotives and steam boats. During Watt's partnership, which terminated in 1800, some five hundred engines had been applied. As Boulton had predicted, the rotary engine proved very popular, and had accounted for 60 per cent of the applications.

With the expiration of his patents in 1800, Watt retired from the business, in comfortable circumstances, and spent the remainder of his life at his estate at Heathfield, near Birmingham. There he resumed his early skills as a mechanic, and with an interest in sculpture, devised a proportional sculpturing machine for replicating statues in various sizes. Watt died at Heathfield in 1819.

The new steam engines gave great impetus to the science of heat, or thermodynamics. This, in turn, revealed the exact properties and the potentialities of steam as power. With the expiration of Watt's patents in 1800, new inventors and builders entered the field and the development of engines proceeded rapidly for higher fuel efficiency. With better steels, better machining, new types of boilers and higher pressures, engines could be built for higher speeds and larger sizes. Full utilization of the steam pressure was achieved by development of compound engines of the double, triple and quadruple expansion types.

**THE AGE** of the stationary reciprocating engine flourished from about 1850 to the early 20th century. The Boulton and Watt engines and those that followed became the pre-eminent power for all types of drive in their time. In addition to their industrial uses, steam engines made possible for the first time the development of railway locomotives for mass transport, and the steamship for world navigation.

But the days of steam engines were being numbered. Development of the modern steam turbine in 1884 by C.A. Parsons in England, brought a revolution in the age of steam. The high-speed turbine with its connected electrical generator launched the era of electrical energy on a vast scale. With the availability of this more versatile power, came the 20th century conversion to electric drive. And, paralleling the electric motor, came the rise of the internal combustion engine. The reciprocating steam engine, once the epitome of power had, as its last survivor, the steam locomotive. All are now a thing of the past.

Fig. 3.7- WATT'S ROTARY engine (1782). It employed his principles of double-acting cylinder, separate condenser, governor for regulating the throttle valve for speed control, and cut-off for admitting steam during only part of thestroke. Flywheel evened out pulsations of the engine. Science Museum, London.

That Watt's name will always be associated with power, was assured when the Second International Electrical Congress, meeting in Paris in 1889, named the unit of power the *WATT.*

Power, in the physical sense, is the rate of transferring *energy.* Energy, in the International System of Units (SI), is measured in *joules.*

The **WATT** is that power which in one second gives rise to the energy of one joule. In electric power: 1 *watt* =1 *volt-ampere* = 1 *joule/second.* Electric motors have traditionally been rated in horsepower. However they are alternately rated in watts, or multiples of watts. One horsepower is equal to 746 watts, approximately.

The watt is not specifically an electrical unit, since power has been associated with mechanics long before there was any use of electricity as a practical power-producing means. Thus the unit of power belongs originally to mechanical power, and hence has its derivation in the discoveries of James Watt. Nevertheless, the watt, with its multiples the: microwatt, milliwatt, kilowatt and megawatt, is far more well known today in relation to electric power.

**A GREAT CATALYST** of industrial change was not only the availability of power, but also in the increase of rotative speeds. Whereas the old beam engines rocked along at the rate of a few strokes per minute, Watt's double-acting, crank and flywheel engines immediately made possible higher rotative speeds. Power-using machines could then run faster to increase the volume and efficiency of output. Use of higher speeds generally reduced the bulk, space requirements and costs related to production. In time, engine speed ratings were expanded over a wide range to suit the needs of driven machinery that could advantageously use higher speeds. This range in speeds involved a corresponding range in the size of the engines.

The acme of slow speed, with corresponding huge bulk, was the 36 rpm steam engine used to drive all the machinery at the Centennial Exhibition in Philadelphia in 1876. Its two cylinders were 40 inches in diameter, with a 10 foot stroke. The flywheel was 30 feet in diameter. Contrasted to this was the development of high-speed engines providing the same output at one-tenth the bulk and weight, and with correspondingly lower cost.

TWO OF WATT'S INVENTIONS ARE ESSENTIAL TO THE MODERN STEAM TURBINE . . .

Fig. 3.8- Turbine Electric Generator Set - Steam turbines use the principles of two Watt inventions: a condenser for providing the exhaust vacuum, and a speed regulating governor. Photo - Electric Machinery Division of McGraw-Edison Company.

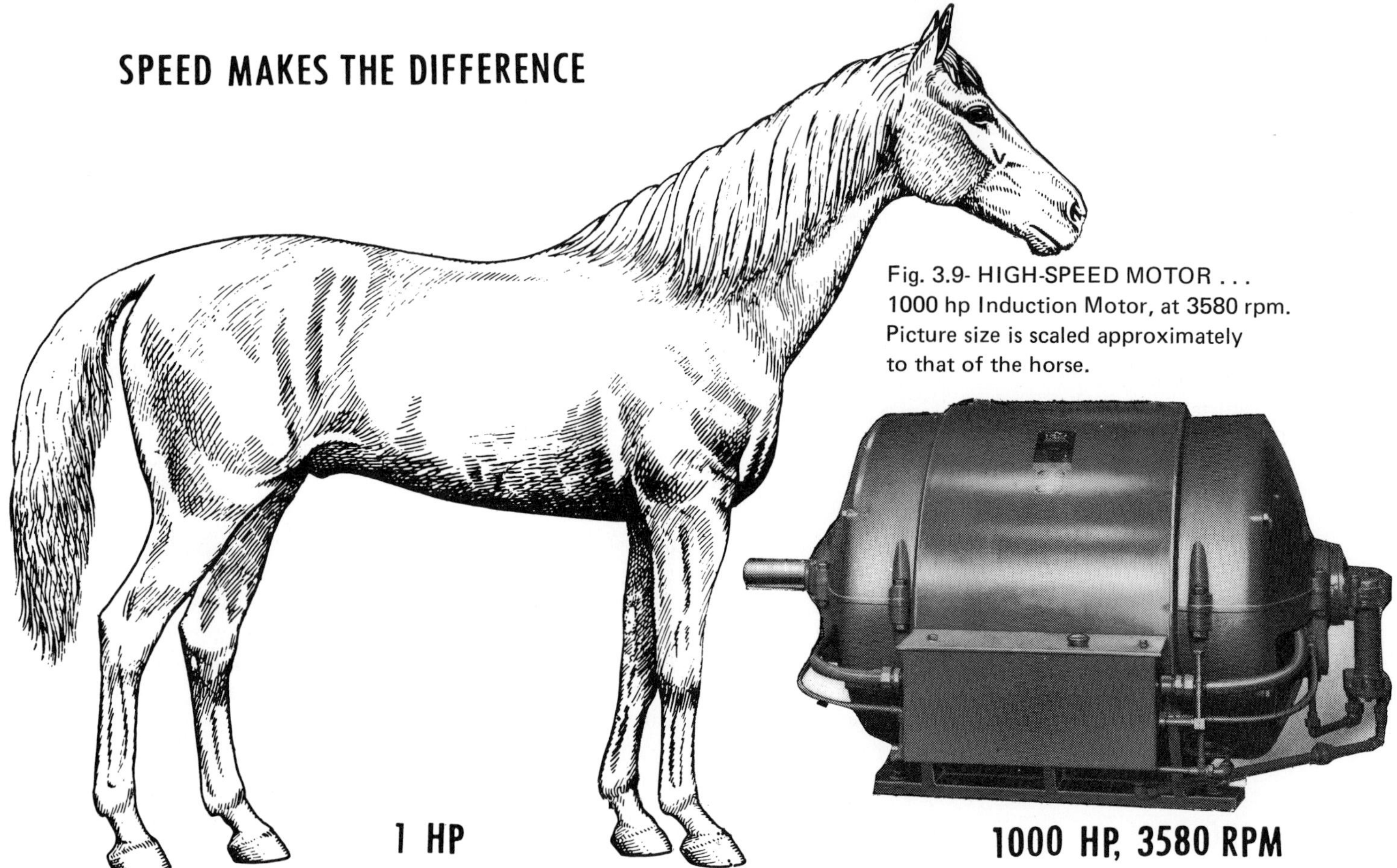


Fig. 3.9- HIGH-SPEED MOTOR . . . 1000 hp Induction Motor, at 3580 rpm. Picture size is scaled approximately to that of the horse.

**ELECTRIC DRIVE** has made possible great gains in efficiency, adaptability and convenience not available with the steam engine. Electric motors have also provided a wider range of speeds. But as was the case with steam engines, there is also a dramatic difference in size and weight for the same horsepower output, depending on the speed.

The electric motor is a force-multiplying device. The force is delivered as a rotative torque at the rated speed of the motor. The motor torque and and the motor speed are related, for a given horsepower, by Watt's rule that one horsepower equals power at the rate of 33,000 ft-lbs per minute:

$$\text{Torque} = \frac{33.000 \times \text{Horsepower}}{2\pi \times \text{RPM}}$$

From this formula it is apparent that, for a given motor horsepower, the torque the motor must deliver will vary inversely with the speed of the motor. Thus more torque is required at low speed and a larger motor structure is necessary to produce that torque, as shown by comparing the illustrations of the two motors above.

The 1000 hp induction motor, revolving at 3580 rpm, for example, has a torque at full horsepower of 1466 ft-lbs. The 1000 hp synchronous motor at the right, however, at 180 rpm, or one-twentieth the rpm, must deliver a torque of nearly twenty times greater, or 29,166 ft-lbs. To obtain the higher torque the motor size, electrically, magnetically and physically must be much larger.

Because of this size-speed relationship, one of the progressing developments in machinery for pumping, compressing and other purposes has been a movement to use of higher speeds by a change from low-speed reciprocating types to high-speed centrifugal type to accomplish the same operation.

**THE EVOLUTION OF SOCIETY** in the 19th century was accelerated at an unprecedented rate by the advent of Boulton and Watt's steam engines, and those of others that came later. Following the industrial languor of the pre-steam world,

Fig. 3.10- LOW-SPEED MOTOR . . . 1000 hp Synchronous Motor, at 180 rpm (40 pole, 60 Hz) to drive cement grinding mill. Picture size is scaled approximately to that of the horse. Motor photos courtesy Electric Machinery Division, Mc Graw-Edison Company.

**1000 HP, 180 RPM**

the new, ample source of power released an unparalleled crescendo of human inventive and productive activity. Steam power gave, for the first time, a mighty new force at the disposal of man to remove the old limitations of human and animal power, and the caprices of power from the wind and water.

The new engines for powering mills and machines brought about a vast restructuring of society. People were attracted from their age-old village and rural existence to the new occupations opened up in the industrial centers. Masses of people moved from the farm to the city. And the industrial age brought with it a host of social problems.

It brought both wealth and squalor, both beauty and pollution. The use of steam power to produce goods and food, and to transport them around the world, also helped to multiply the earth's population - more than qunintupling the world's numbers in the two centuries since Watt's first engines.

**NOWADAYS,** with the disappearance of the steam locomotives and their memorable expression of the power of steam, and with the replacement of the reciprocating steam engine by the internal combustion engine and the electric motor, it may be sometimes felt that the day of steam power is past. This of course is not true. Steam is still the prime medium for converting heat into other forms of energy. The steam turbine provides the power to generate electricity for the bulk of the world's needs. Steam, for the indefinite future will be the prime mover in supplying the energy to relieve the incredible drudgery that once was the lot of mankind, and which still persists in large parts of the world today.

**WATT'S INVENTIONS** live on in the operation of modern steam turbines. His separate condenser and speed-regulation principles are essential to their performance.

# VOLTA

the
volt
is named for
the
discoverer
of
continuous
current

ALESSANDRO
VOLTA

VOLTA'S "crown of cups," one of his arrangements of an electric battery. From the Philosophical Transactions of the Royal Society, London, Vol. No. 90.

**NAPOLEON BONAPARTE** was in the audience on the opposite page. He was listening intently to one of the foremost savants of the day, the Italian physicist, Alessandro Volta, who was in Paris demonstrating his amazing invention - the "electric pile."

France had just come out a great Revolution. The bloody conflict had left the scholars and the leaders of France skeptical as to religion, but with a great curiosity about new knowledge and a consuming desire for progress through science. Napoleon was a daring and ambitious product of the Revolution, and his military skill and administrative prowess had made him First Consul of France.

After his many victories on the battlefields of Europe, Napoleon had visions of making all the countries of that continent into one nation with Paris as the capital. Napoleon had a keen, practical interest in science. He was promoting a national pursuit of scientific excellence as an essential element in progress and superiority of his growing French empire. He had been made an honorary member of France's top-ranking scientific body. the French National Institute.

In 1800, in a letter to Sir Joseph Banks of the Royal Society of London, Volta had announced his novel assembly of an apparatus . . . *for the endless circulation of electric fluid.* Quick to recognize the implications of this important new scientific discovery, the National Institute had invited Volta, who spoke fluent French, to lecture

# his invention of the battery put electricity on the move

in Paris on his great creative accomplishment that had excited the scientific world. Now, in 1801, he was showing Napoleon an astonishing capability of his new current - it could decompose water into its gaseous elements - hydrogen and oxygen.

Napoleon was captivated with Volta's presentation. After the meeting he asked Volta many questions, and sensing the importance and potential of Volta's invention, he instructed the Institute to form a committee of its foremost members to investigate the operation and the possibilities of Volta's creation.

**VOLTA'S CREATION** was the electrochemical cell. For the first time it provided a means for delivering a continuous current. Instead of the Leyden Jar of the previous century, with its instantaneous, quickly exhausted and variable discharge of electricity, Volta's electric pile gave, with a simple assembly, a constant and controllable flow of current. The way was now opened for an immense new field for investigation and use of electricity.

Volta's invention of the electric battery was the final link in a chain of discoveries prompted by curiosity about the effects of metals on the sensibilities of muscles and nerves, and on the muscular contraction that followed stimulation from electrical machines. The possible association of these phenomena with electricity had been hinted at by investigators dating back into the 18th century. The stage for Volta's work was set by a contem-

VOLTA AMAZES NAPOLEON . . .

Volta's discovery of the electric pile so impressed the French scientists that he was invited to lecture at the National Institute of France in Paris in 1801. Napoleon Bonaparte was a member of the Institute, in the audience, and was an enthusiastic listener as Volta demonstrated that his electric pile could dissociate water into its elements. From a painting by Bertini, Italy, 1899. Burndy Library.

porary countryman, Luigi Galvani (1737-1798), a gifted physician and experimenter. His research in "animal electricity" provided a clue that set the genius of Volta on the path leading to his experimental triumphs. And Volta's final achievement was stimulated by a great controversy between he and Galvani on the nature of their findings.

**ANIMAL ELECTRICITY,** as experienced by the powerful shock from contact with an electric eel, has been known from antiquity. A few physicians had speculated that the muscular contractions in the body might be explained by some similar electrical stimulus. 18th century experimenters were familiar with the muscular spasms of humans and animals subjected to the discharge of electrostatic machines. From evidence like this some physicians advocated electric shock as a muscular stimulant.

The search for an explanation of muscular con traction had prompted various anatomical experiments on the possible relationship of metallic contact to the functioning of animal tissue. An early investigator pursuing this possibility was Johann Sulzer (1720-1779), a professor of physiology at Zurich. He described in 1750 a chance discovery that when the tongue was put between two strips of different metals, such as zinc and copper, whose ends were in contact, an unpleasant acid taste was felt. With the ends separated, there was no such sensation. Sulzer ascribed the taste phenomenon to a vibratory motion set up in the metals which stimulated the tongue. Sulzer used other metals with the same results but his reports went unheeded for a half century until new developments called attention to his findings.

The next fortuitous and remarkable discovery in "animal electricity," one of the most significant in the annals of science, was a chance observation that while a freshly prepared frog's leg was being probed by a scalpel the leg jerked convulsively whenever a nearby electrical machine gave off sparks. This occurred in 1750 in the laboratory of Luigi Galvani, at Bologna Italy.

**GALVANI** was a descendant of a very ancient Bologna family. He became a gifted scholar of medicine and at age 25 was made Professor of Anatomy at the University of Bologna. There he became a skillful and persevering experimenter. In keeping with the spirit of the new science of the times, Galvani developed an ardent interest in electricity and its possible relation to the activity of the muscles and nerves.

Newer electrical machines and the discovery of the Leyden Jar gave experimenters a convenient source of electricity. Dissected frog's legs were convenient specimens for investigation, and in his laboratory Galvani used them and other animal parts for studies of muscular and nerve activity. In these experiments he and his associates were studying a variety of irritational responses of the animal tissue from various stimulations.

Galvani, in writing of his experiments with the jerking of frog's legs from electrical stimulation

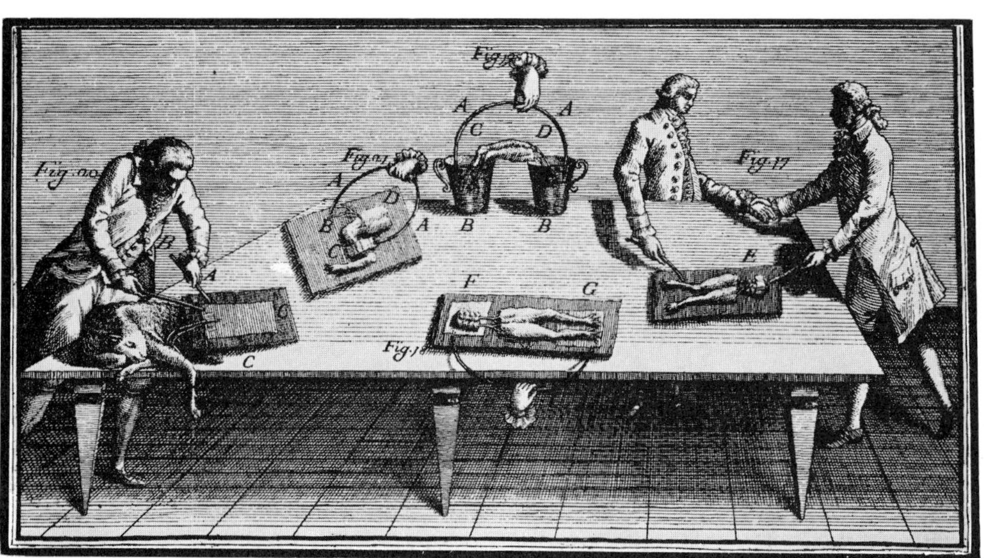

A FROG'S LEG GIVES A CLUE

Fig. 4.1- Convulsions of a frog's leg when touched by metals stimulated a search for animal electricity and provided the first clue to eventual development of the electric battery. From Luigi Galvani's "De Viribus Electricitatis in Motu Musculari," by Margaret Foley at the Burndy Library, Norwalk, Connecticut said that: "He had dissected and prepared a frog, and laid it on a table, on which there was an electrical machine. It so happened by chance that one of my assistants touched the point of his scalpel to the inner crural nerve of the frog; the muscles of the limb were suddenly and violently convulsed. Another of those who were helping to make the experiments in electricity thought that he noticed this happening only at the instant a spark came from the electrical machine. He was struck with the novelty of the action. I was occupied with other things at the time, but when he drew my attention to it I immediately repeated the experiment. I touched the other end of the crural nerve with the point of my scalpel, while my assistant drew sparks from the electrical machine. At each moment when sparks occurred the muscle was seized with convulsions."

With an alert and trained mind, Galvani, amazed at the incident, set out on an extended series of experiments to resolve the cause of the mystifying muscle behavior. On repeating the experiments, he found that touching the muscle with a metallic object, and having the specimen lying on a metal plate, were the conditions for the contractions.

Having heard of Franklin's experimental proof that a flash of lightning was of the same nature as the discharge of a Leyden Jar, Galvani set up to determine whether atmospheric electricity might produce the same results as his electrical machine. By attaching the nerves of frog's legs to aerial wires and the feet to ground, he got during a thunderstorm the same response as with the electrical sparks in the laboratory. But it was another chance observation during this experiment that indicated the future direction of inquiry and discovery.

**GALVANI** noticed that prepared frogs which were suspended by brass hooks through the marrow, and which rested against an iron trellis, showed occasional convulsions regardless of the weather. In adjusting the specimens he pressed the brass hook against the trellis and saw the familiar muscle jerk occurring each time he completed the metallic contact. To check whether this jerking might still be from some atmospheric effect, he repeated the experiment inside the laboratory.

He found that the specimen, laid on an iron plate, convulsed each time the brass hook in the spinal marrow touched the iron plate. He now recognized that some new principle was involved, and he varied his experiments to find the actuating factor. Substituting glass for the iron plate gave a negative result; using a silver plate restored the muscle reaction. He then joined equal lengths of two different metals and bent them into an arc. On touching the tips of the bimetallic arc to the frog specimens he got the familiar muscular convulsions. Not only was metal contact a contributing factor, but also the intensity of the convulsion varied according to the kind of metals joined in the arc pair.

Fig. 4.2- LUIGI GALVANI (1737-1798), Professor of Anatomy at Bologna, Italy. His observations of strange convulsions of frog leg specimens in response to the sparks from an electrical machine, and from metallic probe contacts led to his theory of "animal electricity." Volta however, proved that the electricity was not "animal" but from the metallic contacts. Portrait - The Burndy Library.

**GALVANI WAS NOW FACED** with trying to explain the phenomena he was observing. He had encountered two electrical effects for which his specimens served as indicator - one from the sparks of the electrical machine, and the other from the contact of dissimilar metals. The electricity responsible for the action resided either in the anatomy of the specimens, with the metals serving to release it, or the effect was produced by the bi-metallic contact, with the specimen serving only as an indicator.

Galvani, primarily an anatomist, seized on the first explanation, ascribing the results to "animal electricity" residing in the muscles and nerves of the organism itself. He took his discovery as one in physiology. He compared the body to a Leyden Jar, in which the various tissues developed opposite electrical charges. These charges flowed from the brain through nerves to the muscles. Release of charges by metallic contact caused the convulsions of the muscles. "The idea grew," he wrote, "that in the animal itself there was an indwelling electricity. We were strengthened in such a supposition by the assumption of a very fine nervous fluid that during the phenomena flowed into the muscle from the nerve, similar to the electric current of a Leyden Jar." Galvani was carrying on in the tradition of his day which ascribed the body motivation to a flow of "spirits" residing in the various body parts.

In 1791 Galvani published, in the proceedings of the Academy of Science in Bologna, his paper: *De Viribus Electricitatis In Motu Musculari,* which set forth his experiments and conclusions. Galvani's report created a sensation. It implied to many a possible revelation of the mystery of the life force. The implications challenged men of science and laymen alike, both in Italy and elsewhere in Europe. Investigators proceeded, wherever frogs were available, to repeat the work of Galvani. No one prosecuted Galvani's findings more assiduously and used them as a stepping stone to greater discovery, than Allesandro Volta, a fellow scientist and a professor of physics at the University of Padua.

**ALESSANDRO VOLTA** was born on the 18th of February in the ancient Lombardy city of Como, then a part of the Empire of Austria. Although of noble birth, Alessandro's father, Filippo, was a poor manager of finances so that the children's education had to be taken care of through relatives in the church. As a child Alessandro was so slow in learning as to be of concern to his parents, who thought he might become a mute. But by a seeming miracle young Volta had evolved at the age of seven into a bright and industrious student of exceptional promise.

Within a few years Volta had obtained an excellent classical education, and he was under pressure from relatives to enter the priesthood. But Volta had developed into an interested and discerning observer of the physical world around him and had become far more intrigued with the study of natural phenomena. Encouraged by an older friend who gave him some experimental equipment, Volta had by his mid-teens decided to become a physicist. He made his first experiments in electricity, and by the age of 18 was in correspondense with Abbé Nollet, a well-known French electrical investigator.

**BY VOLTA'S TIME** experimenters in Europe, and Benjamin Franklin in America, had begun to accumulate a large body of facts on the electricity produced by friction, and had speculated on the nature of the electric charges. They had defined the mutual attraction and repulsion of charges and had distinguished between conductors and nonconductors. Positive and negative charges had been identified as representing an excess or deficiency of an electric force or "fluid." The Leyden Jar had devised as an electric accumulator.

With this background of electrical science Volta began his career as a researcher. At age 24 he had published his first scientific paper: *On the Attractive Force of the Electric Fire.* In it he showed an adventurous mind by speculating on the identity of the electric force with that of gravity. Engaged with studies of physics and mathematics, and busy with experimentation, Volta's talents were so evident that before the age of 30 he was named the Professor of Physics at the Royal School of Como. Here he made his first important contribution to science with the invention of the *electrophorous* or "bearer of electricity." This was the first device to provide a replenishable supply of electric charge by induction rather than by friction.

**VOLTA'S ELECTROPHORUS** worked on the principle that a charge may be repeatedly induced on a conductor and transferred to a storage device such as a Leyden Jar, without depleting the original inducing charge. In the electrophorous, Fig.4.3, (1) a charge on the resined surface, A, of plate, B, induces charges on the plate, C, when place on A.

(2) upper negative charge on C is then dissipated, leaving C positively charged. (3) Plate C is removed and the charge transferred to storage. The process can then be repeated without renewing the charge on A. The electrostatic energy is provided by the work required in removing plate C.

Volta's wide-ranging curiosity led him next to investigations of the gases he observed bubbling up from the marshy areas of Lake Maggiore where he liked to fish. He found the gases flammable. In trying to determine the composition, he collected the gas in a closed tube and exploded it with an electric spark. This was a precursor of the spark-fired automobile engine. Experimenting with air, he found that at constant pressure it expanded in the same proportion as the rise in temperature. This gas phenomenon is now called "Charles Law."

**VOLTA** had, in 1782, been called to the professorship of physics at the University of Padua. There he made his next invention, the *condensing electrophorus,* a sensitive instrument for detecting electric charge. The earlier method of charge detection was the "electroscope," which consisted of an insulated metal rod to one end of which was suspended pairs of silk threads, pith balls or gold foil. These pairs diverged by repulsion when the rod was touched by a charge, the amount of divergence indicating the strength of the charge.

By combining the electroscope with his electrophorus, Fig. 4.4, Volta provided a detector for minute quantities of electricity. Later Volta made his condensing electroscope a part of a mechanical balance that enabled measuring the force of an electric charge against the force of gravity. This instrument was called an *electrometer,* and was of great value in Volta's later investigations of the electricity created by contact of dissimilar metals.

Volta had become conversant in several languages and traveled widely on sabbatical leaves to call on other scientists for an exchange of ideas. Volta's electrical inventions, and his speculations and research as communicated to fellow scientists, began to give him international recognition. Dr. Joseph Priestley, a noted British investigator, brought Volta's electrometer to the attention of the Royal Society of London, which elected Volta a Fellow in 1791, and awarded him the Copely Medal for his electrometer in 1794.

**ON LEARNING** of Galvani's report of 1791 to the Bologna Academy on the "Forces of Electricity in Their Relation to Muscular Motion," Volta expressed immediate interest. Galvani responded

VOLTA'S ELECTROPHORUS . . .

Fig. 4.3- A - resin slab, B - metal plate, C - metal plate. D - insulating handle for plate C. Plate C is lifted and resin slab A is negatively charged by wiping it with a piece of flannel. Plate C is replaced on top of A. This induces a positive charge on the under side of C, and a negative charge on the upper side of C. (2) Negative charge of C is discharged to earth by touching C, thus leaving it positively charged. Plate C is removed (3) and its positive charge is transferred to a Leyden Jar for storage. C is then replaced. The operation can then be repeated without restoring the original charge on the resin, B. The energy of the charge is provided by the work done in removing the plate C.

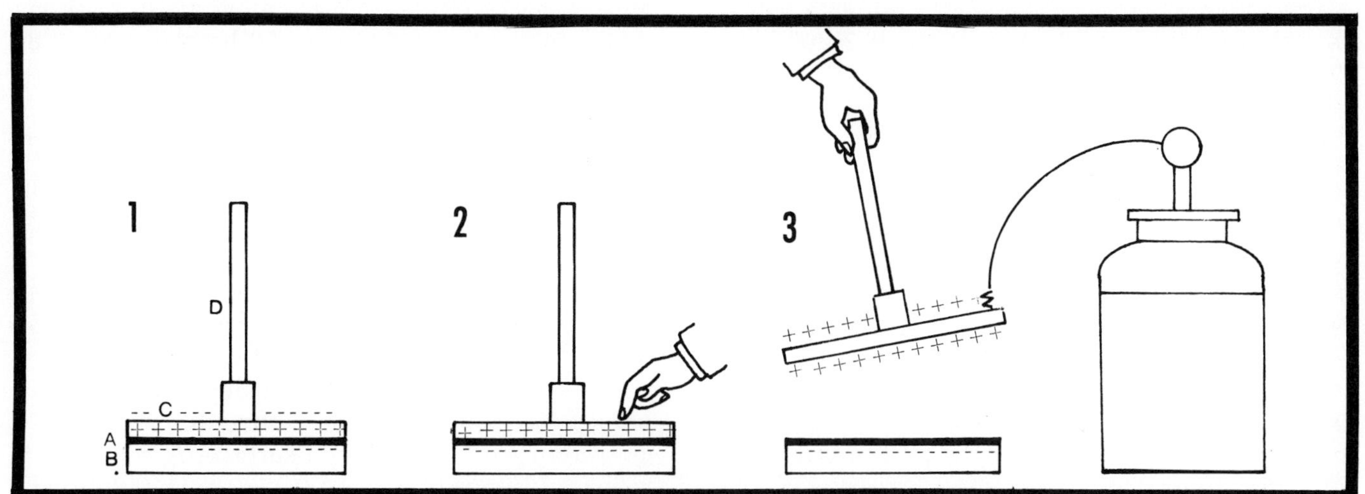

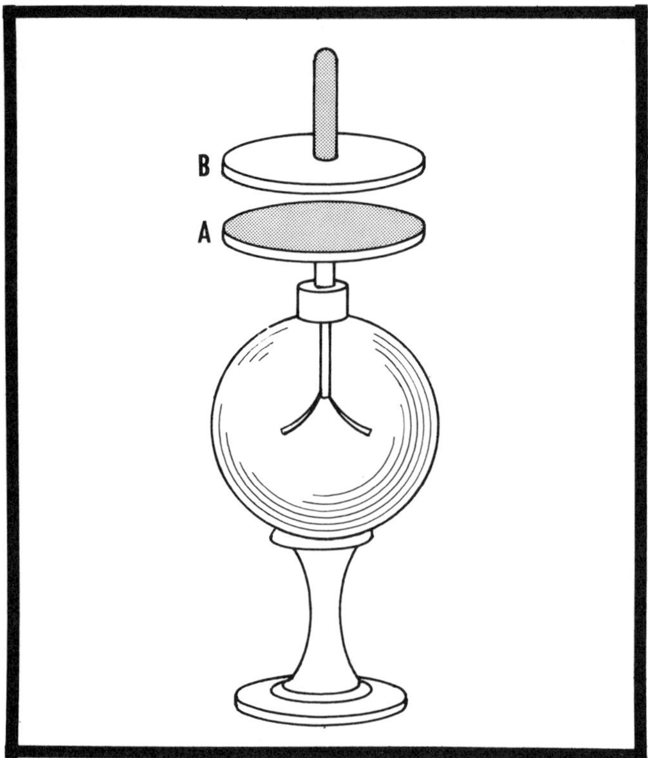


VOLTA'S CONDENSING ELECTROSCOPE . . .

Fig. 4.4- A highly sensitive instrument for measuring electric charge. The gold-leaf electroscope has, instead of the usual knob, a flat brass disk, A, on which can rest a second brass disk, B, of the same size. Plate B has an insulating handle by which it can be raised. Plate A is thinly coated with shellac to form a condenser of large capacity when the disks are in contact. When disk B is raised, the condenser charge is set free on disk A. The charge then diverges the leaves of the electroscope.

by sending him a copy of his paper, which had been privately published.

Volta set up quickly to repeat Galvani's experiments and he confirmed Galvani's conclusions on "animal electricity" as the cause of the muscular reactions. Along with Galvani he ascribed the activity to an imbalance between electricity of the muscle and that of the nerve, which was restored to equilibrium when a metallic connection was made. On continuing his investigations, however, Volta began to have doubts about the correctness of that view. He found inconsistency in the balance theory: in his experiments muscles would convulse when only the nerve was in the electrical circuit. Thus he questioned Galvani's likening of the muscles and the nerves to the coatings of a Leyden Jar, because convulsion would take place when there was an electrical conduction in only one element.

Volta, in an effort to find the cause of the activity, went back to Sulzer's experiment of many years previous. Placing a piece of tinfoil on the tip of the tongue and a silver coin at the rear of the tongue, and connecting the two with a copper he got a sour taste. Substituting a silver spoon for the coin, and without the copper wire, he let the handle of the spoon touch the foil and got the same result. Also, using dissimilar metals to make contact between the tongue and the forehead he got a sensation of light.

From these results Volta came to the conclusion that the sensations he experienced could not originate from the metals as conductors, but must come from the ability of the dissimilar metals themselves to generate electricity. Accordingly, after two years of experimenting, Volta, in 1792, published his convictions. While crediting Galvani with a surprising original discovery, he disagreed with him on what produced the effects.

**BY 1794 VOLTA** had made a complete break with Galvani. He became an outspoken opponent of the theory of animal electricity and proposed henceforth the theory of "metallic electricity." Galvani, by nature a modest individual, avoided any direct confrontation with Volta on the issue and simply retired to his experiments on animals in an effort to maintain his position. But Galvani had a very vocal defender in his own nephew, Giovanni Aldini (1762-1834). Although without a medical education, Aldini entered his uncle's laboratory and carried on enthusiastically with Galvani's experiments in animal electricity directed toward championing Galvani's position and challenging Volta's criticism.

Albini's goal was to prove a case against Volta by showing that muscular reaction could be achieved by stimulus within the specimen itself. By manipulation of nerves and muscle Aldini succeeded in doing this. He and Galvani published an anonymous tract describing the demonstration of muscular action from electrical forces without the use of metals, thus feeling they had cut the ground from under Volta's claims.

Galvani and Aldini, in their efforts to refute Volta's findings, had made discoveries unrelated to metals that would be highly important for future medical science. One of their findings, as shown by later medical investigations, was the reaction due to a body injury current which occurs whenever tissue is irritated or cut. Their work on the presence of body electrical forces was also utilized in the 19th century development of the electrocardiograph which records the electrical impulses generated by contractions of the heart, and by development of the electroencephalograph for measuring brain electrical impulses.

But Galvani's evidence of animal electricity was not proof of the agency which had caused the kicks of the frog's legs produced by the use of metals. By insisting on animal electricity as the sole cause for their results, Galvani and Aldini had excluded the possibility of other causes and missed a great opportunity for electrical discovery.

The claims and counterclaims of Volta of Padua and Galvani of Bologna developed rival camps of supporters and detractors. Scientists swayed from one side to the other in their opinions and loyalties. The subject was complex and dimly understood, emerging from old misconceptions of life forces, but on the verge of an era of revelation. Despite the controversy, however, the thrust of nearly all the experimenters who had gone into action following Galvani's discovery, centered not on animal electricity, but on the role of metals in relation to electricity. There were other early investigators who were on the same route as Volta, and some came close to anticipating the function of electrochemical action for the production of electricity.

Giovanni Fabrioni (1752-1822) of Florence, had suggested that Galvani's electricity was of chemical origin. He observed that when he put plates of different metals into water they gave Galvani's effects. Dr.Richard Fowler (1765-1863) of Edinburgh repeated some of Volta's tests with dissimilar metals in contact with the tongue and confirmed Volta's observation that there was marked difference in the kind of metallic taste depending on the combination of metals used. John Robison (1719-1805) an associate of Fowler repeated the tongue experiments of Sulzer with a significant variation: he tried the multiplying effect of a stack of shilling-size alternate pieces of copper and zinc, and he found that the sensation was very strong and disagreeable. His further tests on metals in a series were halted by illness. Had he been able to carry on he might have anticipated Volta's discovery of the electric pile.

**VOLTA HELD THAT THE SOURCE** of the electricity was in the contact of the dissimilar metals only, with the animal tissue acting merely as the indicator. His contact tests differed substantially according to the pairs of metals used. Volta found that the muscular reaction from dissimilar metals increased in vigor the farther the separation in a series of certain pairs of metals and other materials he had been using. This series, recorded by Volta in 1794, consisted, in the order of the dissimilar activity, of zinc, lead, tin, copper, platinum, gold, silver, charcoal and graphite. The arrangement of this kind was a first depiction of what is now called the half-cell potential of the electrodes.

For better quantitative measurements Volta now dispensed with the use of muscles and nerves as indicators. He substituted his invention of the condensing electroscope, Fig.4.4. He was fortunate in the availability of this superior instrument because the contact charge potential of the dissimilar metals was minute, far too small to be detected by the ordinary gold-leaf electroscope. Volta's condensing electroscope used a stationary disk and a removable disk separated by a thin insulating layer of shellac varnish. The thinness of this layer provided a large capacity for accumulation of charge. When the upper disk was raised after being charged the condenser capacity was released to give a large deflection of the gold leaves.

Volta proceeded systematically to test the dissimilar metal contacts. He made disks of various metals and provided them with an insulating handle. He contacted and measured the quantity of the charge on each disk combination by the divergence of his gold foil condensing electroscope. He then determined whether the charge was positive or negative by bringing near the electroscope a rubbed rod of glass or resin which, by its effect on the divergence indicated the charge polarity.

Volta found that for a zinc-copper contact the zinc was left positively charged and the copper negatively charged. The lead-copper, tin-copper and iron-copper contacts also left copper negative and the other metals positive, but the charge was successively weaker. Now, however, the copper

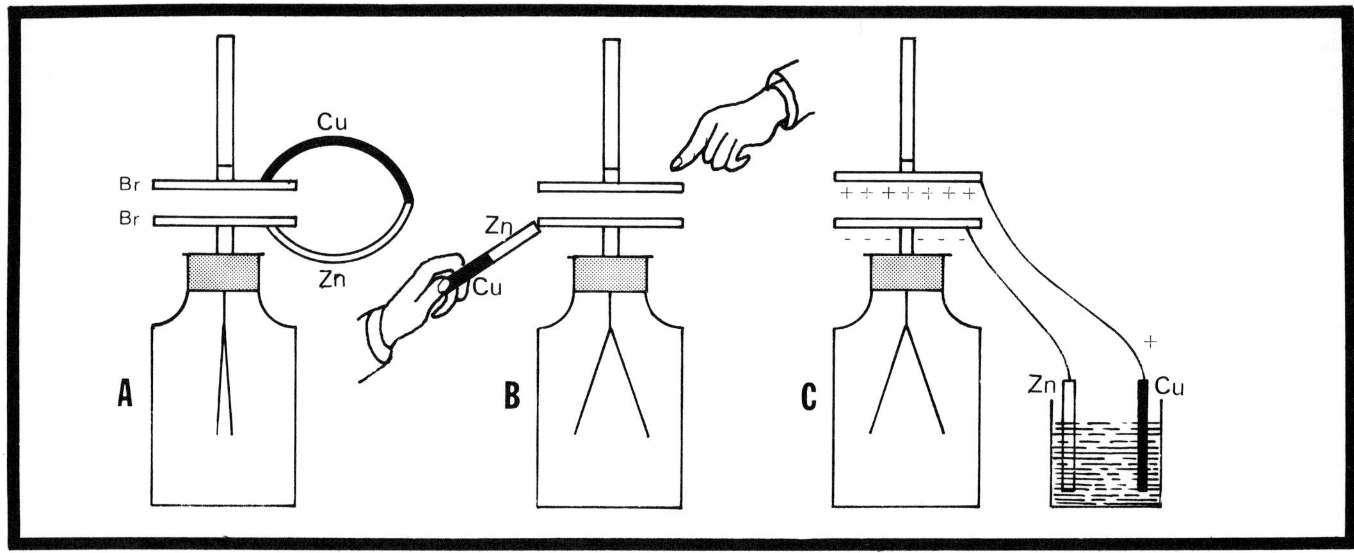


DISSIMILAR METALS AND A FLUID . . .

Fig. 4.5- Volta showed that an electromotive force was produced by the contact of dissimilar metals, and he used his condensing electroscope for measuring the potential. Simple contact, A, however, gave a very small effect. But, B, when the finger tips were inserted in the circuit a much stronger effect was indicated. Volta attributed this to the presence of saline moisture. Although he considered the dissimilar metal contacts primarily active and the fluid only an intermediate "electrifier of the second class," his final arrangement, C, used a fluid, such as salt brine, as an essential element in achieving a constant flow of electricity.

contact with silver or gold reversed the polarity, thus leaving the copper positively charged and the silver and gold negative. Volta concluded that there was a polarity and an electrification relationship, which he arranged as follows:

| | | |
|---|---|---|
| + ZINC | COPPER | |
| LEAD | SILVER | |
| TIN | GOLD | |
| IRON | GRAPHITE | - |

From this arrangement two facts could be predicted: On contact of any two substances, the earlier on the list would be positive and the later negative. Furthermore, the farther the substances were separated on the list the greater the strength of electrification. From his electrometer tests Volta then listed the following charge relationship magnitudes:

| | | |
|---|---|---|
| ZINC/LEAD | = | 5 |
| LEAD/TIN | = | 1 |
| TIN/IRON | = | 3 |

Using the numbers above, the charge for ZINC/IRON was 9, or the sum of the three separations. These results helped establish the *"law of successive contact,"* showing that in any series of contacts around a closed circuit the separations would cancel out.

This series was recognized later by Dr. H. Ritter (1791-1869), in Germany, as showing the order in which each metal will "pre-empt" any in the list below it out of chemical compound - a chemical activity process. However, this conclusion ran counter to Volta'a metal contact theory, hence did not prevail at the time. It was later found correct by Faraday's electrochemical research.

**VOLTA WAS MOVING** toward the idea of an electric force or "potential." This, he assumed, resided in the contact of the dissimilar metals. But as Volta investigated more combinations he found that a potential also existed when there was contact between the metals and some fluids. The early bi-metal tongue tests of Sulzer and the animal tissue tests of Galvani had shown that moisture could also be a participating element in the electric response. Also, in some of his experiments Volta had noted the effect of the moisture of his hands in enhancing the metal contact charge, Fig. 4.5B. Further investigation led Volta to add liquids, such as brine and dilute acids, in his conducting system. He now included in his series of charge-producing metals another dissimilar producer - that of liquid contacts. He classified the metal contacts as "electrifiers of the first class" and the liquids as electrifiers of the "second class."

**VOLTA FOUND** that in a circuit composed entirely of electrifiers of the first class there was only momentary movement of electricity. However when he put two dissimilar metals in contact with a separator soaked with a saline or acidified solution there was a steady indication of potential. Volta was now approaching the decisive and most practical discovery of his career. He was assembling the basic elements of an electric battery - two dissimilar metals and a liquid separator.

He immediately saw that the electric effect, instead of being the result from only two elements, could be enlarged by multiplying the elements. By stacking metals disks and the moistened separators vertically he constructed an "electric pile," the first electric battery, Fig. 4.6. With enough disks in the pile Volta found that by touching the upper and lower disks he received a distinct shock. Unlike the momentary charge of a Leyden Jar, the shock was continuous as long as he maintained the contact. He had made a "Leyden Jar" with a permanent charge! In Volta's words he had produced . . . "an inexhaustible charge, a perpetual action or a perpetual impulsion of the electric fluid."

Volta announced his invention in a long letter, dated the 20th of March, 1800, to Sir Joseph Banks, President of the Royal Society. Volta's letter was read to the Society:

"After a long silence, for which I shall offer no apology, I have the pleasure of communicating to you, and through you to the Royal Society, some striking results I have obtained in pursuing my experiments on electricity excited by the mere mutual contact of different kinds of metal, and even by that of other conductors, also different from each other, either liquid or containing some liquid, to which they are properly indebted for their conducting power. The principal of these results, which comprehends nearly all the rest, is the construction of an apparatus having a resemblance in its effects (that is to say, in the shock it is capable of making the arms,&c. experience) to the Leyden flask, or, rather to an electric battery weakly charged acting incessantly, which should charge itself after each explosion; and, in a word, which should have an inexhaustible charge, a perpetual action or impulse on the electric fluid; but which differs from it essentially by this continual action, which is peculiar to it . . ."

"The apparatus to which I will allude, and which will, no doubt, astonish you, is only the assemblage of a number of good conductors of different kinds arranged in a certain manner. Thirty, forty, fifty, or more pieces of copper, or rather silver, applied each to a piece of tin, or zinc, which is better, and as many strata of water, or any other liquid which may be a better conductor, such as salt water, lye,

VOLTA'S ELECTRIC PILE

Fir. 4.6- The electric pile in one form consisted of pairs of zinc disks (Z) and silver disks (A) separated by cloth or pasteboard disks soaked in brine, lye, sal-ammonica or other liquid.

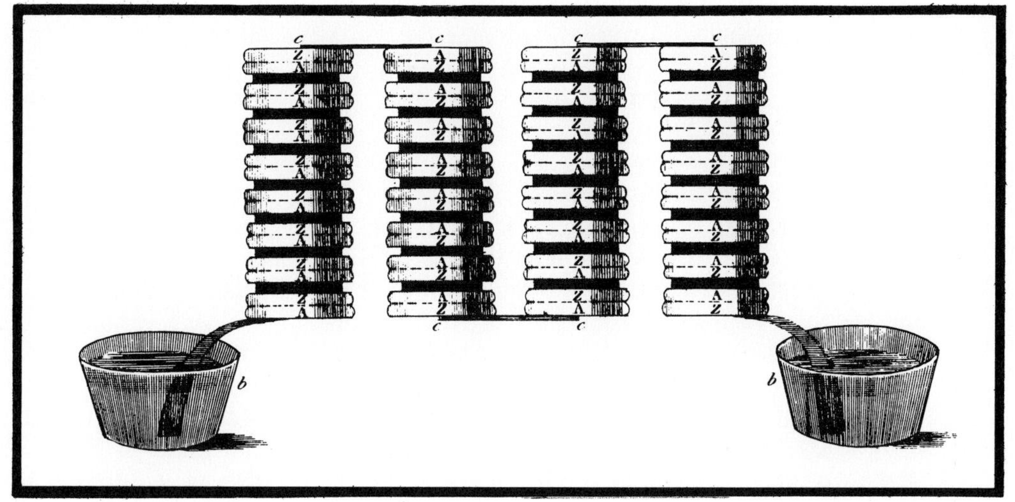

or pieces of pasteboard, skin, &c well soaked in these liquids; such strata being interposed between every pair or combination of two different metals in an alternate series, and always in the same order of these three kinds of conductors, are all that is necessary for constituting my new instrument, which, as I have said, imitates the effects of the Leyden flask . . . although its output was weaker it had the great advantage of needing no outside charge, and instead of a momentary discharge provides a continuous flow of electricity."

**VOLTA DESCRIBED** various assemblies of his pile consisting of stacks of zinc and copper or zinc and silver plates and their liquid-soaked separators. He arranged multiple piles to increase the strength of output.

One disadvantage of Volta's pile was the rapid drying out of the separating disks; another was the interference with action when spillage due to a squeezing of the disks flowed liquid over the pile. Volta, with characteristic ingenuity devised a new arrangement. For the separating disks he substituted tumblers partly filled with brine or lye solution. Into each glass he dipped one plate of copper and one of tin or zinc. The plates were soldered or otherwise connected between the tumblers to form a series of cells. He called this a "crown of cups," see page 44. Volta noticed that the strength of the current varied with the number of cells, with concentration of the solution and with the size of the immersed plates.

In his letter to the Royal Society, Volta reported on the effects of electricity on the senses of touch, taste, sight and hearing. He commonly judged the electric strength by the intensity of shock to the fingers or to the two hands.

He compared the electric producing capability of his cells to the equivalents he conceived of in the body of the torpedo fish or the electric eel:

"To this apparatus, much more similar in its form to the *natural electric organ* of the torpedo or electric eel than to the Leyden flask, I would wish to give the name of the *artificial electric organ;* and, indeed, is it not, like it, composed entirely of conducting bodies? Is it not also active of itself without any previous charge, without the aid of any electricity excited by any of the means hitherto known? Does it not act incessantly, and without intermission? And in the last place is it not capable of giving shocks which produce the same torpor in the limbs as occasioned by the torpedo?"

Volta was amazed at what he had wrought: "this endless circulation of the electric fluid (this perpetual motion) may appear paradoxical or even inexplicable, but it is no less true and real, and you may feel it, as I may say, with your hands."

He continued to distinguish his results from those of Galvani:

"I found myself obliged to combat the pretended animal electricity of Galvani, and declare it external electricity moved by the mutual contact of metals of different kinds,"

**VOLTA'S LETTER** to the Royal Society created great excitement among scientists in England and and France, and stimulated lively interest in further experimentation. Because he was dissatisfied with Volta's arrangement of the metals in the pile, Dr. W. Cruickshank in England in 1800, constructed a "trough" battery. It was a long box with partitioned cells for the liquid in which large metal plates were immersed. The trough arrangement provided a large and more steady output of electric current for investigations.

**USING A TROUGH BATTERY** , Sir Anthony Carlisle and William Nicholson, in London on 30 April 1800, made the first great discovery in electrochemistry. Inserting the platinum wire terminals of the battery into a jar of slightly acidified water, they saw the wires immediately coated with gas bubbles which formed continuously and rose to the surface. The gases were collected and found to be hydrogen and oxygen. When they were mixed and exploded the gases recombined to form water. Thus water, once considered to be an elemental substance, was shown to be a compound of two elements.

Within a few years, Sir Humphry Davy, using a bank of 500 voltaic cells in the basement of the Royal Institution in London, produced a brilliant arc between two pieces of carbon connected across the battery - the first electric light. The arc light was compared to the sun, and proved to be so exceedingly hot that it fused or sublimated all materials put into it. Using the great battery Davy also showed that metal salts could be decomposed

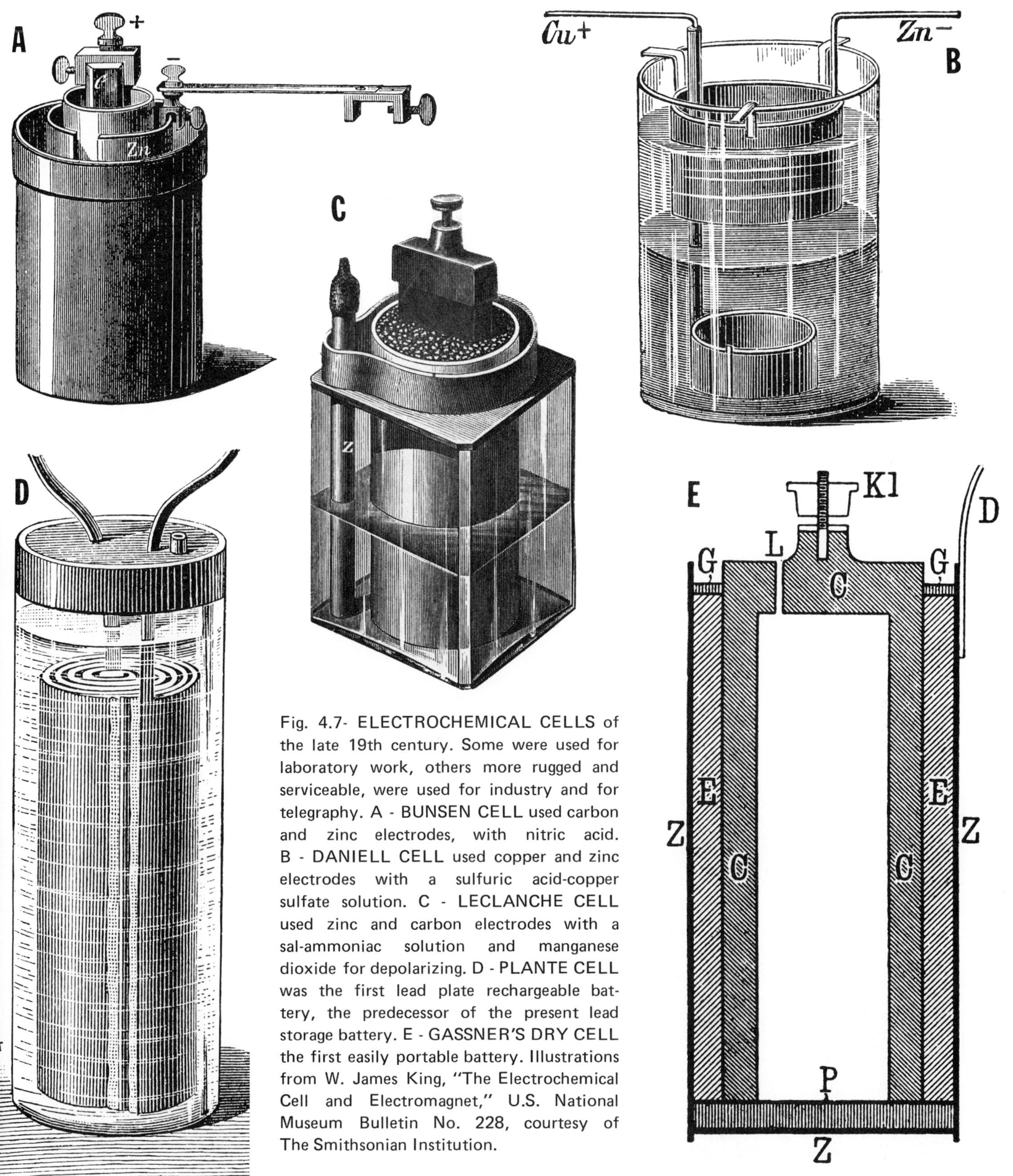


Fig. 4.7- ELECTROCHEMICAL CELLS of the late 19th century. Some were used for laboratory work, others more rugged and serviceable, were used for industry and for telegraphy. A - BUNSEN CELL used carbon and zinc electrodes, with nitric acid. B - DANIELL CELL used copper and zinc electrodes with a sulfuric acid-copper sulfate solution. C - LECLANCHE CELL used zinc and carbon electrodes with a sal-ammoniac solution and manganese dioxide for depolarizing. D - PLANTE CELL was the first lead plate rechargeable battery, the predecessor of the present lead storage battery. E - GASSNER'S DRY CELL the first easily portable battery. Illustrations from W. James King, "The Electrochemical Cell and Electromagnet," U.S. National Museum Bulletin No. 228, courtesy of The Smithsonian Institution.

Fig. 4.8- TROUGH BATTERY of the type used by Carlyle and Nicholson in 1800 to decompose water into its two elements, hydrogen and oxygen, thus opening the field of electrochemistry. Illustration from W. James King, "The Electrochemical Cell and the Electromagnet, U.S. National Museum Bulletin No. 228. The Smithsonian Institution

electrically into their elemental components. By sending a current through molten caustic soda and caustic potash he isolated shiny globules of metal, which he proved to be elements, and which he named "sodium" and "potassium." Thus, in less than a decade after Volta, the age of the electric light and of electrochemistry had begun.

**THE FRENCH GOVERNMENT** came out of their great Revolution with a new and high enthusiasm for science. Impressed with Volta's work, they gave him national recognition by inviting him to lecture before the French National Institute. This he did in November 1801. One of the lectures was attended by Napoleon Bonaparte, who was an eager listener and who engaged in a lively discussion with Volta after the meeting. Later, Volta was showered with honors by Napoleon, who ordered the appointment of a commission of French scientists to repeat and expand on Volta's work.

A leading member of the French commission was Coulomb, who, as related in a previous chapter, had recently remodeled the science of electrostatics. His presence was significant because a great debate was shaping up on the cause of the voltaic currents. Coulomb and his followers, in complex and contrived theories; ascribed the current to multiple electrostatic discharges between the plates and the separators of the pile. Other experimenters, however, began to produce evidence that the voltaic action was of chemical origin. Volta, meanwhile, maintained his rather uncomplicated theory of electric impulsion simply by contact of dissimilar metals in the presence of a fluid. Volta apparently never had any expressed thoughts of the association of chemical change with an electric force. Many years would pass before the nature of electrochemical cell was understood.

Volta had reached the peak of his career. Then came a great change in the pace of his efforts. At the age of 50 he had decided to marry and raise a family. He continued his experimental work but it was quite undistinguished when compared to his earlier triumphs. Volta, in his later life, was apparently content with his earlier accomplishments.

After Napoleon had defeated Austria in the wars of 1796-7, Volta became a senator and eventually a Count in the Napoleonic empire. On the defeat of Napoleon, Italy reverted in 1817 to its former connection with Austria. The Austrians were forgiving, and Volta who was anxious to retire to his home at Canmago, again became a good citizen of Austria. The man who had revolutionized electricity died in the quiet of his country estate in 1827 at the age of 82.

Volta did not participate in the sciences that ensued from his discoveries. He was not part of the growing revelation that electricity and chemistry were intimately related, and that the operation of the voltaic cell was an electrochemical phenomenon. Volta, who had discovered electrochemistry without knowing it, never took part in the great field it opened.

**RECOGNIZING** the greatness of Alessandro Volta's contribution in placing a practical source of electric power at the disposal of mankind, scientists and engineers meeting in Paris, in 1881, at the first International Electrical Congress, saw fit to immortalize him by naming the unit of electromotive force the **VOLT**.

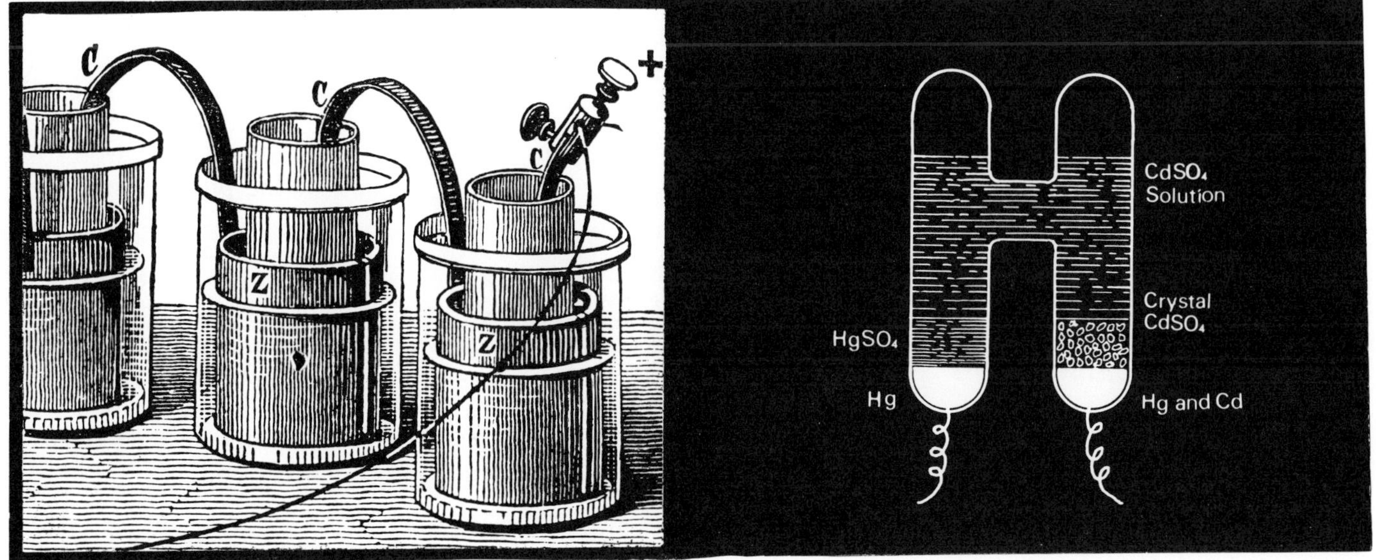


THE OLD VOLTAGE STANDARD CELL . . .

Fig. 4.9 - The DANIELL CELL, 1836, used in telegraphy, was an early standard with a potential of about 1.07 volts. Illustration from A. Niaudet, "Electric Batteries," 1880.

THE PRESENT VOLTAGE REFERENCE CELL . . .

Fig. 4.10 - The WESTON CELL, developed in 1893, is a saturated cadmium sulfate cell, and the standard reference cell, with a potential of 1.01860 volts at 20 degrees C.

**THE VOLT** - An electric current results from the movement of electric charge carriers, such as electrons in a metal, or ions in an electrolyte or a gas. To produce the movement in an electric circuit requires a force. The force results from a difference of electric potential. The difference of potential, which may be provided in many ways, as shown on the following pages, is expressed in *volts,* and designated by V (voltage) or E (electromotive force).

A potential difference, or an electromotive force, implies the ability to do work, and this gives rise to the definition of the volt in the International System of Units (SI).

*The VOLT is the difference of electric potential between two points of a conducting wire carrying constant current of 1 ampere, when the power dissipated between those two points is equal to 1 watt.*

The volt may also be defined as the potential difference such that one joule of work, J, is done in driving one coulomb of charge, Q, from one point to another in the circuit. Thus, $J = QE$.

But, one joule, J, equals one watt of power, P, acting through a second, t, or $J = Pt$.

One coulomb of charge, Q, equals one ampere of current, I, acting through one second, t, or $Q = It$.

Then, $Pt = It \times E$, and $E = P/I$, or a potential difference of one volt produces one watt of power per unit of current.

**STANDARD OF VOLTAGE** - The measurement of voltage has generally been in terms of some primary electric cell. Previous to 1870, the Daniell cell, Fig.4.9, with a potential of 1.07 volts, was used for voltage estimates in operating telegraph lines. In a first effort to provide a stable reference standard for the volt, Sir Latimer Clark, in in 1872, developed "A Standard Voltaic Battery." The Clark cell had electrodes of zinc amalgam and of mercury, and with an electrolyte of mercurous and zinc sulfates. Its potential was 1.434 volts at 15 degrees C.

A still better standard reference cell, Fig. 4.10, was introduced by Dr. Edward Weston (1850-1936). He found that the temperature stability was greatly improved by replacing zinc with cadmium. The Weston cell, with a potential of 1.01860 international volts at 20 degrees C, and reproducible to a few parts in 100,000, has become the universal voltage reference standard.

# through history..... WITH ENERGY CONVERSION TO CREATE A VOLTAGE

## electrostatic

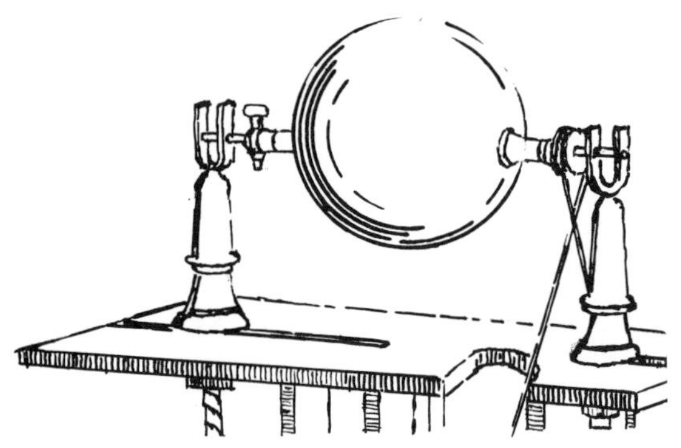

**1705 - FRANCIS HAUKSBEE**
England - Frictional electricity by rubbing glass globe with the hands.

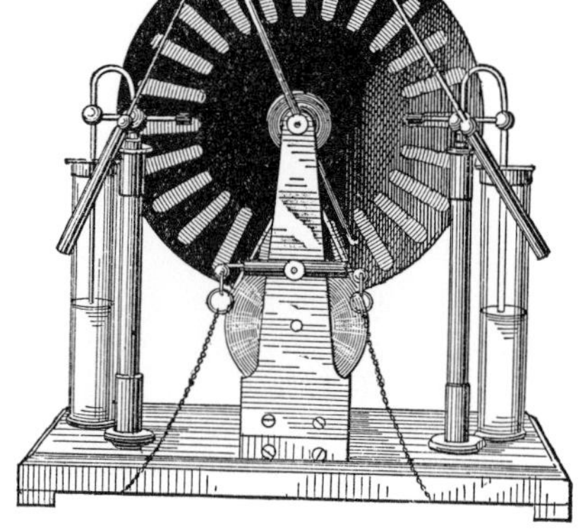

**1880 - JAMES WIMSHURST**
England - Electric charges produced by induction between turning disks.

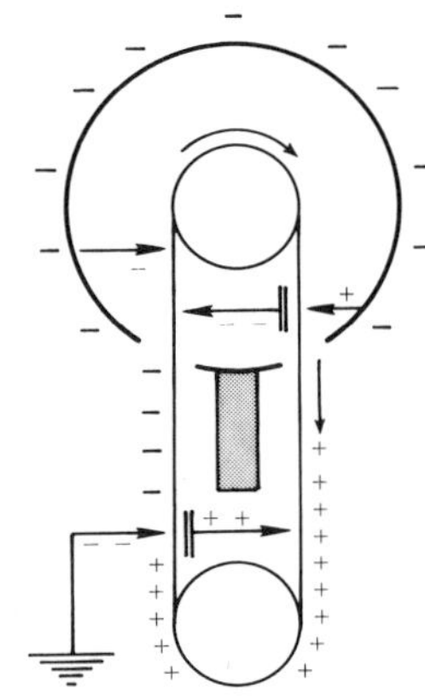

**1931 - R. J. VAN DE GRAAFF**
USA - Electric charges sprayed traveling belt, collected on sphe

## electrochemical

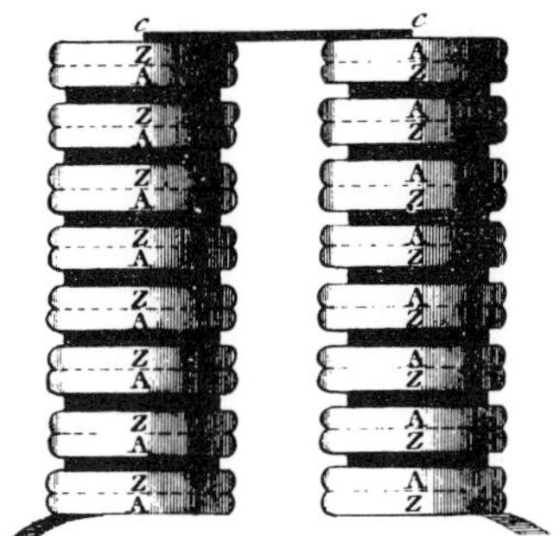

**1800 - ALESSANDRO VOLTA**
Italy - First battery. Zinc and silver disks separated by moistened paper.

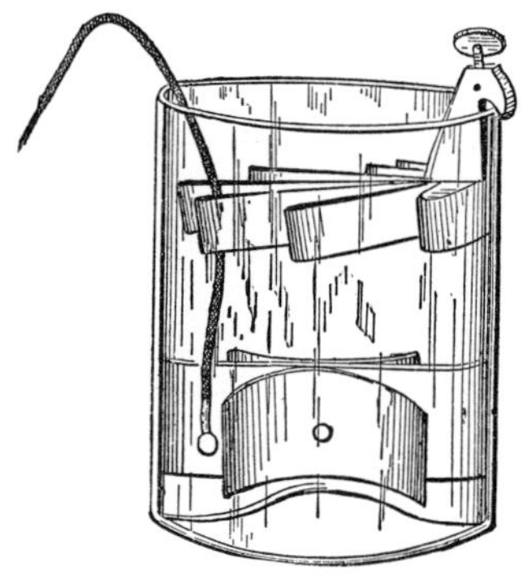

**1836 - JOHN F. DANIELL**
England - First wet cell to elim ate the effects of polarization

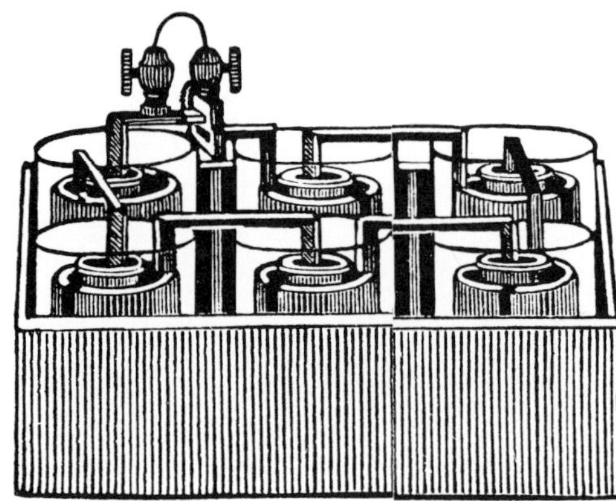

**1839 - SIR WILLIAM R. GROVE**
England - Hydrogen-oxygen electrodes. Predecessor of the fuel cell.

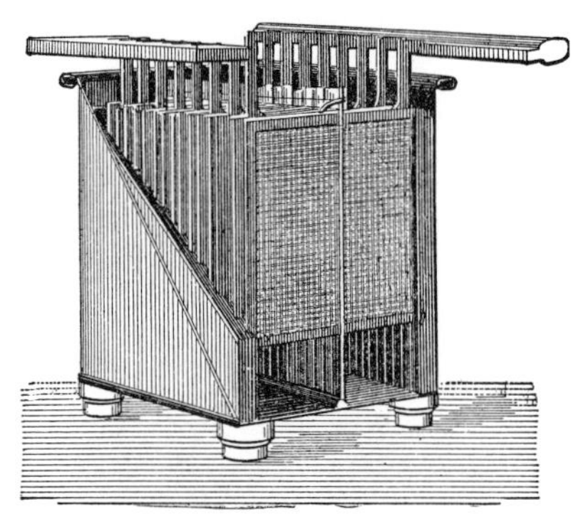

**1859 - GASTON PLANTE**
France - Prepared lead plates. The first rechargeable storage battery.

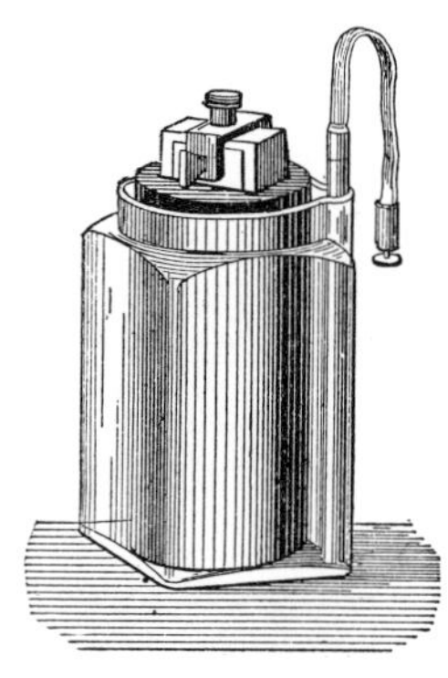

**1868 - GEORGES LECLANCHE**
France - First dry cell. Mangane dioxide absorbed the polarizing ga

## electromagnetic

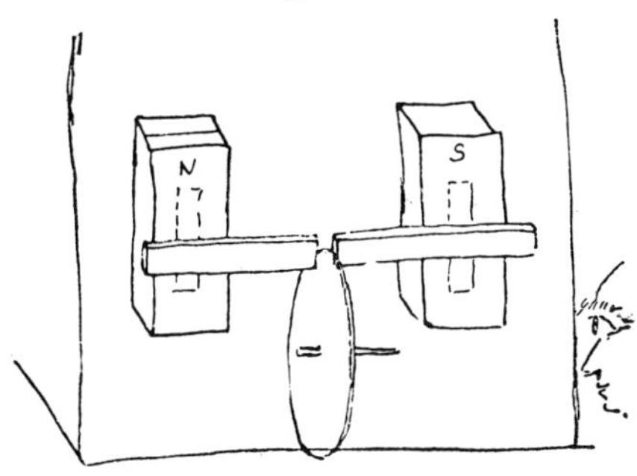


**1831 - MICHAEL FARADAY**
England - The first electromagnetic generator. Sketch is by Faraday.

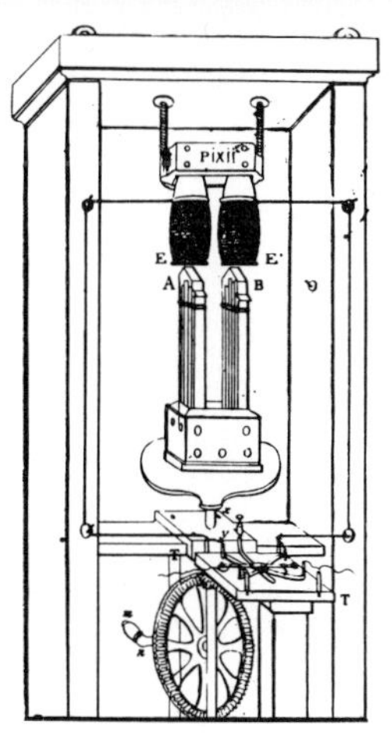


**1833 - HIPPOLYTE PIXII**
France - First generator to rectify the voltage with a commutator.

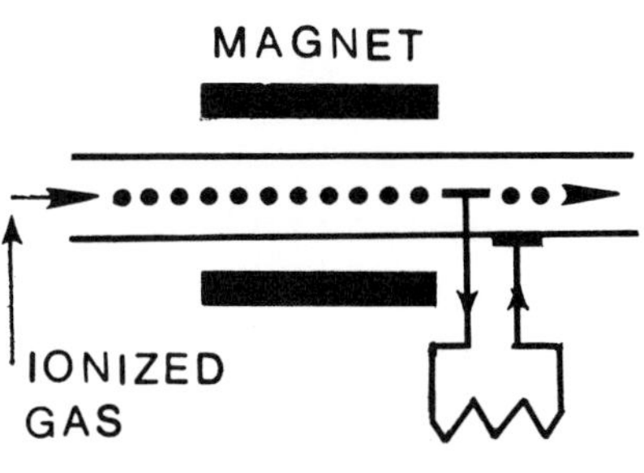


**1840 - MICHAEL FARADAY**
England - Ionized gas moving in a magnetic field generates a voltage.

## thermoelectric

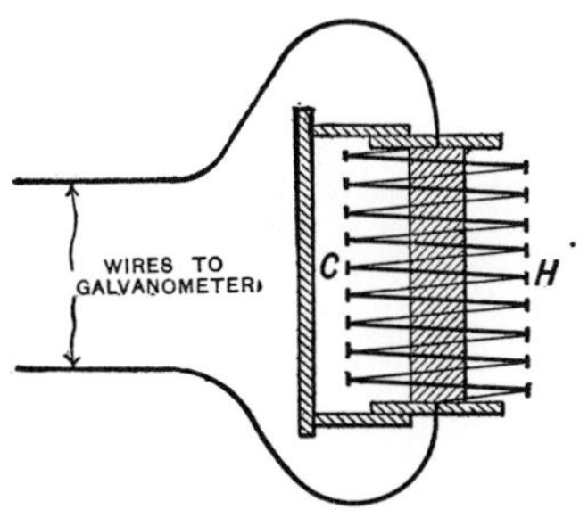


**1823 - THOMAS J. SEEBECK**
Germany - Temperature difference on metal junctions gives a voltage.

## thermoionic

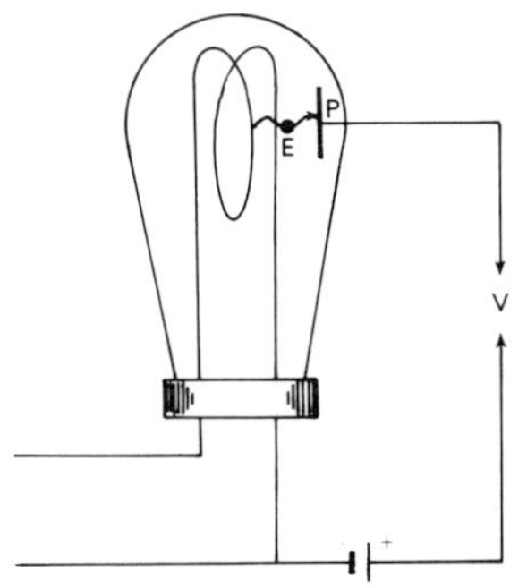


**1883 - THOMAS A. EDISON**
USA - Electrons from hot filament create voltage between it and plate.

## photovoltaic

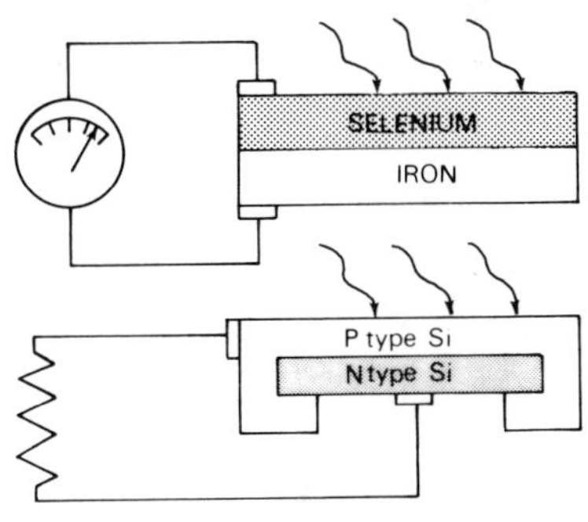


**1869 - A. M. BECQUEREL**
France - Light energy on dissimilar metal voltaic cell creates a voltage.

## photoemissive

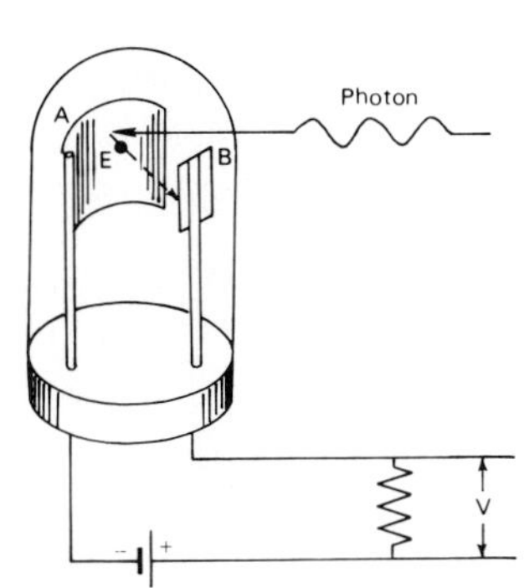


**1887 - HEINRICH HERTZ**
Germany - Photon energy converts to a voltage by electron emmission.

## piezoelectric

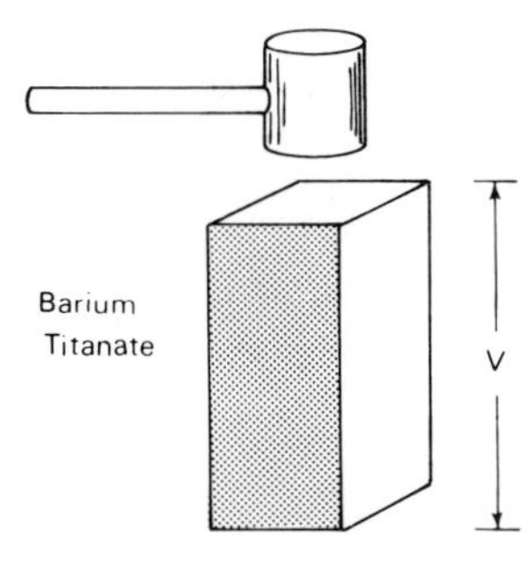


**1880 - PIERRE CURIE**
France - Crystals shift electrons under pressure to create a voltage.

## pyroelectric

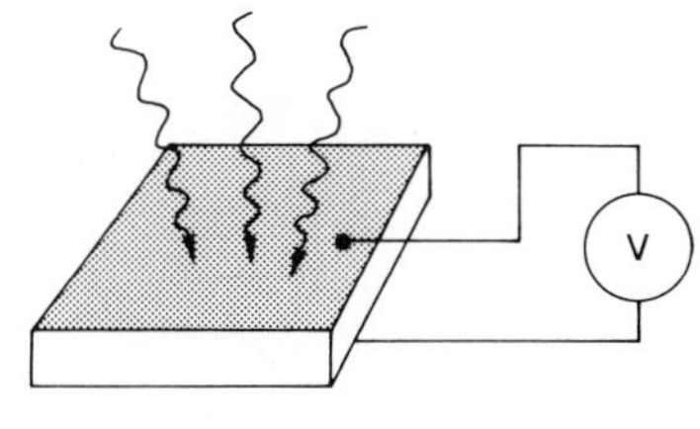


**1824 - DAVID BREWSTER**
Scotland - Temperature change on pyroelectric crystals creates voltage.

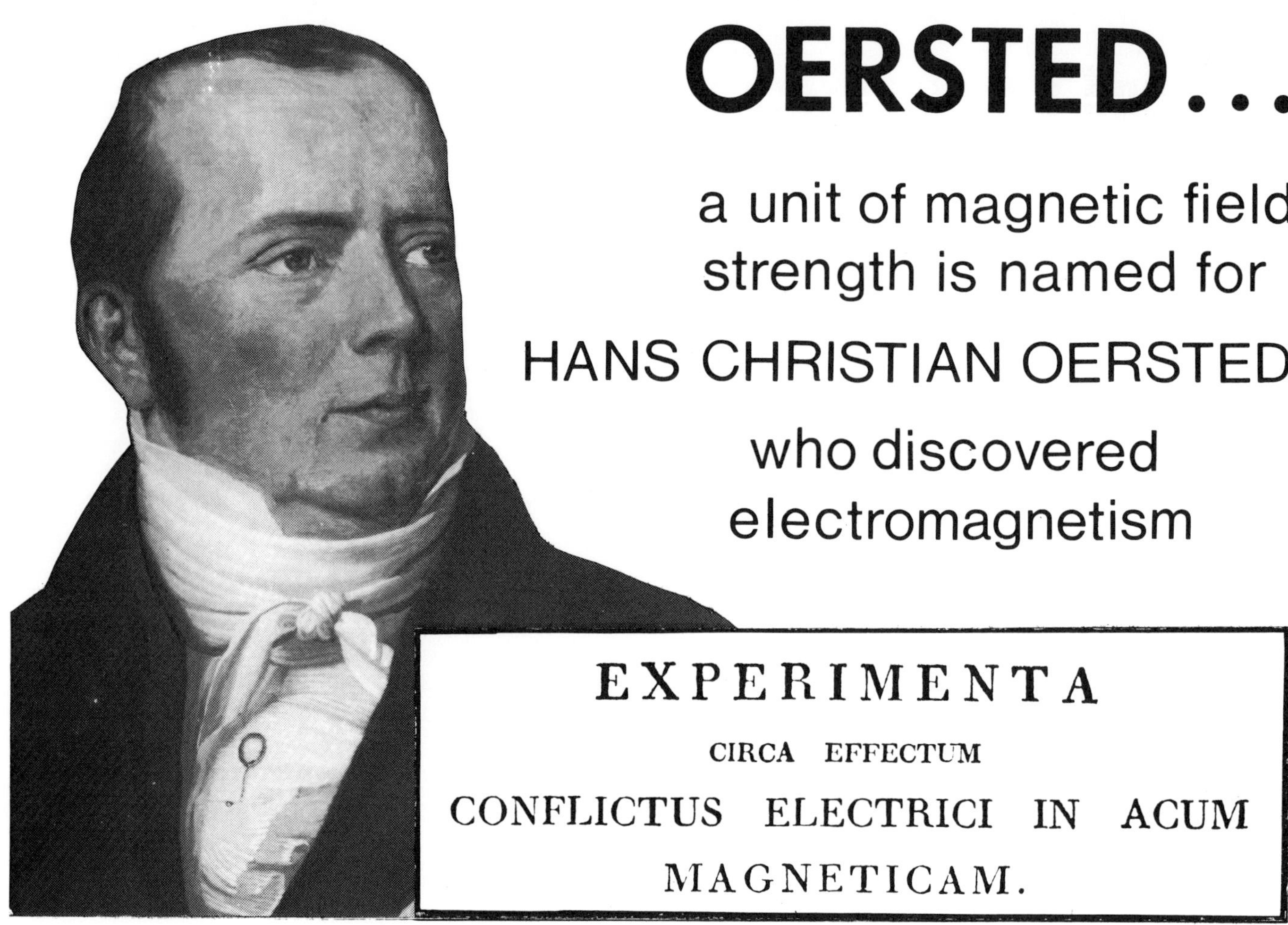


**THE PROFESSOR** beckoned his class in natural philosophy to come closer to the laboratory table. He was about to present a demonstration of an electric current, produced by a voltaic pile on a nearby bench. In particular he would show the ability of an electric current to heat a wire. This current, discovered by Volta twenty years earlier, could make a wire red hot - even melt it if the current was strong enough. This was an amazing oddity. The wire gave off both heat and light.

At a signal for the demonstration, a laboratory assistant connected the battery, and all eyes were riveted on the wire - all eyes except the professor's. Leaning forward he gazed startled at a compass that happened to be near the wire. The compass had come to life. Swinging strongly, the needle quivered as it came to rest pointing to the wire. The professore stared in amazement. Why had the needle swung into action? To what was it responding - to the current that had just been switched on in the wire?

In one of the great intuitive compulsions of discovery, the professor signalled the battery to be disconnected - the needle then quickly returned to its normal north-south position. Again he signalled for the battery to be connected, and again the needle, tugged by an invisible force, swung toward the wire. His amazement turned into conviction: *the needle was responding to a force from the current!* The professor turned to his class. They had witnessed, and the professor, Hans Christian Oersted, a physicist at the University of Copenhagen, had just discovered, the first convincing evidence of a force connection between electricity and magnetism - *a connection that eventually would change the world!*

# he found a connection that changed the world.....

**ONE OF THE GREATEST ANNOUNCEMENTS** in the history of electrical science, dated July 21, 1820, was broadcast in a document with the title bordered on the opposite page: "Experiments on the Effect of an Electric Current on the Magnetic Needle." Oersted, who, along with others, had been on the trail of a connection between electricity and magnetism, fully appreciated the importance of his discovery. He privately printed a 4-page pamphlet describing his results. Published in Latin, the universal language of science at the time, the tract was sent simultaneously by Oersted to learned societies and eminent scientists all over Europe. The importance of the announcement was such that the paper was reprinted within a few weeks into English, German, French, Italian and Dutch.

Oersted was proclaiming a new force - the force accompanying electric current - *electromagnetism.*

Before Oersted, electricity and magnetism had been treated as independent sciences. A few experimenters had noted magnetization of steel needles from electrostatic discharge, or from an association with an electric current. But none had drawn any conclusions from the observations. Some scientists had envisioned an association of electricity and magnetism, but could not prove it. Other investigators maintained that electricity and magnetism had nothing to do with each other.

*Oersted showed the world that they did.*

Volta had invented the battery in 1800. It aroused immediately, throughout Europe, an intense investigation into various effects of electric current. Electrochemical action and heating manifestations had been quickly found. But up to 1820 the relationship of electricity and magnetism had eluded discovery. Then, in the winter of the years 1891-20, Oersted made his famous compass observation to put electricity and magnetism together. Electric current was not alone in a wire - *it was embraced by a cloak of magnetism.*

**OERSTED** considered the magnetic effect in the space around the wire as a "conflict of electricity." He envisioned this conflict as acting only on the particles of magnetic matter. These particles resisted, like those in the compass needle, the passage of the conflict, hence they could be moved by the conflict. Oersted's perception of the action was unique, as was his discovery. It was a first step in explaining the utilization of one of nature's greatest forces - *electromagnetism* - and it opened the age of electric power.

Fig. 5.1 - PROBABLY the first great discovery resulting from a classroom observation occurred to Oersted in the spring of 1820. The compass scene at the left led to his discovery of electromagnetism, which he announced in the title on the opposite page. Courtesy Burndy Library.

**GREAT DISCOVERIES** excite the imagination. Just how did they occur? What train of thought, what circumstances converge at a point in time to reveal some new facet of nature's forces that until then had resisted revelation? What part do luck and chance have in the discovery process?

Oersted's discovery of electromagnetism, as variously told, happened by "lucky chance." He was demonstrating electricity in a lecture to a class of advanced students in natural philosophy. While working with an electric current through a length of wire supported on the laboratory table, Oersted was startled to see the needle of a compass that was near and below the wire suddenly swing to point toward the wire whenever he switched on the battery. The needle was apparently responding to a force exerted invisibly by the current in the wire. Oersted's curiosity was immediately aroused and he proceeded with systematic experiments to confirm and enlarge on the original observation. In a few months came his announcement that the action of the compass needle was the result of a force performing in circles around the conductor. The needle and other attracted bodies were responding to the magnetic effect of an electric current. This was the first authenticated evidence of a relationship of electricity and magnetism.

From the foregoing account it appears that "accident" had the major role in Oersted's discovery. But as William Whewell, the British philosopher, said more than a century ago in his *History of the Inductive Sciences:*

"Men are fond of repeating that such discoveries are most commonly the result of accident; and we have seen reason to reject this opinion, since that preparation of thought by which the accident produces a discovery is the most important of the conditions on which success of the event depends. Such accidents are like a spark which discharges a gun already loaded and pointed."

Fig. 5.2- ONE OF THE SIMPLEST MAGNETIC DEVICES was used for one of the greatest discoveries in electrical science. Laboratory compass at the left was used by Oersted in his crucial experiment of 1820. The compass is in the possession of the Polytechnic Institute of Copenhagen, of which Oersted was one of the founders. The compass rests on a rare surving copy of Oersted's 4-psge Latin tract, sent to scientists in Europe to announce his discovery. Illustration courtesy The Burndy Library.

A review of Oersted's professional career shows that the chance compass observation did not occur in isolation. It was a crucial event in a sequence of of thought preparation and investigation, tenaciously pursued, that led up to the magic moment of discovery. It was Oersted's background of dedicated and purposeful scientific inquiry that was a deciding factor in his "accidental" triumph.

**OERSTED** was born on the 14th of August 1777, in the small Danish town of Rudkoebing, on the tiny island of Langeland, washed by the Baltic sea. Hans, and his younger brother, Anders, were the two oldest sons of Soeren Oersted, the village apothecary.

Rudkoebing had no school, and the father's income was so meagre that the brothers could not be sent away for education. But friendly villagers of some learning took an interest in the boys and gave them rudimentary instruction in reading, mathematics and in drawing. The town mayor taught them French and English. The brothers were apt and eager students. They read every book available, helping each other and developing a thirst for knowledge.

At an early age Hans assisted his father in the apothecary, an experience which sparked interest in science and medicine. When Hans was 16, he and his brother decided to enter the University at Copenhagen. The brothers managed expenses there by dint of tutoring, economizing and with some small assistance from the government. Hans decided to follow his interest in science and medicine. Anders studied law and later became a jurist and statesman.

Like many men entering science at that time, Hans occupied himself not only with the study of natural science and medicine, but also with aesthetics, history and philosophy. Thus, his first university recognition was not in science but for an essay on *Limits of Poetry and Prose,* which got him a gold medal. Hans obtained his degree in pharmacy in 1797, and in 1799 was created a Doctor in Philosophy for a dissertation on *The Reality and Beauty of Nature's Laws.*

In his scientific studies Oersted had been aroused by Volta's discovery of the electric pile. He had already, while at the university, resolved that experimentation with the voltaic current would be a major field of interest. In 1800, after graduation, he had taken over, in addition to lecturing at the university, the management, in the absence of the owner, of an old and large Copenhagen pharmacy. This institution had a collection of physical and chemical apparatus which was lacking at the university, and which became available to Oersted for investigations into the chemical effects of electricity. He had been working with the electric pile in the decomposition of acids and alkalis.

**TO PROMOTE** his interests in science, Oersted decided to enlarge his background of information by discussions with others outside of his country. Post-graduate travel was customary at the time, and Oersted took an opportunity in 1801 for a year of touring to widen his outlook. He visited Germany, France and The Netherlands. There he met prominent philosophers and scientists and was exposed to the ferment of reawakening sweeping the intellectual world as a result of the American and French Revolutions. European science was vibrant with speculation and acitivity in new areas of investigation opened up by the discovery of animal electric current by Galvani, by Volta's invention of the electric pile and by the electrochemical revelations that quickly followed.

One of Oersted's most noteworthy travel associations in science took place during a month's stay in Jena with the physicist J.W. Ritter (1776-1810). Ritter, a visionary, had questioned Volta's metal contact theory for the production of electricity, and proposed that the source of the voltaic current was of chemical origin. He had anticipated electroplating by electrically precipitating metal out of solution, and had shown the possibility of reverse decomposition as a source of electricity, the principle of the storage battery.

These were subjects of vital interest to Oersted. He was intrigued by Ritter's ideas, both those that were sound and those bordering on fantasy. He helped to assemble Ritter's projections into an essay submitted to the Institut de France for the annual prize, but it was not accepted. Ritter, with remarkable foresight, made a prediction, based on an extrapolation of past years of epochal discoveries, that the years 1819-20 would see another great electrical discovery. Ritter died before he could his prediction come true, and that it would fulfilled by Oersted.

Returning to Copenhagen in 1803, Oersted made application for the university's chair in physics, but

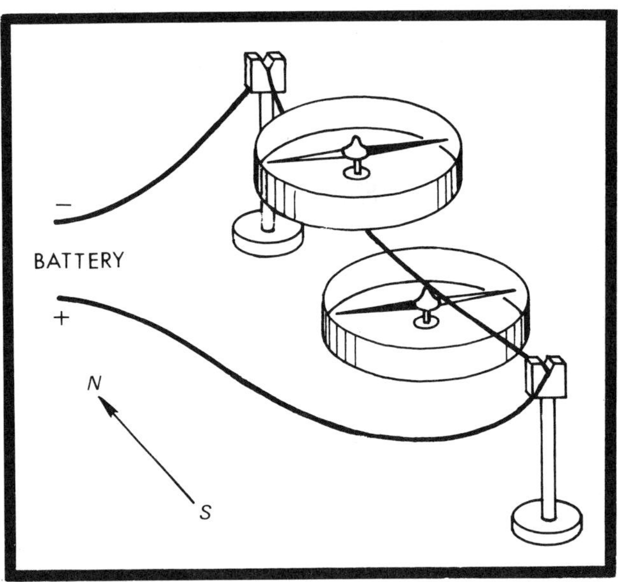


Fig. 5.3 - A COMPASS normally pointing in a N-S direction was observed by Oersted to swing laterally toward the wire, in one direction above and the opposite direction below, thus indicating circularity of magnetism around the wire.

was not immediately accepted. He then returned to lecturing and experimenting, and distinguished himself so well that his early ambition was realized at last by an appointment, in 1806, to the Chair of Physics, a post he held up to his death in 1851.

In his academic duties Oersted was a popular lecturer in electricity, magnetism, and the voltaic current, which was then called "galvanism." At that time these three subjects were independent branches of the electrical science:

*Electricity* referred to electrostatics derived from the ancient lore of amber and the electric charges produced by rubbing rotating glass spheres.

*Magnetism* dealt with the lodestone of antiquity, the compass and the magnetic field of the earth.

*Galvanism* had to do with the effects of the current produced by the voltaic pile. This was the newest branch of electrical science.

**THE INTERRELAITONSHIP** of the above three electrical phenomena had long been a subject of curiosity and conjecture by many investigators and thinkers. It appeared there must be some natural unity of the forces being observed. The German philosopher, Immanuel Kant, in his *Critique of Pure Reason,* published in 1781, proposed that all world matter was an expression of two forces - one attractive, the other repulsive. They were essentially identical, and under proper conditions could be converted into one another. Oersted was very impressed with Kant's concept; identical forces in nature must be convertible. Because of the unity of forces, magnetism and electricity must each be some aspect of the other.

Oersted recalled that it was, for example, common experience that both electric charge and voltaic current could cause shocks. He was aware that there had been occasional incidents of steel needles being magnetized when put in the presence of discharges from electrostatic machines, and of magnetization in the presence of voltaic current.

Could there be any connection between the two - electricity and magnetism? From his experiments he knew that sufficient current through a wire would cause it to glow or even melt - thus light and heat were radiated as an accompaniment of the current. Oersted wondered whether magnetism was also a force somehow "hidden" in the electricity. It was, he thought, and he sought a means of revealing a relationship. The heat and light were lateral radiations of electric current. So, Oersted thought, magnetism might also be a radiation.

In 1813 Oersted, in Berlin, published a book, written first in German, then translated into

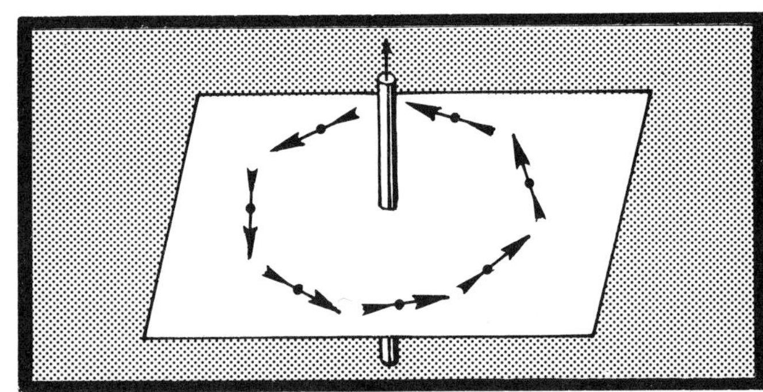

Fig. 5.4 - CIRCULARITY of the magnetic field around a conductor is shown by a tendency of the compass needles to align tangentially to the direction of the magnetism. Needle direction is reversed with reversal of the current.

French titled: "Recherches sur l'identité des forces chimiques et électriques," in which he laid the foundations of the electrochemical system. In one chapter of this book Oersted first recorded his thoughts on the possibility of finding a relationship between electricity and magnetism. After remarking on the great resemblance between electric and magnetic attractions and repulsions, and the similarity of their laws, he concluded by saying: "an attempt should be made to see if electricity, in its most latent stage, has any action on a magnet as such."

**PRECEDENTS TO OERSTED** - Is there a real and physical analogy between electricity and magnetism? This question had intrigued many thoughtful experimenters who reasoned that two phenomenon so closely related must be due to a single agent. The association of electrostatic sparks and magnetism had been frequently observed. Benjamin Franklin had magnetized steel needles by the discharge of Leyden Jars. Giovanni B. Becaria (1716-1781), a noted Italian experimenter, came close to identifying the causal factor in magnetization. He reported that the discharge of a Leyden Jar sent across a length of watch spring produced magnetic poles at the ends of the piece, but the same discharge sent longitudinally along the spring had no similar effect.

With a constant current available from the voltaic pile, new experiments were reported indicating a magnetizing effect in the presence of the current. It was noted in 1801 that two wires connected to the terminals of a battery, and in close parallel proximity, would tend to cling to each other when brought into contact. The same observation was made by others, but the connection between the clinging force of the wires and the force of magnetism due to the electric current was not recognized.

Benedeto Mojon, a chemist of Genoa, reported in 1804 that steel needles placed in a galvanic circuit for several days acquired a definite magnetic polarity. Other experimenters had observed similar results.

The most frequently reported experiment pertinent to an association of electricity and magnetics (Fig 5.5) was performed in 1802 by the Italian physicist Gian Domenico Romagnosi (1761-1835), at the university at Parma. As translated in J.J. Fahie's *History of the Telegraph to the Year 1827,* the account by Romagnosi of his experiment

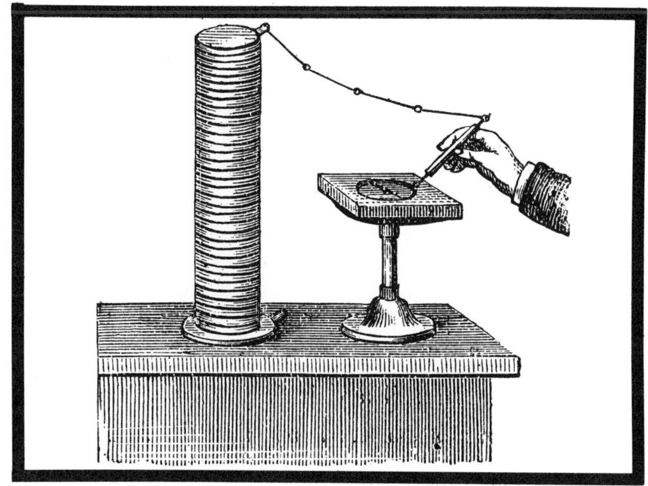

ROMAGNOSI'S EXPERIMENT . . .

Fig. 5.5 - A small silver knob at the end of an insulating handle was energized from a voltaic pile. Romagnosi reported that when the knob was touched to one end of a compass needle, the needle was deflected. This discovery, in 1802, was not followed up by Romagnosi - did not lead to a theory of electromagnetism. Illustration Burndy Library.

showing the action of the voltaic current on magnetism was in part as follows:

"Having constructed a voltaic pile, of thin discs of copper and zinc, separated by flannel soaked in a solution of sal-ammoniac, he attached to one of the poles one end of a silver chain, the other end of which passed through a short glass tube, and terminated in a silver knob. This being done, he took an ordinary compass-box, placed it on a glass stand, removed its glass cover, and touched one end of the needle with the silver knob, which he took care to hold by its glass envelope. After a few second's contact, the needle was observed to take up a new position, where it remained, even after the removal of the knob. A fresh application of the knob caused a still further deflection of the needle, which was always observed to remain in the position to which it was last deflected, as if its polarity were altogether destroyed.

"In order to restore this polarity, Romagnosi took the compass-box between his fingers and thumbs, and held it steadily for some seconds. The needle then returned to its original position, not all at once, but little by little, advancing like the minute or seconds hand of a clock.

"These experiments were made in the month of May, and repeated in the presence of a few spectators, when the effect was obtained without any trouble and at a very sensible distance."

Any of the foregoing investigations, pursued purposefully and under the right conditions, might have yielded the linkage between electricity and magnetism. None were brought to a cause and effect conclusion, and none stimulated the intense activity by others that came immediately after Oersted's announcement, and that led to the immediate formulation, in France, of the laws of electrodynamics.

**IN HIS BIOGRAPHY** *Oersted and the Discovery of Electromagnetism,* Bern Dibner states that five general accounts of the main events relating to his discovery were written by Oersted between the years 1821 and 1830. The narrations vary in detail with the passage of time, but the significant elements show Oersted's clear conception, and his verification and follow-through to a conclusion that set his work apart from that preceding it. Summed up briefly, the main events were as follows:

. . . the accidental compass observation made during a lecture, in the company of students and his assistants.

. . . realization of its significance - current in the wire is attracting the compass needle in the same way a magnet would, therefore the action of the current must produce magnetism the same as a magnet.

. . . planning and execution of subsequent tests using more powerful batteries.

. . . a series of placements of the compass that delineated the direction of the force.

. . . the conclusion that an electric current is accompanied by magnetism that has a circular motion around the current, and that reverses in direction with a reversal of the current.

Oersted never published the exact dates and steps he took for the sequence above. However, by his many experiments he had proved his conclusions in anticipation of the announcement in 1820.

**AS STATED** by Kirstine Meyer in the book: *Prominent Danish Scientists Through the Ages,* by V. Meisen, Copenhagen, the chief contents of Oersted's tract of 1820 were as follows:

"The electric current - in Oersted's language, 'the electrical conflict,' in the conductor - acts on a magnetic needle; direction of the force is sideways and outwards from the conductor, and it has been found in, one might almost say, every possible position of the conductor in relation to the magnet. A simple rule is given for finding the direction of the force. It is shown that the action of the force does not depend on the intermediate substance between the magnet and conductor.

"The magnitude of the force is found to depend on the distance from the magnet, the power of the battery, and the quality of the connecting conductor. He also noticed that a circuit in the form of a polygon carrying an electric current attracts or repels the pole of a magnet. Finally, quite briefly, attention is directed to the fact that judging from these experiments 'the electrical conflict' is not restricted to the conductor, but is communicated to the surrounding space, and must be assumed to traverse circles, the planes of which are at right angles to the conductor.

"The experiments were extended and the results were published soon after, entitled: *Neuere electromagnetische Versuche* in the July number of 1820 of Schweigger's *Journal fur Chemie und Physik.* This work was evidently finished shortly after the first and should be regarded as connected with it, as it forms an important supplement. It contains the following results: (1) The effect of a conducting wire on the pole of a magnet depends on the quantity of electricity and not on its tension - in modern words on the current and not on the electromotive force of the supply. (2) The reaction effect is found by showing that a suspended closed circuit is turned by a magnet. (3) It is established in a new way that a closed circuit has a north end and a south end just like a magnet."

Kirstine Meyer continues: "From the moment that Oersted's discovery became known it created an enormous sensation. The results communicated were so astounding that they were received with a certain distrust, but were stated with such accuracy that it was scarcely possible to entertain doubts. All were soon impressed with the excessive importance of the discovery and the scientific journals were filled with electromagnetical essays.

"In 1821 Faraday wrote a historical survey of the evolution of electromagnetism up to April 1821. Although only eight months had passed since Oersted's communication, so much had already been written that Faraday found it very difficult to make head or tail of the many works, of what had been done and by whom. He then undertook to go through systematically 'with great labor and fatigue' everything that had appeared in journals and elsewhere. Faraday gives full credit to the work of Oersted; about the first communication he writes: 'It is full of important matter and contains in a few words, the results of a great number of observations; and with his second paper, comprises a very large part of the facts, that are as yet known relating to the subject.'

"Faraday acknowledges Oersted's discovery to be not only lucky chance but also the fruit of a deliberate search: 'Mr. Oersted has for many years been engaged in inquiries respecting the identity of chemical, electrical and magnetic forces . . . his constancy in the pursuit of his subject, both by reasoning and experiment, was well rewarded in the winter of 1819-20 by the discovery of a fact of which not a single person beside himself had the slightest suspicion; but which, when it once was known, instantly drew the attention of all those who were at all able to appreciate its importance and value.' "

**RECOGNITION** of Oersted's contributions was attested by his election to a fellowship in many scientific societies. He was given the Copely Medal by the Royal Society of London, and was presented a prize of 3000 gold francs by the Institut de France.

In 1821 Schweigger's Journal proclaimed Oersted's discoveries the most important of the century and that a new era would open with them.

Ampère wrote in the *Journal de physique* in 1822: "M. Oersted has forever attached his name to a new epoch. This learned Danish professor has opened, by his great discovery, a new field of research to the physicists."

Oersted was proficient in several languages. He corresponded and was conversant with the eminent scientists of Europe. He visited England several times and was on friendly terms with Sir Humphry Davy, Sir Charles Wheatstone, Michael Faraday, William Woolaston and other leading British men of science.

Davy visited Oersted in Copenhagen and found a simple, unpretentious man, but with high scientific ingenuity and considerable stature as a philosopher. In 1834 Oersted called on Karl F. Gauss and William Weber, two celebrated German experimenters in electricity and magnetism, and who had first established a basis for systematizing of the electrical and magnetic units.

Although remembered primarily for his discovery of electromagnetism, Oersted, in his time and country, was also highly regarded as a progressive educator and a famous lecturer in natural sciences. He promoted the founding of the Danish Society for the Advancement of Natural Science and the Royal Polytechnic Institute in Copenhagen. Oersted was always concerned with identifying and unifying the manifestations of nature and her laws as evidences of a grand design implying the wisdom and creativity of the overall Diety. Oersted's last published work was: *The Soul of Nature,* a compendium of his lectures and essays in science, and his views on the inner working and outer revelations of nature as expressions of an animate and strictly rational process.

The final honor to Oersted was a jubilee in 1850 celebrating his half century association with the University of Copenhagen. That same year a grateful government gave him, as one of its foremost citizens, a country home near Copenhagen. Oersted died there on the 9th of March 1851.

**THE MAGNETIC NEEDLE** found two new uses immediately after Oersted's announcement - the needle began to be used for measurement and for communication.

For measurement, the juxtaposition of the compass needle and a conductor afforded the first means of monitoring and comparing the strength of the voltaic current. That is, the amount of deflection of the needle in any particular arrangement of the compass and conductor, corresponded to the strength of the current.. This was first

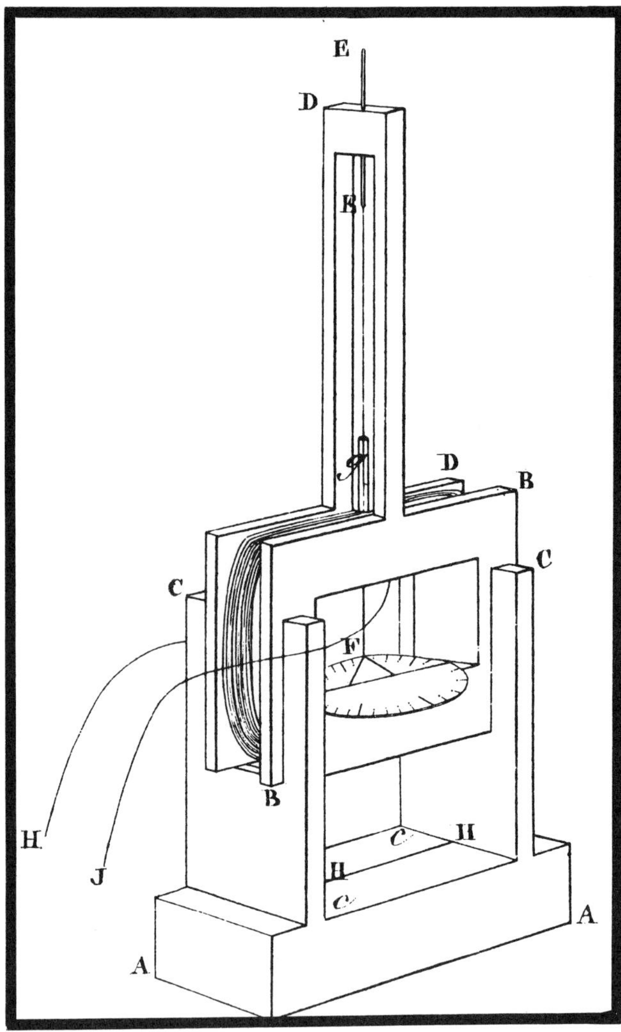


Fig. 5.6 - OERSTED'S DISCOVERY opened the field of electromagnetic instruments. An early type of galvanometer used by Oersted in his work consisted primarily of a coil, C, around a magnetic needle, F, hung over a calibrated disk.

recognized by Ampère, and he applied the name "galvanometer" to an electromagnetic means of measurement. It was obvious from the Oersted experiments that a wire placed above the magnetic needle and looped back underneath it, increased the force on the needle and hence its sensitivity. This prompted the use of multiple turns around the needle to activate it more effectively, and gave the term "multiplier" for early galvanometers.

Priority for the use of electromagnetism as a means of measurement of electricity has been given to Johann S.C. Schweigger (1779-1857), a professor of chemistry at the University of Halle in Germany, and the editor of a technical journal. Instruments of the early type were popularly called "Schweigger Multipliers." However, practically simultaneously, in 1820-21, arrangements of a compass and a coil of wire to measure current were invented by Johann C. Poggendorff (1796-1877), a scientist and editor of *Annalen der Physik,* and by an English professor, James Cumming (1777-1861) at Cambridge University.

The early instruments were little more than detectors, and were made in a variety of coil and needle arrangements, Fig. 5.6. Calibration of the instruments was soon introduced and their sensitivity increased. Availability of new instruments opened wider fields of investigation into the capability of various forms of galvanic batteries and the circuits to which they were connected. Using the early instruments Georg Ohm was able to carry out his experiments on the electrical conductivity of materials that led to his famous law relating electromotive force, current and resistance.

**OERSTED'S** discovery of electromagnetism immediately provided a new way of transmitting information. It was shown that a magnetic needle would respond to a current at a distance to indicate a signal. All that was required was to string the transmitting wires from a battery that could be switched on and off, to a remote terminal where a magnetic needle responding to a current in the wires would indicate a signal from the sending end. Thus messages could be sent beyond the limits of human visibility. This promised a great advantage over the existing mechanical semaphore methods whose range was subject to weather conditions.

One of the first arrangements to use the needle principle for an electromagnetic telegraph is shown in Fig. 5.7. Early devices such as this were considered only novelties. But the advantages of electrical communication for railway traffic and for military maneuvers soon became apparent and promoted development of improved sending and receiving equipment for longer distances. By 1850 the telegraph was the most important electrical technology, and became a major factor in stimulating the development of improved electrical equipment. Telegraphy required improved batteries, suitable all-weather lines, and apparatus sensitive over long distances. These achieved, the telegraph became a major factor in economic expansion.

**THE "OERSTED"**- In the early use of electricity for telegraphy, the chief needs for measurement were those of the electrical circuit, having to do mainly with characteristics of the lines to maintain good signalling. This involved the electromotive force developed by the batteries and the effect of resistance of the line wires in transmitting the battery voltage.

However, with the rapid development, in the 1880s, of dynamo-electric machines such as motors and generators, which worked by the interaction of electromagnetic forces and required large amounts of magnetism, the measurement of the magnetic circuit became as important as that of the electric circuit. This led to the adoption of magnetic quantities measured in units with appropriate names. The magnetic dimensions were those involved in the production of magnetization and the resulting intensity of the magnetic flux for various types of magnetic circuits.

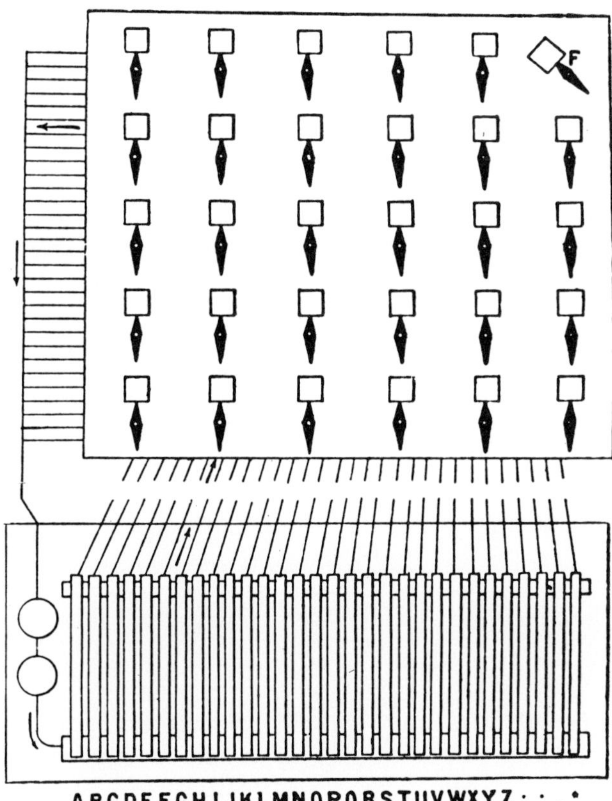


ALEXANDER'S TELEGRAPH - 1837 . . .

Fig. 5.7 - Based on a suggestion by A.M.Ampere, and exhibited in Edinburgh. Each pivoted magnetic needle, when actuated by its key, would uncover a letter. Illustration from W. James King, "The Telegraph and Telephone," U.S. National Museum Bulletin No. 228, The Smithsonian Institution, Washington, D.C. 1962.

Use of Oersted's name for a magnetic quantity was first proposed by the American Institute of Electrical Engineers meeting in Chicago in 1894. The *oersted* was suggested as the unit for the "reluctance" of a magnetic circuit. The reluctance is a proportionality factor relating the amount of magnetic flux that will be produced by a given magnetomotive force, and is analogous to the resistance in an electric circuit. The oersted was defined as the reluctance of 1 cubic centimeter of free space in the centimeter-gram-second system of measurement. However, the oersted as proposed, never received international adoption, and was rarely used outside the United States.

In a revision in 1930, the International Electrotechnical Commission, confirmed the oersted for the unit of magnetic field strength (H) in the CGS system of electromagnetic units.

*The OERSTED is the magnetic field strength in the interior of an elongated, uniformly wound, solenoid that is excited with a linear current density in its winding of one abampere (ten amperes) per 4π centimeters of axial length.*

Other definitions of the oersted are:

*The OERSTED is the force per unit magnetic pole, where the unit pole is defined as one which exerts on a similar magnet pole at a distance of one centimeter a repulsion of one dyne.*

*The OERSTED is the magnetic strength at the center of a place circular coil of one turn, and one centimeter radius, when there is a current of 1/2π abamperes in the coil.*

* * * * * * * * *

For the first fruits of Oersted's great work the scene in the next chapter, shifts to France. There, within a few weeks after Oersted's announcement, A.M.Ampère had established the first laws of a new electrical science - **ELECTRODYNAMICS.**

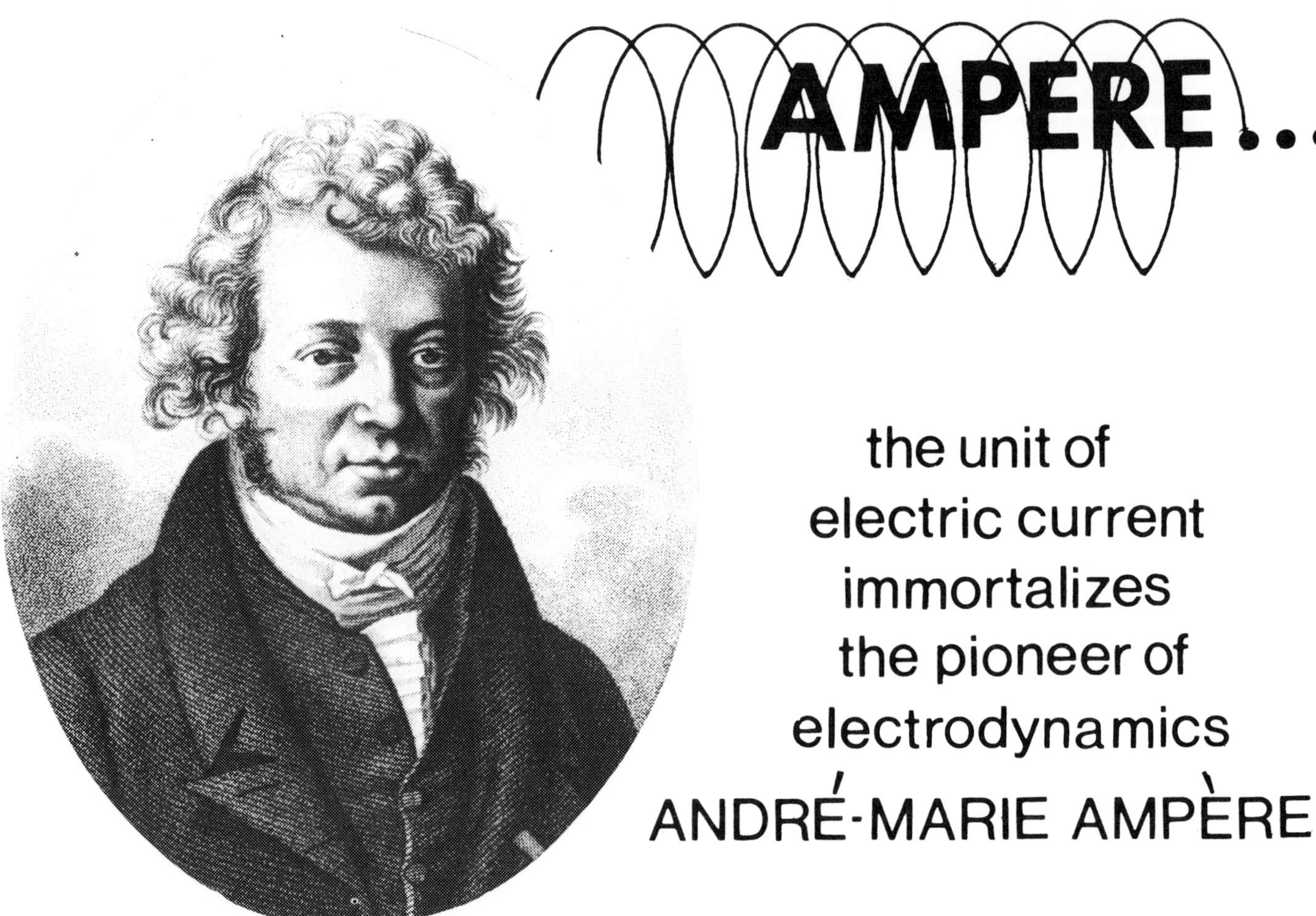

# AMPERE...

## the unit of electric current immortalizes the pioneer of electrodynamics ANDRÉ-MARIE AMPÈRE

**EXCITEMENT WAS IN THE AIR** among the members of the French Academy of Sciences in the latter part of 1820. On 11th September of that year, the French physicist, François Jean Arago (1786-1853), had read to a meeting of the Academy, Hans Christian Oersted's announcement of his experiments leading to the discovery of electromagnetism.

The Academy of Sciences in Paris, funded by and in the service of the French government, was the national assembly for the presentation and discussion of scientific papers. It was the forum for debate on topics and theories of the day as put forth by its members and by the foreign scientists with whom they were in correspondence. Members of the Academy, following the Napoleonic era, were among the most sophisticated and best organized in Europe. The French scientists were institutionalized by the government, and had developed a capability of precise instrumentation for experiment, and the mathematical ability to express their results. The prospect of election to the Academy and a spirit of rivalry among them promoted continuing developments and a speedy exchange of ideas that were recorded in the annals of the Academy. These memoirs form one of the most important records of scientific progress of that period.

To members of the Academy the Oersted accomplishment was sensational, and at first not quite believable. Had not their Coulomb, many years before, proved that there could not be any connection between electricity and magnetism? But Arago stated in his report that he had already repeated and confirmed Oersted's results. A surprising and radically new phenomenon had burst upon the scene. The French physicists were now confronted with a new effect of the voltaic current, and the challenge of the significance of Oersted's revelation that electricity and magnetism were interrelated.

# ..magnetism arises from the circulation of currents......

One member of the Academy left the September meeting highly impressed and motivated. He was André-Marie Ampère, a professor of mathematics at the École Polytechnique of Paris, and also recognized as a chemist and a physicist of great distinction. Ampère immediately followed up on the Oersted discovery and soon broke ground of the greatest importance for growth in understanding the implications of electromagnetism. Within a week of Arago's report, Ampère had repeated the Oersted experiment, and on September 18th informed the Academy of his discovery that electric currents in parallel wires recognized each other's presence by a mutual magnetism that exerted a force between them.

In the next few years Ampère was to present a series of remarkable memoirs on the basics of electromagnetism that established the science of *electrodynamics.*

**ANDRE-MARIE AMPERE** was born in Lyon in south-central France in 1775. Lyon was a city of great silk mills and prosperous shops which supplied products to the nobility and aristocracy of France. But the outbreak of the French Revolution and the beheading of King Louis XVI in 1793, and later his queen Marie Antoinette, brought the court and the monarchy to a violent end, and threw France into the bloody terror that took so many lives and changed the course of world history. Ampère's father, Jean-Jacques Ampère, was a man of independent means - a merchant, scholar and a justice of the peace. In 1782, soon after his son's birth, he moved the family to his country estate in Poleymieux, a few miles north of Lyon. The house today is a national museum, containing books, documents, and a display of models of Ampère's experiments, Fig. 6.1 and 6.2.

Because of a lack of local schools, and because he was influenced by some educational theories of the time, the father took a hand in his son's instruction, by exposing André to the great books of his library, and by encouraging him to educate himself. Among the works Ampère read were Buffon's *Histoire naturelle,* and Diderot's *La Grande Encyclopedie.* These were of great influence in determining his future interests in the natural sciences.

By age 4, Ampère had shown a precocity in

Fig. 6.1 - AMPERE'S boyhood home at Poleymieux, France. The home is now a national museum displaying Ampere's contributions to electricity. Illustration courtesy of the Ampere National Museum.

Fig. 6.2 - AMPERE'S library in the home at Poleymieux, with displays of his works. Illustration courtesy of the Ampere National Museum.

mathematics. Encouraged and tutored by his father young Ampère developed a scientific bent, and at age 12 was studying Laplace's *Treatise on Celestial Mechanics* and Lagrange's *Analytical Mechanics.* Finding that works by the celebrated mathematicians Euler and Bernoulli were in Latin, Ampère immediately learned the language sufficiently well to read the books. Ampère had a remarkable memory, and all who knew him were highly impressed with his intelligence.

Ampère's mother, a devout woman, brought up her son in the Catholic faith. Throughout his life, however, Ampère had to maintain a balance between his faith and his doubts. His religious fervor was often assailed by the abiding influence which the *Encyclopedie* had on him. This great work was marked with skepticism as to theology, a curiosity concerning knowledge, and propagation of the scientific faith applied to human life. It was to have a lasting effect on Ampère's approach to science and his concern with metaphysics.

Ampère's happy boyhood ended when the family was struck with a crushing tragedy. The elder Ampère, as a justice of the peace in Lyon, had ordered the arrest of a revolutionary leader, who was finally put to death. Later, when the terror and destruction of the national revolution had reached Lyon, troops of the new Republic took over the city, and had revenge. In 1793, at the age of 18, Ampère witnessed his father's execution on the guillotine. The father and son had been close companions, and Ampère's grief was such that his mind was temporarily deranged.

After he had regained composure, Ampère again became active, resuming his studies in science, the arts and foreign languages, and carrying on with his experiments. He invented a universal language, and worked with kites to investigate atmospheric electricity. Ampère continued, after his father's death, to live at the estate in Poleymieux, set in the beautiful countryside. In a nearby village Ampère met the young Julie Carron, to whom he was married in 1799. The following four years were the happiest of Ampère's life.

**AMPERE** started his career as a teacher of mathematics in Lyon where, in 1800, his son, Jean-Jacques was born. Needing more money, Ampère moved, in 1802, to Bourg-en-Bresse, to become a professor of chemistry at the École Centrale. With ambitions for the future, Ampère started to prepare himself for a post at a new school Napoleon intended to establish at Lyon. He began work on a paper on probability theory that he hoped would establish the reputation needed for the promotion. Then tragedy struck again with the sudden death of his young wife. Broken in spirit, Ampère decided to leave his native Lyon and its memories, and in 1804 he moved to Paris.

His paper on probability had achieved its purpose and the impression it made enabled him in Paris to be named a lecturer at the École Polytechnique.

But Ampère was still unsettled emotionally. The strangeness of the big city, and a craving for human company and family warmth led Ampère, in 1806, into a second marriage, to Jeanne Potot. This connection soon proved incompatible and unbearable, and after the birth of his daughter, Albine, he obtained a divorce, and then established a household presided over by his mother, and by an aunt who had moved to Paris.

But Ampère's domestic life in the years to come continued in turmoil, with the debauch of his daughter and the utter failure of his son, for whom he had great hopes, to establish a career. These personal miseries had their effect on the trend of Ampère's own development. He had constant recourse to his religion to renew his strength. The many catastrophies that had plagued his life with uncertainty, reinforced a resolve to achieve, in his scientific investigations, a goal, not of probabilities but of unassailable truths.

**AMPERE** early came under the influence of the philosopher Immanuel Kant (1724-1804), who held that the content of knowledge comes from sense perception, with its form determined from effect to cause by the mind. From this, Ampère interpreted two levels of knowledge of the external world - phenomena presented through the senses, and objects understood by intuition without aid of the senses. The relations or rapport of the two was a springboard for extension of scientific knowledge - a scientific idea that was not necessarily supported with evidence, could be hypothesized - the criterion was only whether or not it worked to explain the cause of some phenomenon.

Kant also declared God could not be denied, but neither could he be proved. Ampère's philosophy permitted him to achieve balance between his belief in God and an acknowledgement of the essential existence of objective nature that could be tested. It was from these holdings that Ampère based his resolve to determine what could be found out about the workings of physical nature.

Ampère's excursions into science had no bounds. From 1800 to about 1814 he devoted his time mainly to mathematics. His ability in this field was his means of obtaining an election to the prestigious French Academy of Sciences. Ampère was also fascinated with chemistry, arising from his early teaching days at Lyon. Ampère followed closely the work of Sir Humphry Davy at the Royal Institution, who had isolated sodium and potassium by electrical means. Ampère had anticipated the elements chlorine and iodine, which Davy had separated, but he did not carry through with the proofs. Ampère's work in chemistry led to his involvement in the evolution of molecular physics, which was to permeate his thinking and theories in electricity. He also envisioned a necessary relationship between the growing list of elemental bodies, and set up a classification which anticipated the periodic table of elements devised by the Russian chemist Mendeléev.

In 1808 Ampère was named the inspector-general of a newly formed government promoted university system, a position which he held to his death. In 1814 he was appointed a member of a group of mathematicians at the Imperial Institute. In 1819 he offered a course in philosophy at the University of Paris, and in 1820 was named assistant professor of astronomy. In these positions, and from his many other activities, carried on with diligence, Ampère established himself as a competent member of the scientific community of the city.

Then came the latter part of 1820 and news of the discovery of electromagnetism. This was the event that would give Ampère the opportunity to show his own talent for discovery and formulation that would catapult him into the first rank of science, and give his name immortality. In accordance with his philosophy, Ampère immediately accepted Oersted's discovery and proceeded to build on it. In a few feverish weeks from September 18 to October 9 he established the science of electrodynamics.

**THE FRENCH REVOLUTION,** with the increased devotion to science in its aftermath, had opened the floodgates of French genius. Sponsored by the government, science was becoming a profession. Scientists, many educated by the state, were provided incentives for a wide variety of research careers. Paris became the center for teaching mathematics and its applications to experimental physics.

Scientists of the capital city at the time of the opening of Ampère's career, were among the best organized and equipped for electrical research, which was then concerned with the charge and discharge of electricity produced by friction. Following the lead of Coulomb, who had invented the sensitive torsion balance to formulate the laws of electrostatic forces, Paris had become a focal point for study of static electricity and for development of precise instruments for electrostatic measurement. Thus, when in 1800 Volta announced his invention of the electric pile which gave for the first time a continuous current, French investigators proceeded to look upon the new phenomenon and to explain its operation in terms of their refined science of electrostatics.

Early in the 19th century, at the direction of Napoleon, a special committee of Academy members had been formed to examine and to enlarge on the Volta experiments that had led to his development of the electric pile. The committee focused their investigation on the pile itself rather than the potentially more revelant study of the total circuit. Furthermore, under the influence of Coulomb, the leading member, the committee proposed an electrostatic theory for operation of the pile.

**JEAN BAPTISTE BIOT** (1774-1862), a young physicist member, in a report to the Academy, dismissed Volta's simple theory of metallic contact in the presence of moist separators as the cause for the continuous generation of current by the pile. Instead he proposed that the pile produced current by instantaneous electrostatic discharges that were alternately stored in the metal contacts and then were released by the liquid of the separators. A multitude of these discharges made it appear that the current was continuous. Because of the prestige of the committee, Biot's theory of electrostatic discharge was the dominant one in France for many years.

Before long, however, Volta's metal contact theory and Biot's electrostatic explanation faced other considerations. Across the channel the British investigators had quickly exploited the chemical effects opened by voltaic current, such as the decomposition of liquids and the separation of metals from their salts. This gave them a chemical view of the operation of the pile itself. Sir Humphry Davy of the Royal Institution,London, had suggested that chemical force in the pile was somehow the cause of the electrical force of the pile. This view was taken up by some of the French chemists. One of these was Ampère, who, although he had done no extensive work on the pile itself, was, nevertheless, because of his chemical background, brought into conflict with Biot's electrostatic theory.

As the century moved on, Biot's descriptions of electrostatic operation were modified to accomodate the chemists, and the ideas and presentations on the subject at the Academy had reached a kind of equilibrium.

Then, the conceptions of electricity were suddenly disrupted. In 1820 news of a revolutionary new phenomenon had arrived. Oersted's discovery of magnetism produced by an electric current aroused tremendous interest, and was quickly picked up and the experiment of the electromagnetic attraction repeated by the French scientists. The impact of this new manifestation of current called into question the whole of the electrostatic approach, which was soon to be outmoded by new thinking stimulated by the advent of electromagnetism. The spotlight now shifted from the electric pile to what was going on in the electric circuit. And Ampère, with remarkable rapidity and sagacity proceeded to illuminate the strange forces of magnetism and how they originated.

BEGINNING in December 1820 with his first paper on Oersted's discovery, Ampère, in the next several years, continued to present his views, experiments and investigations in a series of memoirs and treatises in the transactions of the Academy. These revealed and clarified the attributes and potentialities of the electric and magnetic circuits. Ampere's classic experiments with current-carrying conductors were original, and at first considered unorthodox. They made him a great and independent ground breaker whose discoveries are basic to electrical science and to electrical engineering.

**ELECTRIC TENSION & CURRENT** - Ampère suspected that forces exerted by the electric current, or by electricity in motion as produced by the voltaic pile, were entirely different from those of static electricity, or electricity at rest, as produced by the electrostatic machines. Hence, he proceeded to define explicitly two characteristics of the electromotive action.

"Electromotive force is manifested by two kinds of effects which I believe I should first distinguish by precise definitions : I shall call the first 'electric tension,' and the second 'electric current.' "

"The first," Ampère stated, "is observed when two bodies, between which the action occurs, are separated from each other by non-conducting bodies at all points of their surfaces except those where it is established, and is the only one which can arise when the electromotive action develops between the different parts of the same non-conducting body . . ."

By "non-conductor" Ampère meant the source of electricity, the voltaic pile, which produced the electric tension before being connected to the conductor.

"The second case," Ampère then continues, "occurs when bodies make a part of a circuit of conducting bodies . . . there is no longer any electric tension . . . nevertheless electromotive action continues; for if, for example, water, or an acid or saline solution forms parts of the circuit, these bodies are decomposed, and also the magnetic needle is turned from its direction when it is placed near a portion of the circuit, but these effects cease when the tensions are re-established."

Ampère was wrong in assuming the tension disappears when the circuit is completed, but he correctly presents the new idea that the current produced by the tension is responsible for the effects, such as decomposition, etc., that he lists. Ampère had established the concepts of voltage and current. But the exact relation between them had to await the law derived by Ohm in 1826, which is described in the next chapter.

**ELECTRICAL "REUNIONS"** - Ampère had conceived of electricity as an *electric fluid,* the condition of which determined its properties: static effects resulted from the fluid at rest, while the fluid in motion produced the effects of a current, such as electromagnetism. The electromotive force segregated electric fluids and held them apart. Then, in the case of static electricity, the separated fluids were discharged instantly. On the other hand the current electricity was active, uniting the electric fluids in the conductor, whereupon they returned upon themselves for another separation, and so on, to make a continuous flow. From this Ampère hit upon the idea of *reunions* to explain what went on in the conducting wire.

Reunions occur in such a way, Ampère stated, that . . . there results a double current, one of positive electricity, the other one of negative electricity, starting out in opposite senses from points where the electromotive action arises, and going out to reunite in the parts of the circuits remote from these points. The currents of which I am speaking are then accelerated until the inertia of of the electric fluids and resistance which they encounter because of the imperfection of even the best conductors make equilibrium with the electromotive force, after which they continue indefinitely with constant velocity so long as this force has the same intensity, but they always cease on the instant that the circuit is broken . . ."

It will be noted from the above that Ampère had introduced the factor of *resistance* in an electrical circuit. He imagined it to result from a frictional impediment to the otherwise smooth flow of the electricity.

Ampère then set up a sense of direction for the current from the pile. He designated as "positive" that direction in which current flows from the hydrogen-releasing side to the oxygen side in a cell for the decomposition of water.

The **"GALVANOMETER"** - To differentiate between electric tension and electric current, Ampère proceeded with a means for measurement. Whereas the ordinary electrometer, commonly used for static electricity, could measure electric tension, there was no instrument to recognize the presence of electric current, and to indicate its energy and direction.

Going back to Oersted's compass needle discovery of magnetism, Ampère found in its action a ready-made instrument for indicating electric current. He describes it:

". . . all that is needed is that the pile, or any portion of the conductor, should be placed horizontally, approximately in the direction of the magnetic meridian, and that an apparatus similar to a compass, should be placed above the pile or either above or below a portion of the conductor. So long as the circuit is interrupted, the magnetic needle will remain in its ordinary position, but it will depart from this position as soon as current is established, and so much the more as the energy is greater, and it determines also the direction of the current . . ."

"Knowing the direction of the current, the compass arrangement also provided a handy way of determining direction of the magnetic circulation:

" . . . if one places oneself in thought in the direction of the current in such a way that it is directed from the feet to the head of the observer and he has his face turned toward the needle, the action of the current will then always throw toward left that one of the ends of the needle which points to the north . . ."

Ampère gave a name to the instrument:

"I think that to distinguish this instrument from the ordinary electrometer we should give it the name *galvanometer* and that it should be used in all experiments on electric currents, as we are accustomed to use an electrometer on electric machines, so as to see at every instant if a current exists and what is its energy."

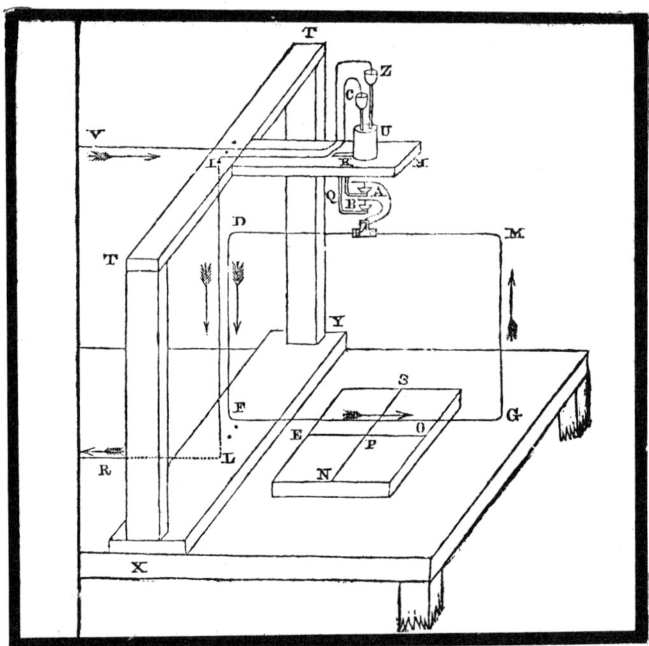


FORCE BETWEEN CURRENT-CARRYING WIRES . . .

Fig. 6.3 - The dynamics of electric currents began with Ampere's discovery that there was a force between parallel current-carrying wires that was a function of strength and polarity of the currents. Illustration is from an Ampere memoir of 1822, showing how a swiveled wire, DF, was attracted, and a wire, MG, repelled by a parallel wire, IL.

The first use Ampère made of his galvanometer was . . ."to show that the current in the voltaic pile, from the negative end to the positive end, has the same effect on the magnetic needle as the current in the conductor which goes on the contrary from the positive end to negative end."

**ACTION BETWEEN CURRENTS** - From the Oersted experiments and from his own, Ampère soon hypothesized that *electric current must be the reality behind magnetism* - that the properties of magnets are the result of the electric fluids in motion within them. He noted that in Oersted's experiments the compass needle responded to the magnetism produced by an electric current. The same needle was also oriented by the earth's magnetism, which might also be the result of electric currents circulating within it. From this he made the deduction that, like affecting like, one circuit carrying an electric current must have an action on another. Ampère proceeded immediately to demonstrate that electric currents must act on each other. One of his experimental arrangements is shown in Fig. 6.3, on which he reported:

"From the time when I noticed that two voltaic conductors interact, now attracting each other, now repelling each other, ever since I distinguished and described the actions which they exert in the various positions where they can be in relation to each other . . . I have been seeking to express the value of the attractive or repellent force between two elements, or infinitesimal parts, of conducting wires by a formula so as to be able to derive by the known methods of integration the action which takes place between two portions of conductors of the shape in question in any given conditions."

"The impossibility of conducting direct experiments on infinitesimal portions of a circuit makes it necessary to proceed from observations of conductors of finite dimension and to satisfy two conditions, namely that the observations be capable of great precision and that they be appropriate to the determination of the interaction between the two infinitesimal portions of wires.

"It is possible to proceed in either of two ways: one is first to measure values of the mutual action of two portions of finite dimension with the greatest possible exactitude, by placing them successively, one in relation to the other, at different distances and in different positions, for it is evident that the interaction does not depend solely on distance, and then to advance a hypothesis as to the value of the mutual action of two infinitesimal portions, to derive the value of the action which must result for the test conductors of finite dimensions, and to modify the hypothesis until the calculated results are in accord with those of observation. It is this procedure which I first proposed to follow.

Early in 1820 Ampère had experimentally deduced a force law between steady electric currents. His law took the general form:

$$F = \frac{I_1 dl \quad I_2 dl}{r^2} K$$

where $I_1 dl$ and $I_2 dl$ were infinitesimal current elements at separation distance, $r$, and $K$ was a factor dependent on the angular relationship between the conductors.

**THE EQUATION** sums up in one form Ampère's theory that electricity in motion is the source of magnetism. Wherever, powered by an electric tension, there are swift reunions of the positive and negative fluids, which he envisioned as whirling in infinitesimal circles in the conductor, there was also a force of magnetism. Hence the magnetism was electrical.

Ampère's electromagnetic force law is a direct analog of Coulomb's electrostatic force law, in which force between two point electric charges $Q_1$ and $Q_2$ is proportional to their product and inversely as the square of their separation. This implied a relation between electric charges and electric current. But it was not apparent at the time of Ampère that a charge put into motion, and a current in a wire, were essentially alike in their production of magnetism.

Ampère's force equation usually appears in a modified form known as the Biot-Savart law.

**FOUR EXPERIMENTAL FACTS** and one assumption are the basis on which Ampère's theory of mutual action of electric currents is founded. As explained in Maxwell's *Treatise on Electricity and Magnetism,* the fundamental experiments made by Ampère are all examples of what is called the "null method" of comparing forces.

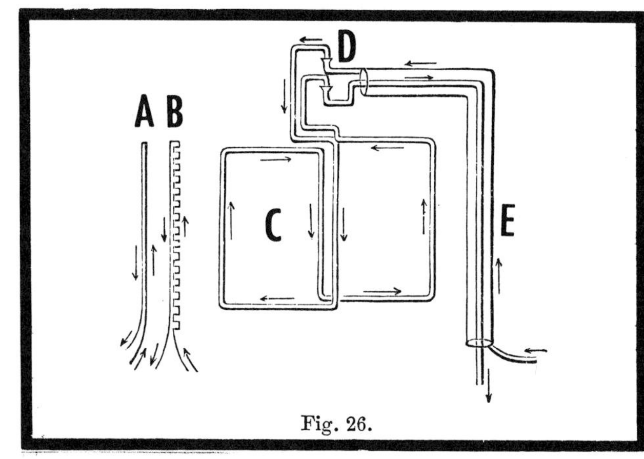


AMPERE'S ASTATIC INDICATING INSTRUMENT

Fig. 6.4 - A coil, C, of two halves, wound in opposite directions to neutralize the effect of the earth's magnetism, is pivoted at D, to permit it to swing as an indicating balance. Current to the balance is supplied axially, at E, by one wire inside a hollow conductor to neutralize the magnetic effect. Ampere used the instrument for the experiments on looped conductors, A and B, as described in the text. Illustration is from Maxwell's *Treatise on Electricity and Magnetism.*

"Instead of measuring the force by the dynamical effect of communicating motion to a body, or the statistical method of placing it in equilibrium with the weight of a body or the elasticity of a fibre, in the null method two forces, due to the same source, are made to act simultaneously on a body already in equilibrium, and no effect is produced, which shows that these forces are themselves in equilibrium.

This method is peculiarly valuable for comparing the effects of the electric current when it passes through circuits of different forms. By connecting all the conductors in one continuous series, we ensure that the strength of the current is the same at every point in its course, and since the current begins everywhere throughout its course almost at the same instant, we may prove that the forces due to its action on a suspended body are not at all affected by starting or stopping the current."

**AMPERE'S FIRST EXPERIMENT,** Fig. 6.4A, was to determine whether two currents, in opposite directions, and in closely adjacent spacing, would cancel out their magnetic effects in respect to the surrounding space. When the pair of wires,

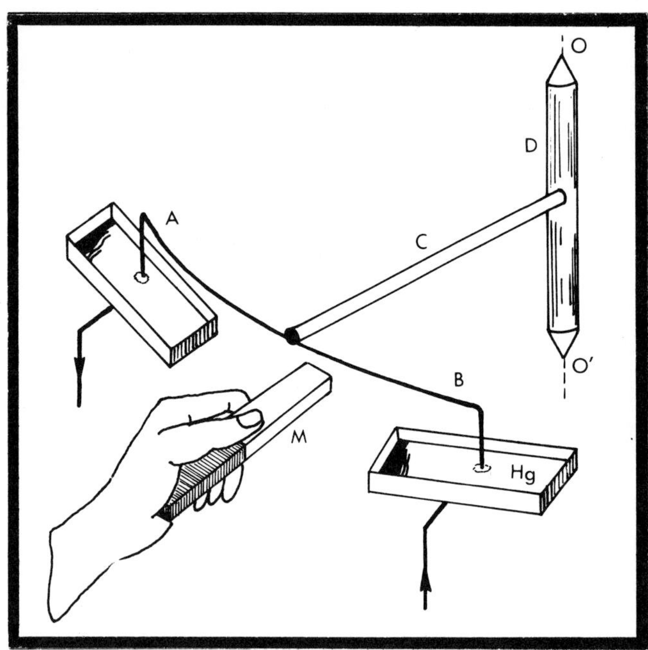


FORCE IS AT RIGHT ANGLES TO THE CONDUCTOR

Fig. 6.5 - Circular current-carrying wire arc, AB, is rigidly fastened to rod, C, and is pivoted so that it can move only in direction of its length. Ampere found that regardless of any external magnetic field, M, relative to AD, there was no movement of the wire arc in the direction of its length.

straight and parallel as in A, were placed next to the balance, there was no reading, indicating that the magnetic effect to the surrounding space of opposed currents neutralized each other.

If, instead of the two wires side by side, a wire be placed coaxially in a metal tube, and if the current pass through the wire and back by the tube the magnetism surrounding the tube is null. This principle, E, was used by Ampère in conveying the current to the balance, C, and is used in modern practice in the construction of coaxial conductors.

**THE SECOND EXPERIMENT**, Fig. 6.4B, used a wire as in A, except that the conductor next to the straight wire was still adjacent to it but of a corrugated shape. The current flowing through the crooked wire and back through the straight wire had no influence on the balance. This proved that the effect of the current flowing through the crooked wire was equivalent to that of the straight wire, and that only the magnetism at right angles to the balance was effective.

In **THE THIRD EXPERIMENT**, Fig. 6.5, a wire n the form of a circular arc was substituted for the ıstatic balance. The arc, AB, was pivoted and its center coincided with the pivoted axis. The ends of the arc dipped into troughs of mercury to provide connections for the current and for freedom of movement. that is, the arc was able to move only along the direction of its length.

Ampère found that another energized conductor could be brought near the movable arc without the slightest tendency of the arc to move in the direction of it length. Neither was there movement when the energized arc was approached by either end of a permanent magnet. The conclusion was that the action of a closed circuit on another element of an electric current is perpendicular to that element. That is, the movement is only sidewise, at right angles, to both the direction of the current and to the direction of the magnetic lines.

**THE FOURTH EXPERIMENT**, Fig. 6.6, used an arrangement consisting of the coil, B, counterbalanced, and pivoted on a support so it could rotate in the horizontal plane. Coil, C, on one side of B, had dimensions a certain amount larger than coil A on the opposite side of coil, B. The coils, C, and, A, were placed with relation to coil, B, such that the distance of, C, from, B, was proportionately larger than that of, A, from, B, as the relation of their sizes. The purpose of the experiment was to prove that the moving coil, B, must remain in equilibrium when the ratio of distance, B-A, and B-C, is the same as the ratio of the diameters.

Ampère found that, B, would remain in equilibrium under the action of, A, and, C, whatever were the forms and distances of the three circuits provided the relations above were maintained. From this experiment Ampere inferred that the magnetic forces varies inversely as the square of the distances.

From the results of the foregoing four experiments Ampère was able to show that:

*(1) The effect of a current is reversed when the current is reversed. In a conductor that is doubled on itself, for example, the outward force of the conductor is neutralized.*

*(2) The effect of an element of a straight conductor may be replaced by that element of a crooked*

*conductor of equivalent length if the same current flows in both cases.*

*(3) The force exerted by a closed circuit on an element of another closed circuit is at right angles to a line joining them, that is, the action is transverse.*

*(4) The force between two elements of circuits is unaffected when all linear dimensions are increased proportionately, the current through them being the same.*

In all the experiments, Ampère used the concept of infinitesimals of electric current, each behaving as though it generated its own magnetic field - the fields acting on each other along straight lines. The actual picture was, however, the summation of these infinitesimals, since an isolated current element could not be measured.

Maxwell, in his *A Treatise on Electricity and Magnetism,* comments on the limitations of the Ampère tests as related to infinitesimals, pointing out that the true picture of magnetic interaction can actually only be indicated by the information obtained in the integration of the infinitesimals into an actual closed circuit:

"It may be observed with reference to these experiments that every electrical current forms a closed circuit. The currents used by Ampère, being produced by the voltaic battery were of course in closed circuits . . . No experiments on the mutual action of unclosed currents have been made. Hence no statement about the mutual action of two elements of circuits can be said to rest on purely experimental grounds. It is true we may render a portion of a circuit moveable, so as to ascertain the action of the other currents upon it, but these currents, together with that in the moveable portion, necessarily form closed circuits, so that the ultimate result of the experiment is the action of one or more closed currents upon the whole or a part of a closed current."

However, Maxwell went on the say that another interpretation could be put to the concept of the current element:

"In the analysis of the phenomena, however, we may regard the action of a closed circuit on an

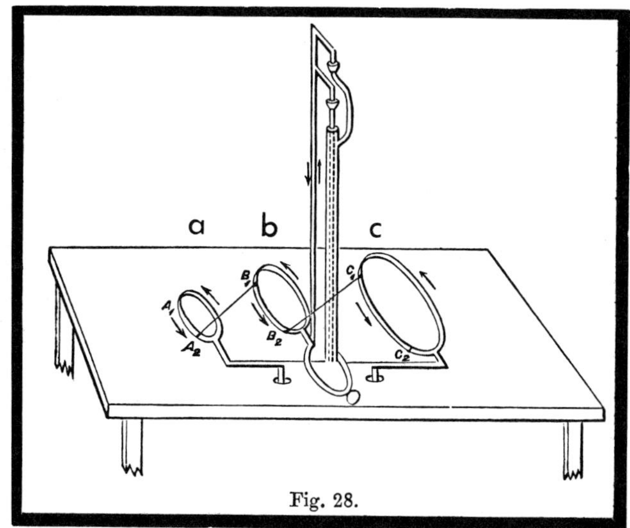


Fig. 28.

CURRENT - FORCE RELATIONSHIPS . . .

Fig. 6.6 -- Movable coil, B, was subject to the magnetic forces from coils, A, and C, all carrying the same current. Ampere found that coil, B, remained in equilibrium when the ratio of distance, BA and BC, is the same as the ratio of the respective diameter of the coils.

element of itself or of another circuit as the resultant of a number of separate forces, depending upon the separate parts into which the first circuit may be conceived, for *mathematical* purposes, to be divided. This is merely a mathematical analysis of the action, and therefore perfectly legitimate, whether these forces can really act separately or not."

**THE SOLENOID** - Oersted's discovery showed that a conductor carrying current in one direction above the compass needle, and one carrying current below the compass needle, acted in concert to swing the needle toward right angles with the conductor. The next step was to complete a loop and to continue winding the conductor into a larger number of turns in the same direction, thus enhancing or multiplying the effect on the needle. This was done by Ampère in 1820. He gave the many-turn coil the name *electrodynamic solenoid,* from the Greek word which denotes "that which has the form of a canal, that is to say, the surface of this form in which the currents lie."

A year later, in 1821, J.C.S. Schweigger (1779-

1857), a German chemist, made the first use of a solenoid, together with a compass, for indicating a current. Using a compass placed within a coil with one hundred turns of wire, he found that the effect on the needle was greatly increased, and that the arrangement could serve as a ready means for indicating electrical intensity. Experimenters who formerly could only guess at the energy of a current now had a visual device for observing their operations. Furthermore, the instrument had the possibility of calibration to provide for numerical comparative readings. Thus, building on Ampère's search for a means of electrical measurement, the age of electromagnetic instruments was born.

Ampère had also observed that the current in the solenoid gave it polarity, exhibiting north and south poles just like a permanent magnet. It was soon found that by winding the solenoid wire over an iron bar the magnetic strength was greatly increased. Using this type of solenoid, William Sturgeon (1783-1850), of Woolwich, England, made the first lifting electromagnet. Taking a bar of soft iron a foot long and one half inch diameter, he bent it into a horseshoe shape to bring the poles near each other. The bar had been given an insulating coating, and on this he wrapped 16 turns of copper wire. When it was excited by a voltaic cell, the magnet was able to hold a weight of 9 pounds. This was the first demonstration of the power of electromagnetism, and a memorable scientific novelty at the time.

**MOLECULAR MAGNETS** - Ampère had noted, from the Oersted experiments, that the compass needle is affected by electric current in a nearby wire. Likewise, the compass needle is oriented by a magnetism generated within the earth. Does it follow then, Ampère conjectured, that the earth's magnetism could be the result of electric currents circulating in the earth's interior? And if so, these currents would have to circle in curves concentric to the earth's axis to provide the north and south magnetic poles. Furthermore, Ampère reasoned, if circulating currents produced magnetism on an earth scale, might not this principle apply on a smaller scale to explain the properties of ordinary steel magnets?

At the time Ampère turned his attention to the subject, popular theories of magnetism attributed its action to a kind of "polarizing fluid." This fluid permeated the structure of the magnet, and its polarity interactions were responsible for the attraction and repulsion observed between the poles of magnets. Coulomb had originally discussed magnetism in terms of a general concept of the existence of "magnetic matter" which had polarity characteristics. But further considerations in accounting for magnetic properties led him to narrow the source of magnetism to the constituents of matter - the molecules. The magnetic fluid, he believed, moved only within the molecules. The fluid, flowing in each molecule, gave it polarity and made each molecule an elemental magnet. Magnets thus derived their properties from the totality of their molecular magnetic constituents.

**AMPERE** had a different idea. Instead of a piece of iron being magnetized by the agency of some imaginary "fluid" which imparted magnetism to the molecules, he hypothesized that the molecules were already magnetic.He based this on his conception that electric and magnetic phenomena were identically associated - that where there was one there was the other. The question was - which one was fundamental? Magnetism accompanied electricity in motion, and this accounted for the force exerted by an electric current in one conductor on that in another conductor. But what accounted for the properties of a permanent magnet? It was only a bar of steel, but it also exerted the same force as a current in a conductor. Ampère's reasoning projected the steel bar's magnetism to the same source - electric currents - which therefore must exist in the structure of the magnet.

Influenced by Volta's theory that the source of voltaic current was in the contact of dissimilar metals, Ampère originally ascribed the source of magnetism to the currents produced by a voltaic pile action between the molecules. The currents generated by these molecular batteries provided an internal electromagnetism that accounted for the external poles of the magnet, Further consideration showed, however, that such internal conduction currents derived from this source would be subject to resistance losses which would generate heat. No heating effect had ever been experienced in a permanent magnet.

Ampère then abandoned the voltaic pile theory, and proposed a bold new explanation. Assume, simply, that each molecule is accompanied by an electric current whirling forever around it without resistance, making each molecule a tiny magnet.

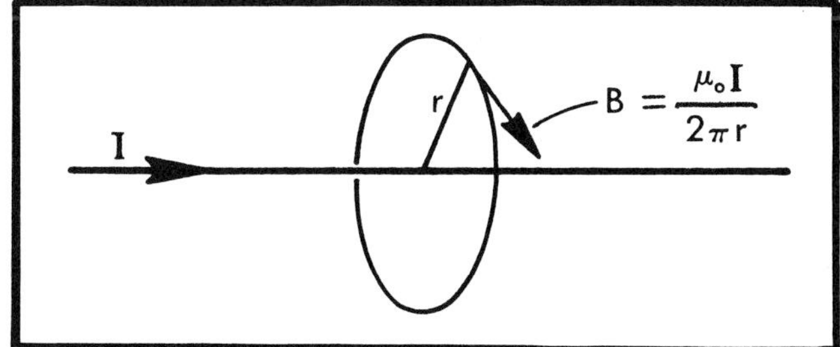


Fig. 6.8 - AMPERE'S CIRCUITAL LAW - The magnetic field lines, B, go around the current-carrying conductor in closed circles. The equation for the magnetic field, B, in the illustration to the left is based on the summation of Ampere's circuit equation (1) below for distance, r.

By Ampère's reasoning the molecules of the steel magnet were already magnetic, with their polarities randomly oriented. By subjecting the bar to an external magnetic field, the magnetic axes of the molecules would tend to line up in one direction to provide the north and south poles. Ampère could offer no proof of his theory other than that the behavior of a bar magnet is equivalent in effect to that produced by current in a long wire-wound solenoid, Fig. 6.7.

The present-day view agrees with Ampère that the fundamental property of magnetic materials is not the magnetism, but the internal currents that produce it. The magnetism is now understood to come from the circulating currents in the electron structure of the atoms - originating in the spin of the electrons combined with their orbital motion. The electronic currents are sometimes called "Ampèrian currents."

**AMPERE'S CIRCUITAL LAW** - His work with the solenoidal characteristics of the magnetic field encircling a conductor led Ampère to his derivation of the circuital law. This law states that the magnitude of the magnetic field in any closed path around a conductor is equal to the current in the conductor.

Thus, for a steady-state current:

$$\frac{1}{\mu_0} \oint Bdl \cos\theta = I \qquad (1)$$

where:

$I$ = current enclosed by the magnetic path.

$\mu_0$ = a permeability constant for free space which has been assigned a value of $4\pi \times 10^{-7}$ weber/ampere meter in the MKSA system of units.

$Bdl \cos\theta$ is the product of, $dl$, an element of the path around the conductor, and $B \cos\theta$ the component of magnetic field, $B$, parallel to $dl$. $B$ is in webers/meter. The line integral, $\oint$, indicates the summation of $Bdl \cos\theta$ around the magnetic loop.

Referring to Fig. 6.8, for the magnetic field near

MAGNETISM FROM MOLECULAR CURRENTS . .

Fig. 6.7 - Each molecule of a magnet was considered by Ampere as a tiny magnet produced by an electric current whirling around it forever in a path of zero resistance. These elemental magnets can be visualized as at the right, looking at a cross-section of a magnet rod. Within the circular area the tiny currents oppose each other and the net current is zero. But at the surface they combine into a circular surface current, making the rod equivalent magnetically to a long, wire-wound solenoid carrying and electric current.

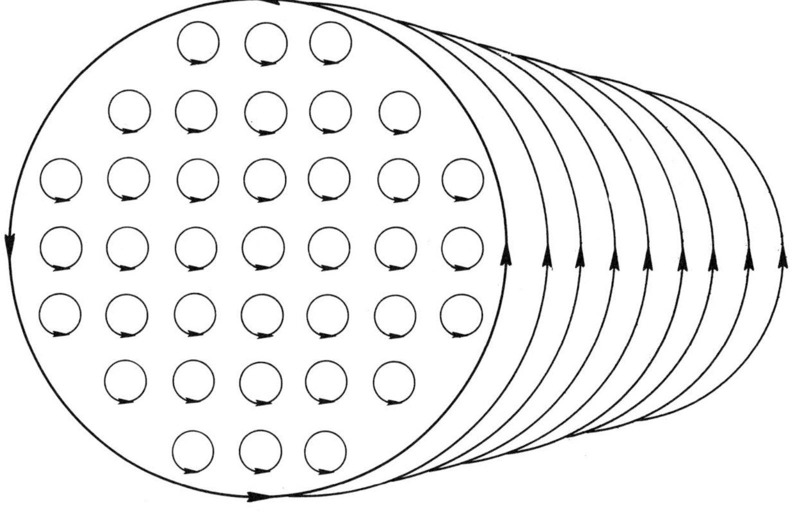

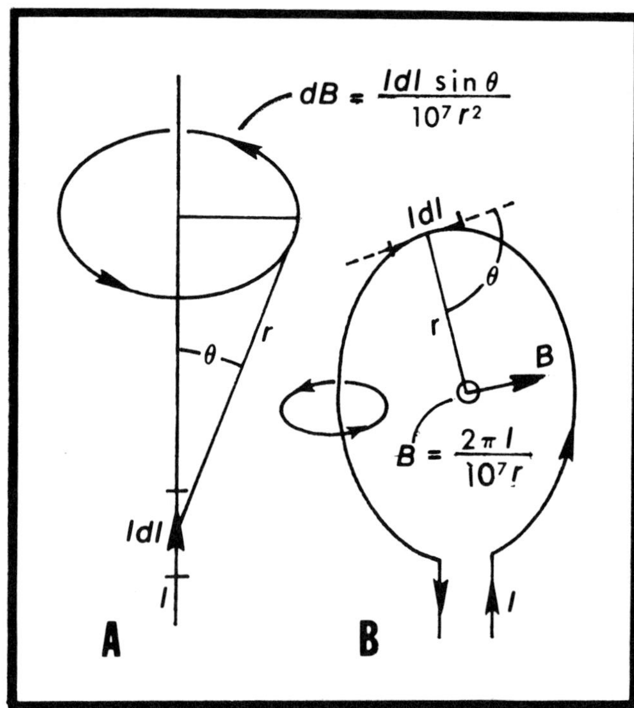


Fig. 6.9 - AMPERE'S CURRENT ELEMENT LAW relates the magnetic field to a current element of a conductor.

(A) Current element, Idl, produces a magnetic field, dB, at a distance, r, and angle, $\theta$, with the current, indicated by the equation above.

(B) Ampere's law applied to the magnetic field at the center of a circular current loop of radius, r. Magnetic field B, at the center, O, is the integral of the equation for (A). The direction of the magnetic field is at right angles to the plane of the coil.

a conductor carrying current, $I$, the magnitude of $B$ can be determined from equation (1).

The magnetic field lines are centered on, and perpendicular to the wire, hence $\cos\theta = 1$. Since $B$ is tangent to and uniform in magnitude around the circular field, the summation of the circle of $B$ will be $2\pi r$, or, from equation (1):

$$\frac{1}{\mu_0}(B)(2\pi r) = I, \quad \text{and}$$

$$B = \frac{\mu_0 I}{2\pi r} = \frac{2I}{10^7 r} \qquad (2)$$

where $B$ is in webers/meter$^2$.

**AMPERE'S LAW** for the Current Element, also known as the *Biot-Savart Law*, for Ampère's associates at the Academy, J.B. Biot and F. Savart, is a general method of determining the magnetic field at any point in the vicinity of a wire carrying a steady current.

Derived from their experimental evidence, it was shown that an element, $dl$, of a conductor with a current, $I$, produces at a point in free space with a radius, $r$, making an angle, $\theta$, with the current, a magnetic field, $dB$, which will be in accordance with the equation:

$$dB = K\frac{Idl \sin\theta}{r^2}$$

In the MKSA system of units, $K = \mu_0/4\pi$, and has the value of $10^{-7}$ weber/ampere-meter. $I$ is in amperes, and $r$ in meters. Thus, as in Fig. 6.9A:

$$dB = \frac{\mu_0 Idl \sin\theta}{4\pi r^2} = \frac{Idl \sin\theta}{10^7 r^2} \qquad (3)$$

For a real circuit the equation must be integrated, and this is usually difficult in any but the simplest cases, such as in Fig. 6.9B, which shows Ampère's current element law applied to the calculation of a circular loop of radius, $r$. The distance of each element of the current, $Idl$, is, $r$, and the angle, $\theta$, is 90°, so that $\sin\theta = 1$. The net field at O due to all the current elements, $Idl$, at distance, $r$, is the summation of these elements around the circle, or $2\pi r$. Thus, substituting in (3):

$$B = \frac{2\pi I}{10^7 r} \quad \text{webers/meter}^2 \qquad (4)$$

**AMPERE'S LAW** for Parallel Wires, Fig. 6.10, derives from the fact that the current in a wire, a, of length, $l$, provides a magnetic field on wire, b, of the same length, and with separation, $d$, in accordance with equation (2), or $B = \mu_0 Ia/2\pi d$.

This magnetic field exerts a force on the magnetic field of current, $I$b, in wire, b, equal to: $F$b $= B$a $I$b $l$.

Thus, the force between the two wires is:

$$F\text{ab} = \frac{\mu_0 I\text{a} I\text{b}}{2\pi}\frac{l}{d} \qquad (5)$$

where $F$, is in newtons, $I$, is in amperes, and, $l$, in meters.

Equation (5) is used to define the value of the ampere in the International System of Units.

**IN 1827** Ampère published a summary of his papers on magnetism and electricity, entitled: *The Mathematical Theory of Electrodynamic Phenomena Solely Deduced from Experiment.* This great work, with its updating revisions. covered the papers Ampère had presented to the French scientific community, starting with his original theories in 1820. Maxwell said of Ampère's accomplishments, as presented in his papers:

"The experimental investigation by which Ampère established the laws of the mechanical action between electric currents is one of the most brilliant achievements in science.

"The whole, theory and experiment, seems as if it had leaped, full grown and full armed, from the brain of the 'Newton of electricity.' It is perfect in form, and unassailable in accuracy, and it is summed up in a formula from which all the phenomena may be deduced, and which must always remain the cardinal formula of electrodynamics."

Maxwell was critical, however, as to whether Ampère's volume represented the true genesis of of his discoveries:

"The method of Ampère, however, though cast into an inductive form, does not allow us to trace the formation of the ideas which guided it. We can scarcely believe that Ampère really discovered the law of action by means of the experiments which he describes. We are led to suspect, what, indeed, he tells us himself: that he discovered the law by some process which he had not shewn us, and that when he had afterward built up a perfect demonstration he removed all traces of the scaffolding by which he had raised it."

Maxwell then compares this to the method of Michael Faraday, Ampère's contemporary:

" . . . who shews us his unsuccessful as well as successful experiments, and his crude ideas as well as his developed ones . . . Every student should therefore read Ampère's research as a splendid example of scientific style in the statement of a discovery, but he should also study Faraday for the cultivation of a scientific spirit, by means of the action and reaction which will take place between the newly discovered facts as introduced to him by Faraday and the nascent ideas in his own mind."

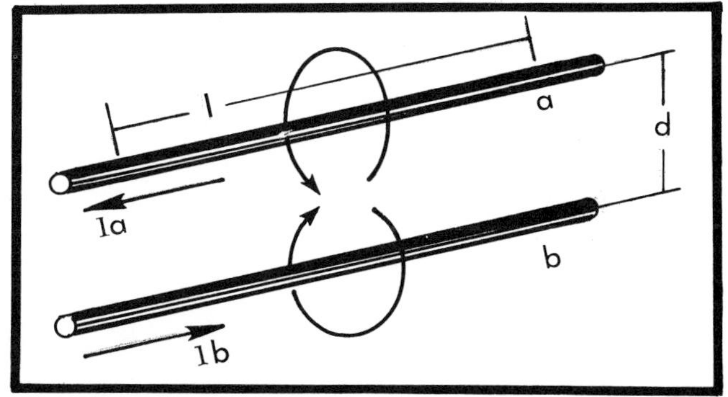


MAGNETIC FORCES BETWEEN CONDUCTORS . . .

Fig. 6.10 - Ampere's experiments showed that the force between parallel conductors is proportional to the currents, length of the conductors, and inversely as the separation. The end-turns of motor and generator coils require anchoring to prevent destructive motion of adjacent conductors that would occur from the opposing magnetic forces.

**AMPERE'S TELEGRAPH** - Ampère was one of the first to see in the Oersted relationship between current and magnetism the possibility of signalling at a distance, since magnetic effect accompanied the current everywhere. A magnetic needle would be able to indicate a current no matter how long the conductor or how far the needle was away, because the current was everywhere the same. Ampère suggested that by using a pair of wires and a magnetic needle for each letter of the alphabet, one could communicate by opening and closing the circuits for the desired letters.

Ampère did not follow up on his suggestion. But, William Alexander demonstrated in 1837, in Edinburgh, how such a telegraph might be set up, and for a short distance it provided successful transmission. Ampère had taken the first step in the invention of the telegraph.

**AMPERE** was international in his correspondence and recognition, both in the fields of physics and chemistry. There was a lively competition between the French and British scientists, and they watched each others work closely. He communicated, as early as 1810, with Sir Humphry Davy, mostly on matters of chemistry, and continued until the latter's death in 1829. He also collaborated with Oersted in Copenhagen in the exchange of scientific information.

Ampère carried on a lively correspondence with Michael Faraday, who was then doing some of his most brilliant work in electromagnetics at the Royal Institution in London. Ampère, who expressed results in elegant mathematical language, and Faraday who sought demonstration only by the results of experimentation, did not see eye to on interpretations.

Ampère's two-fluid theory of electric current, his vision of molecular currents and his conception of magnetism as the product of electricity in motion, were questioned by Faraday for lack of experimental evidence. Ampère, on the other hand, was inclined to look on Faraday's discoveries in electromagnetism as vindication of his own views on the nature and identity of electricity and magnetism.

Ampère's electrodynamics opened a whole new set of concepts and techniques applied to electric currents, immediately pursued by scientists in France and elsewhere. Thus the electrostatic techniques championed by Biot and others of the French Academy became out of date, and scientific communities began to accomodate themselves to the new theories of the electric circuit.

Ampère became a member of the Royal Institute of France in addition to the French Academy of Science, and was made a Fellow of the Royal Society of London. He was honored by institutions in Edinburgh, Cambridge, Geneva, Brussels, Stockholm, Berlin and Lisbon. He was a revered professor at the École Polytechnique and at the Royal College of France.

Ampère died in Marseille in 1836. Nearly a half century later his name was perpetuated by identification with the currents with which he worked. Meeting in Paris in 1881, the First International Electrical Congress, adopted a unit of current and called it the *ampere.*

**The AMPERE** as the unit of current was first defined at the 1881 Paris meeting. The centimeter-gram-second (CGS) electromagnetic unit of current called the *abampere,* was that which, flowing in a circular loop of one centimeter radius, would act on a "unit magnetic pole" at its center with a force of one dyne for every centimeter of wire in the loop. The practical unit of current, the *ampere,* was defined as one-tenth abampere.

To provide a practical and reproducible determination of the ampere, the 4th IEC, meeting in Chicago in 1893, proposed a measurement based on Faraday's discovery that the mass of a substance deposited in an electrolytic cell in any given time is strictly proportional to the amount of current. The ampere was defined as "that current which liberates 0.0011183 grams of silver from a silver nitrate solution in an electrolytic cell in one second," This basis for the ampere was adopted in 1908. Although in use for forty years, complications in the accurate measurement of the silver deposit made a new method desirable.

The International Committee on Weights and Measures, in 1946, set values for a complete revision of previous so-called "International Units." A new "absolute" system of units was established and became effective in 1948. In this system the ampere was named as one of the seven basic units in the International System of Units (SI) along with the meter, kilogram, second, mole,kelvin and candella. The ampere became the fundamental unit in the electrical system of measurement, and its determination was based on Ampère's law of the magnetic forces exerted between parallel current-carrying conductors.

**The SI Ampere** is *that constant current which, if maintained in two straight, parallel conductors of infinite length, negligible cross-section and placed one meter apart in a vacuum, produces between those conductors a force equal to* $2 \times 10^{-7}$ *newtons per meter of length.*

**THE SCIENCE** of Ampère's era was dominated by the theories of motion and gravitation developed by Sir Isaac Newton (1642-1727), the English natural philosopher and mathematician. Newton's mechanics were based on the view that all matter is composed of elementary, changeless particles, or point masses, which have the power to act on each other directly, through empty space. Newton's law of gravitation stated that the force between these mass bodies was proportional to their product and inversely as the square of the distance between them. Furthermore, the attractive forces acted instantly, and in straight lines through the center-to-center of the masses.

Followers of Newtonian theory were led to expect that the law of gravitation would apply to other forces, such as those of electricity and magnetism. Coulomb showed that electrostatic forces between point charges, and magnetic forces between point poles, were the product of their intensities, and inversely as the square of their separation. Coulomb's laws thus fitted exactly into the Newtonian framework.

Ampère's theory of electric currents, like the Newtonian theory of gravitation, was also one of instant, direct-action, at a distance. He hypothesized that the electrical fluid consisted of infinitely small elements, whose motion constituted the electric current. The current elements acted like the Newtonian point masses. The electromagnetic force between electric currents was as the product of these current elements, and inversely as the square of their separation, and thus fitted into the Newtonian model. The measurement of interacting electromagnetic forces of these current elements could not be done individually, however, but only in the aggregate, that is, by working with real currents in closed circuits. These circuits, in various forms, were actually what constituted the experiments by which Ampère developed his electrodynamic circuit laws.

But the Newtonian explanation of electromagtism came under criticism. There were new force elements not covered by Newtonian precepts. The electromagnetic forces had direction as well as magnitude. There was a right-angle relationship between the current and the resulting magnetism, that introduced the possibility of rotation. Also, instead of instant, direct-action-at-a-distance, the electromagnetic action implied the possibility of a time element in expanding into space. These new facts could not be explained by Newton dynamics.

**CONTINUING DEVELOPMENT** of electromagnetics brought a search for new theories and experiments. The most anti-Newtonian of these were by Michael Faraday, a contemporary of Ampère. He rejected the particulate, instant action-at-a-distance theory and introduced a revolutionary supposition that the forces themselves were the sole physical substance. This led to his *field-of-force* theory. Faraday envisioned all space filled with a universal medium - the fields-of-force, which converged at points of mutual action, and which accounted for all the manifestations of electromagnetism.

Faraday pictured the force-field by *lines-of-force* which connected the sources of action. The force-field concept was one in which the physical processes were determined by field lines which had a reality in terms of a well-defined value and direction at each point in space. Put into mathematical form by J. Clerk Maxwell, the Faraday fields-of-force took over, after Ampère, to provide the main flow of scientific thought in the stream of development in electromagnetic theory.

ERECTED IN HIS HONOR by the French Academy of Sciences, this statue of Ampere stands at the ancestral home of the Ampere family at Polymieux, near Lyons.

# OHM . . .

## the unit of electrical resistance is named for the German professor

## GEORG SIMON OHM

## whose investigations resulted in the famous law that bears his name . . . . .

**OF ALL LAWS** relating to the applications of electricity, one of the most pervasive and practical is that symbolized in the circle at the right. Called *Ohm's Law,* it immortalizes the name of its discoverer, the 19th century German professor, Georg Simon Ohm.

Working against the odds of poverty, isolation and obscurity, he laid the foundations that established the relations between the potential, the flow of current and the resistance in an electric circuit. Unrecognized, and derided by some at first, the soundness and basic importance of his findings at last achieved an international acclaim, giving him an eternal niche in the gallery of founders of the electrical science.

To review the conditions that led up to Ohm's crucial law, the following paragraphs present briefly the state of electrical knowledge in the early nineteenth century.

OHM'S LAW ($I = V/R$) - The current in an electrical circuit is directly proportional to the voltage in the circuit, and inversely proportional to the resistance in the circuit. The law applies to linear, constant-current circuits.

The circle above symbolizes the Ohm's law relationships. When a finger tip covers the factor that is wanted, the other two factors of the Ohm's law equation are shown in the relation for the required calculation.

Up to about the year 1800, the electricity that was experimented with, and had been known from antiquity, was the electricity from friction. This involved the electrical charges produced by rubbing on a suitable material such as glass or hard rubber. Electrification was a popular science, and also a pseudo-science for medical quackery. The early workers made large, powerful electrical machines, and experimented with sparks, shocks and the strange purple glow given off by the high intensity electrostatic discharges in evacuated vessels.

Eighteenth century investigations with frictional electricity exhibited the differences between electrical conductors and non-conductors, and showed that charges could flow long distances in wires, leading some to believe that electricity was a fluid. It was also evident that charges, though separated, could affect each other across the intervening space by an inductive action. Experimenters showed heating and chemical effects from electricity, and there were glimmerings of a reciprocal relationship between electricity and magnetism.

Charles A. Coulomb, showed, in Paris, in 1785, that electrical charges attracted or repelled each other in proportion to their magnitude, and that the force between point charges varied inversely with the square of their separation. With this quantification of charge by Coulomb the science of electrification had reached a plateau. The 18th century closed with the work of Luigi Galvani, in Italy, whose investigations with animal tissue had developed a new effect which he called "animal electricity."

**FROM ITALY,** in 1800, came an announcement of a revolutionary new kind of electricity. Alessandro Volta, inventing new principles from the pioneering work of Galvani, revealed that he had produced, for the first time, a "continuous and inexhaustible source of electricity." He used alternate interconnected disks of zinc and copper, separated by pasteboard soaked in brine, to create the "electric pile." In contrast to the spasmodic spark-type discharges from the frictional electrical machines, the scientific world now was provided with a new and ready means to produce sustained and large electrical currents conveniently and at a controllable potential.

The value of Volta's contribution was immediately recognized, and the availability of electric power from batteries soon opened new fields of discovery. Shortly after Volta's invention it was found, in England, that the wires from a battery, dipped into acidified water, decomposed the water into its elements hydrogen and oxygen. Then, in 1807, the separation of the new elements, sodium and potassium, from their salts was achieved with an electric current by Sir Humphry Davy, in England.

In the twenty years following Volta's great discovery, experimentation with the chemical and heating effects of electric current were vigorously pursued. But one important effect, the interrelationship of electricity and magnetism, escaped clear observation until 1820. In that year a new era of electricity began with the startling announcement to the scientific world by Hans Christian Oersted at the University of Copenhagen, that a wire carrying an electric current is accompanied by a magnetic field of force. He found that the magnetism surrounded the wire, and circulated at right angles to the direction of the flow of current.

Oersted's discovery sparked a flood of new investigations. Now, in addition to the electrical effects, the new phenomenon of electromagnetic action was opened, so that the dynamic forces of magnetism could be studied and used.

**WITHIN A FEW MONTHS** after the Oersted announcement, André Marie Ampère, in Paris, demonstrated that parallel conductors carrying current would exert magnetic forces on each other. There was a mutual attraction or repulsion, which depended on the direction of the current. He showed that the force was proportional directly to the intensity of the current, and inversely proportional to the separation of the conductors.

Then Ampère revealed that wire wound into a coil, which he called a "solenoid," acted like a bar magnet, having definite north and south poles. The solenoid arrangement, would in the future, become basic to operation of electromagnetic apparatus..

Ampère further conceived that a tension, which he called "electric potential," existed between the terminals of a battery, and that the "intensity" of the current provided the energy of the output.

In the early nineteenth century, the concepts of electric potential and intensity were largely descriptive. The quantitative relationships had not yet been formulated. There was no clear distinction between the electromotive force of a battery, and the resulting flow of current, although it was apparent that they were being consumed in the circuit to produce heat or mechanical or chemical effects. Very importantly, Ampère had noted a property of a conductor which he called *resistance.* This appeared as an impediment to the flow of the current. He saw this as an "imperfection" in a conductor. The resistance due to the imperfections in a conductor was a factor limiting the current in the conductor.

Some studies of electrical resistance as related to size and shape of conductors had been made. But the significance of resistance as related to the electromotive force and the current in the circuit had not been defined. Progress of electrical investigation was being retarded by lack of exact analysis and formulation of these fundamentally related factors. Bringing order to this situation, and putting circuit measurement on a precise foundation, was to be the epochal work of Ohm.

**GEORG SIMON OHM** (christened as Johann Simon) was born 16 March 1789 at Erlangen, Bavaria, then a principality in southern Germany bordering on Austria. He was the older of two sons of Johann Wolfgang Ohm, a master locksmith, a man of ability, and with interests in philosophy and science. Although the family was not well-to-do, the father had ambitions for the scholarly advancement of his sons. He, in his youth, had been instructed in mathematics, in return for rent, by a student lodged in the parental home, and this gave him a strong feeling for the value of a formal education. It also enabled him to teach his sons the rudiments of mathematics and some physics and chemistry.

George, and his brother Martin, were also indoctrinated into the writings of Kant and Fichte, contemporary German philosophers who were wrestling with the morality problems in an era of advancing science. Kant believed that man's inherent rationality would prevent a social-moral conflict: that along with his power to develop pure and applied science, man had the inborn ability to guide his conduct in the growing age of reason.

While it was the father's thought that the two sons would eventually join him in the locksmith business, he was ambitious for their further formal education, so they were sent to the local Gymnasium, a preparatory school for a university. Five years there gave them a basic classical training. The brothers displayed ability, Georg leaning to science and his brother Martin toward mathematics. The talents of both brothers were particularly remarked on by Prof. Langsdorff, head of the mathematics department at the Gymnasium, who predicted that history would repeat itself in them the brilliance of the 18th century Bernouilli brothers, who were famous scientists and mathematicians.

In view of their outstanding abilities the brothers were enrolled at the University of Erlangen. Georg's first stay there was short. A family disagreement over his school social conduct led to a kind of exile in Switzerland. He supported himself there by tutoring and was able to save enough to continue his education.

In 1811 Ohm returned to the University of Erlangen where he received a Doctor of Philosophy degree and a modest position teaching mathematics. However, the low salary and poor prospects for advancement led Ohm to seek employment elsewhere. Anxious to repay his father for the sacrifices he had made, he petitioned the King of of Bavaria for a new appointment. This was granted, but the best he could obtain was a disappointing and poorly paid post in a secondary school at Bamberg. Even this position came to an end in 1816 with the closing of the school as an aftermath of the general unsettlement that followed the overrunning and disruption of Germany in the Napoleonic wars.

While at Bamberg Ohm had written an *Essay on Geometry,* which embodied his ideas on the usefulness of a mathematical education, and the self-reliant attitude a student should have in the pursuit of knowledge. Ohm sent published copies of his work to the reigning monarchs of Germany with a plea for help in securing a more suitable employment. The Prussian King, Friedrich Wilhelm, looked favorably on Ohm's request and he was appointed a principal teacher in mathematics and science at the reformed Jesuit Gymnasium of Cologne. There Ohm found a friendly accomodation to his abilities and interests. He remained at Cologne for nine and a half years.

**THE SCHOOL AT COLOGNE** had well-equipped laboratories, and his involvement in teaching science stimulated in Ohm a new and ardent interest in physics. To get into the mainstream of the current science he immersed himself in the classic books of the time in science, particularly the works of the French mathematicians and physicists: Lagrange, Laplace, Legendre, Biot, Fourier, Poisson and Fresnel. The discovery of electromagnetism by Oersted, and the work of Ampère, Biot and Davy particularly captivated Ohm's imagination. He began a program of experiments in electricity and magnetism, leading to his study of the galvanic circuit that was to be a major part of his life's work.

**OHM** was a successful teacher. But his ambition was for a university post that would give him more time for physical research. As the years at Cologne moved on, Ohm saw that he was reaching a dead end. His teaching duties were heavy, and although his consciencious work was appreciated, there was little possibility of promotion that would allow more time for personal pursuits. The same disappointments of earlier teaching positions were being repeated. His only hope was to seek an environment of greater promise. Desperateness of his situation now prompted the same solution that had worked before to get a favorable notice. He must publish, and this must be based on some worthy research.

Although burdened with teaching, and with slender resources, Ohm threw himself into a series of experimental efforts on the question of electrical relationships that were pressing for a resolution. His accomplishments were slow. But persistence and a perceptive mind, plus the mechanical skills acquired in his days as a locksmith, produced over a period of several years, the experimental results and mathematical formulations that were finally to make him famous.

What were the conducting powers of wires and other conductors? What was the effect of different metals, and the various lengths and cross-sections of wires in the electrical circuit? How did they relate to the output of a battery as the current coursed through the conductors of the circuit? These were questions that had occupied investigators before and during the time Ohm began his researches.

**THE FIRST EXPERIMENTS** in conductivity were those of Henry Cavendish (1731-1810), a reclusive English investigator whose notes were unknown until published by J. Clerk Maxwell in 1879. Cavendish's work was done before the availability of voltaic current. Using the discharges of Leyden Jars for power, he tested the relation of potential drop to current flow through glass tubes of various lengths and diameters containing a salt solution. Lacking instruments, Cavendish measured the "degree of electrification" by the amount of shock he received in grasping the terminals with his hands to complete the circuit through his body. Maxwell commented that Cavendish's results were remarkably accurate.

Cavendish expressed the outcome of his tests with a statement that the "resistance is directly as the velocity." He meant by "resistance" the whole force which resists the current, and by "velocity" the strength of the current through unit of area of the section of the conductor. In the estimation of Maxwell, Cavendish's results were an anticipation of the law of electrical resistance discovered by Ohm, but obviously Ohm knew nothing of the work of Cavendish.

Sir Humphry Davy (1778-1829) and Peter Barlow (1776-1862) in England, and A. Becquerel (1788-1878) in France, experimented with wires of various metals, using voltaic cells for their source of electricity. They were primarily interested in the conducting power of different wires. This experimenting was being done at a time when the quantitative circumstances of the electric current was being designated by such terms as "intensity of electricity," "excitation force," "quantity of electricity," and "tension," to which terms no accurately defined meaning was attached.

**THE DISTINCTION** between electrostatics and and the electric current was not always clear. The concepts of electrostatics, the older science, left it uncertain whether flow of electricity was a surface phenomenon, as with electrostatics, or a function of the interior structure. The term "resistance" was used for different quantities, and did not have its modern meaning. Instruments were primitive, yielding only rough results. The gold-leaf electroscope was used to measure electric charge; the magnetic torsion balance and other early forms of galvanometers were used for measuring current. The early voltaic cells were subject to polarization and gave an unpredictable output. The cell's inner resistance was a factor not well understood.

Davy found that the load imposed by wires of a given material having the same ratio of length to section was the same. Barlow also sought a relationship between current intensity and the length

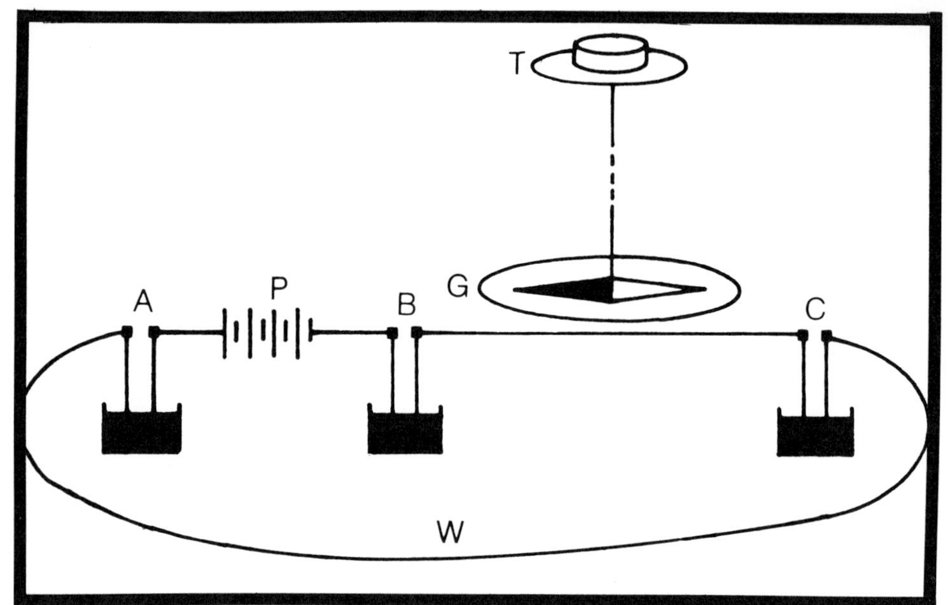


Fig. 7.1- OHM'S APPARATUS for testing conductivity relationships between metal wires of various dimensions. Terminals of the voltaic pile, P, dip into mercury cups, A and B. An "invariable" conductor ran between B and C, under the needle of a Coulomb-type torsion balance, G, to measure the electromagnetic force when the "variable" test wires connected between A and C, completed the circuit.

and diameter of a wire. He concluded that the current intensity was approximately proportional to the square root of the length, and for wires of the same length the current intensity increased with the diameter of the wire.

Becquerel approached the problem from the point of view of the electrical flow in the circuit. He considered the potential, or tension, of the voltaic cell to result from two fluids, originating equally from the two poles of the cell, and diminishing toward some central point in the circuit. He pictured the diminution of tension as an arithmetrical decline from both poles, which would indicate that something to do with the current would remain constant in the circuit. Becquerel showed that the conductivity of wires of the same metal, having the same ratio of length to cross-section, was the same. This was equivalent to finding that the conductivity of the metal, per unit volume, remained constant, and was independent of the current.

Ampère had first drawn the distinction between electric tension and the intensity of the current. This opened the possibility that one of these elements might be the constant factor and the other a varying factor in the resolution of the flow around the circuit.

Ohm was familiar with the reports of these contemporary investigators. He noted the differences in their results, and set out to reconcile them with experiments of his own. He took up his researches in the effort to sift out the true situation.

**OHM'S FIRST SCIENTIFIC PAPER,** published in 1825, dealt with the relationship between the length of a wire carrying a current, and the loss of magnetic force due to that length. The apparatus, Fig. 7.1, used voltaic cells for the electric source. The upper circuit, A,B,C, including the cells, P, and the wire from B to C for energizing the compass needle, G, Ohm called the "invariable conductor."

Using this fixed source, he then added to the circuit the test wires, W, which were the "variable conductors." When the circuit was completed, the needle, G, suspended from torsion head, T, over wire, BC, was deflected. The deflected needle was then returned to zero by twisting the torsion head. The torsion reading was a measure of the force of the current.

For the series of tests, Ohm used a short, thick wire to give a "normal" or reference reading. Compared to this were the lesser readings of six thinner wires ranging in length from one to seventy five feet. Ths loss in force was expressed as the difference between the normal force of the reference wire, and the smaller force exerted by the test wires, divided by the normal force. The loss in force was actually proportional to the change in current corresponding to the length of the test wire. It was an indication of a variable in the circuit related to the dimensions of the test wires.

Ohm found that a relationship between the force and the length of the test wires could be expressed empirically by the equation:

$$v = m \log \left( 1 + \frac{x}{a} \right)$$

where, $v$, is the decrease in force, $x$, the length of the conductor, and, $a$, the length of the reference wire. For the coefficient, $m$, Ohm reported: "The coefficient, $m$, is a function of the standard force, of the thickness of the conductor, of the quantity, $a$, and, as I have reason to believe, of the electric tension of the force. I am at the moment, still engaged in making quite sure, through more exact experiments, of the exact nature of the function."

**CONDUCTION** of wires differed not only with dimensions, but also with the metal composition. The relative conductivity of various metals had been investigated by Davy, Barlow and Becquerel, but their results differed markedly.

Ohm proceeded with a series of new tests to determine the conductivity of a variety of metals. His investigations involved three factors: 1- the length of the conductor, 2- the conductor section, and 3- the conductor material. Using the same apparatus, Fig. 7.1, as in his previous work, he tested the relative conductivity of nine different metal wires. As a reference metal he used a copper wire of a certain length and section. Its reading he arbitrarily took as 1000. Then, using wires of other metals, he maintained the same section but varied the length to get the same reading as the copper reference. The ratio of length for the various metals thus indicated the proportionality to copper.

One series of comparisons reported by Ohm is

| | | | |
|---|---|---|---|
| copper .... | 1000 | iron .... | 174 |
| gold .... | 574 | platinum .... | 171 |
| silver .... | 356 | tin .... | 168 |
| zinc .... | 333 | lead .... | 97 |
| brass .... | 280 | | |

shown in the table above. Ohm later found an error in the reading for silver, due to faulty measurement of the wire. Silver proved to be a better conductor than copper. Except for this, Ohm, in the table, had determined the correct comparative order of conductivity for the metals. His results were later corroborated by many others.

From his experimental work it was apparent to Ohm that an increase in cross-section, and a reduction in length, increased the conduction. Furthermore, the conductivity was a unique property of each individual metal. Ohm formulated his data into a law to the effect that:

*Electrical conductors of the same substance but different diameter, have the same conductivity in their lengths in proportion to their cross-section.*

This is expressed for copper in Fig. 7.2.

Ohm's results appeared in two German publications, one edited by J.S.C. Schweigger, and the other by J.C. Poggendorff, both of whom had been active in the development of early galvanometers. The articles did not, however, attract much atten-

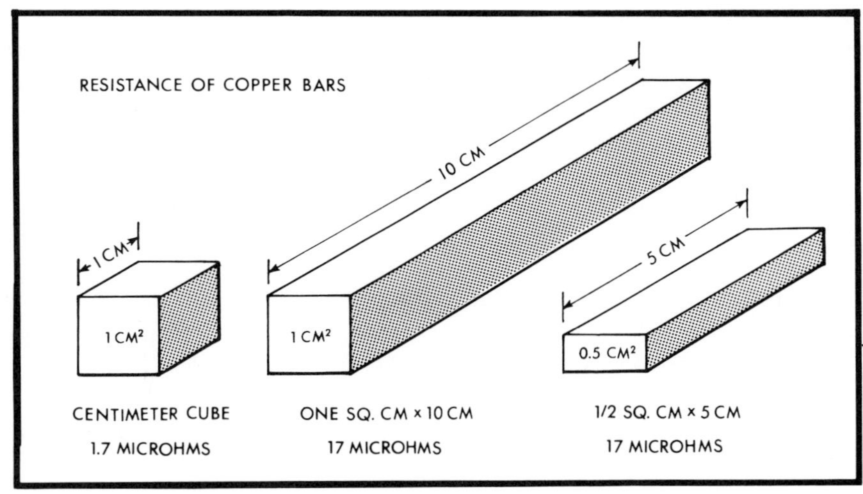


OHM'S PROPORTIONAL RULE

Fig. 7.2 - Resistance of a conductor is not altered if the cross-section is kept proportional to the length. Two bars at right have the same resistance. Cube at left is the ohm-centimeter resistance of copper.

tion, and Ohm's work did not impress his contemporaries. Despite this Ohm forged ahead with his experiments to resolve the problem that had confronted and long intrigued him: *the fundamental forces in an electric circuit.* He was seeking a law to define the relationship between the electric force, the resulting electromagnetic intensity, and a length of the connecting wire. What was the nature of the attenuation of the battery's tension as the conductor was lengthened or its section reduced? Did conductibility vary with the strength of the current? At this point Ohm found the beginnings of an answer in the work of a French scientist.

**JOSEPH FOURIER** (1768-1830), the French mathematician and physicist, published in 1822 a work entitled: *The Analytic Theory of Heat* that attracted the attention of the scientific world. One of its subjects was a discussion of the flow of heat in materials, that is the conduction transfer of heat from a higher temperature extremity to a lower one. Fourier summarized his results in a mathematical law stating that the quantity of heat flowing in time in a given direction is the product of the conducting area normal to the path, the temperature gradient along the path, and a property of the material known as the thermal conductivity. Fourier assumed the conductivity of a material to be constant for all temperatures.

Ohm was struck with the analogy to an electric circuit. He saw similarity in the heat flow in a metal rod to the electrical flow in a conductor. He was the first to grasp the association of the flux of heat to the transfer of electricity from one particle to the next in a wire. And, in analogy with heat, this transfer should be in an amount proportional to the difference of the electrical force between succeeding particles of the wire. The analogy set him thinking of the electric flow as being proportional to the tension, and the possibility of Fourier's heat transfer equation in an equivalent form for electricity.

**OHM** opened his insight into the galvanic situation by a geometric thought experiment, using the simplest possible circuit, one of uniform material and a uniform size conductor:

"Imagine," he wrote, "a ring everywhere of equal thickness and homogeneous, having at any one place one and the same electric potential, i.e., inequality in the electrical state of two surfaces situated close to each other; from which causes, when they have come into action, and the equilibrium is consequently disturbed, the electricity will in its endeavor to re-establish itself, if its mobility be solely confined to the extent of the ring, flow off on both sides.

"If this tension were merely momentary, the equilibrium would very soon be re-established; but if the tension is permanent, the electricity by virtue of its expansive force, which is not sensibly restrained, produces in an almost inappreciably brief space of time, a state which approximates closely to equilibrium, and consists in this: that by the constant transmission of the electricity, a perceptible change in the electric condition of the parts of the conductor through which the current passes is nowhere produced.

"The peculiarity of this state, which occurs also frequently in the transmission of light and heat, arises from the fact that each particle of the conducting medium situated in the circuit of action receives each moment just the same amount of the transmitted electricity from the one side as it gives off to the other, and therefore constantly retains an unchanged quantity.

"Now, since by reason of the first fundamental position the electrical transition only takes place directly from the one particle to the other, and is under otherwise similar circumstances, determined according to its energy by the electrical difference of the two particles, this state must evidently indicate itself on the ring uniformly excited in its entire thickness, and similarly constituted in all its parts, by a constant change in the electric condition, originating from the point of excitation, proceeding uniformly through the whole ring, and finally again returning to the place of excitation; whilst at this place itself, a sudden spring in the electric condition constituting the tension, is, as was previously stated, constantly preceptible. In this simple separation or division of the electricity lies the key to the most varied phenomena.

"The mode of separation of the electricity has been completely determined by the preceding observation; but the absolute force of the electricity at such point by the length of a perpendicular line erected upon it; that directed upward may represent a positive, but that downwards a negative

OHM'S THOUGHT EXPERIMENTS . . .

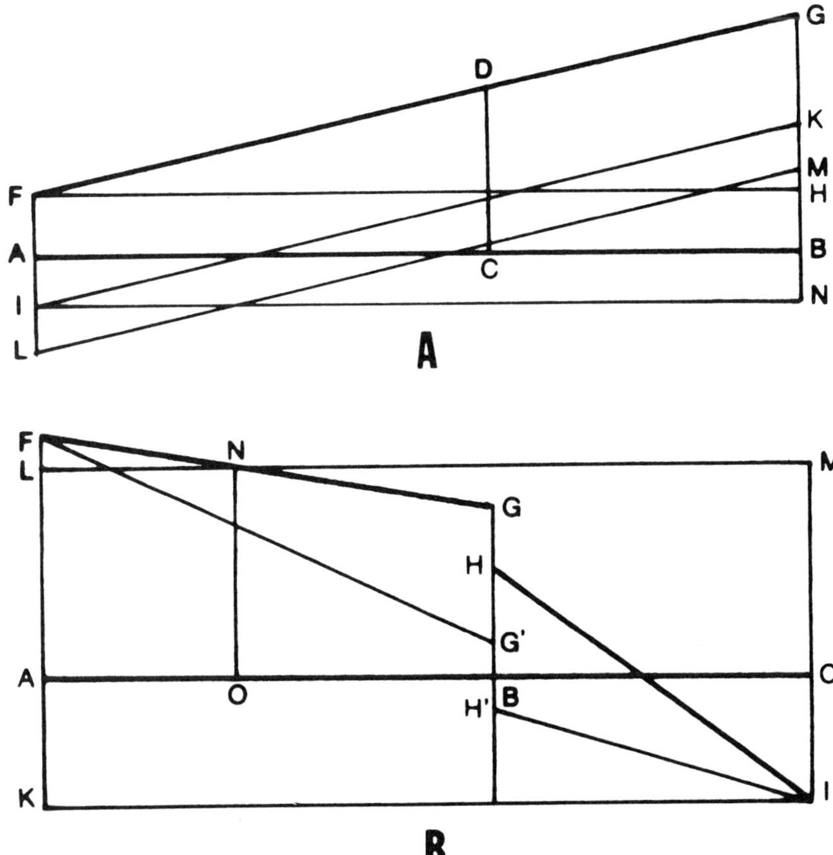


Fig. 7.3 - These graphical representations were part of the reasoning used by Ohm in arriving at the relations in the galvanic circuit. The length of the circuit is shown by the horizontal lines, the tension by the vertical lines, and the fall in tension by the sloped lines. "A" is a circuit of uniform material and uniform size of the conductor, and "B" is a circuit of two differenct metals of the same thickness. Ohm assumed the current to be constant, the tension (potential) falling in proportion to the conducting powers of the circuit.

From *The Galvanic Circuit Investigated Mathematically* by G.S. Ohm, translated by William Francis at the University of Berlin. Van Nostrand Company, New York (1891 edition), the Kraus Reprint Company, New York (1969 reprint).

electrical state of the point.

**POTENTIAL or TENSION** and its fall may be graphically represented:

"The line A-B, Fig. 7.3A, may accordingly represent the ring extended in a straight line, and the lines A-F and B-G perpendicular to A-B may indicate by their lengths the force of the positive electricities situated at the extremities of A-B.

"If now the straight line F-G be drawn from F to G, also F-H parallel to A-B, the position of F-G will give the mode of separation of the electricity, (or, as we should now express it,'the difference of potential that exists between the extremities or the total electromotive force at work in the circuit.') and quantities B-G, -A-F or G-H the tension occurring at the extremities of the ring; and the force of the electricity at any other place, C, may easily be expressed by the length of C-D drawn through C perpendicularly to A-B.

"But from the nature of the voltaic excitation, the absolute magnitudes of the lines A-F and B-G are not determined, but merely the amount of the tension, or the length of the line G-H; consequently the mode of separation may be represented quite as well by any other line parallel to the former; e.g., by I-K, for which the tension still constantly retains the same value expressed by K-N, because the ordinates situated at present below A-B assume a relation opposed to their former one.

"Which of the infinitely numerous lines parallel to F-G would express the actual state of the ring, cannot be generally stated, but must in each case be separately determined from circumstances which occur. Moreover, it is easily conceived that as the position of the line sought is given, it would be completely determined for one single part of ring by the determination of any one of its points, or, in other words, by knowledge of the electric force.

"If, for instance, the ring lost all its electricity at the place, C, by abduction, (if the point, D, of ring, that is to say, were reduced to zero potential by connecting it with a conducting body such as the earth, whose potential is assumed to be zero) the line L-M drawn through, C, parallel to F-G, would in this case express with perfect certainty the electrical state of the ring.

"It is to this variability in the separation of the electricity that the changeableness of the phenomnon peculiar to the voltaic circuit is to be attributed."

**OHM** then repeated his thought experiment with a circuit composed of any number of sections of varying size and material, and another circuit conductor formed of two parts of like cross-section but different materials, Fig. 7.3B. In each case he predicated a law that each section of conductor must receive from one side the same quantity of electricity that it gives off from the other side, i.e., that the current is of equal strength in all parts of the circuit. The variable must thus be the tension. Proceeding in this manner with his deductions, Ohm arrived at the following:

"In a galvanic circuit consisting of any indefinite number of prismatic parts, there takes place in regard to its electrical state at each place of excitation a sudden transition, from one part to the other, forming the tension there prevailing and within each part a gradual and uniform transition from the one extremity to the other; and the dips of the various transitions are inversely proportional to the products of the conductivities and sections of each part."

Ohm commented in one of his papers that he was not happy with the operation of the voltaic cells in his experiments. Due to the nullifying chemical action of polarization and because of local corrosion, the electromotive force of the early batteries declined rapidly under load, and they had a limited life. There were fluctuations in the output depending on the condition of the cells. J.C. Poggendorff, editor of *Annalen der Physik und Chemie,* and familiar with Ohm's work, suggested that he substitute a *thermoelectric generator* for his electric source.

**THERMOELECTRICITY** had been discovered by the Russian-born scientist Thomas J. Seebeck (1770-1831). Thermoelectricity operated on the principle that the junctions of dissimilar metals, such as bismuth and copper, when maintained at different temperatures, would produce a flow of current through a circuit. The generator gave a very low electromotive force, but it also had a low internal resistance, consequently delivered a large current. The electromagnetic effect from the current in the circuit was steady as long as the temperature differential was maintained on the junctions.

Although Ohm had accumulated many tests using voltaic cells, he found thermoelectricity preferable for his concluding investigations. Resistance of the thermoelectric circuit was measurable and stable, contrasted to the unpredictable internal resistance of the voltaic cells. Ohm found that the electromotive force of the generator was approximately proportional to the temperature difference of the thermoelectric junctions.

Ohm used the test arrangement of Fig.7.4. But, instead of calculating his readings with reference to the standard copper conductor, he used the figures shown by the torsion head directly for the dependent variable.

In January 1826 Ohm carried out one of his series of tests, using different metals and various lengths of wire for each. In one test he prepared eight samples of copper wire, of the same diameter, but lengths 2, 4, 6, 10, 18, 34, 66 and 130 inches. The current source was a thermoelectric generator, one junction being heated in a steam jacket and the other junction was in an ice pack. Ohm obtained the following results:

| Length of copper conductor,inches | Restoring angle of torsion head in divisions |
|---|---|
| 2 | 326¾ |
| 4 | 300¾ |
| 6 | 277¼ |
| 10 | 238¼ |
| 18 | 190¾ |
| 34 | 134½ |
| 66 | 83¼ |
| 130 | 48½ |

**OHM'S CONCLUSION,** from the above data and from previous tests of a similar nature, was that the

HOW OHM TESTED RESISTANCE . . .

Fig. 7.4 - Ohm's conclusions regarding the relations between electromagnetic force of the current, the potential, and length of the connecting test wire, were verified in a test arrangement such as this. Magnetic force of the current was measured by a Coulomb torsion balance galvanometer. Ohm found that resistance of a circuit could be determined by an equation involving the ratio of voltage to current.

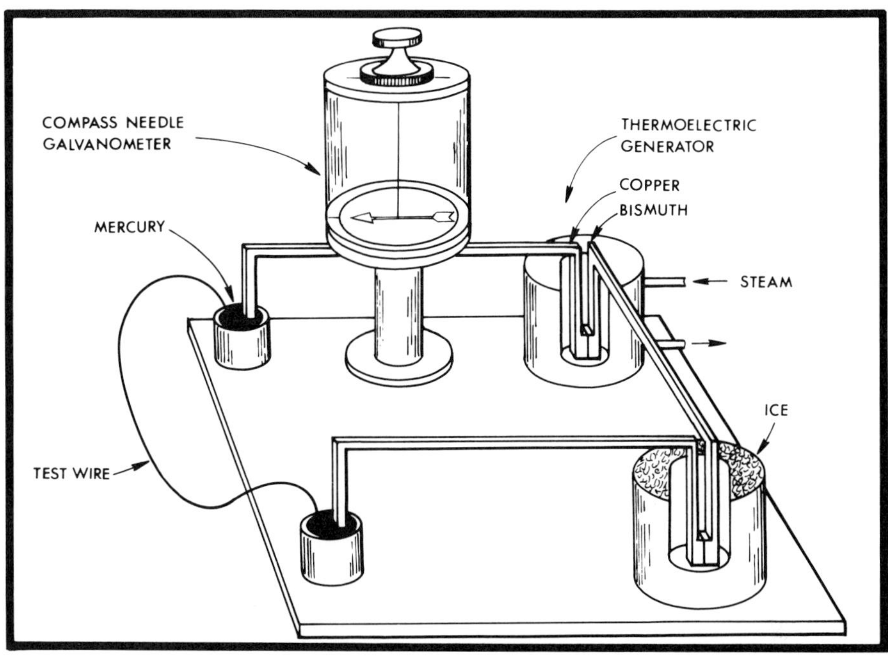


results could be represented by an equation of the form:

$$X = \frac{a}{b + x}$$

where, $X$, the reading of the torsion head, is a measure of the electric current in a conductor of length, $x$. The quantities, $a$ and $b$, were constants of the circuit whose values were to be determined by an additional series of experiments.

Ohm observed that, $b$, remained constant for all the series of tests, while, $a$, varied with temperature of the thermoelectric generator. He concluded that $a$, depended only on the electromotive force of the source, while, $b$, represented the resistance of the circuit.

Ohm gave a value of 20¼ for, $b$, and for the experiment in the Table on the previous page, a value of 7285 for, $a$. Thus, for example, for the conductor length of 18 inches, $X = 7285/20¼ + 18$ which works out to 190.72 compared to the measured value of 190¼. Other test lengths also agreed.

Ohm varied the experiments by using brass test wires and by changing the electromotive force of the thermoelectric generator by using ice in one junction and room temperature on the other.

The results of the formula held up.

**OHM PUBLISHED** in February and April 1826, two papers dealing with aspects of his theory of a galvanic circuit as derived from his extensive experiments. Together, the papers served to resolve the previously confused situation pertaining to open and closed circuits, and presented precisely for the first time the notions of the relations between the electric tension and the quantity of electricity which was delivered to the circuit.

Ohm's second paper contained the equation:

$$S = k\omega\left(\frac{a}{l}\right)$$

where, $S$, was the intensity of current produced by the tension, $a$, in a length of wire, $l$, cross-section, $\omega$, and conductibility, $k$. By reducing the actual length of the wire and its cross-section and conductibility to an equivalent length of wire, $L$, chosen as a standard, Ohm simplified the equation to:

$$S = \frac{A}{L}$$

where, $S$, is the quantity of electricity passing through a given cross-section of the conductor in a unit time, $A$, is the difference of electric force between the extremities of the length of wire under test, and, $L$, is the "reduced length" of the

A
2164
B
0
5
10
5
20

**APPARATUS USED BY OHM** . . . *now in the Deutsches Museum, Munich*

A - **BATTERY** - voltaic pile using copper and zinc plates

B - **GALVANOMETER** - using a compass needle mounted above a coil of wire

C - **THERMOMULTIPLIKATOR** - A thermogalvanometer containing two rectangular bars, probably one of bismuth and the other of antimony, one above the other at a distance sufficient to allow the lower magnet of an astatic needle to swing between them. Ends of the bars are soldered together, and the upper bar has a longitudinal saw-cut through which the lower suspended magnet can pass to its proper level.

circuit, which was equivalent to the sum of the resistances in the circuit.

The equation showed, as stated by Ohm . . . *that the force of the current is as the sum of all the tensions, and inversely as the entire length of the circuit.*

This became known as Ohm's Law, which in its usual equation form is:

$$I = \frac{V}{R}$$

When current, $I$, is produced in a conductor by an electromotive force, $V$, the ratio of the electromotive force to the current is independent of the strength of the current, and is called the resistance, $R$, of the conductor.

Some doubts had been raised with respect to the mathematical accuracy of Ohm's law, and as to the exactness of the proportionality between the electromotive force and the current in the same conductor. The subject was taken up by the British Association for the Advancement of Science, and in 1874 experiments were made by which the exactness of the law, as it relates to metal conductors, was tested by currents of every degree of intensity. The tests upheld Ohm's law, showing it was unambiguous and correct.

**OHM'S WORK,** theoretical and experimental, clarified the following principal points:

*1 - Resistance is a unique property of a given conductor.*

*2 - The current flowing in a given circuit is directly proportional to the difference of potential at the two ends of the circuit.*

*3 - The current is directly proportional to the section of wire, and the current passing through the wire is everywhere the same.*

*4 - The current is directly proportional to the specific conductivity of the material of which the conductor is composed.*

*5 - The potential of a circuit connected to the source of electricity is highest at the positive pole of the source, and lowest at the negative pole.*

*The potential falls uniformly through the whole circuit, when the conductor is of uniform size, and otherwise falls uniformly through uniform resistance.*

**IN 1827,** to complete his experimental work and to have access to better libraries, Ohm was given leave to go to Berlin. Here his great book: *The Galvanic Circuit Mathematically Worked Out* was published. It was a comprehensive volume, encompassing all of his earlier investigations, and was divided into three parts: "Introduction," "The Voltaic Circuit" and "Appendix."

To expound his theory of the galvanic circuit to the mathematically unsophisticated, Ohm, in the Introduction used an essentially geometric presentation. Some excerpts of his method and his conclusions have been given in previous pages.

Ohm' most important postulate, upon which his theory was based, was that the flow of electricity within a body depended on a particle to particle transfer proportional to the difference of electromotive force between them. Conductivity was thus a measure of the quantity of electricity transferred per unit of time across a unit of distance. The postulate had originated in an analogy with the theory of heat flow, and Ohm pointed out its similarity to the work of Fourier. From this postulate Ohm arrived at his general law that the current in a galvanic circuit is constant across all cross-sections, and is equal to the sum of all the electromotive forces divided by the total reduced length (resistance) of the circuit. This implied a change in electromotive force across each portion of the closed circuit.

The second part of Ohm's book was devoted to a mathematical presentation of the results he had deduced from his theory and experiments. The Appendix was a discussion of circuits in which the current was accompanied by a chemical change.

Ohm prefaced his book as follows:

**"I HEREWITH** present to the public a theory of galvanic electricity, as a special part of electrical science in general, and shall successively, as time, inclination, and the means permit, arrange more such portions together into a whole, if this first essay shall in some degree repay the sacrifice it has cost me.

"The circumstances in which I have hitherto been placed have not been adapted either to encourage me in the pursuit of novelties, or to enable me to become acquainted with works relating to the same department of literature throughout its whole extent. I have therefore chosen for my first attempt a portion in which I have the least to apprehend competition.

"May the well-disposed reader receive the performance with the same love for the objects that which with it is sent forth!"

Ohm's book appears not to have been fully appreciated by his contemporaries, and his law came into use only after considerable hesitation and delay.

Ohm was accused of offering in his book purely theoretical deductions without substantiating experimental proof. This accusation has been considered questionable, however, in view of the wide circulation of Ohm's previous papers reporting his experimental evidence, all available before the appearance of his book. Some described the hesitancy of immediate acceptance to the fact that many physicists of the time were lacking in the mathematical approach to science, whereas Ohm presented his theories primarily in that form. Ohm's conductibility equations seemed to some so simple as related to complex considerations, and to reality, that they were not acceptable. The obscurity of some of the definitions in the book, and the complication of presentation and logic of some of Ohm's writing may have been a hurdle to understanding and appreciation.

**RECEPTION** of Ohm's book by some critics was hostile. One of them called Ohm's theories a "web of naked fancies." Furthermore, the critic said, "he who looks upon the world from this book as the result of an incurable delusion, whose sole effort is to detract from the dignity of nature." A doctrinaire school official who was opposed to the mathematical school of thought, and to Ohm's book, stated his objections to the Minister of Education at Berlin. Accepting this report uncritically, the Minister pronounced that "a professor who preached such heresies was unworthy to teach science!"

The cause of this harassment and Ohm's failure to get his aspired university post has been ascribed to ideological and political situations. The Berlin academic establishment was in the grip of the Hegelian philosophy which was sweeping Germany at the time. Hegel's philosophy held that Nature is a realm of law and order, open to the mind without the aid of matter - an attitude against the need for experiment, and somewhat anti-science. On the political side, Ohm's brother Martin, with whom Ohm lived in Berlin, had clashed with Berlin officialdom because of his criticism of the educational system, and was considered a dangerous revolutionary, and this situation contributed to the rejection of Ohm's work.

John Tyndall

Die

**galvanische Kette,**

mathematisch · bearbeitet

von

**Dr. G. S. Ohm.**

Mit einem Figurenblatte.

Berlin, 1827.

Bei T. H. Riemann.

Fig. 7.6 - TITLE PAGE of Georg S. Ohm's great work *The Galvanic Circuit Mathematically Investigated,* in which appeared the fully developed presentation of his theory on the relations of the electric circuit. This title page is from a copy of Ohm's book in the possession of the Burndy Library. The signature at the top is that of John Tyndall, the English scientist and lecturer, who succeeded Michael Faraday as Director of the Royal Institution, London.

In the face of this rejection, Ohm took the only course he considered consistent with self-respect. He resigned his position at Cologne, and was forced into a kind of academic exile through the years 1827-31, during which he taught mathematics at the Military School in Berlin. Then came a break, and Ohm's discouragement was assuaged by a decree from King Ludwig I of Bavaria, giving him a professorship at the Polytechnic School of Nuremberg. But Ohm had ambitions for a university post, and this was not the first-rank institution to which he aspired. From here, however, he was at least able to press his claims for the recognition of his work and its significance.

Ohm's electrical work was based on the soundest of scientific effort - experimental pursuit, and discovery formulated into mathematical language for universal application. His laws were to have a profound importance to the electrical art. Ohm's publications could not therefore be long obscured by wilful neglect, bureaucratic obstruction, or the lack of use. The progress in electrical theory made Ohm's law a necessity in the understanding of and calculation of electrical phenomena. Scientists working with electrical circuitry were meeting the same problems of relationships that Ohm had solved. They gradually learned of Ohm's research and appreciated the elegance and utility of his fundamental law.

**JOSEPH HENRY**, in America, Wheatstone in England and Lenz in Russia, among other active investigators in electricity, admired Ohm's work and began to give it publicity. Then, the younger physicists, dealing with and writing on electricity, subjected Ohm's law to test and confirmation, and began to include it in textbooks. However, it was not until the 1840s that the scientific world generally became aware of the great implications of Ohm's writings.

The crowning accolade came in 1841 when the Royal Society of London awarded him the Copely Medal, their highest and most coveted honor, in recognition of his law.

In 1849 Ohm received the first mark of substantial recognition and merit from his home country. At the end of 1849 Ohm was appointed by the Emperor Maxmillian II, as Professor of Physics to the University of Munich, and the Akademie of Science selected him Conservator of Mathematical Physics. These positions fulfilled one of the great ambitions of Ohm's life. His name was now being honored in all scientific circles, and Ohm was inspired to renew and broaden his scholarly and experimental efforts. His later work was mostly in fields of acoustics and crystallography, and he was an early investigator in molecular theory.

In addition to research, Ohm was an innovative and effective teacher. His brother, Martin, who taught at the University of Berlin, was among the first to stress the importance of a training in mathematics as prerequisite for the understanding of science. This flew in the face of tradition, however, particularly in the military where promotion had been by the privilege of birth rather than from superiority achieved by an imposed scientific training. Hence, adoption of mathematics in the curriculum lagged in Germany. Holding his brother's views, Ohm insisted that mathematics be coordinated with physics as the best basis for achieving practical scientific knowledge. He personally prepared mathematical material for his students, and pioneered the "example" method of demonstrating the use of mathematics in physics.

Ohm's last years were spent in the realization of his dreams finally come true - a first-rank university post, an honored place in science, the respect of colleagues, and the admiration of his students. He carried on his University duties until his death in 1854 at the age of 65. Ohm never married.

Ohm's scientific writings and research notes were collected and published in Leipzig in 1892, under the editorship of Dr. Eugene Lommel, Professor of Physics at the University of Munich. The book of 855 pages covers the results of Ohm's efforts extending for a period of over thirty years.

**RESISTANCE** was the first electrical property to come under concerted scientific attention for standardizing. The British Association Committee on Standards of Electrical Resistance, appointed in 1861, adopted a standard unit of resistance in 1862. It was first named the "Ohmad," and this was finally abbreviated to "Ohm." Meeting in Paris in 1881, the International Electrical Congress paid the ultimate homage to Ohm by confirming his name for the practical unit of electrical resistance.Thus was initiated the practice of naming units after the founders of the science, a practice which has continued to this day.

RANGE OF RESISTIVITY

Fig. 7.7- Resistivity of various conductors and insulators is shown on a logarithmic scale. This reveals the enormous spread in resistivity values, about 10 to the 25th power, between the best conductors and best insulators. This characteristic is due to the nature of the electron bonding in various materials, to provide the availability or absence of free electrons for conduction.

The enormous range of resistivity is a fact of nature that makes possible all of our present electrical technology.

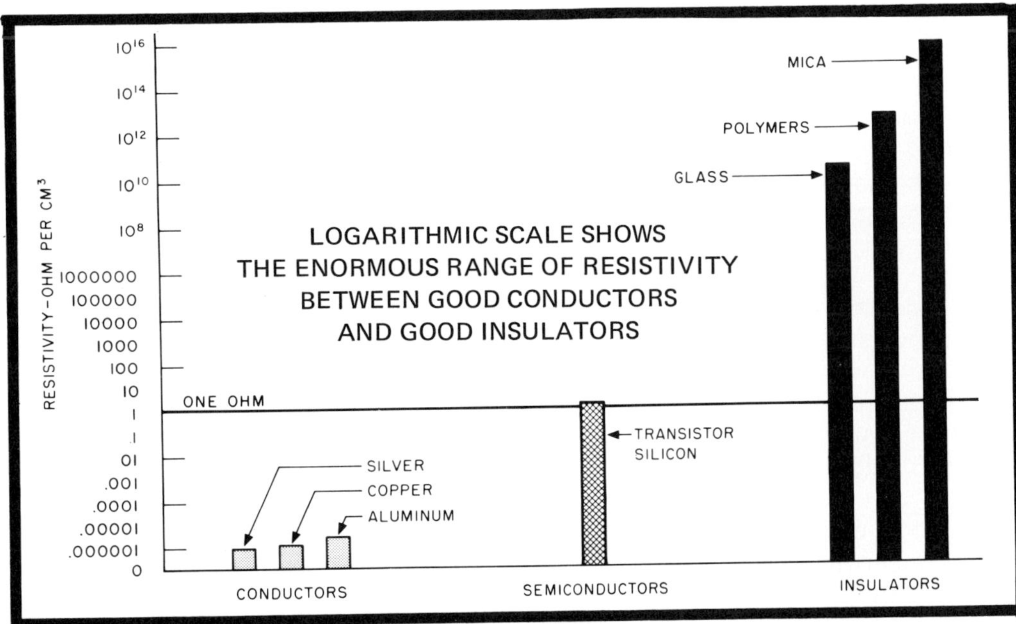


**OHM'S WORK** was a link bridging the time-gap existing between the static electricity of previous ages, and the new era of voltaic current. Whereas static electricity resided on the surface of the conductor, Ohm saw the circuital current as electricity in motion, penetrating the interior and utilizing the volume of the metal wires. For this to occur, Ohm visualized an elemental, particulate structure of matter. The current in the wire, he conjectured, was a transfer of the electrical fluid from particle to particle. The transfer was accompanied by a uniform drop in potential, particle by particle, in forcing the current against resistance in the conductor. Nothing was known of the nature of the electrical "particles" in Ohm's time. But later discoveries would reveal their structure.

Investigations in the 19th and early 20th centuries revealed the atomic elements of matter, and the fact that the atoms had a composition of electric charges. The protons of the atomic nucleus provided positive charges, and the atomic electrons provided negative charges. Electrons were the electric charge carriers in solids and gases. Ions, which resulted when atoms or groups of atoms gained or lost electrons, provided either positive or negative electric charge carriers in solutions and plasmas. The flow of these charge carriers constituted the electric current.

The metals are the best conductors we know of, and Ohm was the first to determine the order of their conductivity. To account for the high conductivity of metals, the "free electron" theory was developed early in the 20th century. This theory was based on a view of the electrons in discrete motion within the crystalline structure of the metal conductor. In the metallic atomic bonding, certain of the outer electrons, about one per atom, become loose in the sense of no longer being localized in an individual atom, but belonging to the crystal as a whole. Freedom of these electrons leaves a residual lattice of positive ions in the metal, within which the free electrons can wander.

The atomic structure is in a constant state of vibration, the intensity of which is determined by the temperature. The atoms impart some of this vibrational energy to the free electrons, putting them into a mazy motion. The electrons randomly collide with the ionic lattice and recoil in a manner similar to the molecular collisions in a gas. The motion of the free electrons provides the mechanism for the conduction of electricity.

When an electric potential is applied to the conductor, a force will be superimposed on the random motion of the free electrons, causing them to accelerate directionally. By convention the direction is opposite to that of the electric field. The acceleration acts for a very brief interval until the electrons collide with neighboring atoms, giving them an impact, then recoiling to speed off again until other atomic collisions occur.

Thus the total electron motion is a series of accelerations directed preferentially by the force of the electric field. The net effect, acting on the totality of the free electrons, is to superimpose a drift velocity on their normal random motion. The drift velocity constitutes the electric current.

The amount of current will depend on a combination of the velocity and density that makes up the charge carrier transport. The net drift velocity, because of the scattering action of the atomic lattice, is exceedingly small, only an infinite fraction of the random velocity of the free electrons. But the number of electrons traveling at drift velocity is enormous.

A cubic centimeter of copper, assuming one free electron per atom, provides about $8.5 \times 10^{22}$ conduction electrons. When it is considered that one ampere consists of the flow of $6.26 \times 10^{18}$ electrons per second, the availability of a sufficient number of charge carriers for very large currents is apparent.

**NET DRIFT VELOCITY** varies with the strength of the electric field. For most metals the drift velocity, and hence the current, is directly proportional to the voltage that is applied. The proportionality factor between the current and the voltage can be expressed as the *conductance, G.*

The current will then be a product of the voltage and the conductance, or, $I = GV$. Since resistance is the reciprocal of conductance, that is, $R = 1/G$, there is an alternate equation for the current, using resistance, that is, $I = V/R$. Thus the drift velocity theory operates in conformity with Ohm's law.

The ratio of the voltage to the current, where Ohm's law applies, is constant, and can be plotted as a straight line, or linear relationship. Slope of the line is determined by the resistance, expressed by Ohm's law, or, $R = V/I$. This straight-line relationship has been shown to be valid up to the highest range of practical current density for most metal conductors. Hence, resistance is a definite property of the metal.

Imparting a drift velocity to the free electrons in a conductor results in much more vigorous collision impacts with the atoms of the conductor, thus doing work on them, and this increases their thermal vibration. The voltage and energy of the conductor are consumed in this process in proportion to the resistance of the conductor. Thus the resistance in a conductor manifests itself by a drop in voltage proportional to the current and to the resistance in the circuit. This is expressed by the Ohm's law equation: $V = IR$, commonly called the "voltage drop" equation.

A power loss is also involved in overcoming the resistance, and this loss is expressed by Joule's law: $P = I^2 R$. The power loss is converted into heat, in the quantity equivalent to joule-seconds, or, $I^2 Rt$, expressed in watts. The resulting temperature rise will depend on the rate of dissipation of the heat from the conductor. Any temperature rise will in turn affect the resistance of the conductor.

Ohm found that the resistance of metals he tested varied with temperature. There was an increase in resistance with increase in temperature, and vice versa. Thus. resistance was not only a characteristic of the individual metal, but was also dependent on the temperature. A resistance value must therefore include the temperature at which it holds. Much research has gone into development of metal alloys for minimum resistance variation with variation of temperature.

**SUPERCONDUCTORS** - Ohm could never have imagined that the resistance of a conductor could disappear completely. But his discovery that the resistance was affected by temperature was a first step that culminated, nearly a century later, in the achievement of zero resistance.

H. Kamerlingh Onnes (1850-1926), a Dutch physicist, succeeded, in 1908, in Leyden, Holland, in liquefying helium, and thereby had attained a temperature of 4.2 degrees Kelvin. Later, by boiling helium regeneratively, he was able to get even closer to zero degrees. From the known facts of the drop in resistance with a drop in temperature, it was apparent to Onnes that if temperature could lowered to absolute zero, resistance would tend to vanish. With the means of achieving a temperature near to zero, Onnes was now able to pursue an investigation into the possibility of zero resistance.

He first experimented with platinum, but was unable to get the resistance below a certain critical value. He blamed this on supposed impurities in the metal. Assuming that it could be had in a much purer state, he next tried mercury, and found in

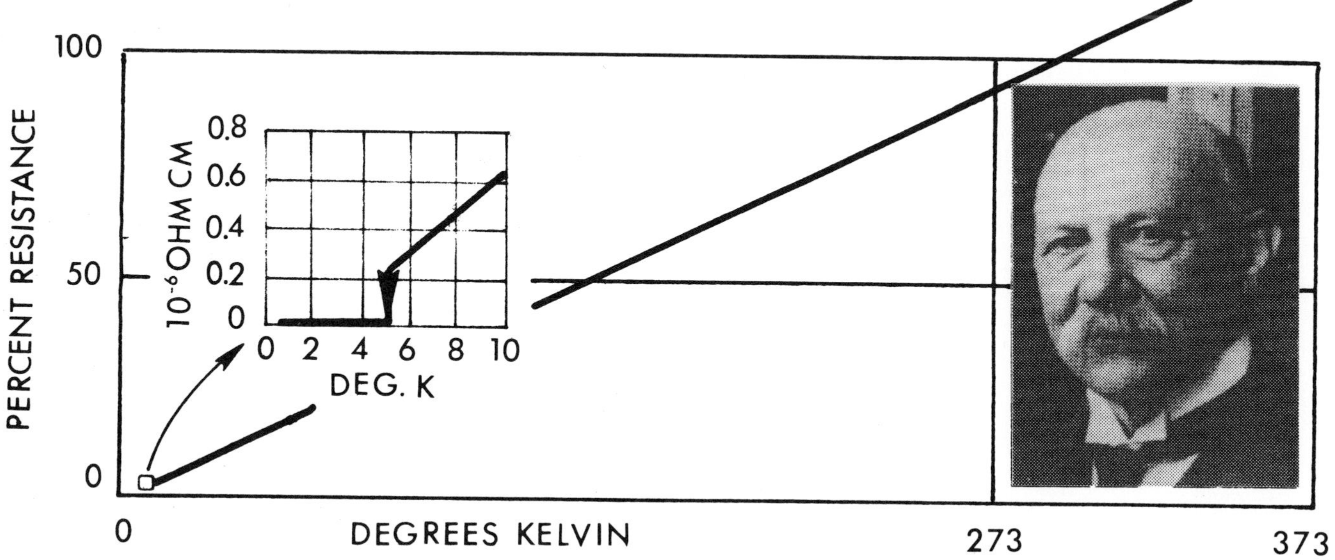


1911, that at a temperature of 4.2 degrees Kelvin the resistance of the mercury fell sharply to zero. Fortunately, the transition to zero resistance had occurred within the limits of low temperature refrigeration available at the time. Onnes called this extraordinary electrical state of a conductor - *superconductivity.*

Onnes later found that lead reached the zero-resistance state at 7.2 degrees Kelvin, and tin at 3.7 degrees. Other metals entered the state of superconductivity at their certain so-called *critical* temperatures. Onnes saw at once that conductors of zero resistance could carry large current without the usual heat dissipation. This would be accompanied by magnetic fields much larger than was possible with conductors at usual temperatures. But he found that at even moderate magnetic field strengths, the superconductivity would suddenly be quenched, and the normal resistance would be restored.

The possibilities of improving performance of electromagnetic apparatus by use of superconductors did not emerge until 50 years after Onnes had made his original investigations. A break-through came after World War II with the discovery that conductors using alloys of niobium, vanadium and other "transition" elements, could carry current densities of thousands of amperes, and very large magnetic fields, without quenching. The remarkable properties of these alloys stimulated research for new types of superconductors which were

ONNES DISCOVERS SUPERCONDUCTIVITY . . .

Fig. 7.8 - The sudden fall to zero of the electrical resistance of some metals at very low temperature was first observed by the Dutch physicist, H. Kamerlingh Onnes (above) in Leyden, Holland, in 1911. The first metal to go into the zero-resistance state was mercury, at 4.2 degrees Kelvin. Onnes called this electrical phenomenon *superconductivity.*

operative at higher critical temperatures, and for materials that could satisfactorily be drawn into wires.

Superconducting alloy materials with critical temperatures up to 25 degrees Kelvin have been developed. Metal-clad alloy wires can now be wound into electromagnets to operate with possible current densities up to $10^6$ amperes, and yielding fields up to 50 teslas ($50 \times 10^4$ gauss).

Now, with fabrication of transition-metal alloys into wires, and with developments in cryogenics, the field of low-temperature refrigeration, wide areas of application for high-intensity magnetic fields have opened. Present-day technology is built around the copper-wound, iron-core magnetic solenoid, which saturates at about 2 teslas, or, 20,000 gauss, and this constitutes the upper limit of conventional magnetic fields. Superconducting magnets promise to multiply this many times.

Superconductivity, with its ability to produce intense, large-volume fields, is now being pioneered for electric power generation and transmission, and

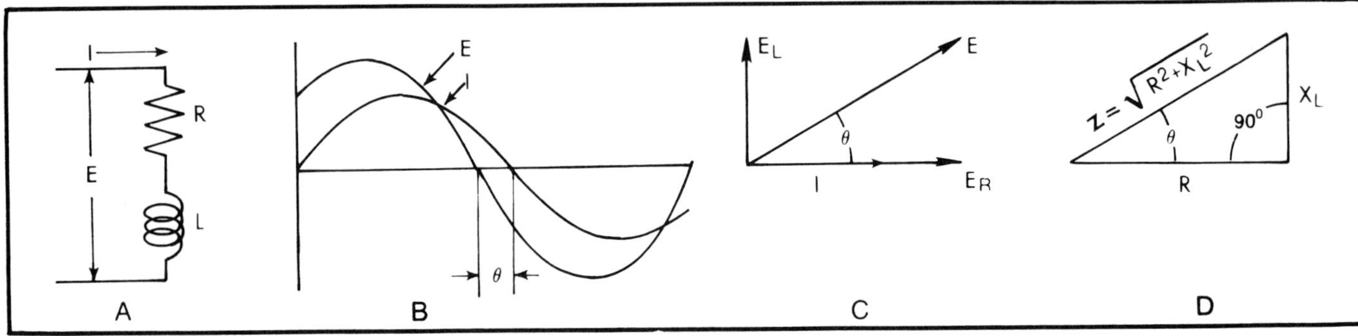


Fig. 7.9 - OHM'S LAW for an a-c circuit, Fig. A, containing ohmic resistance, R, and inductance, L. Inertia effect of the magnetic field, L, causes the current to lag the voltage by and angle, θ, Fig. B. Thus the voltage, E, and the current, I, are related geometrically, Fig. C. The impedance, Z, of the circuit, Fig. D, is the vectorial addition of the resistance, R. and the inductive reactance, $X_L$, which are at a right-angle relationship to each other.

for application to nuclear fusion systems, high-energy particle accelerators, and magnetohydrodynamic apparatus. There is a continuing search for better materials and methods of fabrication that will yield higher intensity magnetic fields.

**CONVERTING OHM'S LAW** for Alternating Currents - Ohm's law assumes a steady electrical state, and applies very simply to direct-current circuits, that is $I = V/R$. With the introduction of alternating-current systems, where the voltage and current change direction rapidly and periodically, this simple relationship no longer holds. Ohm's law was then thought to be on forbidden ground. However, in the late 19th century, the needs of a-c calculation brought forth, by mathematical manipulation, a new variety of Ohm's law that was applicable to alternating-current circuits.

In alternating currents, Ohm's law assumes the new form: $I = V/Z$, where, $Z$, is the "apparent" resistance, or *impedance,* and is no longer constant, but depends on the frequency of the circuit.

The impedance has two components: *resistance, R,* and the *reactances, X.* The principal sources of the reactances are electromagnetism and electrical capacity of the a-c circuit.

The periodic alternation of voltage and current in a-c circuits introduces two frequency-dependent effects. The first effect operates on the *magnetic inductance of the circuit, L,* because of the periodically changing magnetic fields, to produce an *inductive reactance,* $X_L$. The second effect operates on the *electrical capacitance of the circuit, C,* because of the periodically changing electric fields, to produce a *capacitative reactance,* $X_C$.

The voltage and current consumed in heat losses, due to ohmic resistance, $R$, are in phase with each other, independent of frequency, and related to each other by Ohm's law. The current of the magnetic fields, because of magnetic inertia, *lags* 90 degrees behind its voltage, and is related to it by the inductive reactance, $X_L$ . The current of the electric fields, because of electric elastance, *leads* its voltage by 90 degrees, and is related to it by the capacitative reactance, $X_C$.

Because of the quadrature relationship of these voltages and currents, the magnetic and electric fields represent merely a surging of energy back and forth in the circuit, and are called "wattless."

The reactances, $X$, can be expressed for any frequency, $f$, and for any value of inductance, $L$, or capacitance, $C$, by Ohm's "laws":

$$\text{Ohms Inductive Reactance} = 2\pi f L = X_L$$

$$\text{Ohms Capacitative Reactance} = \frac{1}{2\pi f C} = X_C$$

All a-c circuits will be some combination of the *resistance, R,* and the difference of the *reactances,* $X_L$ and $X_C$ in vectorial addition, giving the total:

$$\text{Ohms Impedance} = \sqrt{R^2 + (X_L - X_C)^2} = Z$$

**OHM'S LAW** is thus re-established for a-c circuits by replacing resistance, $R$, by the impedance, $Z$.

**THE VARIATIONS** in the manipulations of Ohm's law, and the uncountable ramifications of resistance in all manner of devices, are the essential pillars supporting the electrical technology of today. Resistance, its kind and control, wanted or unwanted, enters into the design, fabrication and efficiency of every electrical element from the microscopic to the gargantuan.

**ALTERNATING - CURRENT MOTORS** are an example of the use of resistance to achieve desired performance. Resistance is an essential element in the production of torque in induction and synchronous motors. Control of this resistance provides the fundamental method of designing the shape of the torque curve from standstill to full motor speed. Torque of the induction motor is at its maximum point when the rotor resistance, and the rotor inductive reactance, are equal. The amount of resistance with relation to the reactance affords the means of varying the torque curve to meet the starting and running requirements of the motor.

The Design D starting torque curve for an induction motor, shown in Fig. 7.10, develops maximum torque at standstill, where the high-resistance cage winding matches the inductive reactance at zero speed. The Design B motor, on the other hand, develops maximum torque near full speed, where the low-resistance rotor cage winding matches its inductive reactance.

The wound-rotor induction motor provides a means of matching the resistance to the reactance at all speeds, by use of an external resistor in the rotor circuit. Successive shorting out of steps of the resistor supplies a series of high starting torque curves as shown in Fig. 7.11.

The synchronous motor starts as an induction motor by use of cage bars in the tips of the rotor poles. With bars of appropriate resistance, and with bar placement in the poles for proper reactance, a wide variety of torque curves can be obtained to meet a wide range of drive requirements.

The double-cage rotor construction shown in Fig. 7.12, provides a combination of starting and running torques. The upper, high-resistance cage delivers a high breakaway torque. The deep, low-resistance cage matches the reactance at full speed, providing high torque during the starting period.

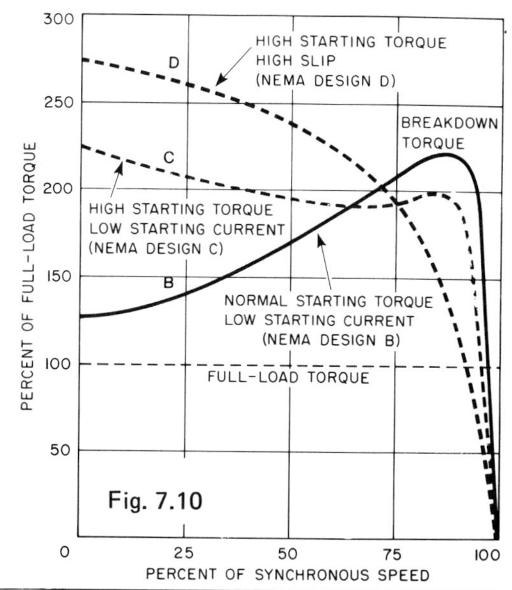

Fig. 7.10

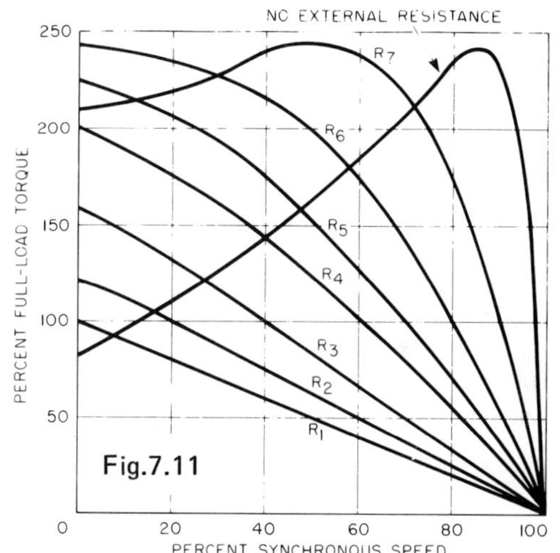

Fig.7.11

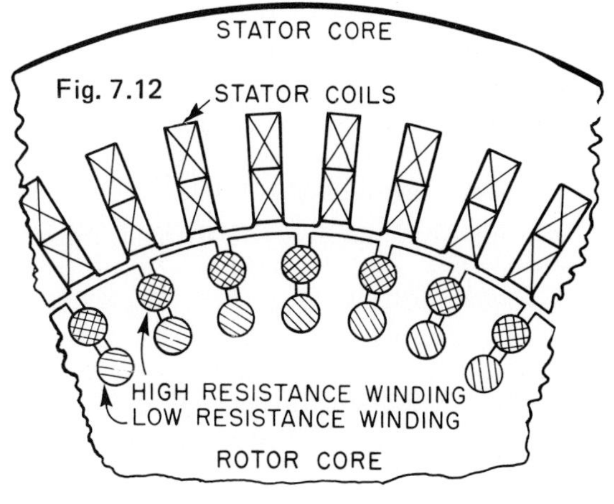

Fig. 7.12

**OHM'S LAW IN OPERATION** - A conductor obeys Ohm's law where the curve *V/I* is linear, that is, if *R* is constant over the range of its application. It is a law whose applicability is determined by the material, and the operational situation to which it is applied. Although Ohm's law is an experimental relationship, it holds over a wide range of conditions. With temperature taken into account, it is obeyed, in metals, within a broad spread of current densities.

Ohm knew nothing about electronics, nor about devices where resistance might be non-linear. For some materials it is possible for a change in voltage to cause a change in resistivity by the release of more charge carriers. In this case an increase in voltage produces a disproportionate change in the current, and Ohm's law applies only incrementally.

The existence of non-ohmic devices is fortunate. Otherwise, much of electronic technology would not exist, and modern life would be vastly different. The vacuum tube and semiconductor diodes are typical non-ohmic devices: the current vs voltage is non-linear. In the diode, the resistance characteristic is asymmetrical, passing the current in one direction and blocking it in the other. Some devices have a negative resistance in part of their operation, and in these cases Ohm's law needs special interpretation.

**WHAT IS AN OHM?** The ohm is defined in the International System of Units (SI) as:

*The ohm is the electrical resistance between two points of a conductor when a constant difference of potential of 1 volt, applied between the points, produces a current of 1 ampere, the conductor not being the source os any electromotive force.*

The SI defines the ohm in terms of volts and amperes. In practice there is need for a reproducible physical standard of the ohm.

In 1851, Wilhelm Weber (1804-1891), an experimental scientist in Germany, proposed an electromagnetic system of measurement in which the electrical resistance would be expressed as a velocity, equal to the distance of one earth quadrant, or $10^9$ cm/sec. The British Association for the Advancement of Science, responding to the needs of the telegraph engineers, set up, in 1861, a pioneer committee to establish electrical units and standards. They adopted the system proposed by Weber. For the standard of resistance, the BA committee prepared, in 1864, a metal-alloy, wire-wound ohm as a standard to represent Weber's $10^9$ electromagnetic velocity units. The BA resistance standards of this type were in use for about twenty years.

In 1860, to resolve the differences in the various resistance measurement standards existing in Europe at the time, and to provide a reproducible ohm, Werner Siemens, the German scientist, proposed independently a standard for the ohm. It was a column of mercury of one square millimeter section, one meter length, at the temperature of melting ice. This found wide acceptance as the standard ohm.

The column length was later increased to 106 cm for greater accuracy. In 1884 the "International Ohm" was standardized as the resistance, at zero degrees C, of a column of mercury 106.300 cm in length, of constant section, and weighing 14.4521 grams. This ohm was the standard for many years until development of newer precision methods of establishing the standard for the absolute ohm.

Since 1948, the legal standard absolute ohm has been defined in terms of the reactance to an alternating current of a calculated standard inductor or capacitor. If an inductor of known impedance: $z = 2\pi fL$, or a capacitor of a known impedance: $z = 1/(2\pi fC)$, is placed, at a known frequency, in one arm of an alternating-current bridge, it can be balanced against a resistor in the other arm to establish precisely the value of the resistor in ohms.

A computable capacitor for the alternating-current bridge method of ohm determination, is shown in Fig. 7.13, being assembled at the National Bureau of Standards. It uses a new form of a segmented capacitor tube, enabling the precise measurement of the absolute ohm.

**PRIMARY STANDARDS** for resistance measurement are made in various forms, one of which is shown in Fig. 7.14. This type of one-ohm standard was developed at the National Bureau of Standards by Dr. J.L. Thomas for use as a reference standard in laboratories requiring extremely high precision. The resistance element uses a wire alloy of copper, manganese and nickel having a low temperature coefficient of resistance. The wire is wound bifilar, hermetically sealed, and mounted in the outer case. In use, the standard is placed in an oil bath, maintained at a constant temperature, usually 25 °C.

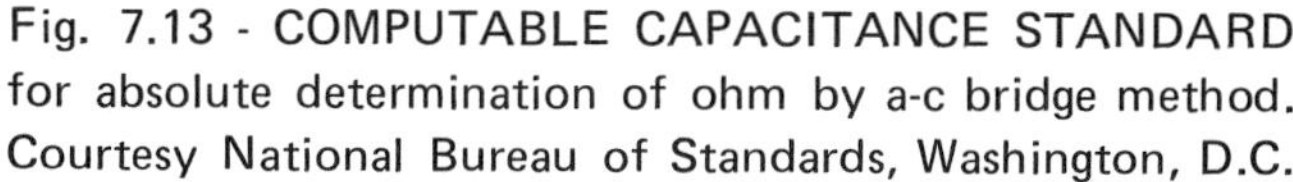
Fig. 7.13 - COMPUTABLE CAPACITANCE STANDARD for absolute determination of ohm by a-c bridge method. Courtesy National Bureau of Standards, Washington, D.C.

Fig. 7.14 - PRIMARY RESISTANCE STANDARD of the wire-wound, hermetically sealed Thomas type, provides precise one ohm resistance. Courtesy Leeds & Northrup Co.

---

**THE ONE HUNDREDTH ANNIVERSARY** of Ohm's birth was commemorated, in 1889, by a group of scientists, meeting at the Royal Bavarian Academy of Sciences, at Munich. The following passage from Professor E. von Lommel's address at the meeting, as quoted in *Makers of Science*, by Ivor Hart, Oxford Press, pays tribute to Ohm:

*"The deeds of a man of science are his scientific investigations. Truth once discovered does not remain shut up in the study or the laboratory. When the moment arrives, it bursts its narrow bonds and joins the quick pulse of life. That which has been discovered in solitude, in an unselfish struggle for knowledge, in pure love of science, is often fated to be the mighty lever to advance the culture of our race. When, nearly a hundred years ago, Galvani saw the frog's leg twitch under the influence of two metals touching, who could have suspected that the force of Nature which caused those twitchings would transfer the thought of man to far distant lands, with lightning speed, under the water of the ocean and would even render audible at a distance the sound of the spoken word?"*

*That this force of Nature - after man by ceaseless investigations had learned vastly to increase its strength - would illuminate our nights like the sun. This enormous development of electrotechnology could only be accomplished upon the firm foundation of Ohm's law. Ohm, by wresting from Nature her long-concealed secret, has placed the sceptre of this dominion in the hands of the present."*

**AND NOW,** more than a century and a half after its formulation, Ohm's law stands, unchanged and unambiguous. as perhaps the most basic and the most widely used of all the laws in the electric science.

## the farad of electrical capacity honors the 19th century genius

## MICHAEL FARADAY

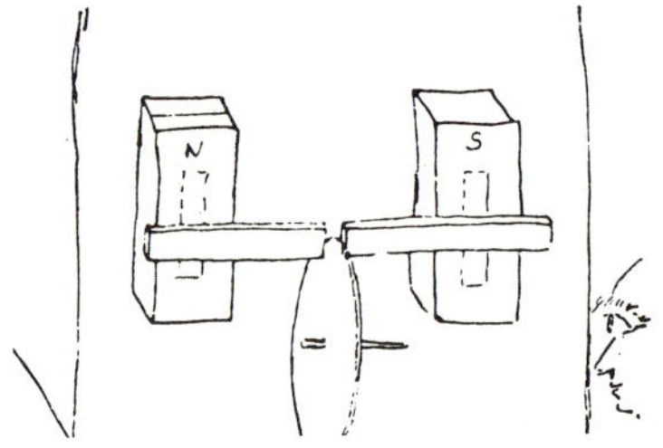

SKETCH from Faraday's Diary shows his rotating disk for producing electric current. This was the first electromagnetic generator.

**IN ALL THE ANNALS** of achievements one name stands out for everlasting wonder - that of Michael Faraday. A nineteenth century experimental genius he is considered pre-eminent as the most gifted discoverer of his time in the fields of electricity and magnetism. And his many other accomplishments in science have given him fame as one of the founders of modern physics.

Born in London in 1791, the son of an impoverished blacksmith, and educated largely by self-instruction, Faraday achieved in his brilliant career, through original thinking and rare experimental skill, a monumental series of fundamental discoveries. In a few inspired weeks in the fall of 1831, Faraday demonstrated electromagnetic induction and the principal laws governing it; he built the first electrical transformer, and the first electromagnetic generator, which were the embryonic beginnings of the mighty electric power industry of today. Faraday also pioneered in electrochemistry and in magneto-optics, and he laid the foundations of field-of-force theory which revolutionized previously held concepts, and which is now a vital part of scientific thinking.

During his long scientific career as experimenter, discoverer, consultant and a popular lecturer, Faraday, by his diaries, his vast experimental notes, scores of published papers, and an international correspondence, left one the most lucid, complete and revealing records of the problems and progress of the electricity and magnetism of his time.

**MICHAEL FARADAY'S** forebears were of Irish extraction, having fled that country for England in the 16th century to escape the bloody Irish rebellion that was put down by Queen Elizabeth. The family settled first in Lancashire on the west coast of England, and then moved eastward to a small, rural village in Yorkshire. There the family joined a new church group called "Sandemanians," founded in the mid-18th century by John Glas and his son-in-law Robert Sandeman. A fundamentalist and strictly Bible-oriented creed, its doctrines demanded humility, reverence, and a faith in the Divine origin of nature. Its teachings, passed down the families, were eventually to become an inspiration and sustaining factor in the life and future work of Michael Faraday. He was throughout his

# ... his discoveries opened the age of electric power ...

life, to be a profoundly religious man, under the Sandemanian precept that the pursuit of worldly endeavor was a participation in the overall design of the Divinity.

The Faradays were small farmers and craftsmen. Michael's father, James, worked and prospered as a blacksmith, but then fell on hard times in the upset and depression that accompanied the French Revolution in 1789. With the hope of doing better in the city, the James Faradays moved to Surrey on the outskirts of London. Here Michael was born on 22 September 1791.

Some years later the father's health began to fail. Reduced income forced the family to shift to cramped quarters over a coach house in Manchester Square, London. Here, at times, Faraday said later, they had to subsist on little more than bread and water. But the family adjusted and remained intact despite misfortune, and Michael occupied himself much as other boys in the neighborhood. His early schooling was minimal. "My education," as Faraday wrote later, "was of the most ordinary description, consisting of little more than the rudiments of reading, writing and arithmetic at a common day school."

**AT AGE 14** occurred one of the key events in Faraday's destiny. He became bound for seven years as an apprentice to a bookbinder and bookmaker, a kindly and understanding Mr. G. Riebau and his wife. Michael took well to the work, and in a few years had become an industrious and well-liked assistant to Riebau.

No job and surroundings could have been more suited to awakening and stimulating the curiosity and intellectual hunger that the young and unusual novice soon displayed. His daily tasks of binding books, and the volumes on the shelves of the store, exposed Faraday to the new and wonderful world of learning. He soon developed a consuming interest in reading, and the articles on science had an irrestible attraction for him. He immersed himself in the kinds of books that were to be stepping stones to the furtherance of his talents.

Faraday developed an early habit of making copious notes on items of interest and fascination from the assortment of store reading material that came to hand. With unabated curiosity, he had accumulated such a body of facts that finally he had to ask himself what use he might make of them in shaping a direction for the future.

**THE ANSWER** came in a book entitled *The Improvement of the Mind,* a self-help manual for ambitious young men, written by the clergyman Dr. Isaac Watts. To Faraday, who lacked a formal education, the book by Dr. Watts provided a mine of guidance for self-improvement - by study, by observation, by conversation and by acquiring a facility and precision in language. All these disciplines, cautioned Dr. Watts, were to be pointed to selected and purposeful ends.

Faraday became an enthusiastic follower of Dr. Watts precepts, and soon found the direction for his efforts. "I loved to read the scientific books which came to my hands," said Faraday. He was especially delighted with an article on electricity he chanced to see in the Encyclopedia Britannica, and was fascinated with the prospect of exploration in that lively new field of science. Faraday soon began to experiment with frictional electricity, using Leyden Jars and other simple equipment that he could make, or that his limited income would allow. He had sighted a field of investigation that held promise for his talents, and he was on his way in the science that would eventually bring him immortality.

**FARADAY,** in 1810, was introduced into a company of young men, interested like himself in science and in self-improvement. In 1808 they had organized a discussion group called the "City Philosophical Society." The leader, Mr. John Tatum, a science enthusiast, opened his London house and the use of his scientific equipment for evening lectures held once a week. Tatum, who was knowledgeable in a wide variety of subjects, and a good demonstrator, lectured every two weeks. The other members of the Society would take over on intervening weeks to speak on any subject they chose.

In the two years he regularly attended the meetings, Faraday formed close association and lasting friendships with the other youths keen on mutual improvement. He was exposed to a wide range of topics: static and galvanic electricity, mechanics, optics, hydrostatics, chemistry, astronomy and others. To profit most from these lectures he took copious notes, and then reworked them into a text on the substance of the evening's presentation. Faraday was a critical listener, and included in his reports ideas of his own evoked by the subject. His conclusions often did not agree with those of the speaker.

**LITTLE WAS KNOWN** about the nature of electricity and magnetism at the time. Static electricity was looked upon as the result of one or two electric fluids, corresponding to the old idea of electric effluvia. Galvanic electricity was considered a flow of electric fluid. The discussion of electricity was in continuous contention and this gave Faraday the opportunity for his first lecture to the Society. Tatum was a believer in Franklin's one-fluid theory. James Tyler, author of the article on electricity in the Encyclopedia Britannica, which had launched Faraday into his love for electricity, had proposed a wave or undulatory explanation for the electric current.

Faraday made bold to disagree with both theories, and was invited to express his views. A third explanation, the two-fluid hypothesis, which had been expounded at Trinity College, Cambridge was more convincing to Faraday, and in his *Lecture on Electricity* to the Society he championed it. The separation by friction of the so-called resinous and vitreous electricities, with their attractive and repulsive powers, was one of the evidences that was elaborated by Faraday for superiority of the two-fluid theory. Faraday's lecture was the work of a nineteen year old, and while positive in its assertions, it was made with the realization of very limited background, which Faraday was striving to enlarge upon.

Faraday based his lecture on his own acquaintance with electricity produced by friction. He was shortly introduced to the wonders of another form of electricity - that produced by the voltaic cell. This introduction came through a series of books *Conversations on Chemistry* by Mrs. Jane Marcet. Among topics covered was voltaic current and its use in the new science of electrochemistry.

**FARADAY** was captivated by Mrs. Marcet's presentations and was swept along with new ambitions to be a student of chemistry. He set up a small laboratory in a back room of Mr. Riebau's shop to investigate the nature of electrical decomposition of solutions. By his reading and experimenting, Faraday adduced the correctness of his position on the two-fluid theory of electricity. He speculated on how this duality could account for separation and isolation of elements that took place in the electrochemical cell. He discussed with Society friends the results of the experiments that had attracted his attention, and sought their views.

Mrs. Marcet's volumes on chemistry were a product of a new tide of interest in science. This arose largely from the popularity of lectures on chemistry and electricity given at the Royal Institution by Humphry Davy, a rising, young, attractive and somewhat idolized experimental genius. Davy's investigations were revealing the ultimate relationship between electricity and chemical action. His discoveries in the field of electrochemistry were making scientific history, and forcing a rexamination of older theories of the voltaic cell and of chemical composition. Davy's discoveries were establishing the Royal Institution as a foremost center of science.

**THE ROYAL INSTITUTION** was founded in 1800 through a proposal by Count Rumford (1753-1814), the colorful American expatriate who was born Benjamin Thompson in Woburn, Massachusetts. Coming from an impoverished family, Thompson was endowed as a youth with a passion for science and a high resolve to make a mark for himself in the world. The combination produced a brilliant scientific investigator, and a successful but somewhat unscrupulous international political careerist.

Thompson was a Tory spy during the Revolution and had to flee to England, abandoning his wife and infant daughter. There, as a Royalist, he became favorably known in influential circles for his services for the King during the War. He was made an inspector of supplies sent to America. However, suspected of taking bribes in military purchases, Thompson found it expedient to move to the Continent. There he shunted around during the Napoleonic Wars as a military adviser, intelligence agent and scientific experimenter. His role as kind of professional courtier made him suspect,

THE ROYAL INSTITUTION

Fig. 8.1 - This classical building, Albemarle Street, London, is where Faraday did his life work. Founded in 1800, the Institution was originally intended for the improvement of agriculture. This was to help relieve rural poverty and exploit the land holdings for greater yields by establishing scientific methods, and educating the workers in their use. Later the emphasis shifted to that of a professional organization for the benefit of industry, and to meet social needs. From an 1848 contemporary painting. Courtesy of The Royal Institution.

but this was offset by the value of his military and technological expertise wherever he was engaged.

He found entree to the Court of Bavaria, and was appointed Minister of the Interior. Here Thompson carried out his famous experiments in the boring of canon in which he showed that the production of heat was not due to some imaginary element called "caloric," but to the mode of molecular motion. He was the first to measure the mechanical equivalent of heat. He also made important investigations in the ballistics of artillery. In recognition of his services he was made Count Rumford (the name of his birthplace) by the Elector of Bavaria.

Rumford's use of this position and power made him many enemies, and he was impelled to leave Bavaria. After a short stay in Italy, he returned to England in 1798. He arrived with a large fortune, and with the intent of pursuing a career of private gentleman scientist, an endeavor in which he had already achieved a considerable reputation. Always active in practical research, Rumford's invention of a stove for cooking, an new type of chimney to prevent smoking, and a system of central heating, plus his past inventions, had made him well known and popular in the moneyed and influential circles of London.

Rumford's popularity was also a reflection of an increasing awareness among the wealthy gentry that science and invention could be a driving force in society. Land owners were taking notice of the possible uses of science and invention to the economic and social advantage of their land holdings.

**RUMFORD'S ARRIVAL** in England, and his scientific reputation, had come at a critical time for the wealthy land owners of England. The economic upsets resulting from the social impact of the Industrial Revolution and the French Revolution had brought great impoverishment to the land owner's farm workers, to the extent that posed the possibility of violent rural uprisings.

As one way of alleviating this threat and of relieving the stress on the lower classes, it was proposed that more use should be made of science in agriculture to increase crop productivity and thus improve the income of the farmers. Count Rumford's advice was sought and he recommended an institution for spreading practical scientific knowledge and practical scientific improvements for the benefit of the working classes.

Rumford's idea was enthusiastically received by the landed gentry. They were now perceiving that the new wave of science and technology could be associated with industrial and agricultural advance. Philanthropic education of craftsmen and farm workers was viewed as a means of mitigating social

Fig. 8.2 - HUMPHRY DAVY at age 23 when he was made Director of the Royal Institution, shortly after its formation in 1800. Portrait by Howard at the Royal Institution.

unrest. With the backing of the eminent scientists of Britain, and the wealthy land owners, Rumford organized the Royal Insitution in London in 1800, and Humphry Davy was selected for its director.

**DAVY** was born in 1778, in Penzance, Cornwall, near Land's End, England, the eldest of five children. Davy attended grammar school until he was 17, when the death of his father forced him to find employment to support the family. In 1795 he apprenticed himself to a local surgeon-apothecary, J.B. Borlase, who was a noted physician of the area. The work proved to be congenial, and before long young Davy had set his sights on a career in chemistry. He imposed a rigorous course of self-study that included two outstanding works, the *Elementary Treatise on Chemistry* by Lavoisier, one of the founders of modern chemistry, and the *Dictionary of Chemistry* by Nicholson, who was the first to decompose water with the voltaic pile.

Davy, alert and ambitious, was soon caught in a medical controversy that was to direct the course of his career. Joseph Priestley, the discoverer of oxygen, had prepared, in 1772, one of its compounds, the gas *nitrous oxide,* or "gaseous oxyd of azote," as it was first known. The new gas came to the attentions of an eminent American medical writer, Dr. S.L. Mitchill, who had a general theory that gases were responsible for the spreading of contagion. He advanced the view that the nitrous oxide gas was dangerous to the skin, and that it had the power to carry disease. Learning of this through his employer, Davy became curious, and while still an apprentice in the apothecary shop, he experimented with the gas to check the supposed dangers. He had only small amounts to work with, but he exposed himself and animals to the gas without evidence of injury. He reported to his employer that his tests showed that Dr. Mitchill's fears were baseless. This was a direct challenge to Dr. Mitchill's theories, and an evidence of young Davy's courage and precocity.

The information on Davy's experiments was sent by Borlase to a Thomas Beddoes (1760-1808), who operated the "Pneumatic Institution" in Bristol. Beddoes was a medical radical who had developed the idea that lung diseases could be cured by the inhalation of gaseous air. On learning of Davy's experimental ingenuity, and his work with nitrous oxide, Beddoes invited Davy to join the Pneumatic Institution as a chemist, which he did in 1798. Here, Davy undertook extensive investigations on the physiological effects of nitrous oxide and other gases on the patients of the Institution to provide a scientific foundation for Beddoes's theories.

Davy's report of the hallucinations from inhaling nitrous oxide, or "laughing gas" as it was soon popularly called, created a sensation. His descriptions of the delirium after breathing laughing gas caught the public fancy, and gas sniffing became a popular parlor amusement. But, more seriously, Davy had made a medical advance by discovering that instead of the dangers Dr. Mitchill had predicted for the gas, nitrous oxide had anaesthetic properties. Davy had suggested it could be used to eliminate pain in surgery. But this humanitarian aspect of nitrous oxide was not picked up until late in the 19th century, when Horace Wells, an American dentist first used it in tooth extraction.

**DAVY'S STUDIES,** while at Penzance, had introduced him to the subjects of combustion and heat. These were two mysterious entities that for more than a century had been attributed to an imaginary element called "phlogiston," which was thought to be given off when substances burned. Lavoisier upset the phlogiston theory by showing that combustion resulted from a combination with oxygen of the air. However, to account for the heat and light given off during combustion, Lavoisier had to invent another imponderable element which he called "caloric." Lavoisier's work was covered in Davy's copy of Nicholson's *Dictionary of Chemistry.* In it Nicholson also reported on a conflicting theory that heat was actually a mode of motion of the constituent particles making up substances.

In an early evidence of the effectiveness of Davy's self instruction, and his alertness to scientific opportunity, Davy determined that the conflict could be resolved by experiment. Using a vacuum to keep out the caloric, Davy found that rubbing two pieces of ice together resulted in their melting nevertheless. He then rubbed two pieces of wax coated on one side with ice. The wax melted, but the ice through which the caloric had to pass, did not. These and other tests proved to an exultant Davy that Lavoisier's caloric theory was wrong.

**COUNT RUMFORD,** who had waged his own war against the caloric theory by experiments in the boring of canon, heard of Davy's work in the field. Also Rumford was aware of the wide recognition Davy had received from his investigations on gas inhalation. These evidences of Davy's talents, and other favorable notices, prompted Rumford to bring him to the Royal Institution. There, Davy was to set up a laboratory and provide popular lectures in chemistry.

In the spring of 1801 Davy gave his first public lecture at the new Institution. Thereafter, for a dozen years, his genius set the tone and established the reputation of the place. His lecturing skills created an excitement over science. His presentations had a romantic flair, and a linking of science to the practicalities of life and industry, that attracted not only students of science, but also men and women of the fashionable and moneyed class of London. By convincing his audiences that science was an essential element in the forward progress of the nation, he helped insure patronage for financing continuance of the Institution.

**DAVY'S** Institutional commitments to practical science in fields such as agriculture, food, tanning, mining and other applications of chemistry, had left him little time to pursue pure research. But news of Volta's discovery of the electric cell quickly attracted Davy's attention, and was the spark that set him off, in all the time he could spare, into electrical experimentation. His work in this field, although not initially recognized as important, eventually was to bring Davy enduring fame.

Nicholson and Carlisle in England in 1801, had discovered that a voltaic current could be used to decompose water into its elements, thus revealing a connection between electricity and chemistry. Davy's experiments took over from there. With a powerful battery which he had obtained through the munificence of the Institution membership, he began to exploit the powers of electrochemistry.

In 1807 Davy decomposed potash and soda salts electrically to isolate their metallic elements which he named potassium and sodium. Later, by the same electrolytic process he produced the new metal elements, barium, calcium, strontium and magnesium from their mineral salts.

By the electrical decomposition of muriatic acid Davy obtained a strange, pungent, greenish gas. Lavoisier, who had discovered the combustion properties of oxygen, and who had proposed that all acids were oxides, said this new gas should be a higher oxide of the muriatic acid. Davy disagreed, and went on to prove that the gas was an element, which he named "chlorine."

**DAVY'S RESULTS** were bringing bewilderment to the scientific circles of Europe. Trying to explain them sparked numerous odd theories on how electricity could tear compounds apart. Some suggested that the consituents were carried through the wires and bubbled up at the poles. Others proposed that electricity itself was an acid that could induce electrochemical deposition. Davy brought these speculations to an end in a lecture to the Royal Society in 1806, entitled *On Some Chemical Agencies of Electricity.* Davy summed up his findings and deduced that the decomposition was a property of electrical segregation in solution, and a consequent migration, due to electrical attraction, to the poles, where the elements were released or deposited.

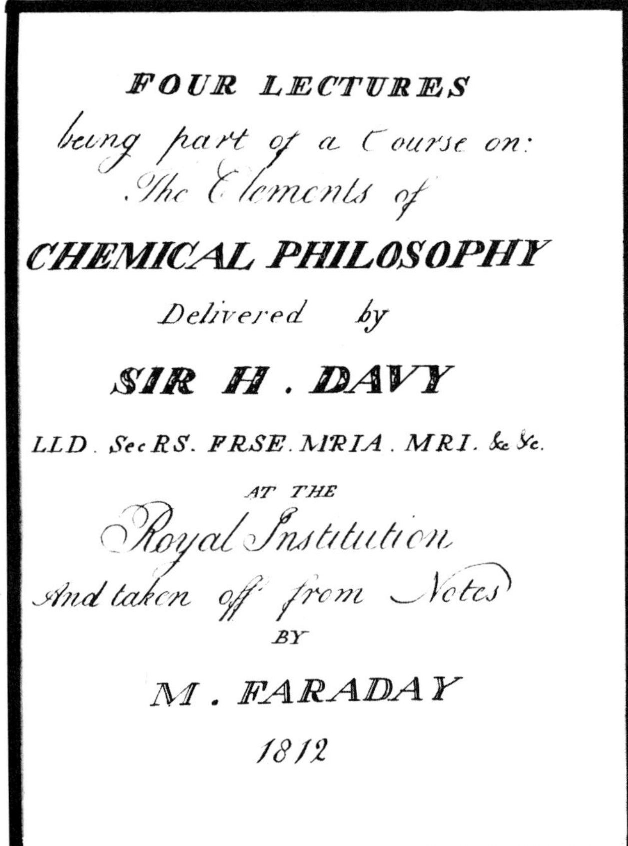
FOUR LECTURES

being part of a Course on:

The Elements of

CHEMICAL PHILOSOPHY

Delivered by

SIR H. DAVY

LLD. SecRS. FRSE. MRIA. MRI. &c &c.

AT THE

Royal Institution

And taken off from Notes

BY

M. FARADAY

1812

Fig. 8.3 - TITLE PAGE of Faraday's notes of Davy's lecture series on chemistry which led to his employment, in 1813, as an assistant in the laboratory of the Royal Institution.

Davy had begun the science of *electrochemistry.* But more than this, as events would show, his work had inspired an even greater genius.

**FARADAY,** of all the members, had profited most from association with the City Philosophical Society. Entering it with little in the way of formal schooling, he had used the opportunities of the education and association to develop his potential. The Society was a training ground for grammar, for the capability of clear and concise speech and writing, and for acquiring self-confidence. He became acquainted with scientific apparatus and disciplines, and kept in touch with scientific proceedings. The Society promoted his consuming interest in science, and projected it as a most promising career. But for the time being his endeavors were profitable only as an attractive hobby.

In 1812 Faraday's bookmaking apprenticeship came to an end and he had no desire to continue in that trade. He wished to carry onin science, but had no place to practice it as a full-time vocation. The outlook was bleak. But Faraday's years of preparation were soon to pay off. As the old adage goes: "When the student is ready the teacher appears." Before long Faraday was to be under the tutelage of the man whose exploits he had followed closely, and who was his inspiration for a future in science.

**THE OPPORTUNITY** came in 1812, when, through the help of a friend, Faraday had the good fortune to attend a series of four lectures on chemistry given by Davy, who had now been knighted, and was president of the Institution. Faraday took extensive and handsomely prepared notes based on his bookmaking skills. Desperate for his future, he hit upon the idea of entering into the service of science by becoming an assistant in Davy's laboratory. To gain attention he decided to make use of his notes.

As Faraday related his action later. "My desire to escape from trade induced me at last to take the bold and simple step of writing to Sir H. Davy, expressing my wishes . . . at the same time I sent the notes I had taken of his lectures." Davy was impressed with Faraday's zeal and replied, "I am far from being displeased with the proof you have given of your confidence."

An opening came in the spring of 1813 when the incumbent assistant quit the job. Faraday was interviewed by Davy, and had caught the eye of the great chemist. Davy's recommendation of Faraday to the managers of the Royal Institution was as follows:

*"Sir Humphry Davy has the honor to inform the managers that he has found a person who is desirous to occupy the situation at the Institution lately filled by William Payne. His name is Michael Faraday. He is a youth of 22 years of age. As far as Sir H. Davy has been able to observe or ascertain, he appears to be well fitted for the situation. His habits seem good, his disposition is active and cheerful, and his manner intelligent. He is willing to engage himself on the same terms that were being given to Mr. Payne at the time of his quitting the Institution."*

Faraday was hired as an Assistant in the laboratory at a wage of 25 shillings a week, and the use of two rooms at the top of the Institution.

Faraday had entered the Hall of Science!

And thus began an association of two great minds that was to be highly productive for science.

Faraday entered happily into his duties, and his skills were soon apparent. He was entrusted with an increasing responsibility in preparing the chemicals with which Davy was working. He also helped in the preparations for lectures, and began his observations on how science could be presented most effectively and interestingly to the public audiences that regularly attended the Institution. He recorded comments and drew up recommendations for best procedures. These were a prelude to his future role as one of the greatest and most popular scientific demonstrators and lecturers of his time.

Davy married a rich widow in 1812, and decided on two years of Continental travel for purposes of science and pleasure. Davy had been given a prize in 1807 by Napoleon, and despite the fact that the Napoleonic War was at its height in 1813, Sir Humphry and Lady Davy were given special clearance by the Emperor to visit France and to travel in other countries under his influence. Faraday was whisked along as a secretary and an experimental assistant for the Grand Tour of Europe.

**TRAVELING** in France, Italy, Switzerland and Germany, Davy practiced science wherever they went, with Faraday as an assistant. Davy, by then was not only an important figure in British science, but was highly regarded in Europe. In Paris they visited distinguished scientists including Ampère, Gay-Lussac, Humboldt and the chemist Courtois.

Davy's visit to Paris coincided with a paper being presented to the French Academy of Sciences, by Courtois, on a strange black substance distilled from seaweed, that gave off, when heated, a beautiful purple vapor. Gay-Lussac, Davy's French rival in chemistry, was on the trail of the vapor's composition. Davy, with his background as the discoverer of chlorine, and suspecting the elemental character of the new gaseous material, joined the search, somewhat to the discomfiture of the French chemists. Both soon determined that the purple vapor was an element. It was christened "iodine" by Gay-Lussac.

In Paris there was also an impressive meeting with Count Rumford. While in England, at the formation of the Royal Institution, Rumford had been knighted by George III, and had been elected to the Royal Society. As a moneyed man with an eye to posterity he had also established several Rumford Medals. He had left England in 1802 under some suspicion of having mishandled funds of the Institution, and had taken permanent residence in Paris, where he married the widow of the famous chemist Lavoisier who had been guillotined in the Revolution.

The party crossed the Alps into Italy in 1814, and in Florence saw Galileo's first telescope. At Naples they made two ascents of nearby Vesuvius. Turning northward they had a long conversation with Volta at Milan, and then crossed into Switzerland where they spent some time at Geneva with the electrochemist Prof. A.A. de la Rive. Faraday was to become a life-long friend of the Professor and his son, carrying on a fifty-year correspondence with them. After touring Germany, and a return to Rome for the winter, further plans to visit Greece and Turkey were cut short and the party returned to England in 1815.

**FARADAY,** who had never previously been more than a dozen miles from London, was now a traveled and intellectually broadened man. He had received a unique education by contact with some of the famous savants and scientists of the time. He had learned French and Italian, and had established friendships in those countries. In their intimate relations some of Davy's insight and analytical know how had rubbed off on Faraday. His love and enthusiasm for a life in science had reached a higher pitch than ever. Reflecting these new attainments, Faraday was elevated by Davy to Superintendent of the Apparatus at the Royal Institution, with increased salary.

Faraday was the right man in the right place at the right time. He was now able to begin to display the remarkable ability at inquiry and discovery that in a continuous half-century career at the Institution would make him a universally recognized genius in the world of experimental science.

**FARADAY'S** duties became more responsible and comprehensive. His skill as an analytical chemist was making him indispensable. He was also sprouting his wings as a fledgling scientist.

In the years up to 1820 Faraday worked mostly in the areas of chemistry, assisting William T. Brande of the Royal Institution, who was regarded as London's leading chemical practitioner. Brande was deeply involved in commercial analysing and in pharmaceutical research. The projects were utilitarian, involving water purification, disinfectants, the distillation of oil for illuminating gas, the analysis of foods and medicines, and various projects for the Admirality.

Faraday's helpfulness and developing skills made it increasingly possible for him to do original research. The discovery of chlorine by Davy led to operations in which Faraday, by new methods, synthesized several compounds of that element with carbon. Working with the distillation of coal tar he discovered isobutylene, which is now used in synthetic rubber, and benzene. In 1815 he assisted Davy in developing his widely hailed miner's safety lamp, a dramatic example of the help science could give industry, In 1816 Faraday published his first paper: *Analysis of Native Caustic from Tuscany.*

Faraday's intensive study, his experimental manipulative skill and pioneering work had made him a professional chemist - but he always preferred the title of natural philosopher. In his new capacity he periodically attended City Philosophical Society meetings, and gave a lecture, in 1816, *On the General Properties of Matter.* This was the first of many lectures on chemistry Faraday presented to the Society.

**FARADAY'S REPUTATION** in chemistry was bringing demands for his services as a consultant and as an expert witness in technical trials. It is interesting that, as a son of a blacksmith, Faraday's first sustained effort as a consultant was in the development of steel alloys. James Stodart, a maker of surgical instruments, wanted to achieve by alloying, a steel that would not rust. Working in Faraday's laboratory they produced the first *stainless steel.* Faraday's pioneering in this field of metallurgy had made him a foremost authority in the alloying of steel with other metals.

Faraday's research, in the years following 1820

FARADAY AT WORK

Fig. 8.4 - Here, in the basement laboratory of the Royal Institution, the industrious Faraday, with his inexhaustible mind of inquiry, probed the secrets of nature to conjure up the greatest discoveries ever made by one man in the fields of science. Scene is from a 19th century painting. Courtesy the Royal Institution.

covered a very wide range of science: chemistry, electrochemistry, liquefaction of gases, geology, metallurgy, mechanics and optics. In each he made fundamental discoveries. But his greatest fame was to come in the field of electromagnetism.

**IN 1820**, Hans Christian Oersted, a professor at the University of Copenhagen, observed during one of his classroom demonstrations that a compass needle in the vicinity of a wire carrying an electric current was being deflected in a pattern showing there was a force on the needle from the wire. He deduced this unknown force to be due to a magnetic effect produced by the current, and indicating, therefore, that there was an interrelationship between the electricity and magnetism.

Oersted announced his discovery in a 4-page tract, in Latin, sent to the leading scientific centers of Europe. The report created a sensation. Rushing to their laboratories, investigators repeated the experiment to see the remarkable effect.

The amazing aspect, which puzzled Oersted and all others who observed it, was the orientation of the force. According to Newtonian science all central forces should act in straight lines. Here, however, the force was circular around the wire, acting transversely to the current. A circular force was new and different from any that had previously been encountered, and the implications of the force gripped the minds of the scientific world.

Ampère, in Paris, was astonished at the discovery and in a few weeks had elaborated on the Oersted revelation. He reasoned that magnetism was not an entity in itself, but an electrical action. Deflection of the compass needle was the result of electrical action upon itself, and therefore electric currents should act on each other. Ampère went on to demonstrate that the electric current-carrying wires would attract or repel each other depending on the direction of the currents, thus establishing the science of electrodynamics.

**NEWS** of the Oersted discovery had reached the Royal Institution in October 1820. Davy and Faraday, although absorbed in chemical researches, immediately set up to repeat to their satisfaction the Oersted results. In his experiments on Oersted's work, however, Davy seems to have missed the point of the tangential action of the magnetic force around the wire. He interpreted Oersted's force as simply a case of straight-line attraction between compass needle and wire. This, for a time, misled Faraday's conception of the forces.

Meanwhile, William Hyde Wollaston (1766-1828) a noted English physician-scientist, assumed the correctness of the circularity of a force around the wire. To prove it he suggested to Davy that a current-carrying wire ought to rotate on its own axis when a bar magnet was brought close to it. Davy tried the experiment but the expected rotation did not occur. Faraday was busy at the time with other work, and with his courtship of Sarah Barnard, whom he married in 1821. He was not a party to the Wollaston experiment, but he was present at the demonstration, and this would cause him future embarrassment.

Faraday's entrance into electromagnetism began with an assignment that he prepare, for publication in the *Annals of Philosophy,* an historical summary of this new branch of science. In preparing the survey he reviewed the existing literature and repeated the experiments made by Oersted, Ampère and others. He was soon aware, in analysing the forces, that the action between the current and a magnet was not the simple push or pull effect usually experienced, but rather that the magnetism was setting transverse to the current. The action was tangential. Wollaston's experiment, he decided, showed that the effect of the bar magnet had been only to move the conductor from side to side rather than make it rotate. Actually, he concluded, the wire was trying to circle around the magnet.

**TO TEST** this hypothesis, Faraday, in late 1821, devised the ingenious apparatus shown in Fig. 8.5. He fixed a magnet in an upright position in a cup of mercury, with one pole of the magnet exposed. A suspended wire, with its lower end in the pool of mercury and free to circle around the magnet, was connected to one terminal of a battery, and the mercury to the other.

Faraday's reasoning was right! The free end of the wire whirled around the magnet as long as the current was maintained. Then, reversing the set up, he pivoted one end of the magnet rod, which thereupon reacted to circle around the wire, following the circle of magnetism around the wire.

*Electrical energy from the battery was being converted to mechanical energy. This was the world's first electromagnetic motor!*

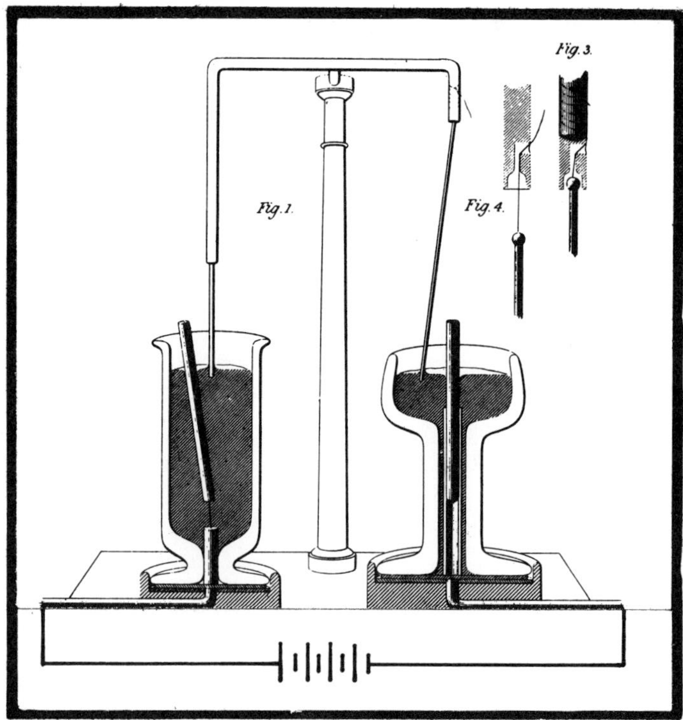


Fig. 8.5 - ELECTROMAGNETIC MOTOR ACTION was devised by Faraday in 1821. Magnet rod, left, pivoted in mercury cup, rotates around the conducting wire, and swiveled wire, right, dipping into mercury cup, rotates around the magnet. Both rotations are due to the interaction between the magnetic field around the wire and the field of the magnet. Faraday had created the first electromagnetic motor, which in modern form, is serving mankind by the millions. Illustration the Royal Institution.

**FARADAY,** where Davy had failed, was able to force rotation from a current-carrying wire. Knowing the importance of his discovery, he published his results immediately in a paper: *On some new Electro-Mechanical Motions, and on a theory of Magnetism.* The news created a sensation, and put Faraday overnight into the top rank of European scientists. But he was in trouble at home.

Friends of Wollaston, who were aware of his pioneering rotative experiments, accused Faraday of stealing the idea without credit. They were unable or unwilling to assess the originality and novelty of Faraday's success contrasted with the failure of Wollaston's idea. Faraday, highly anguished, made every effort to justify his claims for originality. But the affair did not die down for Davy, who appeared to be jealous of Faraday's success. The vital discovery had eluded both he and Wollaston. He saw his pre-eminence waning and was not willing to accept Faraday's rising international fame with good grace. When Faraday was proposed as a candidate for the Royal Society, Davy opposed it. But, with only one vote against him, Faraday was elected in 1824.

In the ten years following his discovery of electric motor action, Faraday was occupied with chemical research and other duties that allowed only occasional opportunity for electrical investigations. But his mind kept revolving on Oersted's discovery, and his interest was further excited by the continuing revelations of electromagnetic effects being reported from abroad. Ampère had shown the attraction and repulsion of wires carrying an electric current. Iron bars were being magnetized by insertion in electrically energized solenoid coils. Joseph Henry, America's electrical scientist, had made electromagnets capable of lifting a ton. The galvanometer had been developed, giving experimenters a means of measuring current intensity, and of determining the direction of the resulting magnetism.

**ARAGO,** in France in 1824, had mystified the scientific community with the apparatus shown in Fig. 8.6. By some unexplained force, the compass needle was dragged around and put into circular motion whenever the enclosed copper disk below it was rotated. Faraday's fellow scientists labored at various explanations for the strange action, but missed the true one. Faraday would soon resolve the mystery by the discovery of electromagnetic induction.

To Faraday, one of the basic implications of a magnetic phenomenon was the possibility of the converse action. In his mind's eye there was a uniformity in the forces of nature. His scientific training and philosophy led to a belief in symmetry and convertibility. He had once stated that of the various natural forces . . . "we cannot say that any one is the cause of the others, but only that they are all connected, and due to one common cause." He reasoned that if electricity could produce magnetism, as in the Oersted experiment, why not the converse? Could not magnetism produce electricity?

**AS EARLY AS 1821,** in his *Historical Sketch of Electromagnetism,* prepared soon after Oersted's discovery, Faraday had already given the following arguments for the conversion of electricity from magnetism:

"In consequence of M. Ampère's theory which attributed the powers of magnets to electrical currents, and also the views taken of the manner in which it was supposed that currents of electricity in the connecting wire induced current in steel bars placed near them, as in M. Arago's experiments in the discovery of electromagnets; it was earnestly hoped and expected that such an arrangement might be made of magnets, wires, etc. as to produce the decomposition of water, or some other electrical effects; for as electricity produced magnetism, it was considered that magnetism might produce electricity.

In Faraday's words: . . . *it appeared very extraordinary that inasmuch as every electric current was accompanied by corresponding intensity of magnetic action at right angles to the current, good conductors of electricity, when placed within the sphere of this action, should not have current induced in them.*

And thus Faraday went into action in his search for the evolution of electricity from magnetism. His methods, progress and the results on this and other electrical investigations over the years, are beautifully noted in his *Experimental Researches in Electricity.* In this journal he numbered each entry, and beginning with No.1 in November 1831, they run consecutively to over three thousand. They cover some of the greatest discoveries in electromagnetism.

Faraday was not alone in his quest for the secret of electricity from magnetism. Arago and Ampère in France, were trying to find a way to produce electricity magnetically, but had failed to find any essential principle. And, in America, Joseph Henry had come close to the critical factor in the conversion process.

In 1824 Faraday had experimented with magnets close to current-carrying coils, but could detect no effect on the current. Later he further investigated the association of magnets and coils of various shapes and in various degrees of proximity without success.

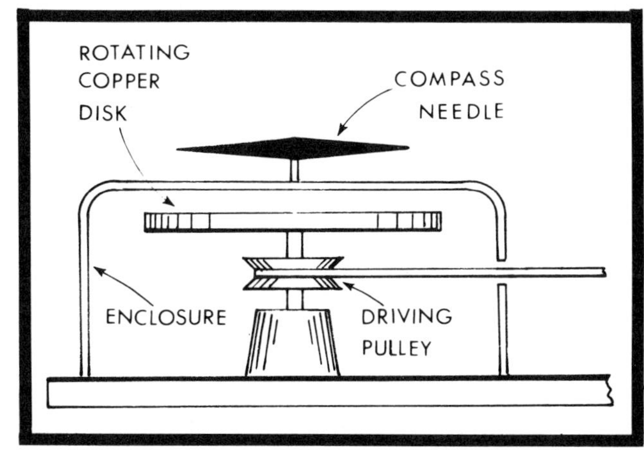


FARADAY SOLVED THE PUZZLE OF ITS OPERATION

Fig. 8.6 - "Arago's Disk" (1824) mystified scientists of the time. The compass needle was dragged around whenever the copper disk in the enclosure below was rotated. Explaining this effect was one of the factors prompting Faraday in his quest of magnetism as a producer of electricity. Faraday's discovery of electromagnetic induction revealed that rotation of the compass needle was the result of interaction between the magnetic compass needle and the magnetic poles it induced when the disk rotated.

Faraday could not give the conversion effort further attention for several years. In 1825 he was appointed Director of the Laboratory under Brande. From then until 1830 he was occupied with the development of optical glass for improved lenses used by the Admiralty. The work was very successful but Faraday regretted the time lost for personal research.

**LATE IN 1831,** further developments in his procedural thinking were so urgent that Faraday dropped utilitarian activity to resume his magnetic investigations. This was the great turning point in Faraday's future. In two brilliant experiments he made two major breakthroughs. He showed that the answer was yes - electricity could be produced from magnetism. However, and this was Faraday's crucial contribution, it was not merely the mutual, steady-state presence of magnetism near a conductor that produced electricity. Something else was involved.

Faraday had reoriented his thinking to the question: if the flow of electric current, that is the

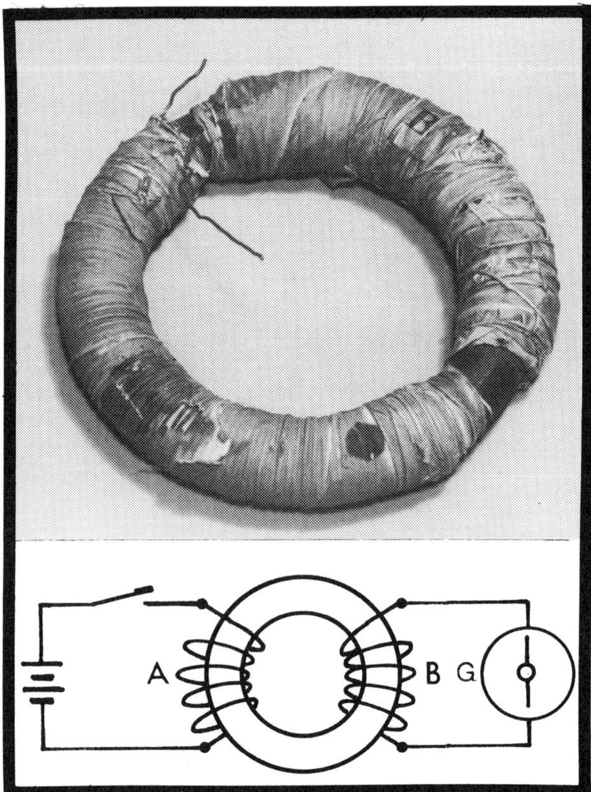


THE FIRST ELECTROMAGNETIC TRANSFORMER

Fig. 8.7 - FARADAY'S RING, with which he produced the first electromagnetic induction. Rise and fall of magnetism in the iron core of the ring, with the making and breaking of the current to coil, A, on one half of the ring, induced current in coil, B, on the other half of the ring, as was indicated by the galvanometer, G. Motion of the magnetism was essential for induction. Photo the Royal Institution.

movement of electric charges, causes magnetism, will movement of magnetism produce electricity?

**IN HIS FIRST EXPERIMENTS** on electricity and magnetism, Faraday based his procedure on Ampère's findings of the forces between mutually associated conductors. How would activation of one circuit affect an adjacent circuit in terms of a possible transfer of magnetic induction? Would the sudden creation of a magnetic force, such as by energizing an electromagnet, produce a special strain or vibration in an adjoining wire that might be an electric current? To test this Faraday prepared two interwound circuits which he described as follows:

"Two hundred and three feet of copper wire in one length were coiled round a large block of wood; other two hundred and three feet of similar wire were interposed as a spiral between turns of the first coil, and metallic contact everywhere prevented by twine. One of these helices was connected with a galvanometer, and the other with a battery of one hundred pairs of plates four inches square, with double coppers, and well charged."

"When the contact was made, there was a sudden and very slight effect at the galvanometer, and there was also a similar slight effect at the galvanometer when the contact with the battery was broken. But whilst the voltaic current was continunig to pass through one helix, no galvanometer appearances nor any effect like induction on the other helix could be perceived."

"Repetition of the experiments with a battery of one hundred and twenty pairs of plates produced no other effects; but it was noted both at this and the former time, that the slight deflection of the needle occurring at the moment of completing the connexion, was always in one direction, and that the equally slight deflection produced when the contact was broken, was in the other direction; and also, that these effects occurred when the first helices were used,"

"The results which I had by this time obtained with magnets led me to believe that the battery current through one wire, did, in reality, induce a similar current through the other wire, but that it continued for an instant only, and partook more of the nature of the electrical wave passed through from shock of a common Leyden jar than of a current from a voltaic battery, and therefore might magnetise a steel needle, although it scarcely affected the galvanometer."

**IT WAS APPARENT** to Faraday that an induction effect, a transfer from one winding to another, was taking place. But the meter indications were weak, and the action was not quite what he had expected. It was associated only with the making and breaking of the battery contacts.

To achieve more positive results Faraday decided to construct an electromagnet whose shape and winding arrangement would provide a very strong magnetic force on an adjacent coil. He chose a thick iron ring six inches in diameter with separate helices wound on the two sides of the ring, as shown by A and B, Fig. 8.7.

Now, when winding, A, was energized by the battery, the galvanometer needle connected to winding, B, spun around several times - an amazing difference due to the greater magnetic receptivity of the iron core compared to the wood core used in the previous experiment. But, as previously, the effect was not permanent. The needle came to rest and was indifferent to the steady current. The needle responded in a similar way in a reversed direction when the battery was disconnected.

Since the magnetization with the iron ring had been so effective, and the induction took place only when magnetization was changing during the making and breaking of the circuit, Faraday hit upon the idea of using a permanent magnet to test the possibility of induction by changing of magnetization in the coil by the movement of the magnet itself. Using the helix of Fig. 8.8, whose terminals were connected to the galvanometer, he thrust the magnet into the coil. He got action - the needle responded. Quickly withdrawing the magnet reversed the swing of the needle. The results were thus duplicating those of his previous experiments.

The actions were all of one piece!

The conclusion was now clear - *the key to the induction had been found.*

**FARADAY** realized that it was the *change* of the magnetic field that was producing the induction. In the iron ring experiment it was the surge of magnetism at the instant the battery was switched on and off that had caused the surge of current in the second coil. Also with the helix and plunger, the magnetism sweeping through the wires had induced the electricity.

Faraday called the action he obtained with the helices *volta-electric induction.* His arrangement of an iron ring with two windings is the fundamental constrcution of our present-day *transformers,* where electric current is transferred magnetically from one coil to another for the purpose of stepping up or stepping down the voltage from one side to the other.

Faraday went on with further experiments to show that whenever there was relative movement between conductor and magnet, an electric current was generated. Faraday's experiments in electromagnetic induction were one of the foundation stones for electromagnetic theory. From them is derived his law that the induced electromotive force in an electrical circuit is proportional to the rate of change of magnetic flux lines in the circuit.

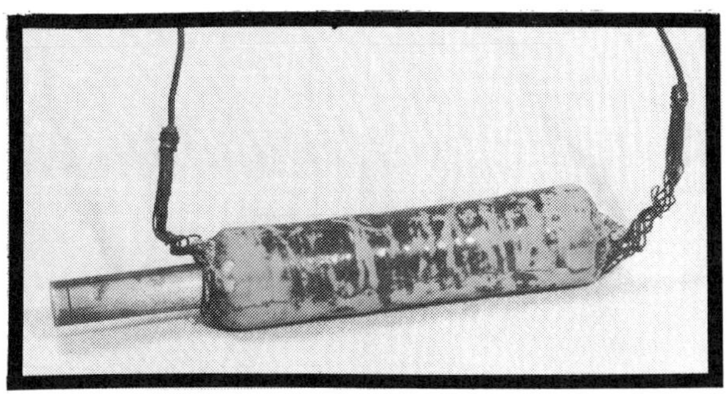

Fig. 8.8 - SOLENOID AND PLUNGER used by Faraday in 1831 to demonstrate generation of electricity. Thrusting the magnet rod into the coil of wire provided the motion of magnetism relative to a conductor to induce a current, as indicated by a galvanometer connected to the terminals. The sweeping of magnetism across coils of wire is the principle used in today's generating stations for producing electric power. Illustration from the Royal Institution.

**FARADAY'S SECOND BREAKTHROUGH,** on 28 October 1831, was monumental. He had thus far been able to produce electromagnetic induction momentarily by making and breaking circuits, and by manipulating magnets in the presence of coils. Now, he was to show that an electric current could be generated continuously by rotation of conductors in a magnetic field.

His apparatus, derived from considerations of Arago's device, Fig. 8.6, used a copper disk rotated in a magnetic field. In Arago's machine, electric current induced in the disk by the magnetic field of the compass needle, obliged the needle to follow the motion of the disk. Faraday inverted the operation. He held the magnetic field stationary and collected current from the rotating disk.

A disk of copper, twelve inches in diameter, and about one fifth of an inch in thickness, fixed on a brass axis, was mounted in frames so as to allow of revolution either vertically or horizontally, its edge being at the same time introduced more or less between the magnetic poles, Fig, 8.9. The edge of the plate was amalgamated for the purpose of obtaining good movable contact, and a part of the axis was prepared in a similar manner.

Two rubbing contacts, one on the axle of the disk, the other on the edge of the disk. were connected to the galvanometer.

Faraday recorded that . . .“when the disk was stationary all was quiescent, and the galvanometer exhibited no effect. But the instant the disk moved the galvanometer was influenced, and by revolving the disk quickly the needle could be deflected 90 degrees or more.”

The deflection continued as long as the disk was revolved. Faraday described the generation as being “currents of electricity in the direction of the radii,” implying that in effect the disk was like an array of radial wires which by turning were cutting across the magnetic field, hence an electric current was generated in them.

Faraday then modified the generation elements, using wire loops for the disk, and the “Great Electromagnet” (Fig.8.10) for the magnetic field. The loops generated an alternating current which he then rectified to direct current by a commutator placed on the shaft. Faraday had devised all the elements of a direct-current generator.

THE FIRST ELECTROMAGNETIC GENERATOR

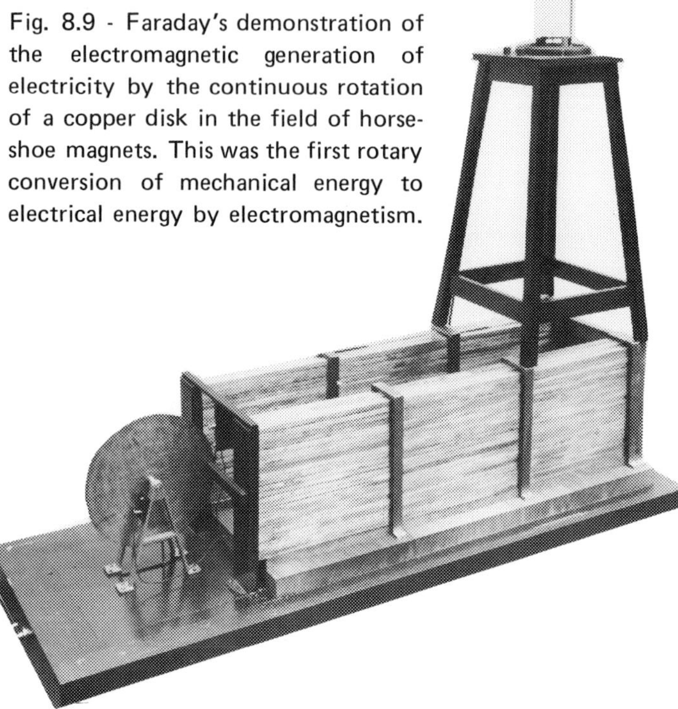

Fig. 8.9 - Faraday's demonstration of the electromagnetic generation of electricity by the continuous rotation of a copper disk in the field of horseshoe magnets. This was the first rotary conversion of mechanical energy to electrical energy by electromagnetism.

**FARADAY** had now achieved a continuous induction, which he attributed to a change in the intensity of the *electrotonic state* of conductors passing through the magnetic field.

But was change in intensity of the magnetism the essential element for induction? To test this further he mounted a copper disk axially on the end of a circular bar magnet, so that the disk experienced a uniform magnetic field. On rotating the combined disk and magnet, a continuous current, collected by wires on the edge and center of the disk, was produced. Thus, electricity could be induced also by a uniform magnetic field. Faraday now altered his theory that there must be a change in the *electrotonic state* for induction. The essential condition was simply motion of a conductor relative to the magnetism in such a way as to intersect, or “cut” the magnetic field.

Faraday had found that an electromotive force could be generated in a conductor in three ways: *(1) by the changing of magnetic flux linkages, produced by one conductor, acting on an adjacent conductor, Fig.8.7, (2) by moving a magnetic field relative to the conductor, Fig.8.8, and (3) by moving the conductor relative to magnetism, Fig.8.9.*

He was soon to find still another inductive principle concerned with the effect of the changing current on the conductor itself. In this effect there was a co-discoverer, who is generally given priority. Joseph Henry, at the Albany Academy, Albany, New York, who like Faraday, was a man with an inspired gift of scientific curiosity, had earlier deduced the principle of “self-inductance.”

Henry noticed that when he separated the ends of the two wires in breaking a long circuit, he would get sparks, and these were intensified when the wire was coiled. He correctly interpreted this as due to the collapse of the magnetic field around the conductor itself when the circuit was broken. The collapsing field generated a reverse voltage sufficient to make the spark. This ever-present property of electric circuits, discovered by Henry and Faraday independently, is called the “self-inductance,” and the unit is named for Henry.

**ELECTRICITY**, to Faraday had now become a fundamental force, holding the possibility of being the principle which accounted for the structure and operation of all natural phenomena. He wrote of this potential in his *Experimental Researches,*

"The science of electricity is in that state in which every part requires experimental investigation; not merely for the discovery of new effects, but what is just now of far more importance, the development of the means by which the old effects are produced, and consquent more accurate determination of the first principles of the most extraordinary and universal power in nature . . . No branch of knowledge can afford so fine and ready a field for discovery as this. Such is abundantly shown to be the case by the progress which electricity has made in the last thirty years: Chemistry and Magnetism have successively acknowledged its overruling influence; and it is probable that every effect depending upon the powers of inorganic matter, and perhaps most of those related to vegetable and animal life, will ultimately be found subordinate to it."

In a short period of inspired thinking and experimenting Faraday had opened the gates of a new domain of science. His discovery of the laws of electromagnetic induction had set the stage for the coming Electrical Age.

Fig. 8.10 - The "GREAT ELECTROMAGNET" of the Royal Society which Faraday substituted for the permanent magnets of Fig. 8.9, to carry on further experiments in electrical generation. Illustration the Royal Institution.

**LEGEND** has it that Faraday was asked by the Prime Minister what use could be made of his discoveries. Faraday is said to have answered, "some day it might be possible to tax them." Faraday made his two discoveries in electromagnetism in a few days of work.But it took more than a half century to make good on his prophecy of taxation.

Faraday, with his many experiments, had demonstrated what he called "production of permanent current by ordinary magnets," and he had given clues on how to accomplish the production by stating that when wires move so as to cut a magnetic field, a power is called into action which tends to urge an electric current through them. But there was to be a 50-year engineering gap before the science was converted into technology. It was not until the late 1880s that really efficient generation of electric power was achieved.

**FARADAY, in 1832,** proceeded to establish that various kinds of electricity, which had been considered of separate origins, were one and the same. These included electricity derived from battery, frictional, static, thermal and animal sources. In Faraday's view there was a unity of world forces; electrical, gravitational, chemical, etc., were forms of a fundamental force. It followed that all the forces were essentially identical and could be converted from one form to another. In 1832, in a new series of experiments, Faraday showed that he could get the same chemical and magnetic effects regardless of the source of the electricity. The identity of all types of electricity resolved a long-standing question.

With the completion of his electromagnetic discoveries, Faraday published the results in the first of a series: *Experimental Researches in Electricity.* Begun in 1831, the series continued until 1860. It is one of the most remarkable documents in scientific history, recording the day to day speculations, theorizing and experimenting of a master investigator, who had a hand not only in the notable advances in electricity and magnetism of the time, but made important contributions in the fields of chemistry, optics and metallurgy. His metaphysical

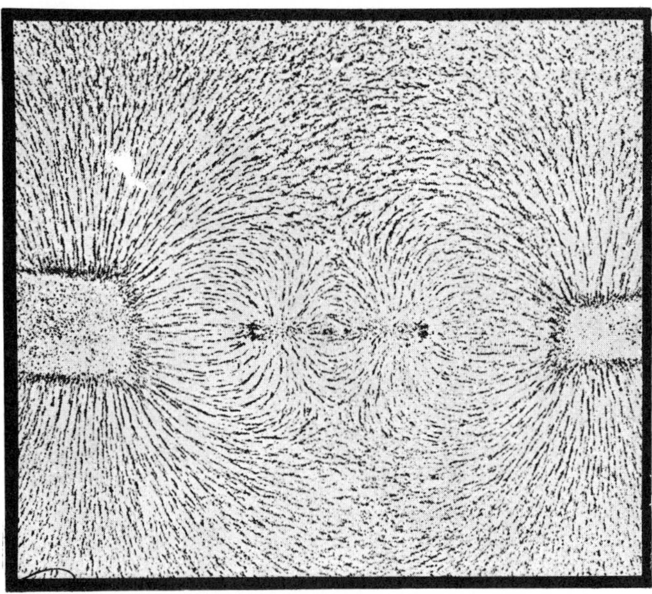

Fig. 8.11 - LINES OF FORCE - Faraday explained the operation of the electromagnetic generator, Fig. 8.9, as due to metal cutting the "magnetic curves." He meant by magnetic curves the lines of magnetic force issuing from a magnet. He pictured the lines by iron filings. The pattern of iron filings above was made by Faraday, and shows the field of force between two magnets with an intervening iron rod. From Faraday's Diary at the Royal Institution.

probings into the ultimate reality of force and matter were to have great influence and repercussion on the future development of physics.

**LINES OF FORCE** - Faraday, who never got beyond simple arithmetic in school, was not a professional mathematician, and was not able to express his ideas and results in mathematical formulations. But he was admirably able to picture his concepts in terms of physical models.

He sought to envision the forces displayed by magnetism. What was emerging from the poles of a magnet that enabled it to attract and repel at a distance, and to generate an elctricity as it swept by a conductor? He saw iron filings arrange themselves in a pattern when he sprinkled them on paper held over the poles of magnets. Why did they form lines; was this merely a picture? Or were they a representation of the real lines of an underlying invisible field of force existing in space? Faraday believed that they were.

One of Faraday's goals, that occupied him until his last days, was to elucidate, and to attempt to prove, that a basal field of force, such as exhibited by magnetism, accounted for the substance and operation of all nature. In this vision Faraday was carrying on in the oldest continuing quest in the philosophy of science - the search for basal unity in nature.

**THE ANCIENT GREEKS** were the first to speculate on what the universe was made of. Thales, in 585 B.C., guessed that the elemental substance was water. Anaximides thought all things were made of air, Heraclitus, of fire, and Empedocles suggested four primary substances - earth, air, fire and water. Democritus, 435 B.C., envisioned a universe of "atoms, eternal and indivisible, operating in the void." In the 17th century came new proposals such as that of Descartes, who pictured the universe as a mechanical system, full of visible and invisible corpuscles, whirling and interacting in vortices. But Descartes' theory could not satisfactorily explain natural phenomena.

The first successful theorizing of a fundamental force of nature was the crowning achievement of Sir Isaac Newton (1642-1727), who in 1867 announced his law of gravitation. Newton's theory stated that bodies consisted of corpuscles, or mass points, between which individually, or in the aggregate, central forces acted instantaneously, and in straight lines, dependent in strength on the inverse square of the distance between corpuscles. Newton's theory successfully formulated the forces of gravity and the motion of planets. It set a standard for future work on the forces of nature.

New discoveries in electricity, magnetism and chemistry, however, made problems for Newton's theory. It could not explain cohesion that held bodies together, nor account for the transverse, circular magnetic force that was produced by an electric current. There was a need for creation of a new theory, and Faraday was the first, compelled by the results of his experiments, to propose a "field of force" theory, that conflicted with the central force theory of Newton.

Newton's theory envisioned forces jumping instantly from one body to another; there was no involvement of the intervening space. But Faraday's immersion in experimental electricity and magnetism made this conception untenable.

In Faraday's view, it did not meet a requirement for continuity. He saw static charge and magnetism acting via a "field-of-force" occupying the whole physical environment, and proceeding with contiguous progression from force point to force point. The convertibility of one force to another, shown by the relationship of electricity, magnetism and electrochemistry, was one of Faraday's supports for his theory.

The attractions and repulsions of static electricity and magnetism, according to Faraday, were a wave of changes in the strain existing between force centers. Faraday's theory abolished the distinction between matter and force; physical substance itself might be considered a convergence or focus of forces. Faraday rejected instant action-at-a-distance in favor of force transferrence by impulsion, through time, of each part impacting on a contiguous neighbor. His theory provided for curved progression, Fig. 8.11, as most likely for the lines of attraction and repulsion, since they involved forces of tension and compression.

**FARADAY'S** most prevasive support for his field-of-force theory was the conception of *lines of force.* These were not merely a pictorial representation, but an actuality that comprised the field of forces. He presented his lines-of-force concept most clearly in magnetic action. Streaming out from magnetic sources, the lines filled the surrounding space and were a dynamical condition of space. Faraday described the situation as follows:

"A line of magnetic force may be defined as that line which is described by a very small magnetic needle, when it is so moved in either direction correspondent to its length, that the needle is constantly a tangent to the line of motion; or it is that line along which, if a transverse wire be moved in either direction, there is no tendency to the formation of any current in the wire, whilst if moved in any other direction there is a tendency . . ."

"A point equally important to the definition of these lines is, that they represent a determinate and unchanging amount of force . . . Though, therefor their forms, as they exist between two or more centres or sources of magnetic power, may vary greatly, and also the space through which they may be traced, yet the sum of power contained in any one section of a given portion of the lines is exactly equal to the sum or power in any other section of the same lines, however altered in form, or however convergent or divergent they may be at the second place. . . ."

"Now it appears to me that these lines may be employed with great advantage to represent the nature, condition, direction and comparative amount of the magnetic forces; and that in many cases they have, to the physical reasoner at least, a superiority over that method which represents the forces as concentrated in centres of action, such as the poles of magnets or needles; or some other methods, as, for instance, that which considers the north and south magnetisms as fluids diffused over the ends or amongst the particles of a bar . . ."

"I cannot refrain from again expressing my conviction of the truthfulness of the representation, which the idea of lines of force affords in regard to magnetic action . . . In a straight wire, for instance, carrying an electric current, it is apparently impossible to represent the magnetic forces by centres of action, whereas the lines of force simply and truly represent them. The study of these lines have, at different times, been greatly influential in leading me to various results, which I think prove their utility as well as fertility. I have been so accustomed, indeed, to employ them, and especially in my last Researches, that I may unwittingly, have become prejudiced in their favor, and ceased to be a clear-sighted judge. Still, I have always endeavored to make experiment the test and controller of theory and opinion; but neither by that nor by close cross-examination in principle, have I been aware of any error involved in their use."

**LINES OF FORCE,** as a concept, was unorthodox. It flew in the face of conventional physics. As Maxwell said, "Faraday, in his mind's eye, saw lines of force traversing all space. where the mathematicians saw centres of forces acting at a distance; Faraday sought the seat of the phenomena in real actions going on in the medium; they were satisfied that they had found it in a power of action at a distance impressed on the electric fluids."

Maxwell was a first strong supporter of the Faraday theory of the force field, and made it the keystone of his electromagnetic theory. He presented a paper before the Cambridge Philosophical

Society in 1865 *On Faraday's Lines of Force* in which he put Faraday's ideas into mathematical language. Maxwell said of Faraday's method:

"It was perhaps for the advantage of science that Faraday, though thoroughly conscious of the fundamental forms of space, time and force, was not a professed mathematician. He was not tempted to enter into the many interesting researches in pure mathematics which his discoveries would have suggested if they had been exhibited in mathematical form, and he did not feel called upon either to force his results into a shape acceptable to the mathematical taste of the time, or to express them in a form which mathematicians might attack. He was left at leisure to do his proper work, to coordinate his ideas with facts, and to express them in natural untechnical language."

**IN 1835** Faraday turned to the study of electrostatics, involving electric charges and their electric fields. Up to this time static electricity had been the subject of curiosity and experimentation for 250 years. Friction machines generated sparks, and Leyden Jars were used for storage of electrostatic charge; various materials were tested for attraction and repulsion, and their ability to hold or transmit electricity. The concept of conductors and nonconductors was finally developed.

The original theory that static electricity was made up of two fluids, was replaced by Franklin's one-fluid proposal. Electricity by friction, he explained, was simply an act of separation which left one body with excess, or positive charge, and the other body with a deficiency, or negative charge - the total charge remaining constant. Coulomb, in 1795, proved that force between point electrostatic charges varied with the strength of the charges, and inversely as the square of separation.

Faraday carried on where Coulomb left off, and his experiments proceeded to put the previous knowledge of static charges and the electric field into its modern prespective. He worked with conductors of all sizes and shapes, and established that there was no such thing as "absolute charge." His lines-of-force theory required that there be tension in two directions, positive and negative. Thus, the creation of separate charges was impossible, i.e., there must be conservation of charge. Every charge was matched by an equal and opposite charge on the neighborhood. The opposite charges were only the different ends of the lines of force.

**FARADAY** concluded the charge must reside on the surface of a conductor. To prove this he had constructed a hollow cube twelve feet on a side. The outside of the cube was covered with a metal foil, and the cube stood on supporters to insulate it from the floor. With Faraday inside, the cube was charged from a powerful electrostatic machine. Faraday reported on the experiment:

"I went into the cube and lived in it, and using lighted candles, electrometers, and all other tests of electrical states, I could not find the least influence on them . . . though all the time the outside of the cube was very powerfully charged, and large sparks and brushes were darting off from every part of its outer surface."

The electric field was all between the surface of the cube and the neighboring bodies.

**THE ELECTRIC FIELD** was the subject of Faraday's next exploration. In his lines-of-force theory the appearance of the electrostatic charge was in the stress of the force field between the charged bodies. The effect of static electricity was a property of the space between the charged bodies, hence it should be affected by the material occupying that space. Coulomb's experiments had led him to conclude that there was no difference in the charge capacity of different materials. Faraday doubted this, and it posed a question to himself which he stated as follows:

". . . suppose, A, an electrified plate of metal suspended in air, and, B, and, C, two exactly similar plates, placed parallel to and on each side of A, at equal distances and insulated; A, will then induce equally toward, B, and, C, (that is, equal charges will appear on these plates). If in this position of the plates some other dielectric than air, as shell-ac, be introduced between, A, and, C, will the induction between them remain the same? Will the relation of, A, and, B, to, A, be unaltered, notwithstanding the difference of the materials interposed between them?

To answer the question Faraday constructed identical spherical condensers, Fig. 8..12, each of which consisted of concentric halves, with an intervening space into which he could put various non-conducting materials. Into the first condenser, with air, he put a measured charge which had a potential indicated by an electrometer. He then connected this to a second condenser, uncharged, in which the intervening space was filled with shellac. Upon connection, the potential, assuming equal distribution of charge, should have halved. Actually Faraday found that the potential was less than half, and that the charge for the shellac was much higher than for air, by a factor he called the *specific inductive capacity.*

**TO FARADAY** the only explanation for the increased charge was a greater crowding of the electrostatic lines of force, hence a more intense electric field. This exerted an elastic strain on the electric charge particles of the dielectric. These particles were normally tightly bound and in a neutral state. However, as a result of the strain produced by the electric field, the particles of the dielectric became slightly displaced, with an accompanying transient current. The displacement Faraday called *polarization.* Upon removal of the potential from the dielectric, the polarization would disappear.

The ability of materials to assume a charge Faraday attributed to lines of tension and compression in the electric field, the density of which was a characteristic of the particular dielectric material.

The concept of polarization in dielectrics led Faraday to further experiments to find evidence

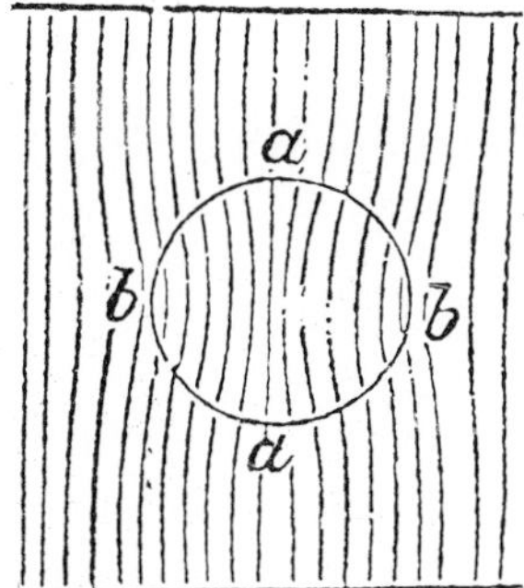


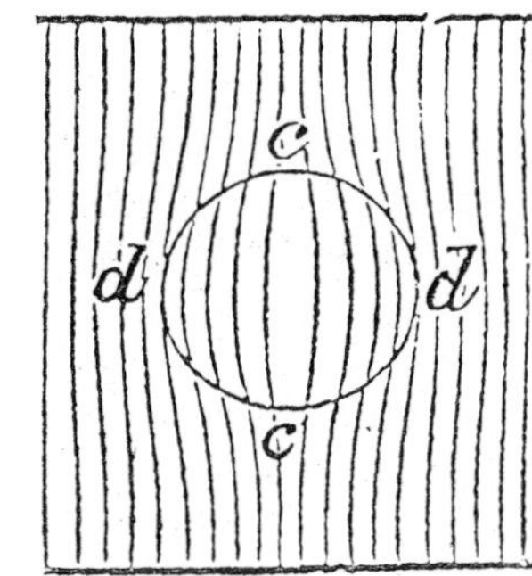


Fig. 8.12 - LINES OF FORCE were used by Faraday to diagram the action, in an external magnetic field, ot two types of materials, which he called "paramagnetic" and "diamagnetic." The former, left, accepted magnetism, and the latter, right, rejected the lines of external magnetism.

Fig. 8.13 - SPHERICAL CONDENSERS used by Faraday to compare "specific inductive capacity" of various materials, that is the comparative ability to accept electric charge. This property is now called the "dielectric constant" of the material. Illustration courtesy the Royal Institution.

of this phenomenon in a magnetic field. He suspended a bar of glass between the poles of the "Great Electromagnet," and found to his amazement that it turned at right angles to the poles. Whereas magnetic materials lined up with magnetism, the glass bar acted as if trying to avoid the magnetic field. Many other materials: wood, lead sulfur and bismuth acted similarly. These materials Faraday called *diamagnetic* because there was an analogy with dielectrics in the electric field. Materials which accepted magnetism Faraday called *paramagnetic.* Faraday, using his lines-of-force concept, indicated, Fig. 8.12, that diamagnetics rejected the lines, and paramagnetics accepted the lines of magnetic force.

**FARADAY** thus continually used the lines of force for theorizing and speculation. In a paper published in 1846: *Thoughts on Ray Vibrations,* he brilliantly foreshadowed the future of electromagnetics as being able to radiate out into space:

"The view I am so bold as to put forth considers radiation as a high species of vibrations in the lines of force which are known to connect particles and also masses of matter together . . . these vibrations take place along the lines of force."

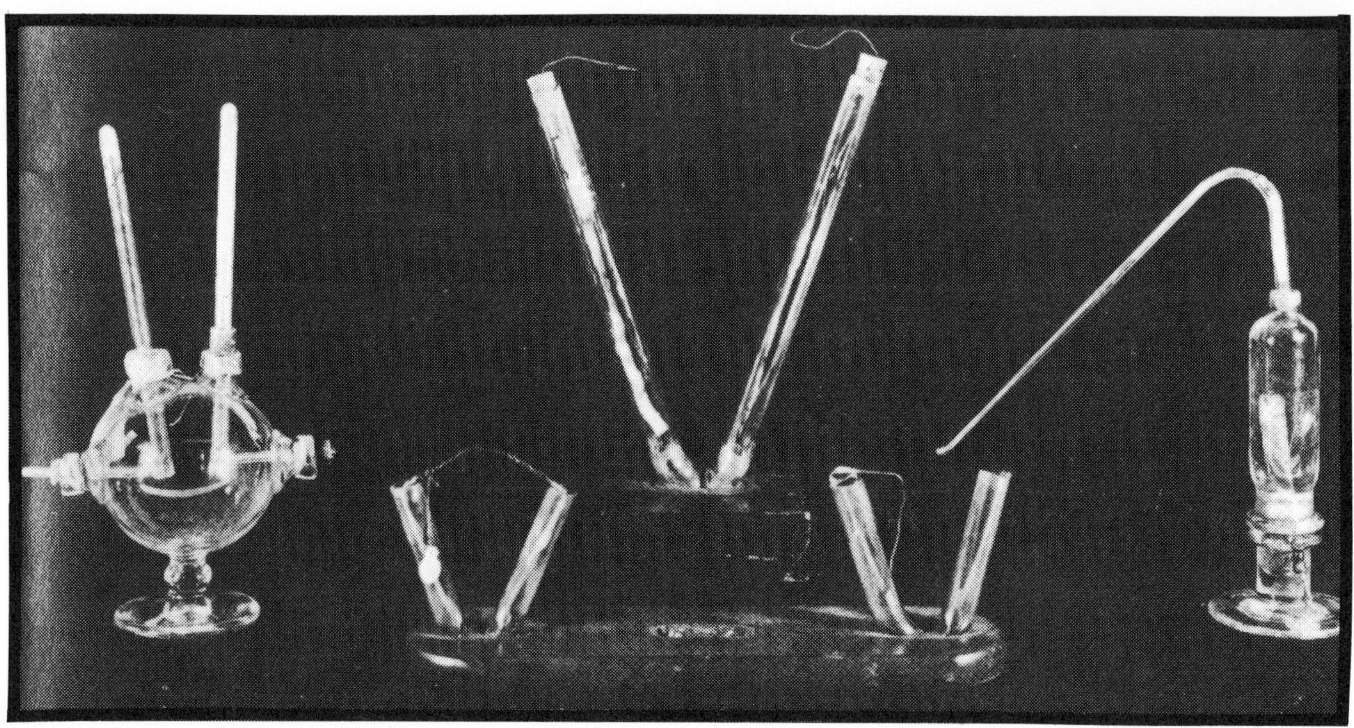

A few years later Maxwell had developed the Faraday vibratory radiation idea mathematically, and presented his electromagnetic theory of light, a revolutionary advance in physics.

**ELECTROCHEMICAL DECOMPOSITION** - An electric current, when sent through a metallic conductor, such as a wire, normally makes no obvious change in appearance or composition of the conductor. When the conductor is a liquid, however, the situation may be different, as was first discovered by William Nicholson and Sir Anthony Carlisle in 1800 in England. Putting the terminals of their voltaic pile into water they witnessed gases bubbling on the wires. The gases proved to be hydrogen from the negative terminal and oxygen from the positive. The water was being torn apart into its elements, revealing that electricity was able to exert a chemical force.

Excited experimenters soon found that electric current not only broke down water, but solutions of compounds of other elements also. It was the work of Davy, however, that provided the most dramatic demonstration of what electricity could do, as has been related previously. A whole new area of possibilities was opened up.

Fig. 8.14 - ELECTROCHEMICAL APPARATUS used by Faraday in his experiments in electrolysis. On the left is a "volta-electrometer" (voltammeter) in which the electric quantity was measured by the volume of gases evolved. Illustration courtesy the Royal Institution.

The mechanics of decomposition by electricity was not well understood, and brought various explanations. Some suggested the gases were coming in through the wires, some appealed to the "phlogiston" theory for the gas formation. Why did the gases bubble up at the wires and not at the center of the solution? How did the current migrate from pole to pole?

**SIR HUMPHRY DAVY**, at the Royal Institution, proposed that the force of the chemical affinity itself was electrical - that it was a property of matter. All matter was a product of its particles, being the union of the opposite charges. The particle charges were normally distributed so that they were neutralized. Under the electrical force, however, the positive and negative charges were separated. Between the poles the charges were rejoined and were neutral. But at the poles the electric current sorted out the charges, depositing negative charges on the positive pole, and positive charges on the negative pole.

It was Davy's brilliant work in electrochemistry that attracted Faraday's attention, and set him, as Davy's assistant, on the path of experimental science. Davy's theoretical framework of electrochemistry was passed on to Faraday, after Davy's death, and he launched the earliest systematic study of electrical conduction in the solutions of various compounds.

Faraday showed that in the solutions of such substances as copper sulfate and silver nitrate, the metal was being liberated at the positive pole and oxygen at the negative pole, when non-corroding electrodes such as platinum were used. Investigating the deposition of these metals quantitatively, with apparatus such as shown in Fig. 8.14, he found two fundamental facts of electrochemical deposition:

*First - The chemical power is in direct proportion to the absolute quantity of electricity which passes.*

*Second - The electrochemical equivalents coincide and are the same with ordinary chemical equivalents.*

"The equivalent weights of bodies," said Faraday, "are simply those quantities of them which contain equal quantities of electricity, or have naturally equal electric power, it being the electricity which determines the combining force. Or, if we adopt the atomic theory or phraseology, then atoms of bodies which are equivalent to each other in their ordinary chemical action have equal quantities of electricity naturally associated with them."

**FARADAY DETERMINED** that a fixed quantity of electricity liberated one gram equivalent of hydrogen, ½ gram equivalent of oxygen, one gram equivalent of silver and ½ gram equivalent of copper. Thus ½ gram equivalent of copper was equal to one gram equivalent of hydrogen. It is now known that the equivalent is the reciprocal of of the valence of the element in a given compound.

The amount of electricity found by Faraday to liberate one chemical equivalent of an element is fixed and equal to an amount now measured at approximately 96,500 coulombs.

This number of coulombs (equals amperes times seconds) deposits 107.88 grams of silver (atomic weight 107.88, valence = 1), 31.77 grams of copper (atomic weight 63.54, valence = 2), or 40.58 grams of antimony (atomic weight 121.76, valence = 3).

**TERMINOLOGY** - Faraday found it necessary to invent new words to describe and clarify the new effects with which he was working. To help in this he consulted his science-philosopher friend, Dr. William Whewell of Trinity College, Cambridge. The fluid conduction process they called *electrolysis,* and the conducting solution the *electrolyte.* The metal conductor was an *electrode.* The two components of the migrating particles in the solution were *ions* (from the Greek "wanderer"). That electrode through which the current entered the electrolyte was called *anode* and the conductor from which it left, the *cathode.* That electronegative constituent of the electrolyte migrating to the anode was the *anion* and the other moving toward, and appearing at the cathode was the *cation.*

Faraday chose his terminology so well that it is in universal use today. It has set a standard which has influenced all later scientific terminology.

Fig. 8.15 - LIBRARY at the Royal Institution as it appears in an 1809 painting. Courtesy the Science Museum, London

**FARADAY'S NAME** lives on in the effects he discovered. He did not formulate mathematically the laws resulting from his investigations. That was done by Maxwell and others.

**Faraday's Law of Induction** - A changing magnetic field produces an electric field.

Whenever a coil of $N$ turns is linked with a changing magnetic field, an electromotive force will be induced in it:

$$e = -N\frac{dB}{dt} \text{ volts, where:}$$

$N$ = number of turns
$B$ = magnetic flux in webers enclosed by the coil
$t$ = time in seconds

The negative sign indicates that the voltage acts to produce an opposing magnetic flux.

Or, a conductor moving perpendicularly to a constant magnetic field and to itself will have induced in it:

$$e = -Blv \text{ volts, where:}$$

$B$ = magnetic flux density in webers per square meter.
$l$ = effective length of conductor in meters
$v$ = velocity of conductor in meters/second

**Faraday's Law of Electrolysis** states that whenever an electric current flows between a metal and an electrolyte a chemical change occurs exactly proportional to the quantity of electricity which passes. The quantity of the change is proportional to the chemical equivalent weights of the substances which separate the electrodes.

96,500 coulombs of electricity will deposit 107.88 grams of silver, or 31.7 grams of copper. This unit is called the *faraday (F).*

**Faraday Effect** - or the optical rotation by magnetism - is the effect produced when a plane polarized light beam is passed through certain transparent materials in a strong magnetic field. The plane of polarization of the emerging beam of light, traveling in a direction parallel to the magnetic lines of force, is different from the original plane.

**Faraday Dark Space** - The relative non-luminous region in a glow-discharge cold cathode tube between the negative glow and positive column. It is associated with another region called "Crooke's dark space."

**Faraday Rotation** - The process of rotation of an electromagnetic wave in a magnetoionic medium, such as the ionosphere.

**FARADAY** spent most of his life in the laboratories of the Royal Institution carrying on his scientific work.After the death of Sir Humphry Davy in 1829, he became the chief lecturer, chief researcher, and scientific head of the Institution. Under his direction the tone and renown of the Royal Institution reached its highest point.

The record of his half-century of experimenting is a catalog of nearly all the areas of physical science under investigation at the time, including many which he initiated. Despite his preoccupation with other projects, he dominated the field of electricity and magnetism in the early decades of the nineteenth century. He developed the relation between electricity and magnetism, and the propagation of their effects. He pursued the themes of electricity, matter and light, and the unity of physical phenomena.

In addition, he found time, and he enjoyed, being a popular educator. He presented hundreds of lectures and demonstrations, and he was noted for his skillful and charming ability in explaining science to the lay public. His talks included many for juveniles, such as his celebrated *Chemical History of the Candle,* series given in 1849. He originated the popular Christmas lectures, which are carried on to this day. Faraday's lectures cover a period from 1825 to 1860.

**IN RECOGNITION** of service to the Empire, Faraday was provided with a house near Hampton Court by Queen Victoria in 1858. The great mind began to fade in 1865, and he died peacefully at home on 25 August 1867.

Faraday could have been a rich man, and while he was reasonably well off for the times, he was essentially indifferent to money, and his tastes were simple. As soon as he had acquired a basic understanding of the phenomenon he was studying he turned to new fields, leaving to others the problems of translating discoveries into useful devices.

**THE VASTNESS** of range of investigations, and the fundamental discoveries wrapped up in the life and accomplishments of one man, sometime defies belief. But Faraday, though fully appreciating the great value of his productivity, was by nature, and through Christian humility, a modest man. He refused the usual titles customarily bestowed by his country for outstanding achievement, and was content to remain plain Michael Faraday. But in his life he received countless medals and scrolls of honor from societies all over the world, including the Copely Medal of the Royal Society. He was held in awesome respect by all ranks of science.

Faraday published only one book: *Chemical Manipulation,* in 1827. But his output of papers published independently was stupendous, totaling: Electrical - 58, Magnetic - 17, Chemistry - 35, Physics - 14, Metallurgy - 26, and Miscellaneous - 21.

Faraday lived to witness the beginnings of the practical application of his discovery of electromagnetic power generation. The first engine-driven magneto-electrical machines were being installed for electroplating and for arc beacons in lighthouses, and Faraday served as consultant in the design and application of some of them.

In addition to the immortality achieved by his discoveries, two units honor his name: the *faraday* for contributions to the field of electrochemistry, and the *farad* for the electrical unit of capacitance. They were recommended by the First International Electrical Congress, meeting in Paris, in 1881.

A farad was defined as . . . *equal to one coulomb per volt;* that is, if one volt will produce a charge of one coulomb on a conductor, the capacitance is one farad.

**UNDERLYING** all of Faraday's presentations was the conviction that the study of science was a discipline that could help liberate the mind and be a factor in offsetting the ills that plagued society. He was one of the first to argue for science to be made a branch of education.

"I am persuaded that all persons may find in natural things an admirable school for self-instruction, and a field for the necessary mental exercise; that they may easily apply their habits of thought thus formed, to a social use; and that they ought to do this, as a duty to themselves and their generation."

Faraday summed up his scientific outlook in a few words:

*The philosopher should be a man willing to listen to every suggestion, but determined to judge for himself. He should not be biased by appearances, have no favorite hypotheses, be of no school, and in doctrine have no master. He should not be a respecter of persons, but of things. Truth should be his primary object. If to these qualities he add industry, he may indeed hope to walk within the veil of the temple of Nature.*

ROYALTY CAME TO HEAR HIM

Fig. 8.16 - FARADAY LECTURING at the Royal Institution. Attending this Christmas lecture in 1856 were Albert, the Prince Consort, and the Prince of Wales, later Edwaud VII. From a contemporary painting by courtesy of the Royal Institution.

FARADAY'S mood and busy life as scientist are well expressed by the letter below which appears in *Faraday Discloses Electro-Magnetic Induction* by Bern Dibner reproduced by permission from the Burndy Library. Fresh from his momentous discoveries in electricity and magnetism, Faraday launches plans for the great series of papers on his experimental researches in electricity, one of the most significant documents of scientific accomplishment. Despite Faraday's inherent modesty in his achievements, he allows himself a moments exultation at the triumph of his experimental method, versus the mathematicians, in the explanation of Arago's disk . . . it lay in the fundamental forces revealed in Faraday's law of electromagnetic induction

## THE LETTER TO PHILLIPS

*Brighton: November 29, 1831.*

*DEAR PHILLIPS, - For once in my life I am able to sit down and write to you without feeling that my time is so little that my letter must of necessity be a short one and accordingly I have taken an extra large sheet of paper intending to fill it with news and yet as to news I have none for I withdraw more and more from Society, and all I have to say is about myself.*

*But how are you getting on? are you comfortable? and how does Mrs. Phillips do; and the girls. Bad correspondent as I am, I think you owe me a letter and as in the course of half an hour you will be doubly in my debt pray write us and let us know all about you. Mrs. Faraday wishes me not to forget to put her kind remembrances to you and Mrs. Phillips in my letter.*

*Tomorrow is St. Andrew's day, but we shall be here until Thursday. I have made arrangements to be out of the Council and care little for the rest although I should as a matter of curiosity have liked to see the Duke in the Chair on such an occasion.*

*We are here to refresh. I have been working and writing a paper and that always knocks me up in health, but now I feel well again and able to pursue my subject and now I will tell you what it is about. The title will be, I think, EXPERIMENTAL RESEARCHES IN ELECTRICITY. I. On the induction of electrical currents. II. On the evolution of Electricity from magnetism. III. On a New electrical condition of matter. IV. On Arago's magnetic phenomena. There is a bill of fare for you - and what is more I hope it will not disappoint you. Now the pith of this I must give you very briefly; the demonstrations you shall have in the paper when printed.*

*I. When an electric current is passed through one of two parallel wires it causes at first a current in the same direction* through the other, but this induced current does not last a moment, notwithstanding the induced current (from the Voltaic battery) is continued all seems unchanged except that the principal current continues its course, but when the current is stopped then a return current occurs in the wire under induction of about the same intensity and momentary duration but in the opposite direction to that first found. Electricity in currents therefore exerts an inductive action like ordinary electricity but subject to peculiar laws; the effects are a current in the same direction when the induction is established; a reverse current when the induction ceases and a peculiar state in the interim. Common electricity probably does the same thing but as it is at present impossible to separate the beginning and the end of a spark or discharge from each other, all the effects are simultaneous and neutralize each other -*

*II. Then I found that magnets would induce just like voltaic currents and by bringing helices and wires and jackets up to the poles of magnets, electrical currents were produced in them these*

*This is a slip in the description; the momentary current induced in the secondary wire on making the current in the primary is *inverse:* it is succeeded by a momentary *direct* current when the primary current is stopped. (THOMPSON)

*currents being able to deflect the galvanometer, or to make, by means of the helix, magnetic needles, or in one case even to give a spark. Hence the evolution of electricity from magnetism. The currents were not permanent, they ceased the moment the wires ceased to approach the magnet because the new and apparently quiescent state was assumed just as in the case of the induction of currents. But when the magnet was removed, and its induction therefore ceased, the return currents happened as before. These two kinds of induction I have distinguished by the terms Volta-Electric and Magneto-electric induction. Their identity of action and results is, I think, a very powerful proof of the truth of M. Ampere's theory of magnetism.*

*III. The new electrical condition which intervenes by induction between the beginning and end of the inducing current gives rise to some very curious results. It explains why chemical action or other results of electricity have never been as yet obtained in trials with the magnet. In fact, the currents have no sensible duration. I believe it will explain perfectly to transference of elements between the poles of the pile in decomposition but this part of the subject I have reserved until the present experiments are completed and it is so analogous, in some of its effects to those of Ritter's secondary piles, De la Rive and Van Beck's peculiar properties of the poles of a voltaic pile, that I should not wonder if they all proved ultimately to depend on this state. This condition of matter I have dignified by the term Electrotonic, THE ELECTROTONIC STATE. What do you think of that? Am I not a bold man, ignorant as I am, to coin words but I have consulted the scholars,* and now for IV. The new state has enabled me to make out and explain all Arago's phenomena of the rotating magnet or copper plate, I believe, perfectly; but as great names are concerned Arago, Babbage, Herschell, etc., and as I have to differ from them, I have spoken with that modesty which you so well know you and I and John Frost** have in common, and for which the world so justly commends us. I am even half afraid to tell you what it is. You will thinking I am hoaxing you, or else in your compassion (you) may conclude I am deceiving myself. However, you need do neither, but had better laugh, as I did heartily when I found that it was neither attraction nor repulsion, but just one of my old rotations in a new form. I cannot explain to you all the actions, which are very curious; but in consequence of the electrotonic state being assumed and lost as the parts of the plate whirl under the pole, and in consequence of magneto-electric induction, currents of electricity are formed in the direction of the radii; continuing for very simple reasons, as long as the motion continues, but ceasing when that ceases. Hence the wonder is explained that the metal has powers on the magnet when moving, but not when at rest. Hence is also explained the effect which Arago observed, and which made him contradict Babbage and Herschell, and say the power was repulsive; but, as a whole, it is really tangential. It is quite comfortable to me to find that experiment need not quail before mathematics, but is quite competent to rival it in discovery; and I am amused to find that what the high mathematicians have announced as the essential condition to the rotation - namely, that time is required - has so little foundation, that if the time could by possibility be anticipated instead of being required - i.e. if the currents could be formed before the magnet came over the place instead of after - the effect would equally ensue. Adieu dear Phillips.*

*Excuse this egotistical letter*
*from*
*Yours Very Faithfully,*
*M. Faraday.*

*This doubtless refers to Whewell, of Cambridge, whom he was in the habit of consulting on questions of nomenclature. (THOMPSON)

**A man of fashion who had, without any claim to distinction, wormed himself into scientific society, posed as a savant, and had delivered a high-flown oration on botany at the Royal Institution. (THOMPSON)

# 50 years after FARADAY

**a half century elapsed between discovery of electro-mechanical energy conversion and development of efficient generators and motors . . . some steps in progress of construction of electrical machines during this fifty year period are illustrated . . . .**

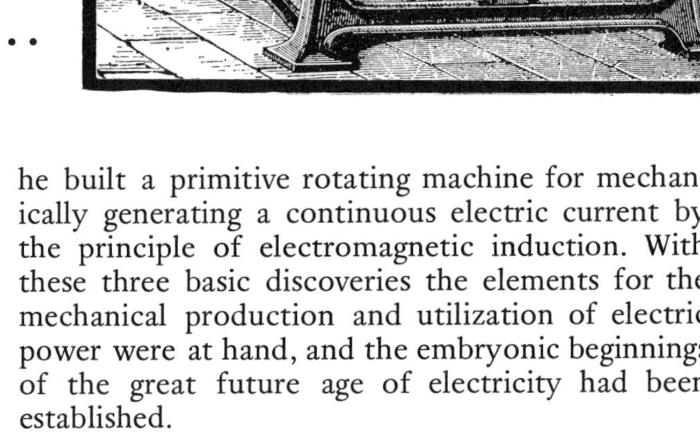

ELEVEN REMARKABLE YEARS, from 1820 to 1831, saw five great discoveries of basic principles that set the stage for the future age of electric power. In the first of these, Oersted, in Denmark, in 1820, demonstrated that magnetism was produced whenever an electric current flowed in a conductor. In that same year, in France, Ampère showed that a current in a coil of wire developed north and south magnetic poles. Then, as related in the previous chapter, Faraday, in England, in 1821, proved that the magnetism that circulated around the current in a conductor could produce rotation as a motor.

In 1831, Faraday performed two crucial experiments. He demonstrated the transfer of electricity from one coil to another by magnetic induction, which is the principle of the transformer. Also he he built a primitive rotating machine for mechanically generating a continuous electric current by the principle of electromagnetic induction. With these three basic discoveries the elements for the mechanical production and utilization of electric power were at hand, and the embryonic beginnings of the great future age of electricity had been established.

When the news of Faraday's discovery that a changing magnetism would cause electricity to flow in a conductor was made public, people wondered, and naturally they asked: "What is the use of it?" The oddity of producing a small current by moving a magnet near a length of wire was seemingly of little importance at that time. Faraday's answer to the question was significant. He is said to have replied: "What is the use of a new-born baby?"

Here, from the pages of the *American Journal of Science,* is how prospects for Faraday's "new-born baby" were viewed in 1837:

*SCIENCE has now placed in our hands a new power of great but unknown energy.*

*It does not evoke the winds from their caverns, nor give wings to water by the urgency of heat, nor drive to exhaustion the muscular power of animals; nor operate by complicated mechanism; nor accumulate the hydraulic force by damming vexed torrents; nor summon any other form of gravitating force, but, by the simplest means, through a power everywhere diffused through nature, but generally concealed from our senses, is mysteriously augmented, a thousand and a thousand fold until it breaks forth with an incredible energy, and there is no appreciable interval between its first evolution and its full maturity, and the infant starts up like a giant.*

*Nothing since the discovery of gravitation and of the structure of the celestial systems, is so wonderful as the power evolved by electricity; whether we contemplate it as the muscular convulsions of animals, the chemical decomposition, the solar brightness of the arc light, the dissipating consuming heat, and, more than all in the magnetic energy, which leaves far behind all previous artificial accumulations of this power, and reveals, as there is reason to believe, the grand secret of terrestrial magnetism itself.*

50 YEARS after Faraday's discovery of electromagnetic generation, electric power was transmitted 57 kilometers at 1400 volts, from Miesbach, Germany, to drive an electric motor operating a centrifugal pump for the artificial waterfall at the International Electrical Exhibition of 1882 at Munich. Illustration courtesy Deutches Museum, Munich

**NOW,** from our twentieth-century vantage point, we can look backward at the countless applications of electric power that rose out of the discoveries of the early 1800s. But, as history has shown repeatedly, discovery did not necessarily mean immediate utilization.

Many years would elapse before Faraday's "newborn baby" could come of age. Compared to the pace of technology today, the evolution of electricity in the 19th century, was a gradual and unhurried process. The electro-mechanical conversion of energy was a brand-new idea. Its implications were remote until the rise of industry provided the outlets for engineering ingenuity in transforming the electrical science into practicality. The growth of the electro-mechanical forces proceeded step by step with the developments of motive power and with the rise in use of electrical energy.

Lighting the darkness, and turning the wheels of industry, were the social and economic factors impelling the large-scale generation and utilization of electricity. But there was to be a gap of a half century between the early science and a resulting effective technology. It was not until the late nineteenth century that the spread and intensity of electrical usage prompted the production of really efficient generators and motors for general use.

**1831** Fig. 9.1 - MICHAEL FARADAY demonstrated the mechanical generation of electrcity. The copper disk produced a voltage when rotated in the field of the magnet. This was the first electromagnetic generator.

Many early experimenters, like Faraday, as soon as they got a basic understanding of the phenomena with which they were working, then turned to newer fields. They left to the future, and to others, the initiative and problems of resolving the discoveries into useful devices. So it was at first for the early tinkerers, and then for the inventors and engineers to develop an interest and the skills for converting the basic ideas into practical utility. And finally, the business enterprise and the commercial applications had to make possible the large-scale promotion and manufacture.

This chapter illustrates some of the early steps in the development of generators and motors, steps that were halting at first, then with a quickening pace, that followed in the fifty years after the first electromagnetic discoveries by Faraday.

**DEVELOPING THE GENERATOR** - In 1831 Faraday had demonstrated the production of electric currents from ordinary magnets. He had also given clues of how a generator could be built by stating that . . . "when wires move so as to cut a magnetic field, a power is called into action which tends to urge an electric current through them." Faraday's experiments showed that an iron core for the coil of wire was better than one of air, wood or other material, because it concentrated the magnetism for the most effective induction of electricity. He reported also that the induction was at a maximum when the magnetism cut directly across the conductors.

Despite these clues there was no immediate large-scale interest in applying Faraday's electromagnetic discoveries for producing electric current. Various types of batteries were the ready and accepted source of current, and for decades they supplied power for the first electrical industry, the telegraph. Until development of electroplating, and the arc light, there was no apparent need for large currents. Hence the earliest electromagnetic generators were hand-cranked, with small output, and mostly for experimental or medical use.

Faraday's rotating machine, Fig. 9.1, for continuous conversion of mechanical energy into electric

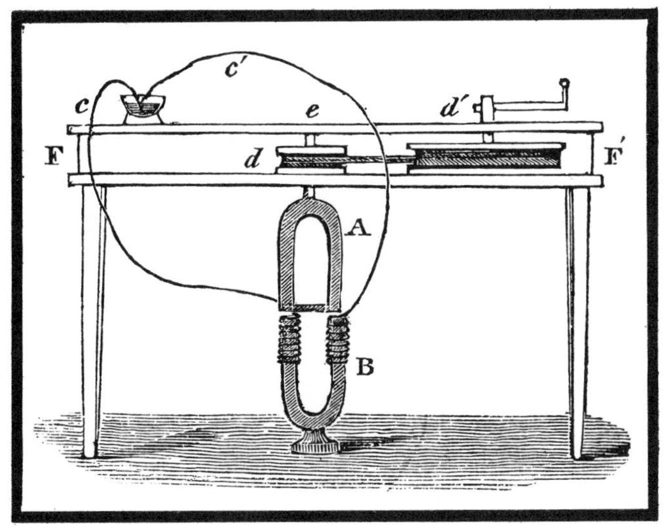


**1832** Fig. 9.2 - HIPPOLYTE PIXII, in France, devised the first alternator. A U-shaped magnet was rotated over a similar armature with wire-wound bobbins. Pixii's machine produced sparks and shocks, and could decompose water. For a second version Ampere suggested a commutator to rectify the alternating current to a kind of unidirectional current. Illustration courtesy the Smithsonian Institution.

energy was invented in 1831. Faraday called it a *magneto-electric* generator, thus distinguishing it from the electrostatic generator.

One year after Faraday's announcement of electromagnetic induction, Hippolyte Pixii, an instrument maker in Paris, made the first model of a generator based on Faraday's principles, Fig. 9.2. A horseshoe magnet mounted on a vertical axis could be rotated over the tips of a U-shaped iron armature with two wire-wound bobbins. When the horseshoe magnet was revolved, the alternating north and south poles passing over the bobbins, generated a current first in one direction then in the other. It thus produced an alternating current for which there was then no use.

For a second version of Pixii's machine, Ampère suggested a cam-operated switching arrangement that reversed the alternations into a kind of unidirectional current. This was the forerunner of the *commutator.* Rectifying provided a more usable current, similar to that of the voltaic cell.

Instead of rotating the magnet, it was of course possible to revolve the coils, and keep the magnet stationary. Also it was possible to use a horizontal arrangement. One of the first to employ this was Joseph Saxton, in 1834, in Washington, D.C. The armature on Saxton's machine, shaped like a cross, had four wire-wound bobbins, which were rotated at the ends of the magnets. This armature provided the option of two voltages, one for sparks and the other for shocks. These were obtained by winding two of the opposite bobbins with many turns of fine wire, and the other two with fewer turns of heavier wire. Saxton's machine was shown at an exhibition in London, in 1833.

Edward Clarke, in 1836, in England, invented a magneto-electric generator, Fig. 9.3, in which the armature bobbins revolved at the side rather than ends of the poles. The machines were used mainly for medical experiments, and lent themselves to quackery and exaggerated claims for bodily cures.

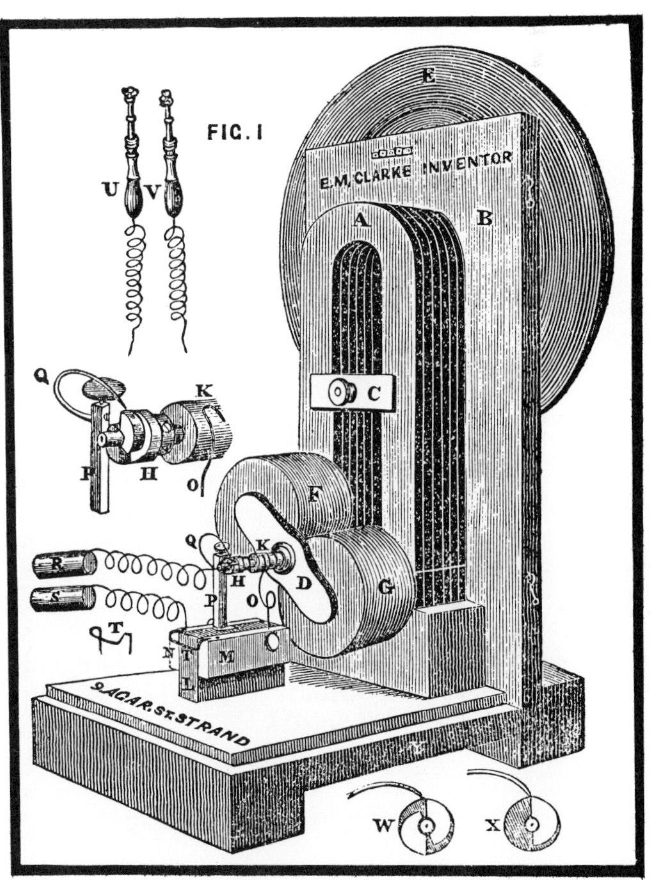


**1836** Fig. 9.3 - EDWARD CLARKE, in England, devised a magneto-electric generator in which he improved output by rotating the armature coils at the side of the magnet.

**1844** Fig. 9.4 - ELECTROPLATING was the pioneering application of magneto-electric generators for industrial use. John S. Woolrich, in Birmingham, England, made the first installations at the Elkington plating works in that city

**THE FIRST IMPETUS** for bigger and more powerful magneto-electric generators came with the development of electroplating. Batteries were the original source of power. Large-scale plating required heavy currents furnished by bulky and and expensive installations of the cells. John S. Woolrich, in Birmingham, England, saw the possibility of great improvement by use of generated power to replace the batteries. In 1844 he made the first application of a magneto generator at the Elkington plating works in Birmingham.

Woolrich's machines incorporated two new features which made them successful. One was radial mounting of the horseshoe magnets to give a multi-polar circular arrangement, an idea introduced by Stohrer in Leipzig. The other feature was segmented metal commutator invented by William Sturgeon, in England. Woolrich went on to increase the power of his machines by multiple magnets, more bobbins and better commutation, and continued building bigger generators, Fig. 9.4.

In the 1850s the carbon arc light had been developed, providing brilliant and popular illumination for large outdoor areas, and for special indoor effects. But the batteries were too bulky and costly for powering the lights, thus limiting their use This presented a new opportunity for electromagnetically generated electricity. By the mid-1850s magneto-electric machines, driven by steam engines were proposed for arc light operation. The first applications were for lighthouses where brilliance and range of the arc light promised to provide greater navigational safety over the existing oil-lamp illumination.

Frederick Holmes, and English engineer, was one of the first builders of generators for this purpose, and after several trials his machines proved successful. Faraday was a consultant in the building of these machines, and was seeing one of the first practical uses of his discovery of electromagnetic induction. Developers in France and Germany then also began producing units of similar design for conversion of their lighthouses from lamp to arc light operation. Alliance machines, made in France, as shown on the title page of this chapter, were widely used.

A typical early Holmes magneto-electric generator, Fig. 9.5, used a bank of three, five-foot diameter disks, each with 20 permanent magnets around the rim, rotating past an equal number of stationary bobbin-type coils. Weighing 2½ tons, and driven by a steam engine at 400 rpm, the generator absorbed 3.2 horsepower and developed about 1000 watts.

**IN THE 1860s** magneto-electric generators had reached the peak of their performance. They were bulky and expensive because of the limited magnetic flux available from the permanent magnets, and the inefficient use of the flux. The magneto-electric machines were being outmoded by new methods of supplying the magnetism for the generator.

The first of these developments was the substitution of electromagnets for permanent magnets to provide more compact and more powerful fields. This principle had been explored by such investigators as Dr. Charles G. Page (1812-1868), an examiner in the Patent Office in Washington, D.C., and an electrical experimenter. He found that by adding a current-carrying coil around a permanent magnet, the strength was greatly increased. Thus it appeared that generator output could be correspondingly increased by putting coils on the field magnets. The Siemens brothers, one in England and the other in Germany, and Charles Wheatstone and C.F. Varley in England, all proposed the use of

**1856** Fig. 9.5 - (right) ARC LIGHTS to replace oil lamps in lighthouses were an early use for magneto-electric machines. Units of this kind, driven by steam engines, were built by F.H.Holmes in England, with the consulting advice of Faraday, for the Dungeness beacons on the Dover Straits.

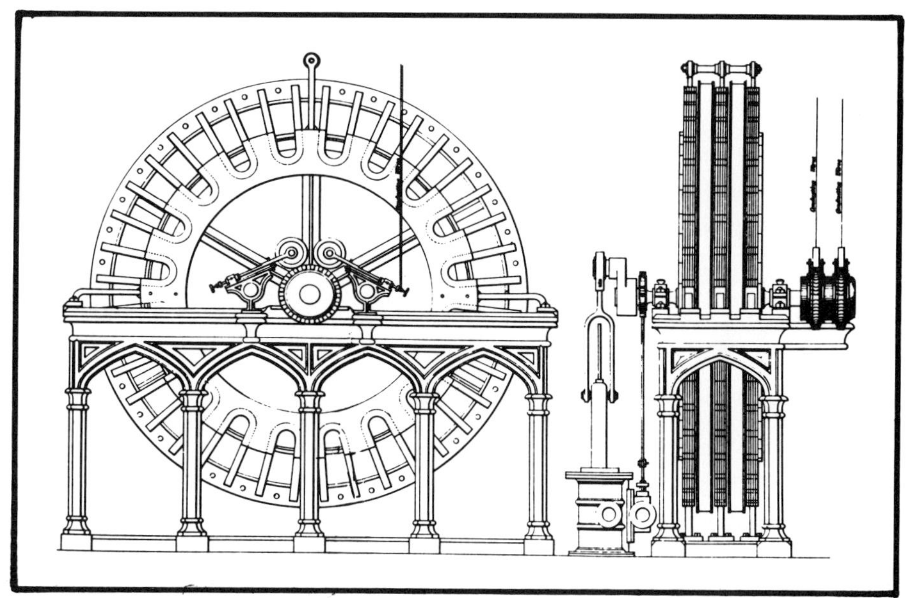

**1864** Fig. 9.6 - (below) ELECTROMAGNET field coils were substituted for the permanent magnets in the machine built by Henry Wilde in Manchester, England. It used a separate magneto, on top, driven from the generator shaft, to supply current for the coils. From W. James King, *The Early Arc Light and Generator,* United States National Museum Bulletin 228, The Smithsonian Institution, Washington, D.C.

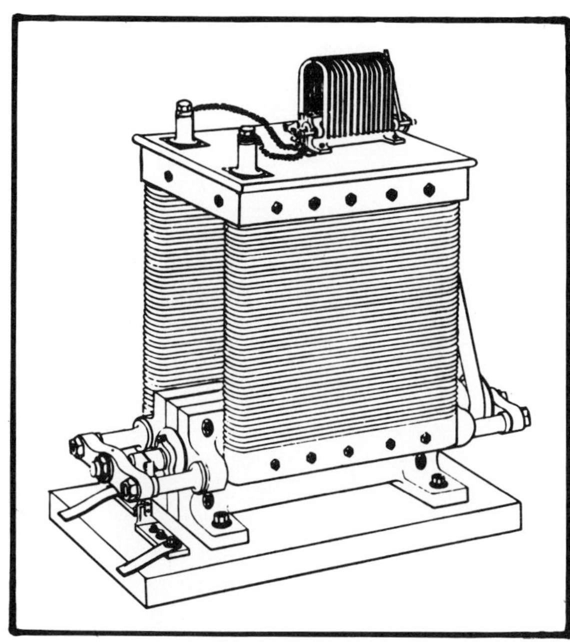

electromagnets in place of permanent magnets for greater efficiency and increased output.

Henry Wilde (1833-1919) in England was one of the first to use electromagnets for field excitation on a line of commercial machines. He patented, in 1864, a generator in which the field magnets were excited by a small auxiliary magneto belted to the machine, Fig. 9.6.

It was then soon recognized that the auxiliary magneto could be dispensed with by the principle of "self-excitation," which simply diverted some current from the armature of the machine to energize its field magnets. When shut down there was enough residual magnetism in the iron core of the field poles so that when restarted the generation would build up automatically without the need of a separate exciting magneto;

The principle of self-excitation was worked out independently, and at about the same time, by C.F. Varley and C. Wheatstone in England, and by E.W. von Siemens in Germany. British and European manufacturers adopted self-excitation almost simultaneously. The Siemens brothers, William and Werner, are generally credited with establishing the name *dynamo-electric,* later shortened to *dynamo,* for this type of direct-current machine.

**ARMATURE MODIFICATION** - After adoption of the self-excited electromagnetic fields for the dynamo, the next step was modification of the armature to improve performance of the machines. In essence this was to achieve what Faraday had called effective "cutting" of the magnetic field by the armature conductors. This was to be accomplished by the proper use of iron for the armature core, the disposition of the coils on the core so as to cut the field magnetism perpendicularly, and by reduction of the air-gap to a minimum.

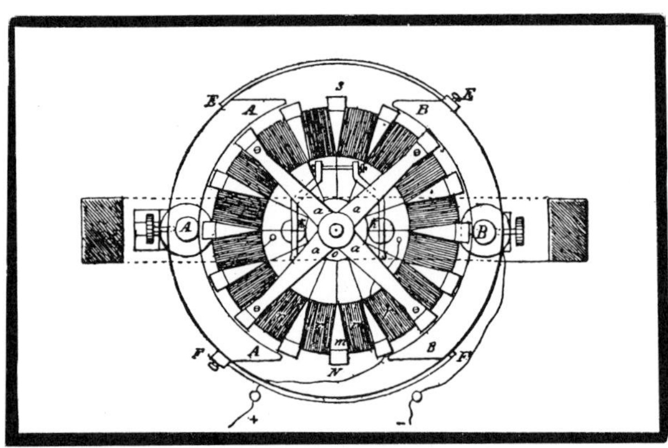

**1863** Fig. 9.7 - PACINOTTI'S GENERATOR was the first to use an iron ring with slots to receive the conductor coils.

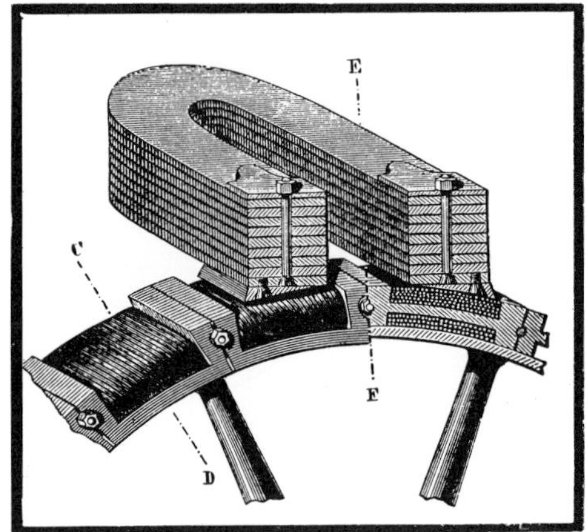

**1868** Fig. 9.8 - DE MERITEN'S (France) ring armature.

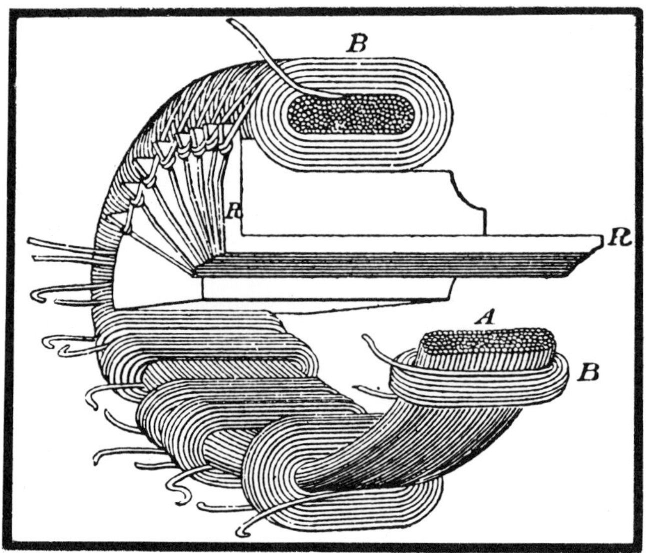

**THE FIRST NEW DESIGN**, departing from the bobbin-type coils, was the "ring" armature, described, in 1863, by Antonio Pacinotti (1844-1922) a young Italian physicist. The coils, Fig. 9.7, were wrapped around a slotted iron ring which rotated under the curved poles of the field magnets. This principle was picked up by De Meritens, in France, who used the ring construction of Fig. 9.8.

Ring armature machines came into widespread use in the hands of an engineer Zenobe T. Gramme (1826-1901), who realized their commercial possibilities, and who associated himself with aggressive business interests in France. Using an armature construction shown in Fig. 9.9, the Gramme machines became very successful. The production of Gramme dynamos for arc lights in factories, lighthouses and naval vessels, and for electroplating, became one of the first large electrical machinery manufacturing operations in Europe.

**1870** Fig. 9.9 - GRAMME ring-type armature (left), and Gramme ring-type armature dynamo (above) for arc lights.

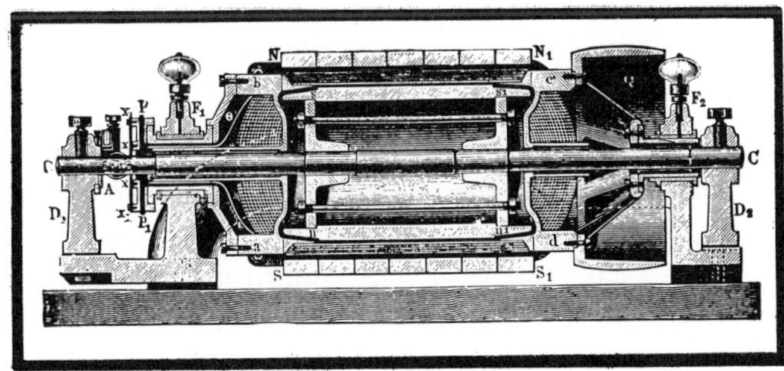

**1872** Fig. 9.10 - DRUM-WOUND ARMATURE used on the 1872 Siemens & Halske dynamo at the right, is shown in the cross-section above. The field coils are wound on iron bars, and connected to form consequent magnetic poles, thus giving a vertical magnetic field through the armature.

The Gramme armature was superior to those that preceded it, but it was still not fully effective, however, because only the outer part of the coils generated a useful voltage. The inner part produced a weak reverse voltage and added to the resistance of the coils. Designers were at work to overcome this deficiency.

**THE FINAL ADVANCE** came in 1872 with the invention of the "drum winding" by Friedrich von Hefner-Alteneck, who was a design engineer with the firm of Siemens & Halske in Germany. Instead of threading the wire around the inside of a ring, as in the Gramme construction, von Hefner-Alteneck used a drum-shaped armature core and wound the wire about the outside of the core, passing the wire directly across the end of the drum to the opposite side. This eliminated the unproductive inner part of the Gramme winding. In the drum winding the maximum length of the conductor was exposed perpendicularly to the magnetic flux of the field. The only non-productive parts of the drum winding were the end-turns. The drum winding is now standard on rotating-armature machines.

Succeeding design steps further improved the efficiency of drum-type machines. These included embedding the armature winding in slots, contouring the field magnets to the armature to maximize flux exposure, and making the air-gap between the armature and field as small as possible. With these accomplishments, the basics of the modern direct-current generator had been achieved.

Some idea of the advance in efficiency can be seen from a comparison made in 1876 between the Holmes magneto-electric machine, Fig. 9.5, and an 850 rpm drum-type dynamo, Fig. 9.10, with the same engine input. The size of the generator was reduced to 1/25th, the weight to 1/15th, and the cost to 1/10th. The power output was increased several times.

Early considerations of generator design soon showed that for the most effective results the magnetic field should be as strong as practical, the air-gap between the rotating armature and the stationary field poles, a minimum, the conductors on the armature should be large for minimum resistance, and speed of rotation should be high.

Early dynamos were built on a trial and error design process. By the late 1800s, however, empirical design was being superseded by calculations based on formulas of electrical and magnetic relationships. The electrical and magnetic properties of materials were being better understood. Increased size ratings and higher voltages speeded up refinements for better efficiency and reliable performance. From this point on, progress in balancing all the design and operational factors to produce the most practical and efficient dynamos would occupy the efforts and skills of many gifted minds to achieve the performance we have today.

**DEVELOPMENT OF MOTORS** - 20th century electric power for turning the wheels of machinery in the factory and appliances in the home, depends on the fact that the electric motor is the natural

**1821** Fig. 9.11 - FARADAY. at the Royal Institution in London, demonstrates the first electromagnetic motor. When the pivoted wire dipping in a pool of mercury is energized, its magnetic field interacts with that of the permanent magnet tangentially to make it whirl around the magnet. Model is at the Smithsonian Institution.

converse of the electric generator. Not long after the development of dynamos, this latent convertibility was discovered, probably accidentally. It was found that when two dynamos were connected to the same line, and the driving engine of one was cut off, it would continue to run as a motor, drawing its electricity from the driven dynamo. This led to some sporadic use of inverted dynamos for motors. However, general usage had to await the availability of suitable sources of electric power.

**FARADAY,** in a brilliant experiment in 1821, first demonstrated rotary motor action based on circularity of the magnetic field around a wire, Fig. 9.11. Rotary motors were not immediately developed, however. The first steps in motor design followed the mechanical tradition of the times. Just as the first automobiles were patterned after horse-drawn carriages, the early electric motors were reciprocating type, patterned after steam engines. The magnetic solenoid served for the cylinder, and an iron plunger for the piston.

Professor Joseph Henry, a pioneer American scientist, was the first to devise, in 1831, a machine for producing reciprocating motion by magnetic attraction and repulsion, that had the possibility of further mechanical development. The machine, described in the next chapter, contained the motor essentials: a magnetic field, an armature, and a current switching device for continuous action.

Henry's machine was quickly followed by other types of reciprocating motors using the attractive and repulsive forces of magnetism. Innumerable experiments and variations in design began to be carried out in America and Europe.

One of the prolific builders of early motors was Charles G. Page, M.D., of Washington, D.C. He was

**1838** Fig. 9.12- Dr. CHARLES PAGE of Washington, D.C., was a prolific electrical experimenter, and the inventor of the spark coil. He developed plunger-type motors of many kinds, and constructed an early railroad motor. This Patent Office model of a reciprocating electric motor is at the Smithsonian Institution.

the inventor of the spark coil, which has been wrongfully credited to Ruhmkorff, a Parisian instrument maker. Dr. Page built a multitude of motors of the plunger type, Fig. 9.12, some with an output of several horsepower.

The reciprocating motor eventually gave way to the motor with a revolving armature. Beginning in the mid-80s the principle of rotary commutation began to be better understood. Inventors then started to work on the more practical rotary motor in which the armature in the magnetic field was kept revolving by a commutator that switched the current to keep pulling the armature around. One of the earliest American schemes for rotary motion is the motor built by Thomas Davenport in 1834, shown in Fig. 9.13.

**THOMAS DAVENPORT** (1802-1851), of Brandon, Vermont, a blacksmith-inventor, is considered to be the first builder of rotary motors capable of doing useful work. In July 1834 he assembled a motor with a 7-inch rotor that revolved within semicircular permanent magnet fields. A patent was issued in 1837 for his motor design. In later models he substituted electromagnets for the field. He continued to turn out bigger and higher speed motors that were capable ot running small wood drills and lathes. Davenport was the promoter of one of the first commercial operations to exploit electric motors. His largest motor used an armature about two and a half feet in diameter, with which he was able to operate a printing press.

**1837** Fig. 9.13 - THOMAS DAVENPORT, of Brandon, Vermont, was granted a patent on a revolving-armature type motor for the propelling of machinery. Shown above in the Patent Office model at the Smithsonian Institution, the motor uses a permanent magnet field and electromagnet armature, with a commutator mounted on the motor shaft.

**1842** Fig. 9.14 - RAILWAY ELECTRIC MOTORS were intriguing to early experimenters. The 4-foot model below was made by Moses G. Farmer of Dover, New Hampshire. This Patent Office model is at the Smithsonian Institution.

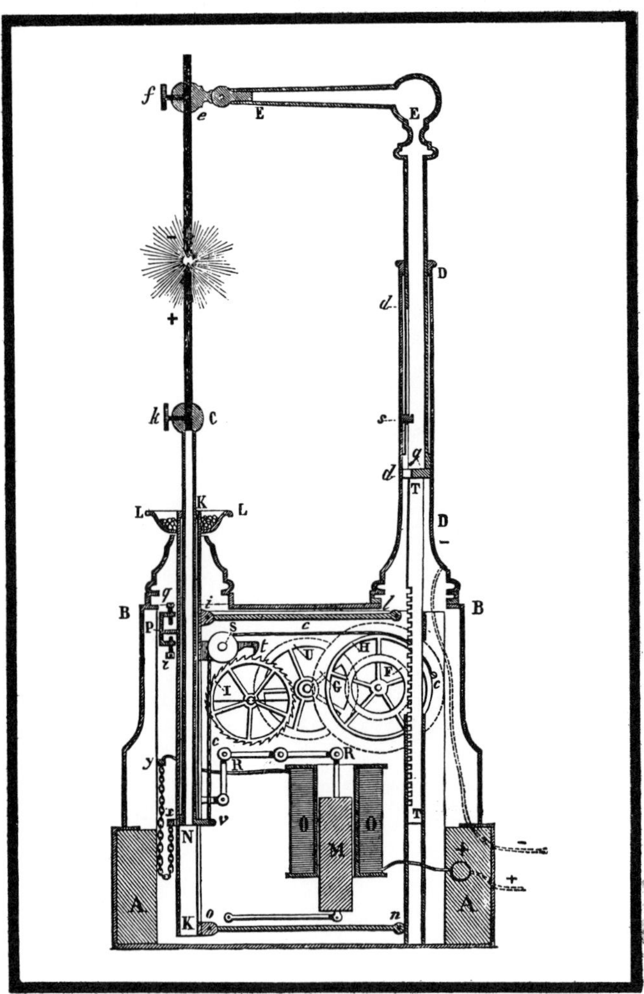


**1862** Fig. 9.15 - ARC LIGHTS created the first large-scale demand for electric power. To maintain proper separation of the carbons for a steady light, many regulating mechanisms were devised, such as this early version developed in France. The arc lamp regulator combined a clockwork with a solenoid. Illustration from the Smithsonian Institution.

Davenport and many other early electrical inventors were intrigued with the electric motor for driving trains. Many model locomotives, using both reciprocating and rotary motors, were built, such as that shown in Fig. 9.14. But the era of practical application for traction was still some decades in the future.

Early motors depended on batteries. The zinc consumed in battery operation made them too expensive for large, continuous power use. Motors thus seemed to have a dim future competing with cheap steam and waterpower. Dr. Page, reporting "On Electro-Magnetism as a Moving Power," in the *American Journal of Science* in 1839, said: "It is not to be presumed that in the present age, or perhaps ever, we shall arrive at a power from electro-magnetism which will supplant the steam engine in the grander operation."

On the other hand, steam and waterpower were practical only for large horsepowers, hence some inventors foresaw possibilities for electric motors in small units of power. However, any sanguine hopes for wider and larger application of motors had to wait for an adequate and cheaper source of electricity, and the means of transmitting it over substantial areas to the point of use. The future thus had to provide the development of suitable electromagnetically generated power, and facilities for adequate distribution. Impetus for expansion of electric motorization came with establishment of the central station system.

**ARC LIGHTING** was the earliest method of electric illumination, and in the last half of the 19th century attained widespread use and became a booming business. It created the first demand for electric power on a large scale, and was a powerful stimulus for developing large and efficient magneto electric machines and dynamos. Decades of talent were expended in devising various mechanisms for insuring constant separation and feeding of the carbons to obtain steady, long-lasting light. One type of arc-light mechanism is shown in Fig. 9.15. Manufacturers of arc lighting systems were established in Europe and America, using various lamp, circuit, regulation and generation methods. This grew into the formation of local electric power companies to provide arc-light service for factories, stores, and for street lighting.

But the arc light was too flaring, hot and current-consuming for any other use than lighting large, open areas. Most important, arc lighting did not lend itself to easy subdivision. There was a great need of an electric light that was subdividable into units of low intensity, long-lasting, free of combustion products, and suitable for use in any location. Two men rose to the challenge. Thomas A. Edison (1847-1931), in 1880 patented his incandescent lamp in the United States and in England. Later that year Sir Joseph Swan (1828-1914) obtained an English patent on a similar lamp. Within a few years the two inventors merged interests for the British market, calling the lamp "Ediswan," while Edison proceeded to apply his invention to the American and world markets.

Light by incandescence of a wire had intrigued experimenters ever since invention of the electric battery. All efforts failed for lack of a durable filament, and this included Edison's first attempts with platinum wires. The story of Edison's final success by using heat-treated, carbon-saturated thread sealed in a unique, highly evacuated, one-piece glass bulb, has been told and retold. Patent No. 223,898, issued January 27,1880, specified "a light-giving body of carbon wire offering great resistance to the passage of electric current, and at the same time presenting but slight surface from which radiation could take place . . . so as to allow practical subdivision of the electric light."

**AN UNIQUE ELECTRIC SYSTEM** - Edison had, however, not only invented an incandescent light,

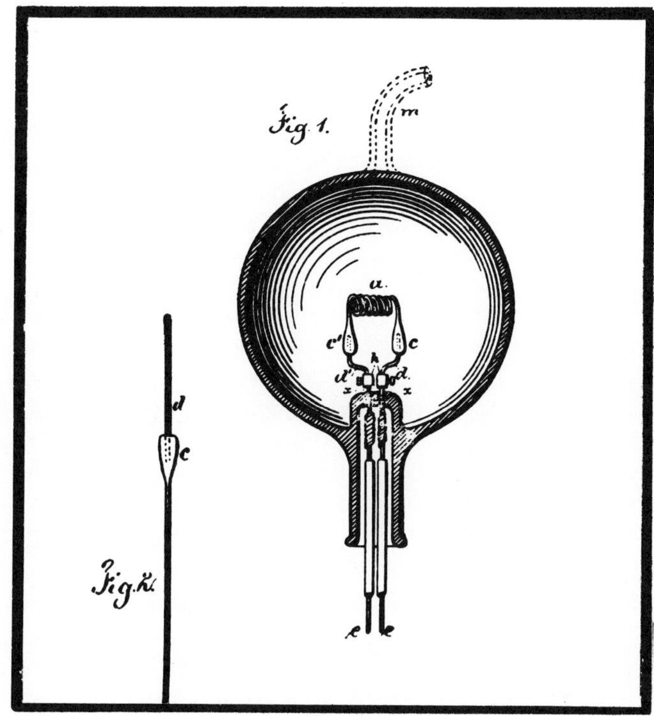

**1880** Fig. 9.16 - EDISON'S PATENT No. 223,898, for an incandescent lamp, was issued Jan. 27, 1880, featuring a carbon filament mounted in a highly evacuated glass bulb.

**1882** Fig. 9.17 - EDISON'S FAMOUS DYNAMO got the name "long-waisted Mary Ann" from the elongated shape of the field coils. In most installations the field coils were disposed horizontally. The dynamo was direct-connected to the horizontal steam engine to provide a compact unit.

but in addition he proceeded to develop a complete system of providing a suitable and adequate source of electric power for it. With his lamp and new power system he not only revolutionized lighting, but also the method of generating and distributing electricity. Power by wire, for lighting houses, factories and institutions, and for running machines in industry, became a practicality.

Edison, along with inventive genius, had also a keen economic insight. He envisioned a goal of providing a complete system of illumination to

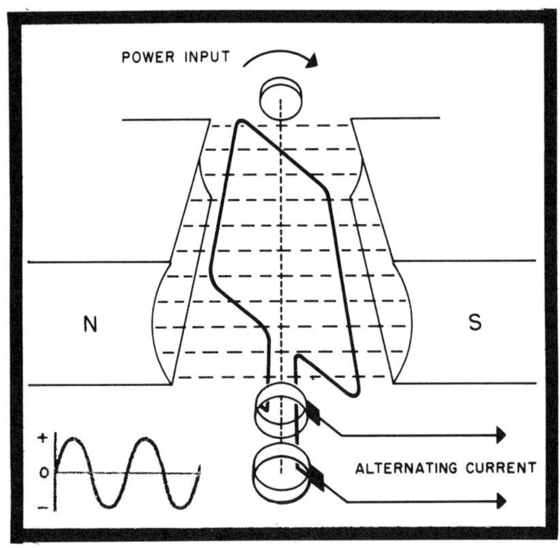


Fig. 9.18 - A DYNAMO is basically an alternator, left, in which by use of a commutator, right, the machine is converted to direct current. By replacing the commutator with collector-rings, left, the machine is reconverted to alternating current.

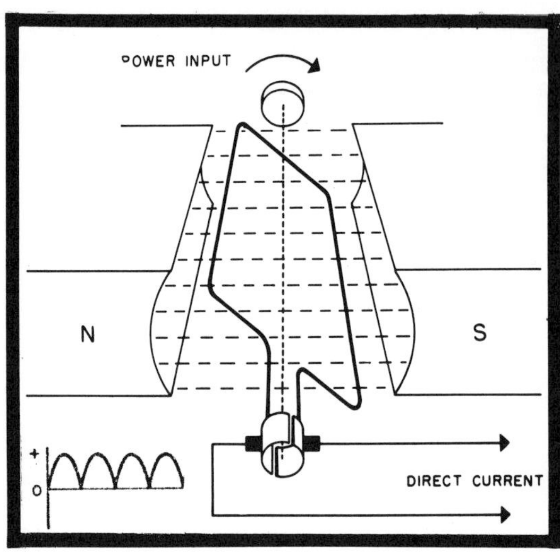


compete with the arc light and the gas light. He wanted to make electric light available, in small units, over a long distance and over a wide area, from electric power generated at a central source. Taking his cue of the public utility concept from that practiced by the gas illuminating companies of the time, he set out to establish a competitive electric utility ststem.

Edison pioneered the central power station as a nucleus of the power distribution system. His system called for large steam engine driven dynamos, centrally located, from which electricity was fanned out to as large an area of use as practical. To minimize the loss of voltage in the conductors, the power was first to be fed out through heavy mains, which then branched out into multiple feeder circuits, and finally into the individual load circuits, each successively using smaller conductors.

Engineering for the system proceeded from Edison's Menlo Park, N.J. laboratory. This was remarkable in representing the first use of planned, progressive research, the prototype of the modern research laboratories. Edison's research for his system developed suitable switches and protective fuses, and a meter for measuring the amount of electricity used. He also pioneered the use of underground conduit wiring.

**EDISON** envisioned supplying all of New York City with lights through a distribution network. His famous Pearl Street Station went into use in 1882, providing electricity for 16,000 eight candlepower lamps in the downtown area. Instead of stringing the 14 miles of wires, he put them into insulating material in iron pipes, and layed them underground. The station had four boilers rated 240 horsepower, each at 120 psi steam pressure, running high-speed 150 hp horizontal steam engines, coupled to 100 kilowatt dynamos delivering 110 volts. Edison adopted for his station the two-pole dynamo with extremely long poles, Fig. 9.17.

It was essential to maintain the proper voltage at the lamps to insure steady light. To accomplish this Edison devised a regulator which automatically actuated a variable resistor in the dynamo field circuit. It increased or decreased excitation with a change of load, to maintain the rated voltage.

Edison's light system gave impetus to improved dynamo designs which provided higher output per unit of weight, and higher operating efficiency. The popularity of the incandescent lights initiated a rapid increase in the size of the generating units, and in the capacity of the power stations.

While the early power plants were built primarily to meet the increasing demands for lighting, there was an even greater load growth potential - the electric motor. Inventors and engineers saw the possibilities for powering street railways, and the opportunities for electrifying industrial machines and mills. As a result, instead of following the early practice of merely inverting dynamos into electric drives, producers began to design motors as such, with the characteristics suited to the widening and diverse fields of application.

**ALTERNATING CURRENT** - Practically all of the electric power in the first three quarters of the 19th century was direct current, supplied by dyna-

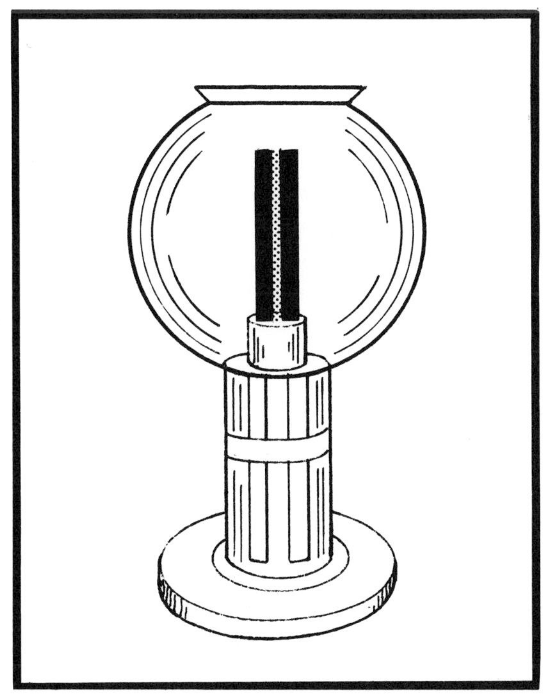

**1876** Fig. 9.19 - JABLOCHKOFF ARC CANDLE eliminated the strike and clearance mechanisms required for the ordinary arc light. The arc candle lasted about two hours.

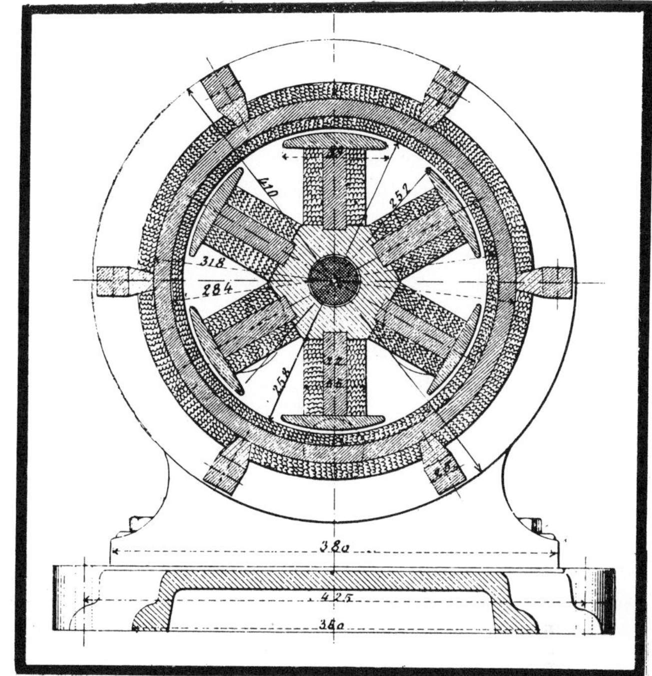


**1878** Fig. 9.20 - GRAMME ALTERNATOR for the arc candle light system. The alternator has a salient-pole field rotating in a ring-type armature, and used a separate exciter

mos and batteries. Latent in the dynamo, however, is a convertibility to alternating current, Fig. 9.18. The dynamo is basically an alternator in which the generated alternating current is converted by a commutator to a direct-current output. By replacing the commutator with collector-rings, the machine is reconverted to an alternating-current output. Events in the progress in electric power generation and utilization were shaping up to produce, by the end of the 19th century, a dominant position for alternating current.

Electric power for arc lighting could be either direct or alternating current, but up to the mid-1870s, the dynamo was the predominant power source, sometimes in the early days being supplemented by batteries. However, a new type of light, the Jablochkoff "arc candle," introduced in 1876, gave an immediate requirement for use of alternating current, and the development of alternators.

Invented by Paul Jablochkoff (1847-1896), a Russian engineer, the arc candle, Fig. 9.19, used long, parallel carbon rods, closely spaced, with an intervening refractory insulating material like kaolin. No mechanism was required for operation; once started the carbons continued to burn until consumed, lasting about two hours. Clusters of candles were arranged so that when one burned down another was automatically started up. The arc candle permitted small units of illumination with a minimum of maintenance expense.

Jablochkoff arc candles became popular immediately. They supplied the first street lighting by electricity on the Avenue de l'Opéra in Paris, creating great excitement. The Paris Universal Exhibition of 1878 was lighted with 68 candles, and scores were installed in London to light streets and bridges.

With direct current one carbon burned twice as fast as the other. Thus alternating current was essential for the arc candle. The Gramme company took the lead in producing alternators specifically designed for arc candle service. The candles were connected in parallel on the line, and operated at constant voltage. Gramme machines of the rotating field type shown in Fig. 9.20, were manufactured in sizes to handle 4, 16 or 32 candles.

For a time it appeared that the Jablochkoff system might be a solution to the problem of electric lighting, and in a few years of its dominance it initiated the use of large generators supplying electric power over distance through distribution lines. But the rise of the Jablochkoff lights came to a sudden halt with the introduction of incandescent illumination, which provided for a subdivision of lighting that was more favorable economically, electrically and environmentally.

Early incandescent light systems were powered by direct current. But as services expanded to meet growing loads in the late 19th century, the limitations of direct current began to be felt. Generated and distributed at low voltage, the large loads involved correspondingly high currents, carried by large and expensive conductors. Distribution costs thus limited the location of the d-c powered plants to the vicinity of the load. It was soon apparent, however, that where favorable for the most economical generation, it would desirable to locate the generating station distant from the area of use, such as at a waterpower site. But to transmit at a distance with minimum power loss would require a much higher transmission voltage than was practical from dynamos. The answer to this situation was provided by the introduction of alternating current where it was possible to transform the voltage to a high value for the transmission of electric power over long distance with a minimum line loss.

**A-C POWER,** transmitted at high voltage, and transformed back to low voltage at the point of utilization, was immediately superior where distance was involved.

But the use and expansion of alternating-current power had a major handicap - the lack of suitable a-c motors. This lack was overcome in 1888. That year marked a milestone in the future of alternating current, with the invention of the polyphase induction motor by Nikola Tesla (1856-1943), an American scientist born in Yugoslavia.

Tesla's several patents covered not only the revolutionary new motor, but a complete system of polyphase power, consisting of the alternator and the transformers in a variety of combinations. Tesla presented his system to the American Institute of Electrical Engineers in May 1888, in a lecture entitled: *A New System of Alternate-Current Motors and Transformers.*

The simplicity and versatility of the induction motor was such that within a decade it was taking over. It provided the most powerful incentive for promoting alternating current as the most practical and economical system of electricity supply. The availability of a suitable motor plus the advantages of high-voltage generation and transmission, gave voice to a growing number of engineers who now advocated that future large power systems should be alternating current.

Tesla's motor arrived at the height of a battle shaping up between those, like Edison, entrenched with direct current, and those promoting the greater merits of alternating current. Edison's low-voltage direct current was popular for local lighting and for small motors, but was very limited in the distance of transmission. Alternating current, on the other hand, could be generated in bulk, and distributed economically. And Tesla's motor completed the system superiority with an advantageous general-purpose drive. But Edison actively opposed the use of altlernating current and waged a vigorous anti a-c campaign based on the alleged dangers of high voltage to human safety.

As the 19th century closed it remained for two great projects to demonstrate dramatically that

**1888** Fig. 9.21 - TESLA'S INDUCTION MOTOR MODEL
It used the rotating field of multi-phase alternating current to induce torque in the closed circuit rotor winding. From the Nikola Tesla National Museum, Belgrade, Yugoslavia.

that alternating current was the system of the future. One was the 1889 Deptford Station in London, built to operate at an unheard of 10,000 volts and designed by a pioneer in a-c technology, S. Z. de Ferranti (1864-1930). The other great project was the 1893 Niagara Falls hydroelectric plant.

There had been great debate as to the best method of harnessing the 100,000 horsepower available at Niagara. There had been serious consideration given to use of mechanical means to transmit the power. Against the advice of authorities such as Edison, and Lord Kelvin of England, it was decided to generate alternating current using Tesla's system. In 1895 two 5000 kVA generators went into operation to send power to the first industrial customers, and in 1896 the first transmission to Buffalo, 22 miles away, was begun.

The success of Niagara was immediate and contagious, and the new system was adopted in other generating plants being constructed, or being changed over. Alternating current was the winner and would be the dominant form of electric power. But direct current would always retain its place where special advantages made its use desirable.

By the dawn of the 20th century, the steam turbine driven generator came into use, and the modern age of electric power had arrived.

---

*CAN WE IMAGINE the shades of Faraday being shown one of today's central power stations? He would doubtless be astonished. He would express amazement at the huge size of the spinning generators with their tremendous output of electricity. He would marvel at the vast strides in engineering and construction that could produce enormous power from such effortlessly running machines.*

*He would then finally smile in contemplation of how his "new-born baby" had grown to such prodigious size and utility. And he would probably recall fondly that this was surely a remarkable outcome of his demonstration, at the Royal Institution in the fall of 1831, that merely moving wires in a field of magnetism would create the electric current that now helped power the world!*

---

THE NIAGARA FALLS POWER SYSTEM went into operation in 1896, setting the stage for alternating current as the dominant system of electric power. The first three vertical a-c generators, driven through long shafts by waterwheels in the pits below. are shown above. The generators, rated 5000 horsepower, operating at 2350 volts, 25 cycles, were by far the largest machines built up to that time. Courtesy Niagara Mohawk Power Corp, Buffalo, New York.

# HENRY. . . .

## America's foremost electrical physicist of the early 19th century

## JOSEPH HENRY

## a pioneer in electromagnetics

### and the first Secretary of the Smithsonian Institution

**THE NEW LIFTING MAGNET**, shown on the opposite page, was ready for testing. It was, in 1831, the biggest that had been made, in America or in Europe, and it was about to demonstrate its capabilities. The magnet had been constructed at the request of Benjamin Silliman, Sr. for the physical laboratory at Yale College. Professor Silliman, who headed the department of natural philosophy and the laboratory, had been alerted to the unusual feats in constructing lifting magnets being achieved by Joseph Henry, an enterprising young instructor at Albany (New York) Academy. Henry was winding electromagnets, for use in his classroom, that had unheard of lifting strength, and Silliman wanted one of them for his laboratory. Professor Silliman was editor of *Silliman's Journal,* which would later become the *American Journal of Science.* He was an enthusiastic promoter for the advancement of science in America.

**EACH ERA** has its marvels that excite the mind of the scientist and layman alike. In the early 19th century a sensation of the day was electromagnetism. Discovered in 1820, by Oersted in Denmark, this strange new force, one of the most impressive ever put into the hands of man, was commanding an attentive interest. Every news of its operation and capability was greeted with fascination as to its scientific and practical implications.

The lifting magnet was one of the first vivid demonstrations of electromagnetism in action. The astonishing new attractive power could magically be commanded instantly into life by a coil of wire energized by a battery, and would as easily disappear instantly when the battery was switched off. In the hands of adventurous investigators like Joseph Henry, electromagnetism was beginning to show its potential for tremendous power, that in time would reshape the destiny of the world.

# he first showed that electricity could travel in space...

Henry had found the answer to a problem: How to wind a horseshoe magnet in a way to achieve the maximum lifting strength from a given battery? He had already assembled a "quantity" magnet with the ability to lift a weight hitherto considered as unbelievable - 750 pounds. Hearing of this Prof. Silliman asked Henry to make an even bigger one to give Yale the most powerful lifting magnet known. Henry proceeded unhesitatingly to build an enlarged design, and with the exception of forging the iron horseshoe, constructed the electromagnet with his own hands. It was delivered to the Yale laboratory for the final assembly in the summer of 1831.

Specifications for the big new magnet were detailed by Henry in a letter to Prof. Silliman: "The horseshoe was forged from a bar of Swede's iron three inches square and thirty inches long, flattened on the edges to form an octagonal prism. The perpendicular height of the exterior arch of the horseshoe - 11¾ inches. Around the outside from one pole to another - 29 ½ inches. Internal distance between the poles - 3 ½ inches. The armature or lifter is nearly 3 inches square, and 9 ½ inches long and weighs 23 lbs . . . a strong iron stirrup, surrounding the armature, fits into a groove in the center of the armature. The weight to be supported is fastened to the lower part of the

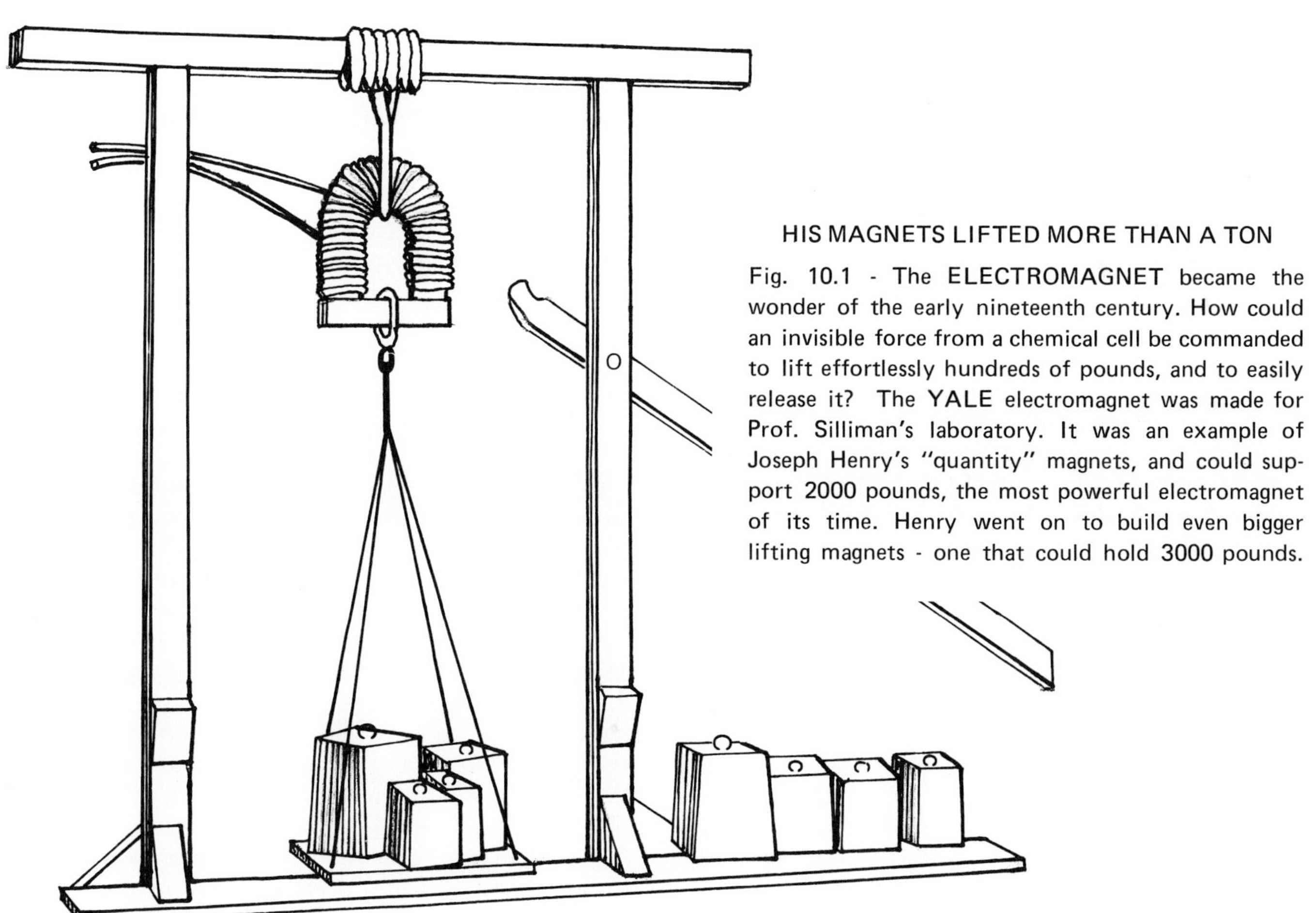

HIS MAGNETS LIFTED MORE THAN A TON

Fig. 10.1 - The ELECTROMAGNET became the wonder of the early nineteenth century. How could an invisible force from a chemical cell be commanded to lift effortlessly hundreds of pounds, and to easily release it? The YALE electromagnet was made for Prof. Silliman's laboratory. It was an example of Joseph Henry's "quantity" magnets, and could support 2000 pounds, the most powerful electromagnet of its time. Henry went on to build even bigger lifting magnets - one that could hold 3000 pounds.

stirrup, and by means of the groove is made to bear directly on the center of the armature.

"The magnet is wound with 26 strands of copper bell wire covered with cotton thread 31 feet long; about 18 inches of the ends are left projecting so that only 28 feet surround the iron; the aggregate length of the coils is therefore 728 feet. Each strand is wound on a little less than an inch; in the middle of the horseshoe it forms three thicknesses of wire; and on the ends or near the poles it is wound so as to form six thicknesses."

How much would the big magnet hold?

Henry first connected a battery containing 3/5 of a square foot of zinc surface, and the magnet could not support more than 500 lbs. He then tripled plate size and the response was more than tripled - 1600 lbs were sustained. Again the plate size was increased - to 4 7/9 square feet - and now the magnet readily supported over a ton! Henry fairly leaped from the floor in witnessing his success. He would later go on with bigger batteries lifting heavier weights. But already Henry had the honor of constructing an electromagnet by far the most powerful ever known.

**THE BIG MAGNET** was, at the time, an unprecedented accomplishment for an American scientist, and an auspicious beginning for a young man who had started his life in poverty. Joseph Henry was born on the 17th of December 1797, near Albany, New York, the son of a Presbyterian minister of Scottish descent, who had come to America at about the time of the Revolution. The limited family means allowed Joseph only a minimum opportunity for schooling. To help with the family income young Henry had to go to work early, first as a farm hand and later as a clerk in a store. But a quirk of fate was soon to widen his outlook and instill an ambition.

Henry had a pet rabbit which escaped one day. In trying to rescue it he accidentally came upon a wonderful world of books. Like Alice in Wonderland who found a new world of fantasy by following a rabbit down its hole, the young boy, in tunneling under a church to get at the rabbit, came up into a library room containing a shelf of romantic novels. He missed the rabbit, but fell to reading some of the books. The unfolding dramas revealed in them aroused a latent feeling in Henry for the wonder of playing a part in a possible life as an actor.

An opportunity came a year later when at age 14 he was sent to Albany to make a living and to help support the family as an apprentice to a watchmaker and silversmith. A stock company was playing in a local theatre and Henry was able to develop his talents by getting side line parts which he performed with an inborn competence. He also was a member of an amateur company and was urged by his associates to seek his fortune in the theatre as a full-time profession. Acting became Henry's passionate interest and he had practically decided on a life on the stage. Then came another chance reading that was to arouse a new enthusiasm and ambition that would change the entire course of his life career.

Confined to his room by a slight illness, he read a book that had been left by a visitor at his mother's house. The book: *Lectures in Experimental Philosophy, Astronomy and in Chemistry,* by George Gregory, an English minister, began by asking questions:

"You throw a stone, or shoot an arrow into the air, why does it stop at a certain distance and then return? On the contrary why does flame or smoke always mount upward though no force is used to send them in that direction? And why shoud not the flame of a candle drop to the floor when you reverse it? . . . again, you look into a clear well of water and see your own face and figure as if painted there. Why is this?"

**YOUNG HENRY'S MIND** was fired by the imagery of the questions, and intrigued by the answers. The effect of Gregory's book on Henry's future was noted in a memorandum on the flyleaf of the volume written when Henry was 46. "This book, although by no means a profound work, has under Providence exerted a remarkable influence on my life. It opened to me a new world of thought and enjoyment, fixed my mind on the study of nature, and caused me to resolve that I would immediately commence to devote my life to the acquisition of knowledge." Acquisition of knowledge was henceforth to become Henry's life passion. He had found a new world in science and had made up his mind to become a natural philosopher. The first step was the need for a scientific education.

Henry began by attending night school. Then, through the efforts of T. Romeyn Beck, the principal of Albany Academy, who became aware of the young man's promise and ambition, Henry was enrolled without charge at the Academy in 1819 to begin his college education. He studied there until 1822, earning his way by a year off to teach in a country school. Albany Academy offered courses in mathematics, chemistry and natural philosophy.

Henry was diligent and capable, and so impressed Beck that he was employed as an assistant in chemistry lectures in 1823 and 1824. Beck was instrumental in convincing Henry to pursue his scientific studies rather than following a dramatic career.

For the next four years Henry worked in a number of capacities to support himself while he was striving for his goal as a developing scientist. He served as a tutor in the prominent Van Rensselaer family in Albany and to Henry James the elder. He participated in a survey for a projected road from West Point to Lake Erie. The engineering on the survey introduced him to geology, meterology and terrestrial magnetism, three sciences which later would occupy much of Henry's activity. The survey also involved mathematics, at which Henry was proficient, and this sharpened his appreciation of its relationship to science. The new field was so appealing to Henry that he might have continued a career in civil engineering had he been successful in his bid for an appointment with the U.S. Army Topographical Engineers.

**HENRY'S SCHOLARLY ATTAINMENTS** and his field experience got him a teaching position, in 1826, at the Albany Academy. In 1828 he was made professor of mathematics and served in this capacity until he received a call from Princeton in 1832. In his years at Albany Academy, Henry made the decision to become an experimental physicist. This, at the time, was a rare profession in America, and his choice illustrated an aspect of Henry's individuality and willingness to pioneer. He was the first man in the country, after Benjamin Franklin, to undertake original investigations in science. In Albany he began the series of experiments in electromagnetism which were eventually to make him famous.

Albany provided a favorable atmosphere for fostering scientific interest and inquiry. Many perceptive leaders in the community were beginning to display a growing belief in the encouragement of American science for its practical utility. This led to formation of the Albany Institute for the discussion of natural history, and the promotion of useful arts. The Institute was originally housed in the Academy and it served as a forum for local gatherings on topics of scientific importance.

Henry participated enthusiastically in the activities of the Albany Institute. It became a second focus for his professional interest. He was made curator, and in 1829 became librarian. His contribution to the scientific community involved many demonstrations and presentations. Henry's training as an actor and a flair for the dramatic made him an effective speaker. He prepared his lecture notes with great care, and was a great believer in the use experimental display. His first public lecture was "On the Chemical and Mechanical Effects of Steam."

**DURING HIS YEARS** at Albany, although engaged in teaching, Henry was vigorously widening his scientific knowledge by self-instruction. He had a range of interests in the field of natural history that included astronomy, chemistry, geology, meterology and terrestrial magnetism. He was an avid and purposeful reader, and associated with those in the community who had an enlightened curiosity in the current topics of scientific interest. In his later years Henry credited much of his progress and success to the rigorous program of self-instruction he had imposed on his early teaching days, and he often asserted himself to be self-educated.

The Albany Academy and the Institute, had a substantial collection of books and periodicals, and as librarian Henry had access to the current scientific literature of Europe. As fledglings in their contributions to basic sciences, Americans looked abroad for news of scientific developments. By about age 30 Henry had become familiar with the work of Faraday, Ampère, Biot, Arago and other European investigators in electricity and magnetism. He had been attracted to this new and lively branch of science, an interest that sprang initially perhaps from his chemical work with galvanic batteries, and his preoccupation with terrestrial magnetism.

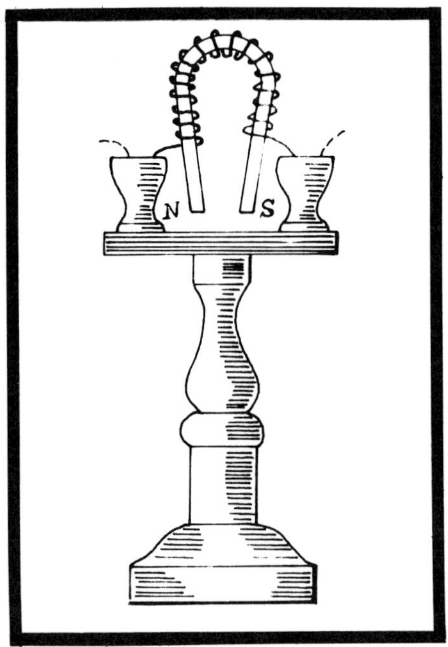


PROTOTYPE OF HENRY'S ELECTROMAGNETS

Fig. 10.2 - STURGEON'S ELECTROMAGNET was able to support 9 pounds. He was the first to use the horseshoe shape to get a more complete magnetic field. Sturgeon's original work in England led Joseph Henry into building lifting magnets eventually supporting a ton and a half.

Before he was thirty years old Henry had begun his electrical investigations - a field he considered "a most interesting branch of knowledge, presenting at this time a most fruitful field of discovery." Henry, pressed with a heavy teaching load, had limited time for experimenting, but he managed during summer vacations to do research in one of the classrooms which was temporarily converted into a laboratory.

Henry had an abiding interest in terrestrial magnetism. It had entered into his measurements for road surveys in New York State, where the calculations where checked with celestial observations, and with the deviations in compass readings that accompanied shifting of the earth's magnetic field. In some of his earliest experiments Henry was attempting to relate electromagnetism with terrestrial magnetism.

**IN 1828** Henry presented his first electrical paper in which he reported on an improved apparatus for illustrating the principles relating to the effect of the earth's magnetism on a freely moving coil of silk-covered wire. Henry's reference to the silk covering on the wire is an indication that he was perhaps the first to insulate a conductor by a compactly wound layer of thread.

In 1828, the year of Henry's paper, an event was taking place across the Atlantic that would eventually impact on Henry's future career. In that year another scientific investigator, James Smithson, an English chemist, had bequeathed his patrimony to "Found in Washington an Establishment for the Increase and Diffusion of Knowledge among Men." Smithson died in Genoa, Italy in 1829. On 10th August 1846, President James K. Polk signed a bill creating the Smithsonian Institution. A year later Henry was appointed its first Director.

**AT THE TIME** Henry entered the field of electricity, experiments in electrodynamics were multiplying rapidly. Numerous science workers across the Atlantic were contriving schemes and apparatus to account for the actions of electricity and magnetism. Arago, in France, had reported his observation that an electric current around a bar of iron made it magnetic. This was put into practical form by William Sturgeon (1785-1850), who was a self-taught scientist and experimenter in Woolwich, England. He built, in 1824, the first lifting magnet, Fig. 10.2, worthy of the name. It was capable of holding 9 pounds. He used a horseshoe of soft iron, insulated with varnish and wound with a single layer of bare wire, keeping the turns separate from each other. Sturgeon's invention made widespread scientific news and was a jumping off point for Henry's experiments in electromagnetics.

Henry learned of Sturgeon's work and that of G. Moll in Utrecht, Holland, who had built some larger magnets along Sturgeon's design. Henry became obsessed with the thought of improving and enlarging on the construction used by Sturgeon, and soon perceived a way of doing it. He proceeded to build larger and larger magnets, including the Yale magnet for Prof. Silliman. He was finally able to support the unheard of weight of a ton and a half. He achieved the electrodynamic force not only by use of insulated wire for large, compact coils, but in anticipation of Ohm's law, hit upon the idea of subdividing the coils and connecting them to maximize the current and hence maximize the magnetic flux.

Henry first found that he could lift more weight

by simply adding to the number of turns. This had been done by Sturgeon and Moll to increase the strength. But adding turns worked only to a point where he then had to add pairs of plates to his battery to keep up the current. More wire reduced the input, requiring more cells. But by using multiple coils connected in parallel he found he could get a superior result with a single battery of large plate area. By using the coils in parallel he was maximizing the current. He called this arrangement a "quantity" magnet.

Henry distinguished his electromagnets by two types. The "quantity" magnet with the parallel turns was one that responded best to a "quantity" battery of large plate area, hence large current output. Since the amount of magnetic flux was proportional to the current, this provided the maximum lift power. On the other hand an "intensity" magnet, using single coils with many turns, was necessary when the magnet had to be operated at the end of a long line. This then required a battery of multiple cells in series, that is, an "intensity" battery producing a high voltage.

Using a "quantity" electromagnet and a "quantity" battery of 132 square inches of zinc, Henry was able to support 3600 pounds. Henry's "quantity" and "intensity" magnet arrangements had anticipated the fact that the maximum magnetic force was obtained when the resistance of the magnet and that of the battery were matched.

Henry reported his results in 1831 in a paper in the American Journal of Science: *On the Application of the Principle of the Galvanic Multiplier to Electro-Magnetic Apparatus, and also to Development of Great Magnetic Power in Soft Iron with a Small Galvanic Element.*

It was not until 1837 that Henry became acquainted with Ohm's law, which would explain the relationships accounting for the operation of his "quantity" and "intensity" electromagnets. But his projections of the possibiities of these arrangements and their extension to his experiments and discoveries of the early 1830s were to establish Henry as a first-rank American scientist, and a pioneer in electrodynamics.

**AMPERE** had shown that the current in a given circuit was everywhere constant. It followed that there would be an accompanying magnetism which could be detected by a magnetic needle. By alternately turning the current on and off, he saw the possibility of sending signals magnetically. Ampère suggested an arrangement using a pair of wires and a needle to signal each letter.

Ampère's suggestion for a telegraph was taken up by various experimenters. Their schemes were not practical however because of the economics and complexity of multiple circuits and because operation was limited to short distances. The necessary relations between power source and receiver for long distance signalling were not yet understood.

It remained for Henry to provide a solution to the problem of sending an electromagnetic signal through thousands of feet of wire. It was accomplished by use of an "intensity" battery, using many pairs of plates in series for the power source at the sending end, and an "intensity" magnet of many turns of small wire at the receiving end. In effect he was matching resistances for the highest efficiency of transmission. To detect the signals he used a magnet with a clapper, Fig. 10.3, to strike a bell.

Henry's development of compact, insulated coil construction, which enabled many turns of wire in

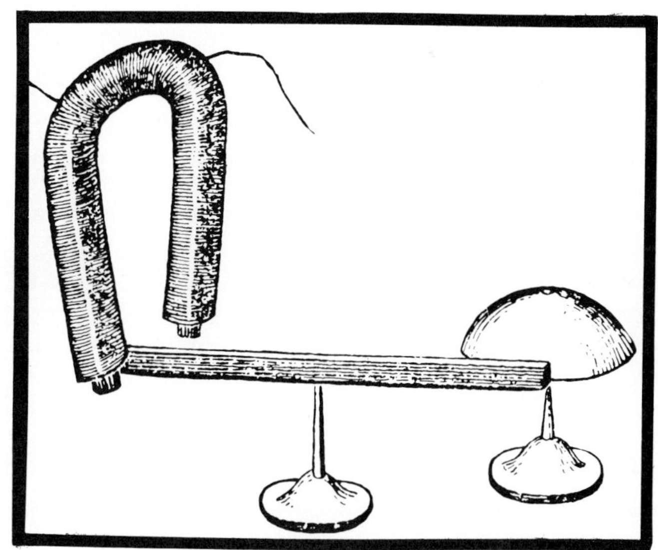

THE FIRST TELEGRAPH RELAY

Fig. 10.3 - BELL - TAPPING RELAY with which Henry received signals over a mile - an early telegraph line. Illustration from a report of the Smithsonian Institution - 1838.

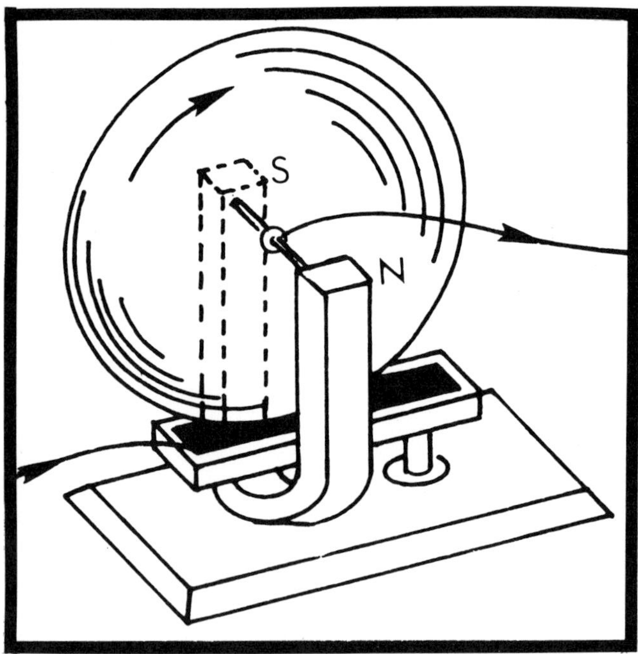


Fig. 10.4 - BARLOW'S WHEEL, built in 1822, consisted of a copper disk pivoted to rotate between the poles of a horseshoe magnet, and dipping into a pool of mercury. The battery connections between the mercury and the axle provided a radial current, the magnetic field of which reacted with that of the magnet to cause rotation. This was one of the first electromagnetic rotating devices.

a small space led to the concept of an electromagnetic relay for multiplying the signal. The relay was a horseshoe magnet wound with many turns of fine wire, and a movable armature which would be pulled against the magnet each time a signal arrived through the sending line. The armature had a contact so that when it moved up and down it could energize a second "quantity" magnet with its own battery to operate a printing mechanism, or to relay the signal on through another length of wire. In one experiment the sensitivity of the relay was such that Henry had no difficulty in sending signals over a mile of wire.

Henry knew that he had assembled all the elements of a successful telegraph system, but he did not make an invention of them. He was content that science had accomplished its purpose, and he left to others the extension of the scientific principles to an utilitarian end. This was to occur within a decade.

**SAMUEL F.B. MORSE** (1791-1872). educated as an artist, became intensely interested in telegraphy. Using Henry's principles he was awarded a patent in 1840 for the first workable electromagnetic telegraph. Morse used Henry's relays, the opening and closing of which he coded into a system of dots and dashes (the "Morse" code) to indicate the letters of the alphabet, numerals and symbols.

During the conception of his arrangement he had sought Henry's advice, which was freely given. In 1846 there was to be a conflict between the two on the question of priority for originating the highly successful system. Henry was anxious to have his place in history of telegraphy recognized and preserved. The dispute highlighted the philosophy held by Henry that the scientific originality by "men of mind" had, for a given concept, a superior role to the technology that resulted by "men of action." In response Morse could point out that his work, too, had been recognized by awards from many scientific organizations.

**AS SOON AS OERSTED** had announced his discovery that electric current is accompanied by magnetism, and that the magnetism encircled the current, Faraday looked for a means of using the circular magnetic forces to provide a continuous rotation. His success in assemblying the apparatus for producing the first electromagnetic motor by revolving a wire around a magnet and a magnet around a wire has been previously described.

Faraday was followed by Peter Barlow (1776-1862), a physicist at Woolwich Academy, England, who in his *Essay on Magnetic Attractions* told of his development of an "Electromagnetic Wheel," Fig. 10.4, which was the converse of Faraday's rotating copper disk electrical generator.

Contrasted to these novel but impractical devices the electric motor constructed by Henry is credited as being the first which comprised the basics that could be developed into a workable machine. In 1831, in an article in the American Journal of Science: *On a Reciprocating Motion Produced by Magnetic Attraction and Repulsion,* Henry told of his device:

"I have recently succeeded in producing motion in a little machine by a power, which, I believe, has never before been applied in mechanics - by means of magnetic attraction and repulsion.

"Not much importance, however, is attached to

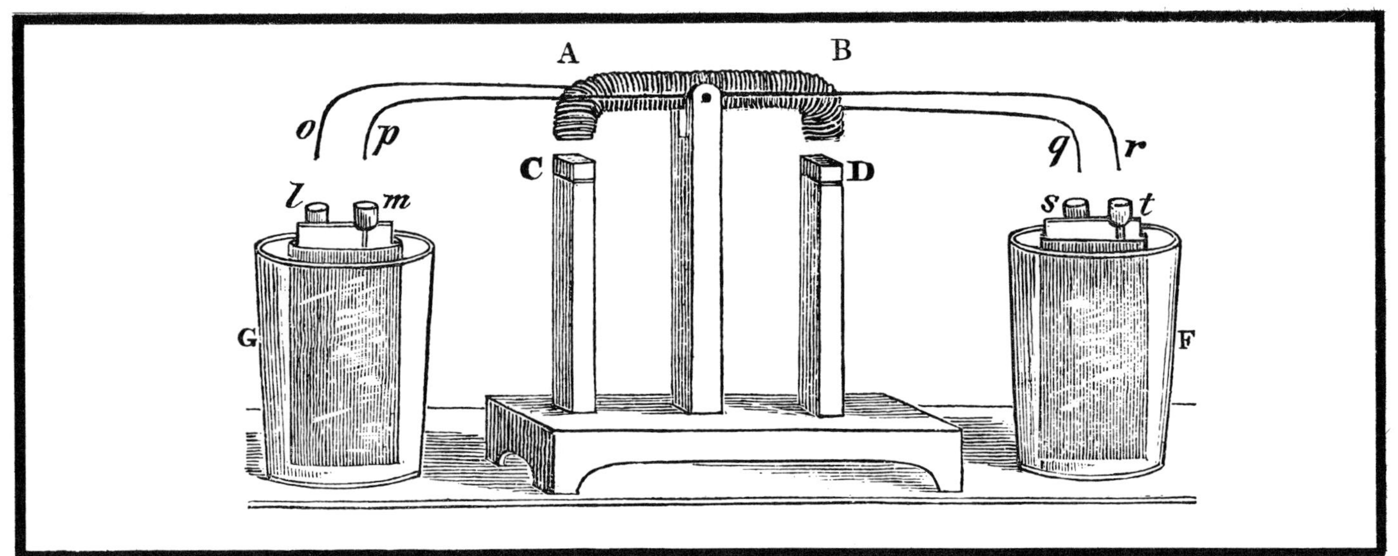


Fig. 10.5 - HENRY'S ELECTROMAGNETIC MOTOR - It had the first essentials of a workable machine. It consisted of a rocking armature, A,B, steel field poles, C,D, and commutator terminals, o,p, and q,r, which dipped into terminals l,m, and s,t, of the batteries, G and F. From the American Journal of Science, Volume 20, 1831.

the invention, because the article, in its state can only be considered as a philosophical toy, although in the progress of discovery and invention, it is not impossible that the same principle, or some modification of it on a more extended scale, may hereafter be applied to some useful purpose. But without reference to its practical utility, and only viewed as a new effect produced by one of the most mysterious agents of nature, you will not, perhaps, think the following account of it unworthy of a place in the Journal of Science.

"It is well known that an attractive or repulsive force is exerted between two magnets, according as poles of different names, or poles of the same, are presented to each other.

"In order to understand how this principle can be applied, to produce a reciprocating motion . . ." Then Henry went on to describe the operation of his electromagnetic motor, Fig. 10.5, based on a rocking bar armature which was alternately attracted and repulsed by vertical magnets.

Although Henry described his motor as a "philosophical toy," it had the essentials that were capable of development into a direct-current motor: a field magnet, an electromagnet for an armature, and a commutator to reverse polarity to provide continuous motion.

Henry concluded: "My friend Prof Green of Philadelphia, to whom I first exhibited this machine in motion, recommended substitution of galvanic magnets for the two perpendicular steel ones. If an article of this kind was to be constructed on a large scale, this would undoubtedly be a better plan, but for a small apparatus, intended merely to exhibit the motion, the plan here described is perhaps the most convenient."

**HENRY HAD FIRST SHOWN** how to turn electrodynamic forces on and off with a commutator. He had assembled the elements that could ultimately be made into devices that would drive machinery. But self-identification as primarily a seeker in science constrained Henry to presenting his apparatus only as a demonstration of basic principles. He, like Faraday, envisioned himself as a discoverer of scientific knowledge rather than one who applied it. He was skeptical of the prospects of his motor for practical use, and expressed a low opinion of those who wasted time and squandered money trying to make electric motors an effective drive. He was impelled in this belief largely by the limitations of cost that would be imposed by batteries as a power source, and the greater effectiveness of the steam engines then coming into use.

Although Henry was unimpressed with his motor except as a scientific toy, his expressed sentiments could not dampen the progress of technology. His "philosophical toy" soon attracted the attention of other experimenters who saw possibilities of further development of the rocking-arm prototype. In a few years they had channeled inventive ability into producing workable plunger type and rotating type motors. Empirically-minded investigators saw the possibilities of electromotive power despite the existing disadvantages that so bothered Henry, and they wanted to give motors a practical trial.

**BY 1838** Thomas Davenport of Brandon, Vermont, had received U.S. Patent No. 132 on his rotary electric motor which had such commercial possibilities that 3000 shares in a joint stock company, organized in New York, were offered to:

" . . . provide an opportunity for public-spirited individuals to become associated in an enterprise which it is hoped, for the benefit of mankind, may prove successful. Surely there are not wanting men, and we trust they are numerous, who will cheerfully pay, and if necessary, cheerfully lose. comparatively small sums, whose considerable aggregate will carry forward this interesting research, until the ratio and extent of its powers are ascertained; and if it should then prove that the limit is far beyond the demands of practical application, so much the better, but neither the ratio nor the extent can be learned without persevering experiments, the expense of making which and of sustaining all who are concerned in making them, will be, we trust, cheerfully borne by the public."

Practical motorization, as time would show, had only to wait for the development of the dynamo.

**HIS WORK ON ELECTROMAGNETICS** led Henry into investigation of electromagnetic relationships, and to the great problem of the day, the possibility of generating electricity from magnetic fields. Since electric current produces magnetism, it was logical, therefore, to suspect the converse. Previous experimenters spent their efforts wrapping wire around around a magnetized bar and looking for an evidence of electricity when the free ends of the wire were touched together. They were misled by the fact that since a steady current produced a steady magnetism, the converse should be true. Henry perceived that the answer lay in a changing magnetic field. As Henry put it: "An instantaneous current in one or the other direction accompanies any change in the magnetic intensity of the iron."

In the crucial experiment that demonstrated his hypothesis, Henry used one of his lifting magnets, Fig. 10.6, which he equipped with an armature wound at its central portion with 30 feet of insulated wire. The armature was placed across the poles of the magnet and the terminals of the armature winding were connected to the terminals of a galvanometer about 40 feet from the magnet.

"When this arrangement was completed," Henry reported, "I stationed myself near the galvanometer and directed an assistant to attach the galvanic battery to the magnet. At this instant the north end of the galvanometer needle was deflected 30 degrees to the west, indicating a current of electricity from the coil surrounding the armature.The effect. however, appeared only as a single impulse, for the needle, after a few oscillations, resumed its former undisturbed position, even though the magnet power was still continued.

"I was, however, much surprised to see the needle suddenly deflected from the state of rest to about 20 degrees to the east, or in a contrary direction, when the battery was disconnected. Motion of the needle appeared to take place only in consequence of the instantaneous development of the magnetic action in one direction, and the sudden cessation in the other.

"This experiment," Henry concluded, "illustrates most strikingly the reciprocal action of the two principles of electricity and magnetism, if indeed, it does not establish their absolute identity."

Henry had singled out the key element leading to an understanding of the dynamic relationship of electricity and magnetism. He had made one ot the world's momentous discoveries - the electromagnetic induction of electricity by magnetism in motion. But alas! Henry was not to have priority on his discovery. While Henry's work, described above, was done in 1831, he failed to publish his findings, because, he explained later, of the press of other work, and a desire to accumulate more data before presenting his results.

In May 1832 Henry read in a British scientific journal Michael Faraday's report of his discovery of electromagnetic induction. Even though he was preceeded by Henry, Faraday's independent and comparable experiments at the Royal Institution, and his superior facility for immediate publication, had established world recognition for Faraday. Henry has lost the opportunity for world distinction of American science, and he was roundly criticized by some of his contemporaries. His despair was profound, but he was eventually persuaded to report on his experiments in the American Journal of Science in a paper: *On the Production of Currents and Sparks of Electricity from Magnetism.*

**BUT WHILE FARADAY** had achieved world credit for electromagnetic induction, Henry clearly had priority on another discovery which had also been made independently by Faraday. In his paper Henry had described experiments showing another phenomenon of electromagnetic reaction: "the sparks of electricity."

Henry reported that . . . "when a short wire is connected across a battery, no spark is perceived when the connection is formed or broken. But if a wire 30 or 40 feet long is used, instead of the short wire, although no spark will be perceptible when the connection is made, yet when it is broken a vivid spark is produced . . . The effect appears somewhat increased by coiling the wire into a helix; it also seems to depend in some measure on the length and thickness of the wire; I can account for these phenomena only by supposing the long wire to become charged with electricity which by a reaction on itself projects when the connection is broken."

Henry had discovered *self-induction.*

He correctly interpreted the spark at the point of interruption as due to the sudden collapse of the magnetic field around the conductor itself; the collapsing field generated reverse voltage sufficient to make the spark. This ever-present property of all electric circuits is called *inductance.*

The amount of inductance, as Henry had noted, was a function of the length of the wire, and was enhanced by coiling. In practice the inductive property is dependent on the contour, the number of turns, and on the kind of material the wire embraces. The effect is also dependent on the speed with which the current changes. Inductance is a factor in operation of all electromagnetic devices.

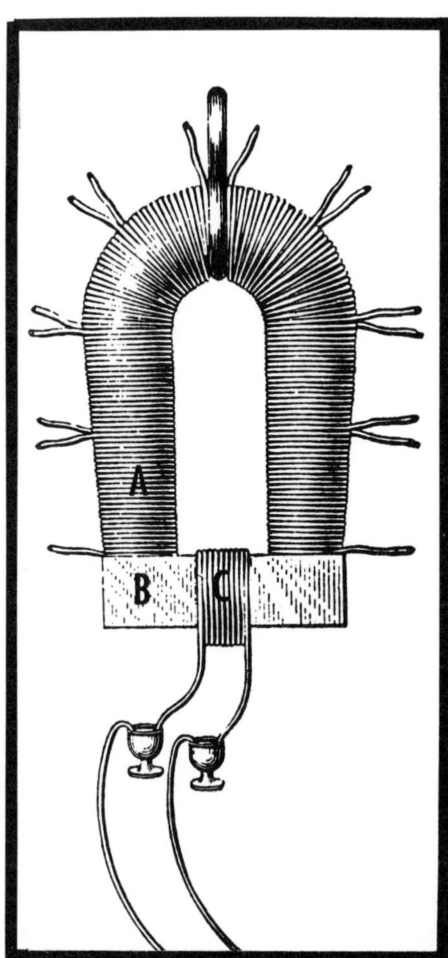


HENRY GENERATES CURRENT MAGNETICALLY

Fig. 10.6 - Henry, in 1831, demonstrated that electricity is produced whenever there is changing magnetism in a coil of wire. Whenever the horseshoe magnet, A, is energized, its magnetic flux sweeps through the armature, B, inducing a current in coil, C, swinging the needle of the meter below momentarily. When magnet, A, was switched off, the meter needle swung the other way. The changing of magnetism was inducing the current.

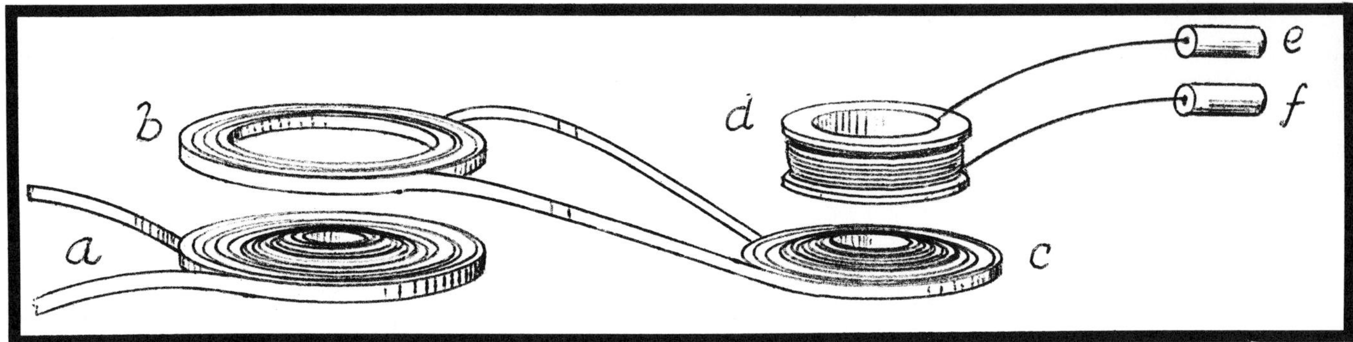


Fig. 10.7 - HENRY demonstrated the principle of the electrical transformer, inductively stepping the voltage up or down depending on the ratio of turns of the several coils.

In May, 1830 Henry married his cousin, Harriet L. Alexander of Schenectady, New York. His talents and interests had now outgrown the environment and the facilities for research at Albany. He wanted relief from a heavy teaching load, and more freedom as a man of science. Henry was thus receptive when Professor John Maclean at the College of New Jersey, now Princeton University, offered him the chair in Natural Philosophy at that institution. Henry accepted the appointment in 1832, and carried on in his new laboratory the research in electromagnetics he had started in Albany.

**PRINCETON,** in 1837, sent Henry to Europe to meet the leading men of science, and to observe methods of European scientific research and education. He was also to buy scientific equipment. This seven month visit was a great opportunity to become acquainted with some of the top talent on the Continent and in England. Henry's reputation had preceded him and he received a warm welcome wherever he went, for an exchange of information on his work, and on the current status of European and American science.

In England Henry met Charles Wheatstone at the Greenwich Observatory, was a guest at the Royal Society Dining Club, saw Charles Babbage's calculating engine, the mechanical precursor of the computer. He discussed the latest discoveries in polarization of light in crystals with Sir David Brewster, inventor of the immensely popular "kaleidoscope."

Henry attended a series of Faraday's lectures on metals, and found them highly instructive. He repeated for Faraday the experiments with wire helices that had been so successful in producing sparks. Faraday, who had not been able to do this as effectively, was reported to have been delighted at what he saw. American science was not widely read in Europe, so that the visit by Henry served to bring first hand some needed attention to the original investigation going on in the new Republic.

Princeton provided a good environment for research and put him closer to Philadelphia, then the scientific center of the country. There he participated in the meetings of the American Philosophical Society and the Franklin Institute. He kept on with experiments in telegraphy, using the relay principle to multiply the distance of transmission. As an offshoot of his experiments on relays and electromagnetic induction he found a means of stepping up or down the intensity and the quantity of battery output.

The apparatus, Fig. 10.7, consisted of coils of flat, insulated copper strip, a, b, and c, with different ratios of turns, and coil, d, with many turns of small wire. When the battery connected to the coil, a, was interrupted, a current was induced in coil, b. This, when also connected to coil, c, induced a further current in coil, d. Instruments for measuring voltage and current did not exist. Henry, by use of hand-held electrodes, e, and f, was able to judge, by the amount of sensation or shock, how much the intensity was stepped up or down by various arrangements and connections of the coils. He had developed the principle of the transformer, but his achievement was little noticed.

Henry's most notable work at Princeton occurred in 1842 when he came nearer to a demonstration of the transmission of electromagnetic waves than anything that had been done before. It began when Henry observed that the effects of sparks in one location could be detected some distance away by their magnetic effect in another location. One experiment showed that the induction from a single spark from the discharge of a Leyden jar was penetrating enough to magnetize steel needles 30 feet away in a cellar, with two floors and ceilings intervening.

To account for the action Henry made the remarkable projection that the electricity, leaping through the space was *undulatory, or wave-like,* in nature, and was made possible by an intervening plenum, or *ether-like medium,* which transmitted the waves.

He ended his report . . . *"It would appear that transfer of a single spark is sufficient to disturb perceptibly the electricity of space throughout a cube of 400,000 feet of space, and that the spark is oscillatory. It may be inferred further that the diffusion of motion in this case is comparable with that of the spark of a flint and steel in the case of light."*

Henry had demonstrated electrical resonance and the *transmission of radio waves.* He had the genius to see, though dimly, that electromagnetism was transmitted across the intervening space with a velocity comparable to light. But further discovery had to wait 50 years for Hertz, in his celebrated experiments to show the propagation of high-frequency waves - a prelude to wireless telegraphy.

Henry's career in research came to an end, and his position as an executive began, in 1846. In that year he accepted the post as secretary and director of the newly established Smithsonian Institution. In this role Henry became the first scientific administrator in America, and the first organizer and promoter of American science research.

**JAMES SMITHSON,** a British chemist, born in 1765, was the illegitimate son of the Duke of Northumberland. Smithson, who spent much of his life in France, died in Genoa in 1829. His will contained a bequest of 100,000 pounds "to the United States of America, to found at Washington, D.C., an Establishment for the Increase and Diffusion of Knowledge Among Men."

Smithson, like Henry, was an enthusiast for the natural sciences. He was a mineralogical chemist after whom "smithsonite," the ore of zinc carbonate, is named. He was a friend of many of the eminent scientists of his time, both in England and on the continent. Whether, being illegitimate he resented the loss of birthright, or lack of recognition in his native country, or was impelled by romantic interest in a new country, Smithson resolved to perpetuate his name so that it would . ." live in the memory of man when the titles of my forebears will be forgotten."

President Andrew Jackson, in 1835, formally recognized the Smithson bequest before Congress. After a decade of proposals and counter-proposals, as to how the money was to be used, President James K. Polk signed a bill creating the Smithsonian Institution. The Act of Incorporation provided for a governing Board of Regents, A Secretary to direct the affairs of the Institution, a suitable building, a museum, an art gallery, a chemical laboratory and a library. It further provided that all objects of art and of natural history which belonged to the Government in Washington be turned over to the Institution. These constituted all the major stipulations contained in the Act, Smithson's intent thus remaining unhampered by detailed instructions. The new Institution was left free to carry out in whatever manner it deemed best the donor's provisions for the increase and diffusion of knowledge among men.

The first responsibility of the Board was selection of a Secretary for the Institution, on whose competence, it was realized, the destiny of the new organization would largely depend. The man finally chosen was Joseph Henry. The choice was deemed fortunate. Henry was a man of high ideals and strong character, and he was to give the rest of his life, without stint, into making the Smithsonian a success.

**HENRY'S FIRST JOB** was to determine the meaning of Smithson's will, and how to realize the intentions of the donor. He presented his proposals in the form of a "Programme of Organization of the Smithsonian Institution," which was approved by the Regents in 1847, and became the foundation of its development. The Institution was to have the U.S. Government as a trustee, not in an establishment in the usual sense, but dedicated to the pursuit of knowledge. Smithson did not restrict the kind of knowledge to be diffused, and Henry concluded that . . . "Smithson was well aware that knowledge not be viewed as existing in isolated parts, but as a whole, each portion of which throws light on all the others, and that the tendency of all is to improve the human mind, and give new source

JOSEPH HENRY WAS ITS FIRST SECRETARY

Fig. 10.8 - "THE CASTLE," the first building of the Smithsonian Institution, whose construction was overseen by Henry, and where he lived and carried on his work. Henry was opposed to the Institution becoming a popular museum but in a compromise, one of the first uses of the building was to exhibit the artifacts collected by the Captain Wilkes Expedition to the Pacific Ocean in the 1830s, and to house a collection of American Indian relics that had accumulated in Washington waiting for popular display. Illustration courtesy of the Smithsonian Institution.

of power and enjoyment."

Henry's original conception of the Smithsonian Institution was idealistic for its time, and partly an attempt to redress the inferior position of science in America. He regarded the prime purpose as the support and encouragement of scholarly publications. To disseminate knowledge he proposed, and there was later formed, an international system for exchange of scientific literature.

Henry resisted efforts to make a national library out of the Institution, or a gallery for popular lectures, or a science museum. But the purity of Henry's conceptions of making the Institution a specialized intellectual font was tempered by practical goals and national needs. Henry had not contemplated the popularization of science. But the Smithsonian ultimately was to become one of the most effective in that field in America - a wonderland of science and technology display. An early program was to set up a system of Meterological Stations across the coutry, centered in Washington, to provide national forecasting of the weather. This led to creation of the U.S. Weather Bureau.

In line with a philosophy developed early in his career, Henry emphasized the discoverer and not the applier. Invention, he felt, was implicit, and would eventually arrive from new theory. Henry was against involvement in applied research, feeling that it had a posterior role to pure science. Application would follow when the time was ripe for the introduction of scientific knowledge, and would be taken care of by others. The diffusion of scientific investigation he proposed to accomplish was to be by publishing the results of original research and exploration in a series of memoirs, and distributing them worldwide.

Henry was to spend a busy 32 years in the development of the Smithsonian, as administrator, advisor to the government, and America's leading scientist. In his later years Henry was a prime mover in organizing the National Academy of Sciences. He took a leading part in formation of the American Association for the Advancement of Science and the Philosophical Society of Washington. Henry died in Washington on the 13th of May 1878.

The impetus given by Henry, and carried on by his successors, was to make the Smithsonian Institution, with its constituent parts: the United States National Museum, National Zoological Park, Astrophysical Laboratory and the National Gallery of Art, one of the greatest repositories of mankind's creations, and the wonders of nature.

**TRUE RECOGNITION** of Joseph Henry's genius and contributions to electricity was to come during the last decade of his life when the telegraph became worldwide, and early development of the dynamo and motor began to utilize the principles he had laid down.

The crowning tribute to his contributions to science came in 1890 when the American Institute of Electrical Engineers proposed . . . "that the name HENRY be given to the practical unit of self-induction, since he was the discoverer and the great investigator of this phenomenon."

The name was confirmed by the International Electrical Congress, meeting in Chicago in 1893.

Henry's name is also memorialized in the geography of the southwestern United States. Major James W. Powell, famed for his first voyage, in 1869, through the Grand Canyon of the Colorado, and at one time Director of the U.S. Geological Survey, named a range of mountains in southwest Utah for Henry.

**INDUCTANCE** - Henry had found that every change of current in a conductor produced a corresponding expansion or contraction of the magnetic flux, which, in its divergence or subsidence, cuts through the conductor, and generates therein a voltage, called the electromotive force of self-induction. This voltage is in opposition to the source voltage, resisting the change of current. In general, the voltage of self-inductance is:

$$V = -L \frac{di}{dt} \quad \text{volts}$$

The symbol, $L$, is a coefficient of self-induction and is measured in henrys. Inductance is basically a dimensional factor, being proportional to the square of the number of turns linked with the magnetic flux, and it depends on the size, shape and material of the magnetic circuit.

*The henry is defined as the inductance of a closed circuit in which an electromotive force of 1 volt is produced when the electric current in the circuit varies uniformly at a rate of 1 ampere per second.*

* * * * * * * * * * * * * * *

**APPRAISING** the work of Joseph Henry, and that of his contemporary, Michael Faraday, Prof. H.S. Carhart made the following comments in an address delivered at the annual meeting of the American Association for the Advancement of Science, as recorded in the Journal of the Franklin Institute in 1889:

*The Physics Section of this Association congratulates itself because it deals with topics of the most lively and general interest, not only from the practical point of view, but still more from a theoretical one. Even popular interest in electricity is now well nigh universal. Its applications increase with such prodigious rapidity that only experts can keep pace with them. At the same time the developments in pure electrical theory are such as to astound the intelligent layman and to inflame the imagination of the most profound philosopher.*

*Of the practical applications of electricity it is not necessary to speak. They bear witness of themselves. A million electric lamps nightly make more splendid the illustrious name of Faraday; a million messages daily, over land and under the sea, serve to emphasize the value of Joseph Henry's contribution to modern civilization. Blot out these two names alone from the galaxy of stars that shine in the physical firmament, take from the world the benefits of their investigations, and the civilization of the present would become impossible. The value of the purely scientific work of such men is attested by the resulting well-being, comfort and happiness of mankind.*

# GAUSS. . .

the units for
magnetic flux
in the cgs and mksa
systems
are named for
the German mathematician

## CARL FRIEDRICH GAUSS

**THE SAILING SHIPS OF EUROPE,** cruising the sea lanes of the globe, were largely dependent, for navigational direction, on the compass, the mariner's first scientific instrument. But the early navigators (including Columbus) knew that the compass did not point to the true north-south of the earth's rotational axis. The compass needle oriented itself along the earth's magnetic field, which deviated about 12 degrees from the true N-S, and which shifted unaccountably with geographic location and with time. To use the compass the mariner had to make an allowance for the angle of deviation, called the *magnetic declination.* This required some knowledge of the declination at a specific location.

Before the earth's magnetic field had been mapped, determining the declination was a matter of experience or guesswork. Improving this situation was one of the motives that led William Gilbert, the 16th century experimenter, to the first scientific effort at understanding the response of the compass to the earth as a giant magnet. It also prompted Gilbert and later investigators to attempt mapping the earth's magnetic field to enable the preparation of charts for navigation.

In the two centuries after Gilbert, sea-borne trade developed rapidly, and with it growing attention was focused on magnetic observations to improve the effectiveness and safety of navigation. Magnetism became a study of first rank, equal in importance to astronomy, which since antiquity had been the prime target of interest. Many capable investigators labored to collect and map data on the declination, dip and strength of the earth's magnetic field. Ideas were developed for picturing the magnetic pattern of the globe, and to try to account for magnetic origins and changes. Where was the north magnetic pole? At one time it was thought to be in the frozen sea somewhere north of Siberia.

Finally, after a two hundred year accumulation of facts on terrestrial magnetism, the time was ripe for reduction of the material to law and order. The process brought to the fore an interdisciplinary power that was becoming an intimate part of scientific process and progress. It was manifested by the early 19th century theorists, men for whom mathematics was the language of science - men with the ability to formulate and measure the results of scientific discovery.

For terrestrial magnetics, the formulation and

# . . .WEBER

## and the German experimental physicist WILHELM EDUARD WEBER who jointly established the absolute system of electrical measurement

measurement of the earth's magnetic field strength fell into the hands of two German scientists of outstanding intellectual genius: Carl Friedrich Gauss, ranked as one of the greatest mathematicians the world has ever produced, and Wilhelm Eduard Weber, his younger contemporary, and an experimentalist of exceptional skill. Joint efforts of the two were published, in 1833, in a pioneering tract on *Intensity of the Force of Terrestrial Magnetism Measured in Absolute Units.* In this presentation, Fig. 11.1, Gauss showed for the first time how to measure the earth's magnetic field strength exactly, and at the same time originated a system of force measurements using the fundamental units of length, mass and time of the metric system. This pioneering established the basis for the future systems of expressing the measure of both mechanical and non-mechanical quantities, and was a turning point in the advancement of science.

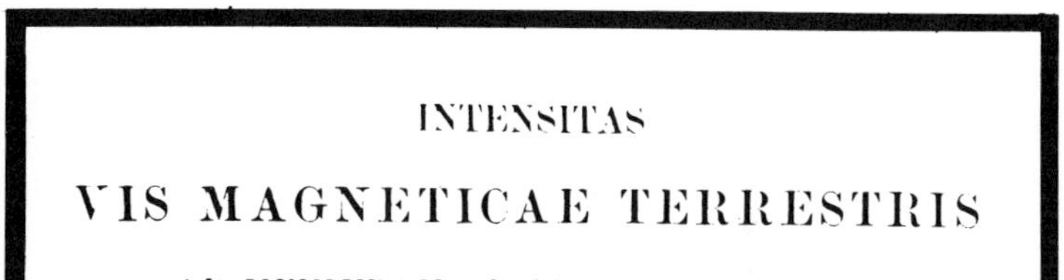

INTENSITAS

VIS MAGNETICAE TERRESTRIS

AD MENSURAM ABSOLUTAM REVOCATA.

COMMENTATIO

AUCTORE

CAROLO FRIDERICO GAUSS

IN CONSESSU SOCIETATIS MDCCCXXXII. DEC. XV. RECITATA.

Commentationes societatis regiae scientiarum Gottingensis recentiores. Vol. VIII.
Gottingae MDCCCXLI.

Fig. 11.1 - GAUSS, in this paper, described the use of the absolute metric units of length, mass and time to measure the force of the earth's magnetic field. This was the first use of absolute units for a non-mechanical force, and it established the basis for other electrical units.

**CARL FRIEDRICH GAUSS'S** parents were poor and unlettered. There was no background to indicate that they had produced a prodigy who would one day be called the "prince of mathematicians," when their son was born on 30 April 1777, in Brunswick, Germany. The father, Gebhard, was a harsh, domineering and untaught middle-class laborer employed as a foreman at various trades. His wife, Dorothea, was said to have had a humorous disposition, strong character and good intellect, but was semi-literate.

Carl Friedrich showed immediate genius in mathematics. Recalling, in his old age, his childhood, Gauss would often say that without help from others he had learned to calculate before he could talk, and that he had learned to read and number entirely by himself. By the age of three his amazing skill at instant calculation appeared. Watching intently as his father, who was foreman for a group of masons, and who was totaling up the weeks wages, and was about to pay them out, Carl piped up: "Papa you have made a mistake," and gave a different figure. He had followed the summing up, and to the astonishment of all present, had corrected an error in the count.

**GAUSS** was enrolled at seven years of age in an elementary school. The young student soon gave more evidence of a self-generated and rapidly growing mathematical skill. When about ten years old he was admitted to a class in arithmetic presided over by J.G. Buttner, a demanding and somewhat inhumane school master, given to caning his charges for the slightest failure. One day Buttner gave his class a problem: "Write down all the numbers from 1 to 100, add them, and give me the sum." Buttner knew the answer, expected no easy solutions from this young group, and looked forward to an hour of little to do while his class labored over the sums.

By custom, the first to finish the problem would put his slate upside down on the schoolmaster's desk, and those following put theirs on top in a growing pile. Buttner had hardly finished giving out the problem when Gauss arose and laid his slate on the teacher's table with the terse "There it lies." For an hour as the other students sweated out their figures, and Buttner sized him up suspiciously, young Gauss sat composed.

One by one the slates were examined; most of the results were wrong and the unlucky boy got a smart rap from the cane. When Gauss's slate at the bottom of the pile was finally turned over there appeared on it a single number - 5050.

**BUTTNER** and the others in the class were astonished at the speed and correctness of the answer, and Gauss was asked to explain how he had got his sum. It was done by a symmetry of pairing numbers: $100 + 1 = 101$, $99 + 2 = 101$, $98 + 3 = 101 \ldots$ to $50 + 51 = 101$, a total of 50 pairs each adding to 101, with the resulting whole sum of $50 \times 101 = 5050$.

Buttner was so impressed with the boy's ability that he broke his usual aloofness and out of his pocket ordered the best book on arithmetic he knew of to help in instruction. Gauss breezed through the text so easily that Buttner gave up. He could teach the young genius nothing more.

Buttner's young assistant, Martin Bartels. was passionately devoted to mathematics, and fortunately then took an active interest in the gifted student, and was of more understanding and help than Buttner in encouraging Gauss's continued mathematical education. The two developed a close friendship, studying and analyzing the proofs and problems of their textbooks together.

**YOUNG GAUSS** now faced an obstacle to his educational progress - his father. Gebhard Gauss was uncomprehending of a future for his son in mathematics, had no funds for financing further schooling, and expected his son to follow in his footsteps as a workman. It took entreaties and representations of both Buttner and Bartels to enable Carl to carry on with a formal education. The father was finally persuaded to allow Carl to enter, in 1788, the local Gymnasium. There it was soon evident that his mathematical talents so exceeded those required by the courses that he was excused from the regular lectures. Carl was thus provided with the time and the opportunity to follow his ambition to study philosophy and the classical languages, in which he soon surpassed his fellow students.

Financial help was now essential to enable Carl to enter a college. His friendship with Bartels, who was acquainted with some of the influential men of Brunswick, brought a happy solution. It was the custom of the locality that some distinguished person of means and willingness to serve as a protector for needy and worthy students who might bring credit to the community. The reigning Duke, Carl Wilhelm Ferdinand of Brunswick, was told of the budding genius, and he agreed to an interview with the fourteen year old student. The Duke was impressed with the precocity and character of young Gauss, and he agreed to become his patron. In 1792 Carl was able to matriculate at the Collegium Carolinum in Brunswick, with the assurance that his bills would be paid by the Duke until completion of the course.

Gauss entered the Caroline College with a background in mathematics and the classics far beyond the student average of the time, largely due to the efforts of his own. He decided first to take advantage of the library now available to learn what was known about mathematics. He studied the important works of Euler, Lagrange, Laplace, Legendre, and particularly Newton's *Principia.*

During his three years at Caroline, Gauss began the independent research and mathematical invention that was to stamp him as the equal of those who were considered the masters. He studied prime numbers, and he evolved an hypothesis concerning their distribution. He investigated repeating decimals to develop an arithmetical nomenclature and notation called *congruences,* which led to a law of quadratic reciprocity.

He invented the method of *least squares,* which today is indispensable in curve-fitting to obtain the probable representation of a set of points observed experimentally. This led Gauss to a study of the theory of errors of observation. He derived the bell-shaped curve of normal distribution of a random variable around a mean value. This probability curve is used in statistical mathematics. It was applied by Maxwell to distribution of velocities in a gas, in his molecular theory of heat.

**IN OCTOBER 1795**, Gauss was admitted to the University of Gottingen, and matriculated as a mathematical student. This institution, one of the foremost in Germany, had been founded in 1837 by King George II of England.

Gauss was still under the patronage of the Duke of Brunswick, but began to realize that this could end at any time and he must look forward to being independent of it. He was now faced with a choice of direction for his life career. He first considered the field of philology, based on his great interest in, and phenomenal ability in, foreign languages, a field which offered a better economic future at the time than science. After some hesitancy, however, the decision came for mathematics. It had been an impassioned and overriding love from his infancy.

**THE POLYGON DISCOVERY** - One of the elements in Gauss's carrying on in mathematics was his triumph, at the age of 18, in the numerical calculations for construction of a regular 17-sided polygon, that could be solved by the Euclidian methods, that is, with only a compass and straight edge. This accomplishment in the Euclidian division of a circle was the first of its kind in 2000 years, and its announcement, his first in public print, immediately made Gauss famous among the mathematicians. In his elation over the feat he told his friends that the 17-sided polygon should be inscribed on his gravestone!

The Euclidian triumph was a corollary to theories of greater content on which Gauss was working. One of these involved complex numbers and how they were to be interpreted geometrically. The complex number arose from the problem of how to solve equations such as, $x^2 + 1 = 0$, and it required the invention of the imaginary rotating operator, $i$, equal to $\sqrt{-1}$ . This led to development of the more general method of expressing a number in the form, $a + ib$, where, $a$, is the real part, and, $b$, with the operator, $i$, the imaginary part of the number. Gauss suggested this general number be called a *complex* number.

Whereas the ordinary number can be represented by a point on a line, the complex number required new treatment because of its two-dimensional character. Gauss's solution was a representation of the complex number, $z = x + iy$, by a point in a plane with coordinates, $x$ and $y$, and the rotator, $i$. This, Fig. 11.2, sometimes called the "Gaussian plane," initiated the use of the general number, and of vectorial methods. Vectors were applied early to the graphical representation and computation of alternating currents.

**ALTERNATING CURRENTS** or voltages can be represented by vectors, or complex numbers, $z = x + iy$, as shown in Fig 11.2. The real part of the voltage or current is, $z \cos \theta$, and the reactive part is, $z \sin \theta$. The phase constant is the angle

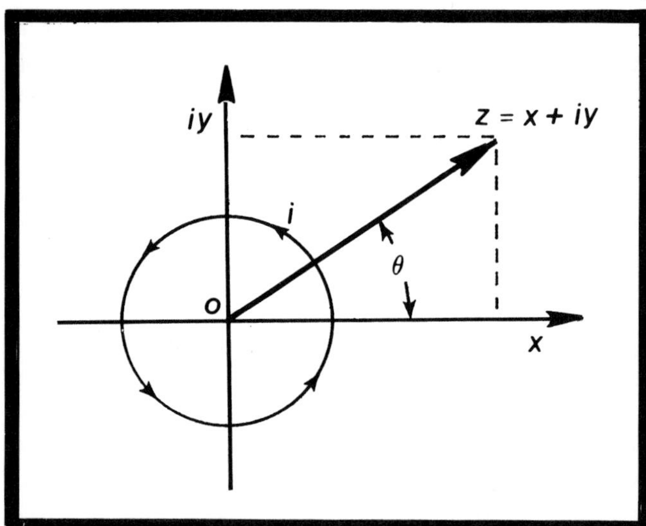


Fig. 11.2 - GAUSSIAN NUMBER PLANE for graphing a complex number. A point, z, is defined by two real numbers, x, and y, and the imaginary operator, i, which is interpreted as the rotation of a vector through 90 degrees.

$\tan^{-1} = y/x$, and the amplitude is $\sqrt{x^2 + y^2}$.

Vectors and their complex algebraic expressions were found exceedingly useful for the resolution and combination of alternating-current waves. Sine waves for example, could be combined, or resolved by adding or subtracting their vector components, and by adding or subtracting their complex algebraic expressions, since the sum or difference of two waves is the sum or difference of their vector representations.

**IN 1798** Gauss returned to Brunswick where he continued intensive work in mathematics, and developed his first interest in astronomy. In 1799, although not present at the occasion, he was awarded a Doctor of Philosophy by the University of Helmstadt for his thesis on the fundamental theorem of algebra, which stated that every integral rational equation in a single variable had at least one root.

In 1801 his work *Disquisitiones Arithmeticae* was published and dedicated to the Duke of Brunswick. The book summed up Gauss's youthful years of creativity in the number theory and was responsible for the development of the language and notations of that branch of the theory of numbers known as the algebra of congruences. The work is usually considered Gauss's greatest mathematical accomplishment. Although the circulation and readership of *Disquisitiones* was initially small, those competent few who could understand the work acknowledged that Gauss was the "prince of mathematicians."

Gauss now proceeded to involve himself in a problem which would make him recognized in the academic circles around the world, and reveal his mathematical superiority over his contemporaries. In January 1801 the Italian astronomer Joseph Piazzi, in the observatory at Palermo, discovered a moving, star-like object in the constellation of Aries. After 40 days of observations he lost sight of it in the rays of the sun and neither he nor other astronomers could locate the heavenly body again. Experts disagreed whether the object was a comet or a new planet. The latter possibility was ruled out immediately by the German philosopher Hegel, who claimed that by pure logic there could be not more than seven planets - the seventh, Uranus, having been found by the famous British astronomer, Sir William Herschel, in 1781.

**HEGEL'S DICTUM** did not deter Gauss, who in early years had already worked with some astronomical questions, such as motion of the moon. The Piazzi problem appeared as a challenge which would give play to his computational skills and and imaginative creativity. The path of a planet is determined by the size and the eccentricity of its solar orbit. Piazzi had been able to plot only a few degrees angle of arc of the path of his star-like object during the 40 days of viewing. This was all that Gauss needed.

"Could I," Gauss said, "ever have found a more seasonable opportunity for testing the practical value of my conceptions, than now in using them to determine the orbit of this new planet, which during these forty days had described a geocentric arc of only three degrees, and after the passing of a year must be searched for in a portion of the heavens very remote from where it was when it last seen?"

Eleven months after Piazzi's discovery, using Gauss's computed positions, G. Zach, an astronomer in Gotha, Germany, again spotted the celestial body. Named "Ceres," it was a minor planet, or asteroid. Later it proved to be the largest asteroid in the solar system, 765 kilometers in diameter, circling the sun between Mars and Jupiter.

After Ceres, astronomers discovered one asteroid after another, and Gauss immediately calculated the orbits. His genius had reduced the computation to a matter of a few hours, and had developed a routine for the process. He needed only three points in the observed arc of an orbit to plot the entire path and the period. Young students of astronomy now try their mathematical prowess by solving orbits by the Gaussian three-point method.

Gauss summed up his new methods of astronomical calculation in the *Theory of the Motion of Heavenly Bodies Revolving about the Sun in Conic Sections.* This work prompted Laplace, who was publishing his own great *Treatise on Celestial Mechanics,* to state that Gauss was the world's greatest mathematician.

**IN 1805** the Duke Ferdinand increased the pension he was providing so that Gauss could marry. His children were born and for a few years Gauss had what he called the happiest time of his life. His joy was shortly saddened by the death of his father in 1808, and the loss of his beloved patron, Duke Ferdinand, who was mortally wounded and captured while leading the Prussian army in the battle of Austerlitz in the Napoleonic wars. Napoleon's cruel treatment of the defeated general was a bitter blow to the man whom the Duke had befriended.

Gauss now found it necessary to look for other income to support his growing family, but there was no difficulty in that. His fame had spread to most of Europe, and he had begun to receive offers of appointments. One of the most flattering was from St. Petersburg Observatory, the most important of the time. Reluctant to see him lost to Germany, however, Baron von Humboldt, the great German naturalist, and his influential friends, arranged for the appointment of Gauss to directorship of the University of Gottingen observatory. This post Gauss quickly accepted because of his high regard for the institution and the freedom he felt the new post would offer for astronomical research.

**GOTTINGEN** was one of the greatest universities in Germany, and Gauss was to be associated with it for the rest of his long life. The appointment marked a new phase in his career. He would successively branch out into new fields in which his mathematical skills and creative imagination would be of use in developing many branches of science.

The first years at Gottingen were upsetting for Gauss. In 1809 he lost his wife after the birth of their third child. Then came a period of war. The French had invaded Germany and were pillaging and extorting the populace to pay for the campaigns. Gauss was forced to contribute 2000 francs to the Napoleonic war chest. The sum was far beyond his means. Among those who helped Gauss meet the demand was Laplace in Paris, a gift which Gauss ultimately repaid. Napoleon's defeat, and Gauss's remarriage in 1811, brought a new era of peacefulness for the pursuit of the many avenues he had laid out for study.

Gauss's early activities at Gottingen were dominated by astronomical duties. He began with the primitive observatory, Fig. 11.3, that was built into the old city walls. Within a few years a new observatory was under way, and a large part of Gauss's time and energy went into its design and construction. Gauss purchased the best equipment available at the time from the noted optical instrument makers of Germany. The observatory was completed in 1821, and was the area where Gauss, with his colleagues and students, was to spend much of his life in the calculating and reporting of observations of the positions and movements of the many heavenly bodies.

Fig. 11.3 - OLD ASTRONOMICAL OBSERVATORY at Gottingen was built into a medieval wall surrounding the city in 1773. Illustration from "Gauss Werke," Volume XI

Fig. 11.4- The NEW OBSERVATORY at Gottingen, started in 1816, and completed in 1821, under Gauss's direction. Illustration from "Gauss Werke," Volume XI.

**DURING THE 18th CENTURY** the mapping of the earth became a matter of important national interest. New political areas were being set up, and maps in terms of the earth's longitude and latitude were essential to the establishment of local boundaries. Maps were also needed on a global scale to determine the surface of the earth.

Mapping required observational skills and theoretical and mathematical abilities similar to those in astronomy, so that it was natural that Gauss should develop an early interest in *geodesy,* the science of earth measurement. Gauss's mathematical reputation was such that at the age of 22 he was consulted on questions of mapping by the Prussian army engineers, who were making a survey of Westphalia. Gauss had previously done some field work of his own, using typical instruments such as the sextant.

Gottingen, in the early 19th century, was in a province of the United Kingdom of Great Britain, Ireland and Hannover, under King George III. In 1820 a grant was provided for a survey of the province of Hannover, with Gauss in charge of the project. Triangulation was started at Gottingen, and carried on to the Dutch and French borders. Gauss took personal charge of the field work, and the entire survey, which took ten years and thousands of calculations, established exact longitudes and latitudes for the area so that accurate maps could be made from the collected material.

**THE TRIANGULATIONS** carried on by Gauss involved the use of trigonometry to establish the reference points and distances over a geographical area. They started with a carefully measured base line, from the ends of which a third point was sighted, and the internal angles measured. This enabled, by trigonometry, the sides of the triangle to be calculated. These sides were then used as base lines for other triangles, a network of which finally enabled distances that were far apart to be determined in length and direction.

Plagued by the limitations of existing lights in sighting points at a long distance in daylight, Gauss hit upon the idea of using a mirror and the reflected light from the sun, an invention he called the "heliotrope." Using this principle, Gauss coupled a mirror with the surveying instruments and was able then to sight easily over 15 to 20 miles. Hence, large-scale triangulation, limited only by the curvature of the earth, could be carried out. The instruments used by Gauss provided the measurement of angles with a precision hitherto not obtainable.

**GEOMETRIC MEASUREMENT** is basic to physics, and for all practical purposes is based on the geometry of Euclid. One of the axioms of this plane geometry is that *the sum of the internal angles of any triangle is equal to 180 degrees,* and this is an element in the triangulation process for terrestrial survey. In the 18th century, however, mathematicians, Gauss among them, began to question whether the axioms of Euclidian plane geometry were applicable to curved space, such as the earth's surface. They proposed the possibility of non-Euclidian geometries of negative or positive curvature, for which the sum of the internal angles would be less or more than 180 degrees.

Gauss speculated on whether it might be possible to detect curvature, indicated by a variation of a triangle from the Euclidian 180 degrees, if the space surveyed were large enough. Using his heliotrope for sighting, he triangulated from three mountain tops in the vicinity of Gottingen - the 4000 foot Brocken, and similar peaks of Hohenhagen and Isenberg, at distance from each other of 69, 85 and 107 kilometers. The sum of the three angles differed from 180 degrees by only 0.68 seconds of an arc, so small a variation that, within the accuracy of observation, Gauss concluded that the space was Euclidian.

Gauss's numerous computations in geodesy and astronomy compelled him to deal most effectively with inescapable observational errors that affected the plotting of results. These were the systematic errors, dependent on external and instrumental conditions, and the random errors still there when systematics were accounted for. Gauss's invention to meet this was his *method of least squares,* published in his paper *Theory for the Combination of Observations Which is Connected with Least Possible Error.*

For a straight line the method of least squares involved taking as the line, $y = mx + b$, of the best fit, that line for which the sum of the squares of of the deviations were a minimum. In the determination of the orbits of planets and comets, the lines were ellipses and parabolas, and the error problems were more difficult. Gauss, however, was able to apply the least square method to them.

**THE YEARS 1830-31** were difficult ones for Gauss, and they marked a dramatic change in the direction of his major efforts. His second wife died after a long illness, and his older son left in a huff for America after a quarrel with his father over the youth's loose life style. In September 1831, as if to assuage both losses, a young experimental physicist Wilhelm Eduard Weber, joined Gauss in an association that was to result in a father-son relationship and collaboration of many years that would have a major impact on the theoretical and practical advance of electricity and magnetism.

In his early twenties Gauss had already developed an interest in physical problems. He had produced a theory of geomagnetism, but did not pursue the subject for lackof experimental facilities. Now, in 1830, at the age of 53, seeking a new source of inspiration and creativity, Gauss again turned to the field of physics and terrestrial magnetism for major investigations.

**A KEY FIGURE** in bringing about Gauss's return to the magnetic science was Alexander von Humboldt (1769-1859), the German world explorer and naturalist. He is generally regarded as the founder of natural geography, and the beginnings of weather analysis. Humboldt, in Berlin, had for twenty years been a leading figure concerned with measuring the earth's magnetism, and attempting to understand and chart its variations.. It was known that the magnetic north pole of the earth did not correspond with the geographic north pole, and that from year to year the mean position of the magnetic pole was gradually drifting westward. Furthermore, there were irregular variations in magnetic intensity, sometimes daily, that were accounted to "magnetic storms" from the sun.

Humboldt's program was essentially the collection of precise data on the intensity of the magnetic forces, on the declination of the compass from the geographic meridian, and on the inclination to the horizontal, or dip, of the needle. These data were necessary for making magnetic charts for maritime use, and for an understanding of the earth as a magnet.

Humboldt was instrumental in encouraging his own, andother governments of the major European countries, to establish more stations for charting the strength and direction of the magnetic field. He also promoted exploratory expeditions around the earth for determining the magnetic configuration and the location of the magnetic poles.

**HUMBOLDT,** for some years, had sought to enlist Gauss's mathematical skills to help in developing better magnetic measurements. He tried to bring Gauss to Berlin to join an academy he was founding there, but there was was a disagreement about meeting the salary set by Gauss, and the bid was unsuccessful. However, Gauss was persuaded to attend a scientific convention in Berlin, where he was invited to stay for three weeks at the Humboldt home. This visit was one of the turning points in Gauss's career. He was exposed to Humboldt's plans for a concerted investigation of terrestrial magnetism, and the hopes of improved methods of measurement. He saw and was aroused by Humboldt's geomagnetic scientific equipment. Gauss was convinced of the importance and challenge of the project, became enthusiastic, and within a short time agreed to participate. As one of the incentives in carrying out the work Gauss was offered the help of an assistant of his choice. The opportunity for the selection of the associate came while Gauss was still in Berlin.

One of the papers presented at the Berlin convention was by a young professor from the University of Halle, Wilhelm Eduard Weber. His talents as an original thinker impressed both Humboldt and Gauss, and they proposed him as a co-worker for the magnetic studies. Weber agreed to the arrangement.. Thus the project was implemented at the beginning with a collaborator who turned out to be a brilliant experimental physicist, who would have six years of association and friendship with Gauss at Gottingen.

**WILHELM EDUARD WEBER** was born in Wittenberg, a city near Berlin which was then in the Kingdom of Prussia, on 24 October 1806. He was one of twelve children of Michael Weber, a professor of theology at the University of Wittenberg. Europe at that time was in the throes of the conquests by Napoleon, who was seeking to establish a grand empire for himself and France. Many principalities of Germany were in the path of the campaigns. In an aftermath of the Napoleonic wars, the city of Wittenberg was bombarded and the Weber lodging was burned. Thereupon the family moved to Halle, a few miles to the south, where the father became a professor at the University of Halle, with which the University of Wittenberg was later merged. The Weber children grew up in a scholastic atmosphere, and in association with academic friends who awakened in Wilhelm and his brothers an early interest in science and medicine. Two of the brothers eventually became famed professors of anatomy.

Wilhelm began his scientific experimenting as a teen-ager, and entered the University of Halle at age 16. There he came under the influence of J.S.C.Schweigger (1779-1857), inventor of the galvanometer, and J.F. Pffaf, a mathematician. He concluded his schooling with a doctoral dissertation on the theory of organ pipes, and was awarded a lectureship, and then an assistant professorship, at Halle.

His major interest was acoustics, in which his investigations of sound and water waves soon made him known in scientific circles. A paper on his experimental work on organ pipes, delivered at the 1828 Berlin convention of the German Society of Natural Philosophy and Medicine, attracted the attention of both Humboldt and Gauss. They saw in Weber a worthy co-worker for Gauss, and this led to his enlistment in Humboldt's organization for the study of terrestrial magnetism. Weber was offered the professorship of physics at Gottingen in 1831.

Gauss and Weber joined forces at a time when the situation in the electrical sciences had been completely transformed. There had been a busy decade of discovery. Electromagnetism was demonstrated by Oersted in 1820; in 1821 Ampère, Biot and Savart had formulated the fundamental laws of force exerted between wires carrying electric current, and between magnetic poles and current elements. In 1827 Ohm had established his law of the electric circuit. Green in 1828 had introduced the potential function into the mathematics of electricity and magnetism. And finally, in 1831, Faraday discovered electromagnetic induction. Upon this foundation Gauss and Weber began their researches.

**HUMBOLDT'S PROGRAM** in geomagnetism was formalized in a newly-founded "German Magnetic Union," under the auspices of which eighteen observational stations were set up by 1835 in many of the nations of Europe. To the operations of the new Union, Gauss and Weber, the one a theoretician and the other a skilled experimental physicist, applied their joint efforts. Two powerful intellects soon refined the methods of observation and the calculation of results. They developed new, sensitive instruments for precise measurement. Together, at Gottingen, they erected a magnetic observatory free from iron that might introduce irregularities in the local field.

Most revolutionary, Gauss recognized the desirability and possibility of representing the results of magnetic measurement in the metric system of units, the first time this had been proposed for a non-mechanical force. Previous to this all magnetic readings were only simple comparisions with an arbitrary quantity of the same kind as that being measured. The readings were thus not subject to reduction to universally comparative standards.

Gauss devised the first rational system involving use of the three fundamental physical entities of length, mass and time to measure the magnetic quantities in absolute units. Gauss's methods and results were presented in his famous paper: *Determination of the Strength of Terrestrial Magnetism in Absolute Units,* written with the assistance of Weber.

**GAUSS'S METHOD** of measuring the earth's magnetism in absolute units required two manipulations to provide equations showing the relationship between $M$, the moment of a reference magnet,

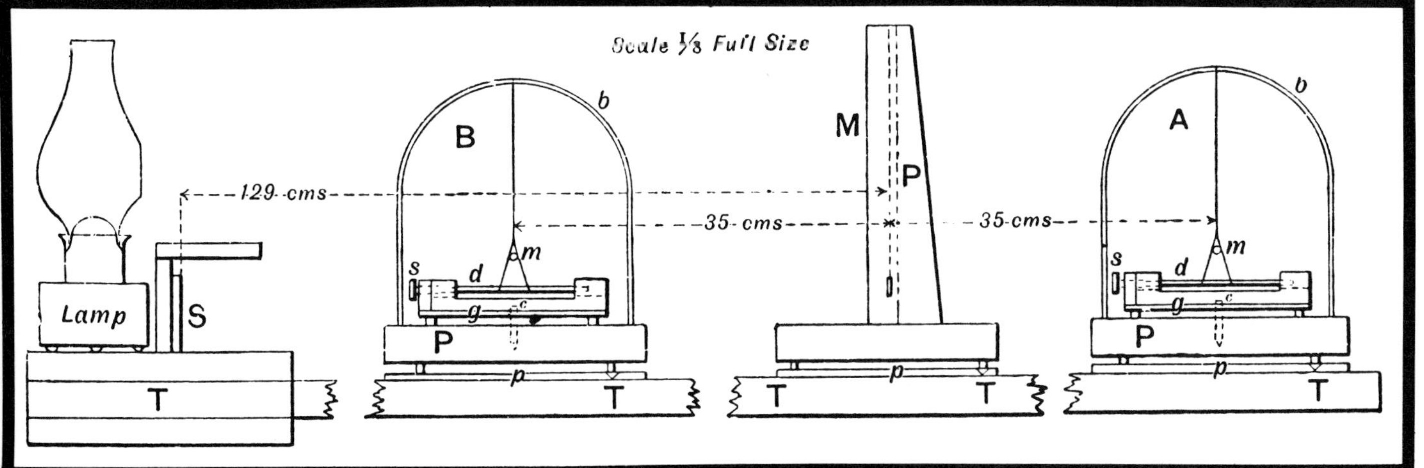


EARTH'S MAGNETISM IN ABSOLUTE UNITS

Fig. 11.5 - Measurement of terrestrial magnetism in absolute units was performed by Gauss and Weber with an apparatus like this. Deflection of the magnetometer, M, due to the bar magnets, A and B, and oscillation period of the magnet in the earth's magnetic field, gave equations yielding the earth's magnetic strength in metric units of length, mass and time. From A. Gray, "Absolute Measurements in Electricity and Magnetism," Dover Publications, N.Y.

and, $H$, the horizontal component of the earth's magnetic field strength as it affects the compass needle. One operation gives the equation for the ratio of $M$ to $H$; the other gives the equation for the product of $M$ and $H$. The two equations are solved simultaneously to yield $M$ and $H$ in terms of physical units. With apparatus as shown in Fig.11.5 the procedure is as follows:

(1) The bar magnet, B, is placed at a right angle, at a suitable distance, to the normal position of the needle of a magnetometer, M, which consists of a light suspended needle magnet with a mirror. The deflection of the magnet, B, is then read.

This yields the ratio: $M/H = 1/2(r^3 \tan\theta)$,

where, $r$, is the separation of the centers of the magnet and needle, and, $\theta$, the angle of deflection of the magnetometer needle.

(2) The bar magnet is then suspended horizontally so as to be free to turn about a vertical axis, and the period of its oscillation in the earth's magnetic field is recorded.

This yields the product: $MH = \dfrac{4\pi^2 I}{T^2}$

where, $I$, is the moment of inertia of the bar magnet, and, $T$, the period of its oscillation.

Solving the two equations gives:

$$M^2 = \frac{2\pi^2 I}{T^2} r^3 \tan\theta, \text{ and}$$

$$H^2 = \frac{8\pi^2 I}{T^2 r^3 \tan\theta}$$

Thus, by the two measurements, the value of $M$, the magnetic moment of the standard bar magnet, and the value of $H$, the earth's magnetic field, have been obtained, both in terms of metric units. Either $M$ or $H$ having been determined, the evaluation of any other magnetic field or magnetic pole can be obtained.

Gauss and Weber had provided a fundamental measurement of the magnetic quantities in absolute units, and this is the basis of our quantitative expression of magnetic phenomena.

With a method of absolute measurement at hand Gauss was now able to put into specific form the theories of earth magnetism that he had originated thirty years previously, but lacked data then to pursue. Now, in 1838, having the measurements collected from a world-wide network of observation stations, Gauss could map the distribution of the earth's magnetism in a general way by expressing the magnetic potential at any point by a series of spherical harmonics.

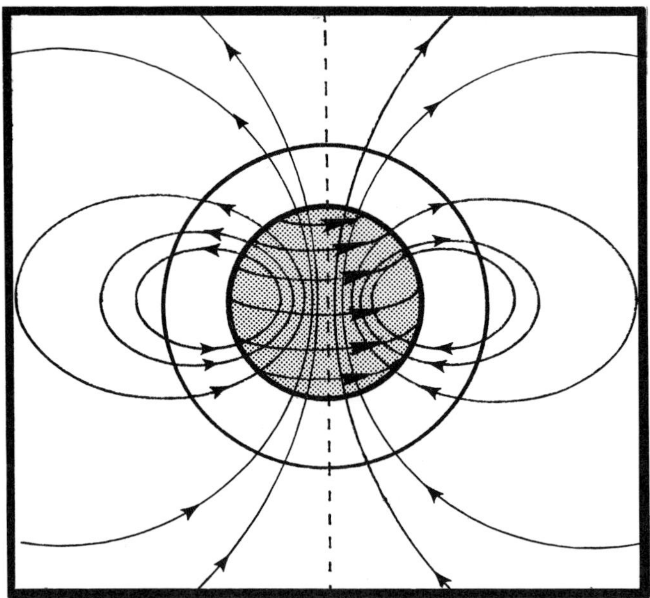

Fig. 11.6 - GAUSS PROVED that the earth's magnetic field must originate in the earth's interior, as a result of its rotation. The electric currents which could produce the field are shown above circling the hot metallic core. Variations in intensity and direction of the earth's magnetic field are due to differential rotative forces between the core and outer mantle, causing local shifts in the current circulation.

Gauss proved beyond question that the magnetic field must originate within the earth. Today there is no doubt but that the earth's general magnetic field is generated by electric currents due to circulation of the hot, plastic metal core of the earth. Interactions between the earth's core and mantle, Fig. 11.6, produce local eddies, giving rise to supplementary magnetic fields which account partly for local variations of intensity and direction of the general magnetic field.

Gauss attempted to evaluate what part of the magnetism that might be due to causes external to the earth. To account for magnetic storms he speculated that these violent variations might be induced from outside the earth by electric currents flowing high in the atmosphere.

Gauss found that there was only one true north and south pole on the earth's surface. The operations of Gauss and Weber provided the first worldwide data on the magnetic movements of the earth, on the shifting of the magnetic poles, and supplied charts showing isomagnetic lines for areas where magnetism was under observation. The work is summed up in Gauss's *The General Theory of Magnetism,* published in 1838.

**THE ELECTROMAGNETIC TELEGRAPH** - Gauss and Weber were well-known in academic circles for their theoretical work, but their public recognition was minimal until news of their electromagnetic signal system got out in 1833. Operating over a mile of wire, from the Gottingen observatory to another station, Gauss and Weber were sending messages to each other in code. Signalling by electricity was in its infancy, and when word got around that the two scientists were in communication with each other at such a distance by wires, it made sensational news. It was an electrical wonder that even the lay public could appreciate.

Transmitting signals by an electromagnetic means was first suggested by Ampère in 1821, but was not carried out at any distance until Gauss and Weber had set up their apparatus. The Gottingen telegraph was a chance product of an experiment originally designed to test the validity of Ohm's circuital law over a long distance. The equipment comprised a battery for the power source, and a galvanometer using a needle about a foot long operating in a coil of about 300 feet of wire. A commutator enabled the current to be reversed so the needle could be swung to the right or left at will. The battery in the Gottingen station, and the galvanometer in the other station were connected by a two-wire circuit about 8000 feet long.

Gauss and Weber soon found out that the experimental setup could be used for sending signals as well as for circuit testing. They first used the swings of the needle to synchronize clocks between the stations; then they began to send words and phrases by coding the swings of the needle. Combinations of swings gave indications of various letters of the alphabet. Transmission was slow, about seven letters per minute.

Over a period of years to 1845, Gauss and Weber experimented with and improved on the apparatus. In one arrangement they substituted a coil and a permanent magnet plunger, Fig. 11.7, as the power source, and they put a small mirror on the galvanometer needle so a light beam could register the swings optically.

Gauss and Weber recognized the utility of their

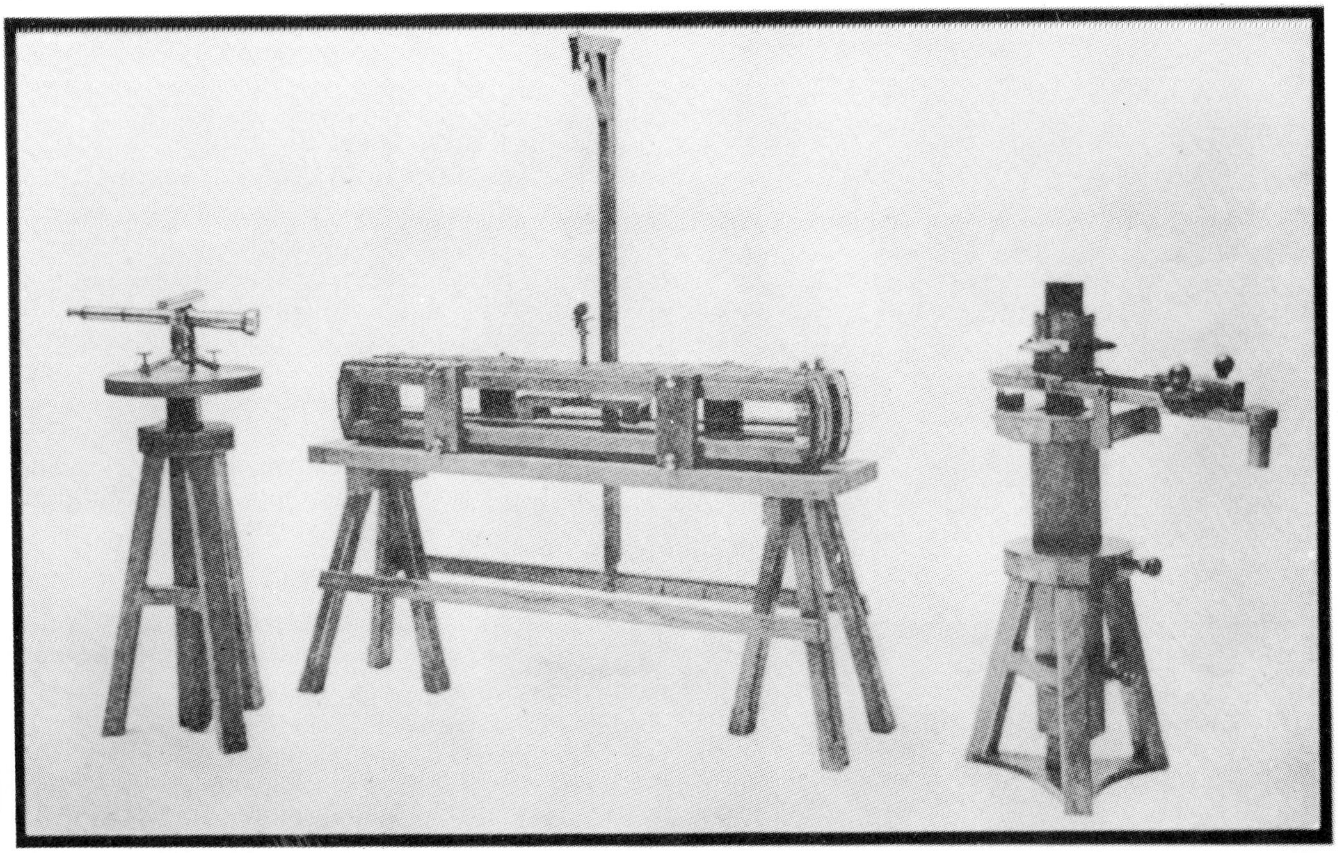

apparatus and the possibilities of future technical development. It was the first electric telegraph then in regular use. Weber in 1835, made a remarkable prediction: " . . . when the globe is covered with a net of railroads and telegraph wires, this net will render services comparable to those of the nervous system of the human body, partly as a means of transport, partly as a means for the propagation of ideas and sensations with a speed of lightning."

Their priority in the field of telegaphy is largely forgotten, but the Gauss-Weber idea soon attracred the attention of scientists and inventors who modified and improved the apparatus. Systems for long-distance, such as developed by Samel F.B. Morse, came into commercial use in the mid-19th century.

**GAUSS'S** close collaboration with Weber came to an end in 1837. Germany, at the time, consisted of a group of kingdoms, each with a ruler whose authority set local politics. In 1837, with accession of Queen Victoria to the throne of England, her uncle Ernst August became the king of Hannover. He immediately abbrogated the existing liberal constitution and demanded from public servants an

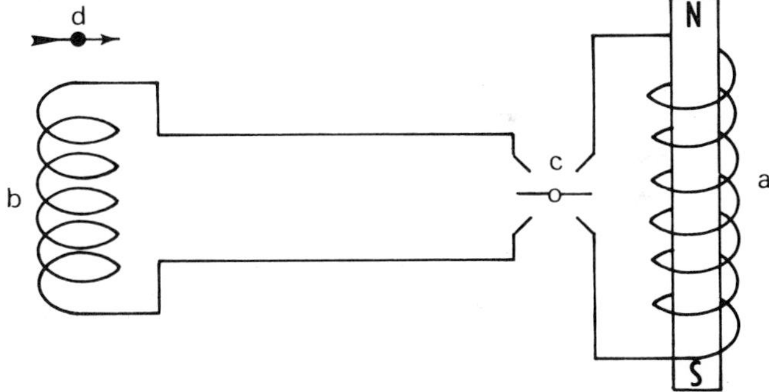


Fig. 11.7 - GAUSS AND WEBER TELEGAPH OF 1836. The galvanometer, with telescope, above, first used for testing Ohm's law on long circuits, was soon found to be suitable for communications. With this apparatus Gauss and Weber made the first use of the binary code for telegraphing. The diagram above shows one arrangement used by Gauss and Weber. When the magnet was plunged into and withdrawn from the coil, a, an induced current was transmitted to receiving coil, b, which deflected the needle of the galvanometer. A commutator, c, enabled direction control of the needle to implement a binary signal code. Illustration above courtesy the Deutsches Museum, Munich.

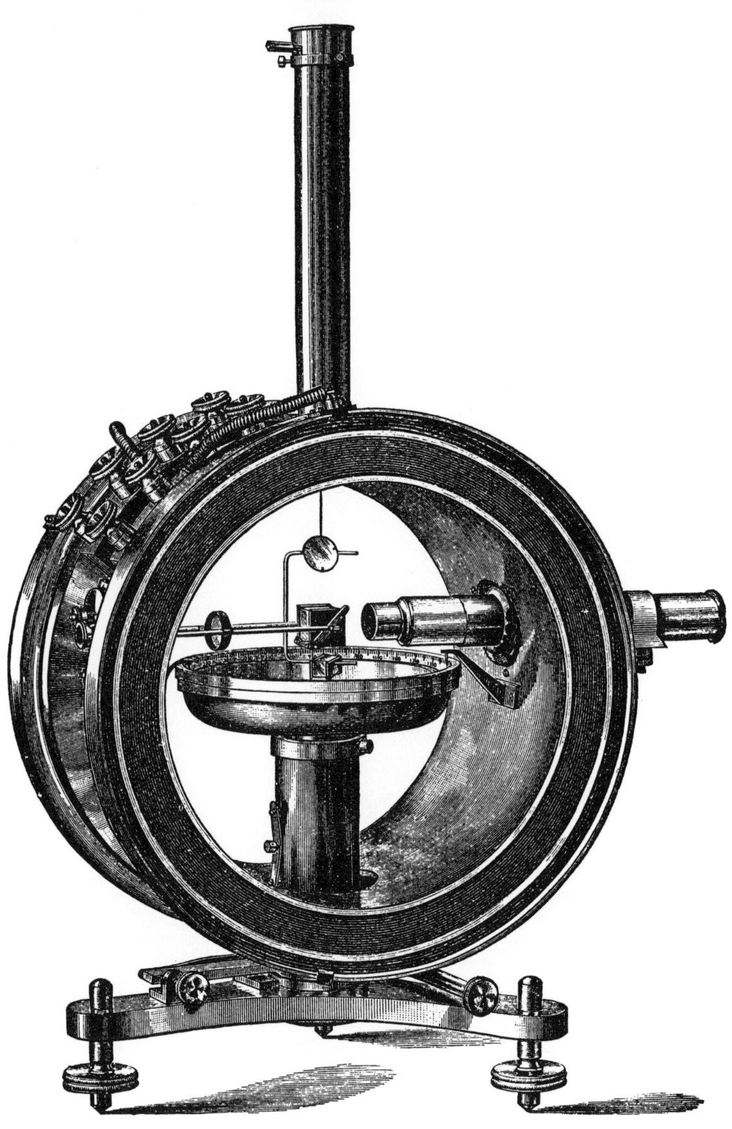

Fig. 11.8 - TANGENT GALVANOMETER such as used by Weber in determining absolute units of current. The instrument consists of a pair of circular coils within which is suspended a magnetic needle, with graduated scale. From Andrew Gray, "Absolute Measurements in Electricity and Magnetism," Dover Publications, New York.

oath of allegiance to himself. Several Gottingen professors, including Weber, protested and were fired. Gauss, who disliked the political radicalism, and did not favor the resistance of his associates, stayed on, but now was deprived of Weber's help, so he abandoned further physical research.

Weber traveled to London and Paris to promote extension of geometric stations, then returned to Gottingen where he drifted for several years until, in 1843, with the help of friends, he received an appointment as a professor of physics at the University of Leipzig. Here Weber began his independent researches in electricity and magnetism. His work over the next quarter century provided the basis for an understanding of electrical and magnetic forces, and the establishment of fundamentals for measurements that would lead to the system of electrical units we have today.

Gauss and Weber had done the pioneering in the procedure of measuring terrestrial magnetism in absolute units. The rationality and usefulness of the method was now evident, and it contained the approach for expansion to other electrical quantities. Weber was the first to grasp the possibilities of a coherent system of units, and to proceed to extend it.

**THE TANGENT GALVANOMETER** - Weber, in 1840, made the first step by showing how an electric current could be measured in absolute units. Since electric current is accompanied by a magnetic field, the numerical measure of a current can be defined by the field intensity it produces at a given point. Weber applied this principle in a simple, direct way by use of the magnetic force of a *tangent galvanometer,* Fig. 11.8.

This instrument compares the field produced by a given current, to that of the earth. The current to be measured flows in two circular coils, on the axis of which is suspended a compass needle. The coils and needle are lined up with the earth's north - south magnetic meridian. When current is applied to the coils, the needle deflects to line up with the coil's magnetic flux. The deflection is opposed by the earth's magnetic field, $H$, which is known accurately. The strength of the current is then:

$$I = \frac{r}{2\pi n} H \tan\theta$$

Since, $r$, the radius of the coil, $n$, the number of turns of the coils, and $H$. are known, the current is equal to the tangent of the angle of deflection. For the current in amperes a factor of 10 is applied.

**THE ELECTRODYNAMOMETER,** invented by Weber, Fig. 11.9, provides another means of measuring current. In this instrument a suspended coil substitutes for the magnet needle. This eliminates the uncertainty as to distribution of magnetism in the needle, and use of the instrument does not any knowledge of the earth's magnetic field. The instrument, although not as sensitive as the tangent galvanometer, had the advantage of being suitable for alternating current.

From the origin of Weber's method, until about 1890, current measurements continued to be made by the electrodynamometer or the tangent galvanometer. Hence a knowledge of the earth's local magnetic field intensity was necessary. Electrical laboratories during this period usually did not use steel in their construction to avoid setting up irregularities in the earth's local field.

**AT LEIPZIG,** Weber collaborated with Gustav T. Fechner (1801-1887), a physicist and the founder of "physcophysics," who had done some theorizing on the nature of an electric current. Fechner had developed the view that current in a wire consisted of positive and negative charges streaming in opposite directions with equal velocity. He pictured the two charges as, *A,* at a higher potential, and, *B,* at a lower potential. When they were connected with a wire, positive charges were transferred from *A* to *B,* and the negative charges from *B* to *A*. The current was thus a counterflow of positive and negative electricities, and this could account for the force between parallel conductors which had been shown by Ampère, and the electromagnetic induction demonstrated by Faraday.

Weber was inspired by Fechner's theory, but his collaboration with him ended when the latter, due to illness, was forced to give up his post in physics and to return to the physcophysics for which he is best known.

Weber, however, carried on with Fechner's ideas, and proceeded to put into mathematical form the properties of electric charges that would account for the effects they produced.

**WEBER POSTULATED** that the electric charges were fundamental, and that electromagnetic phenomena could be explained in terms of the forces exerted by charges, not only when stationary, but also when in motion. He then modified Coulomb's equation to derive the forces:

$$f = \frac{ee'}{r^2}\left\{1 - \frac{1}{2c^2}\left(\frac{dr}{dt}\right)^2 + \frac{1}{c^2}r\frac{d^2r}{dt^2}\right\}$$

where, $f$, is the force acting between the charged particles, $e$, and, $e'$; $r$, the separation, and $c$, is a velocity number relating the static and dynamic condition. The first term $ee'/r^2$, is the Coulomb force, the second term is that of velocity, which represents the inductive forces; the third term is

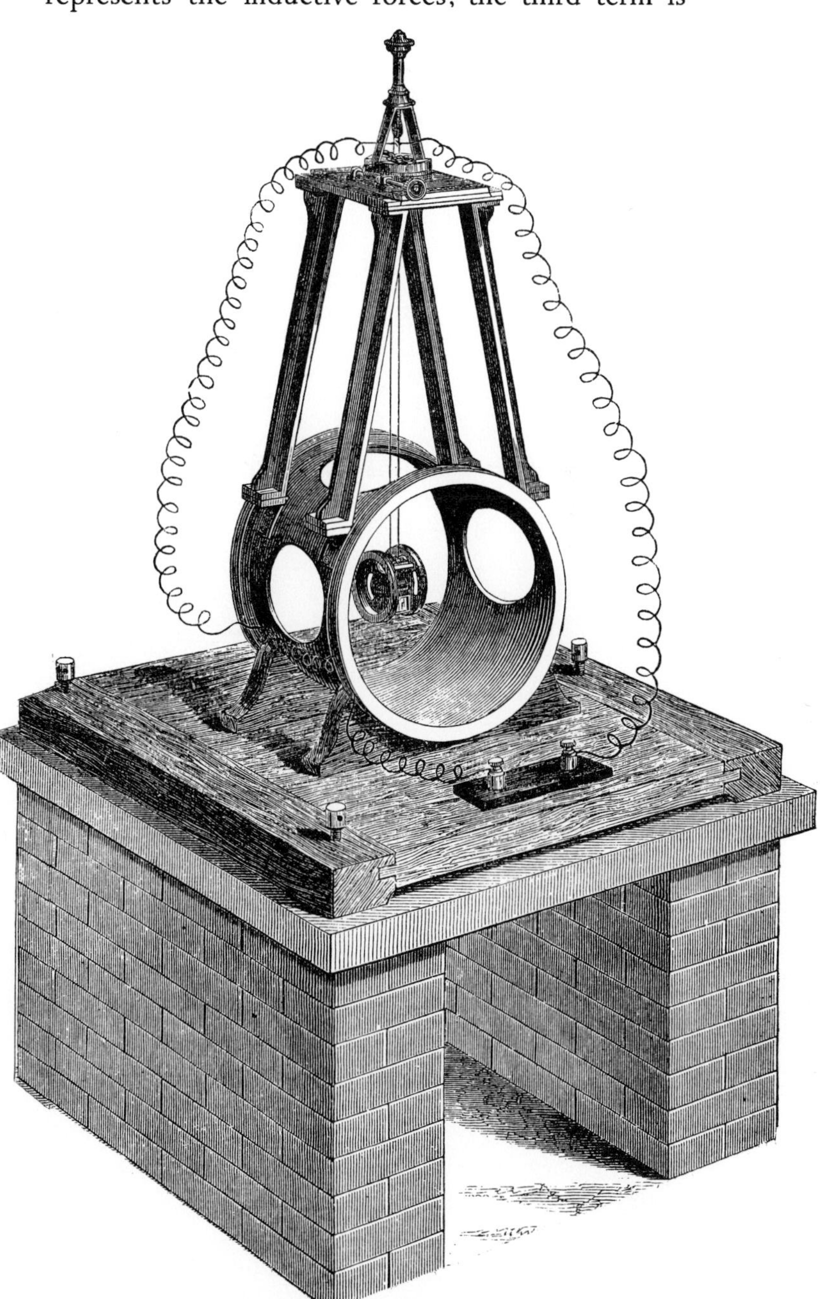

Fig. 11.9 - ELECTRODYNAMOMETER, an instrument invented by Weber. It uses a coil instead of a magnetic needle for the moving element. The bifilar suspension was devised by Gauss. From J. Clerk Maxwell, "Treatise on Electricity and Magnetism," Dover Publications New York.

that of acceleration, representing the flux linking, or transformer action

Weber's law of electrical force appeared in 1846, as one of his first articles on the subject of *Electrodynamic Measurements.* It was one of the first of "electron" theories, in that it attributed the phenomena of electrodynamics to the agency of the velocity of electric charges.

It was soon discovered, after Weber had presented his law, that positive and negative charges did not move with equal velocities in an electrolyte. Other considerations also affected the validity and usefulness of the Weber equation. But the factor, *c,* indicating a velocity ratio between static and moving charges, provided Weber with the next clue for investigation, and its results, later, in the hands of Maxwell and Hertz, would prove to be revolutionary

In 1848 an upheaval in politics reduced the power of King Ernst August of Hannover, and Weber was able to return to his old position at Gottingen, where he became director of the astronomical observatory. He had now established an international reputation for his work in electricity and magnetism. His force equation caught the eye of scientists, and particularly motivated Rudolph H.A.Kohlrausch (1809-1858), a close friend and associate of Weber, who had taught mathematics and physics at several German academies and universities. Kohlrausch was now at the University of Marburg.

**KOHLRAUSCH,** impressed by Weber's force law, which involved a velocity ratio between static and moving electrical charges, proposed to determine experimentally what that ratio was. The existence of a velocity ratio lay in the basic force quantity relations between unit charges and unit currents.

*By Coulombs Law,* the force between unit electrical charges, $q$, and between unit magnetic poles, $m$, is:

$$Fq = qq'/\varepsilon_0 r^2$$

$$Fm = mm'/\mu_0 r^2$$

where, $\varepsilon_0$, is the electrical permittivity constant, and, $\mu_0$, is the magnetic permeability constant, in a vacuum.

From these equations, expressed in physical dimensions of mass [M], length [L] and time [T]:

$$[q] = [M^{1/2}\ L^{3/2}\ T^{-1}\ \varepsilon_0{}^{1/2}] \qquad (1)$$

$$[m] = [M^{1/2}\ L^{3/2}\ T^{-1}\ \mu_0{}^{1/2}] \qquad (2)$$

The Weber theory proposed that a current consisted of moving electrical charges. Thus a current expressed in electrostatic units would have physical dimensions of charge divided by time ($i = q/t$).

Or, using equation (1) in physical dimensions:

$$[i]es = [M^{1/2}\ L^{3/2}\ T^{-2}\ \varepsilon_0{}^{1/2}] \qquad (3)$$

*From Ampere's experiments,* the force of a current element, $i{\cdot}l$, on a unit magnetic pole, $m$, is $F = (m{\cdot}i{\cdot}l)/r^2$, or, $i = (Fr^2)/(m{\cdot}l)$ Thus, using equation (2), a current expressed in electromagnetic units would have the physical dimensions:

$$[i]em = [M^{1/2}\ L^{1/2}\ T^{-1}\ \mu_0{}^{-1/2}] \qquad (4)$$

Equating the currents (3) and (4) above gives:

$$1/\sqrt{\varepsilon_0 \mu_0} = [L\ T^{-1}]$$

Thus the ratio of the electrostatic units to the electromagnetic units is a function of the permittivity constant and the permeability constant in a vacuum, and has the physical dimensions of length over time - a *velocity.*

**KOHLRAUSCH and WEBER** were led, in 1856, in the first experiment of its kind, to find this velocity. The method used was founded on the measurement of the same quantity of electricity, first in *electrostatic,* and then in *electromagnetic* measure. The quantity of electricity measured was the charge of a Leyden jar.

The quantity of electricity in electrostatic measure was found as the product of the capacity of the jar into the difference of potential of its coatings.

To determine the value of this charge in the electromagnetic measure, the jar was discharged through the coil of a galvanometer. The effect was to give the galvanometer a certain angular velocity. By observing the extreme deviation of the galvanometer, the quantity of electricity in the discharge could be calculated.

The value of the velocity obtained by Kohlrausch and Weber was *310 740 000 meters per second.*

This velocity was numerically equal to the number of electrostatic units of electricity in one electromagnetic unit.

The velocity of light had been measured as 314,000,000 m/s by H.L.Fizeau in 1848, and as 298,360,000 m/s by J.L. Foucault in 1850. These velocity figures were such as to make manifest the coincidence with those of Weber and Kohlrausch. The results led G.R. Kirchhoff (1824-1887) to state, in 1857, that electricity was propagated along a wire at the speed of light. And this velocity was one of the foundations of the electromagnetic theory of light, and the propagation of electromagnetic waves proposed by J. Clerk Maxwell, in 1873.

**THE TELEGRAPH INDUSTRY,** in the mid-19th century. was the most important user of electric current. The design and operation of telegraph systems brought to the attention of scientific investigators and practical engineers that there were no satisfactory names or standard units for the electrical factors with which they had to deal. These factors included mostly those for resistance, capacity and electromotive force.

The choice of a unit of resistance was of fundamental importance wherever telegraph engineers were stringing their lines and installing their instruments. A satisfactory unit of resistance was also required to relate to current and electromotive force in Ohm's law. Until the 1850's standards of resistance were local and arbitrary, consisting of some suitable length, section or weight of material, usually wire, and they were unrelated to any absolute unit. In 1843, Wheatstone, in England, proposed a resistance standard equal to one mile of copper wire 1/16 inch diameter; in France a standard was one kilometer of iron wire 4 millimeters in diameter; and in Germany, a German mile of No. 8 iron wire.

Weber was not satisfield with resistance standards based on a particular object or particular material. In 1851 he proposed a system of electrical and magnetic measurements in which the resistance would be expressed, curiously enough. as a velocity, and he performed the necessary experiments to determine the unit of resistance in absolute measure.

The detailed relations leading to resistance as equal to a certain value of velocity are described in a subsequent chapter on the derivation of the

II. *Ueber die Elektricitätsmenge, welche bei galvanischen Strömen durch den Querschnitt der Kette fliefst;*
*von Wilhelm Weber und R. Kohlrausch* [1]).

I. Aufgabe.

Die Vergleichung der Wirkungen einer geschlossenen galvanischen Kette mit den Wirkungen des Entladungsstromes angesammelter freier Elektricität hat zu der Annahme geführt, dafs diese Wirkungen von einer *Elektricitätsbewegung* in der Kette herrühren. Wir denken uns in den die Kette constituirenden Körpern ihre *neutrale* Elektricität in Bewegung, in der Art, dafs deren gesammter positiver Theil nach der einen Richtung sich in geschlossenem, zusammenhängendem Kreise herumschiebt, der negative nach der entgegengesetzten Richtung. Der Umstand, dafs nirgends durch diese Bewegung eine Aufhäufung der Elektricität entsteht, fordert die Annahme, dafs durch jeden Querschnitt in gleicher Zeit die gleiche Elektricitätsmenge hindurchfliefse.

Man hat sich darüber geeinigt, dafs man die *Gröfse dieses Fliefsens*, die sogenannte *Stromintensität*, proportional setzen will der Elektricitätsmenge, welche in derselben Zeit durch den Querschnitt der Kette hindurchgeht. Soll also eine bestimmte Stromintensität durch eine *Zahl* ausgedrückt werden, so ist festzusetzen, welche Stromin-

1) Der Herr Herausgeber wünschte für die Annalen einen Bericht über eine von Prof. Weber und mir gemeinschaftlich ausgeführte Arbeit, deren Resultate unter Prof. Webers wesentlicher und schliefslicher Redaction im 5ten Bande der Abhandlungen der Königl. Sächsischen Gesellschaft der Wissenschaften zu Leipzig unter dem Titel: *Elektrodynamische Maafsbestimmungen, insbesondere Zurückführung der Stromintensitätsmessungen auf mechanisches Maafs* (Leigzig bei S. Hirzel. 1856) niedergelegt sind. Ich gebe hiermit einen kurzen Auszug. Kohlrausch.

WEBER FINDS THE VELOCITY OF ELECTRICITY

Fig. 11.10 - Title page of the investigation by Weber and Kohlrausch: *About the Quantity of Electricity which by Galvanic currents flows through the cross-section of the Cell,* published in the Poggendorff Annals XCIX, August 1856. They describe the first experimental determination of the ratio of the electrostatic unit to the electromagnetic unit, which represented a velocity - about $3 \times 10^8$ meters per second. This figure represents the velocity of light, and is the foundation of the electromagnetic theory of light electrical and magnetic units. The argument involves the force exerted on a conductor carrying a current transversely in a magnetic field, the motion impressed by that force, the work done thereby, the equivalent of that work in terms of the expenditure of current through the resistance of the conductor, and then the equivalency of the resistance to a velocity.

**WEBER'S EXPERIMENT**, as reported in the Poggendorff Annals of 1851, used a large coil capable of being revolved on a vertical diameter, and connected to a tangent galvanometer to form a single circuit having a resistance, *R.* The coil, the sum of the areas of the windings of which, was, *A,* was placed perpendicular to the earth's magnetic meridian, and given a quick half revolution. The quantity of electricity, *Q,* in the current induced by the motion of the coil through the earth's magnetic field was then:

$$Q = \frac{2AH}{R}$$

where, *H,* is the horizontal intensity of the earth's magnetism, and, *R,* the total resistance of the circuit. This current sets the coil of the galvanometer in motion.

Then, from the swing of the galvanometer:

$$Q = \frac{HT}{\pi G} 2 \sin \tfrac{1}{2} \theta$$

where, *G,* is the galvanometer constant, *T,* the time of swing of the needle of the galvanometer, and, $\theta$, the observed angle of deflection of the galvanometer.

From the above equations:

$$R = \frac{\pi GA}{T \sin \tfrac{1}{2} \theta}$$

Weber thus expressed the resistance in mechanical units as a velocity.

In 1863 the British Association for the Advancement of Science Committee for Resistance Standards chose a velocity of $10^9$ centimeters/second for the absolute electromagnetic unit for the *ohm.*

Providing a precise material and reproducible representation of the absolute ohm was the next step. The British Association Committee devised an alloy, wire-wound standard of resistance. Werner Siemens, in Berlin, proposed, in 1860, a column of mercury one meter long, one square millimeter in section, at zero degrees centigrade, as the standard for the ohm. Further refinements in establishment of the absolute ohm, as described later, would occupy scientists for many years.

Weber's wide-ranging experiments covered many fields of electricity and magnetism. He investigated from 1848 to 1852, the properties of bodies in a magnetic field. Weber held, as did Ampère, that the magnetism in magnetic bodies was accounted for by molecular currents, circulating permanently without resistance. Submitted to the action of a magnetic field, the molecular magnets tended to undergo an alignment so that like poles are turned predominantly in the same direction. Due to intermolecular forces and friction, however, not all molecules come into coincidence, so that magnetization can be increased but not indefinitely. The same factors, when magnetizing force is removed, prevent the return to a completely demagneitzed state.

In an effort to explain *diamagnetism,* a property of those substances repelled by an external magnetic field, Weber experimented with bismuth, one of the highly diamagnetic metals. Faraday had originally interpreted this phenomenon as due to a reversal of polarity of the material when it was introduced into a magnetic field.

Weber enlarged on Ampère's theory to explain diamagnetic action. If, according to Ampère, there are molecular circuits in which there is no ohmic resistance, so that currents can flow without dissipation of energy, then currents would be induced in other molecular circuits. By reversing polarities this would confer on a material the properties of diamagnetism. To avoid the conclusion that with this explanation all substances would be diamagnetic, Weber supposed that in iron and other magnetic materials there were permanent molecular currents, not produced by induction, which oriented themselves in a preferred direction when in the influence of an external magnetic field.

Weber's later years at Gottingen were devoted to further investigations and theorizing in electrodynamics, and to efforts in comprehending the electrical properties of matter. He developed the concept of atomic charges, consisting of particles with positive and negative charge. The Amperian currents were now pictured as due to electric charges spinning in orbits about fixed central charges of opposite sign.

**WEBER VISUALIZED CONDUCTION** in terms of positive charges fixed in the lattices of the conductor material around which the negative particles were accelerated with increase in velocity until they penetrated the sphere of influence of neighboring atoms, thus causing a directional migration and hence a conduction.

Weber's work on the interaction of positive and negative particles had a strong influence on contemporary scientists. His orbiting electric charges preceded by only a quarter of a century J.J. Thomson's advancement of the electron theory in 1897. Weber's ideas also led, at the turn of the century, to the classical theory of conduction in metals based on free electrons able to move around in the atomic lattice as do molecules in a gas.

Weber died in Gottingen in 1891 at the age of 86. He never married. His vast accomplishments brought many honors from the scientific circles of France and Germany. He had great influence on the proceedings and decisions of the British Association Committees of Electrical Standards, organized in 1861 to promote a coherent system of electrical units. Weber, along with other notable international scientists, was appointed to help in the determination of a practical unit of resistance. He received the Copely Medal, the highest award of the Royal Society, London.

**MAXWELL,** in his *A Treatise on Electricity and Magnetism,* summed up one of Weber's contributions to the electrical science in these words:

"The introduction, by Weber, of a system of absolute units for the measurement of electrical quantities is one of the most important steps in the progress of the science. Having already, with Gauss, placed the measurement of magnetic quantities in the first rank of method of precision, Weber then proceeded in his *Electrodynamic Measurements* not only to lay down sound principles for fixing the units to be employed, but to make determinations of particular electrical quantities in terms of these units, with an accuracy that previously had not been attempted. Both the electromagnetic and the electrostatic systems of units owe their development and practical application to those researches."

**WEBER'S NAME** was considered, and used for a time, as the unit of electric current. But Hermann von Helmholst (1821-1894), the German physicist, whose relations with Weber, over their rival theories, were often strained, proposed at the First International Electrical Congress, in Paris, in 1881, the name "ampere" for the unit of current.

Weber's name lives on, however, in the terminology of the magnetic units. The Organization of Electric and Magnetic Magnitudes and Units, in Paris, in 1932, proposed, and in 1935 confirmed *weber* as the MKSA unit of *magnetic flux.*

**The WEBER** is now the unit of magnetic flux in the International System of Units (SI): *The weber is the magnetic flux whose decrease to zero when linked with a single turn induces in the turn a voltage whose time integral is one volt-second.*

**GAUSS,** as has been discussed in previous pages, was supplied with material for his investigations in electricity and magnetism by the discoveries of Ohm, Ampère, Faraday and others. Gauss used these not so much as a physicist searching for new phenomena, but as a mathematician seeking to formulate the work of others. He was primarily a mathematician-scientist.

Gauss formulated fundamental laws in electrostatics and electromagnetics.

*"Gauss's law of electrostatics"* relates the electrical field, $E$, from any closed hypothetical surface, $S$, (called a "Gaussian surface"), to the net amount of electrostatic charge, $q$, enclosed by the surface:

$$\varepsilon_0 \oint \mathbf{E} \cdot d\mathbf{S} = q \qquad (1)$$

where, $\varepsilon_0$, is the permittivity constant of free space.

*Coulomb's law* of the inverse square relationship between point electrostatic charges, although it preceded Gauss's law historically, can be deduced as a special case of Gauss's law. Thus, assuming the charge, $q$, centered in a spherical surface, the electric field, $E$, will be constant at all points of the surface, and can be factored from inside of the integral sign. Then, the integral is simply the area of the sphere, and equation (1) becomes:

$$\varepsilon_0 E(4\pi r^2) = q$$

or,

$$E = \frac{q}{4\pi\varepsilon_0 r^2} \quad (2)$$

Equation (2) gives the electric field at any point distant, $r$, from an isolated point charge, $q$.

To compare with Coulomb's law, put a second charge, $q'$, at a point where, $E$, is calculated. The magnitude of the force, $F$, that acts on it is:

$$F = Eq'$$

Combining this with equation (2) gives:

$$F = \frac{1}{4\pi\varepsilon_0 r^2} qq'$$

which is Coulomb's law.

*"Gauss's law for magnetics"* is a formal way of asserting that the net magnetic flux, $\Phi$, through any closed Gaussian surface must be zero. It states that isolated or "free" magnetic poles have not been found; that the lines of magnetic flux have no beginning or end:

$$\Phi = \oint \mathbf{B} \cdot d\mathbf{S} = 0$$

where, $\Phi$, is the flux; **B**, the flux density, and, **S**, is a closed Gaussian surface.

Gauss's equations of electricity and magnetism are part of the basic equations of electromagnetism called *Maxwell's Equations.*

**GAUSS ENDEAVORED,** as Weber was also doing independently, to generalize on Coulomb's law to account for the electrodynamic forces and electromagnetic induction. Gauss had arrived at a force equation similar to that of Weber.

These force manipulations involved consideration of whether action took place instantly over a distance, or was a function of the intervening field as proposed by Faraday. That Gauss was moving away from the former, and toward the latter view, is shown in a letter to Weber in 1845 which stated that the generalizations of Coulomb's law did not satisfy him because the assumed an instantaneous propagation, whereas his real aim had been derivation of the forces from an action propagated in time in a similar manner to light. The work of succeeding scientists would apply this keystone to an understanding of electromagnetic propagation of waves.

Gauss's diary, comprising 19 pages, was first published in 1901. It consists of 146 brief statements of his results from 1796 to 1814. It is a highly important document from which many of his priorities can be verified, and from which the development of his genius can be followed.

All Gauss's publications, and a selection of his unpublished papers and notes, appear in twelve massive volumes, *Gauss Werke,* printed after his death. The unpublished papers printed therein reveal many of Gauss's discoveries and innovations that were later repeated by contemporary scientists independently, unaware of Gauss's priority.

**GAUSS'S LIFE** was one of comparative intellectual isolation, stemming partly from a childhood in which his advanced thoughts in an abstract branch of knowledge could not be shared with others, and from his reclusive nature. Except for collaboration with Weber, he was for most of his life a solitary worker. He avoided emotional involvements and customary ceremonies and formalities, except for the commands of royalty. Only a Humboldt could get him to attend a scientific convention. Gauss rationalized this attitude as necessary to maximize his scientific output.

Gauss had drive and ambition; drive for financial security based on remembrances of the bitter poverty of his youth, and an ambition for great achievement and lasting fame in science. His publications number in the hundreds, and these were exceeded by the ideas on which he worked and were later revealed. Conservatism, a high self-imposed standard of output, and an unwillingness to present something not prefected and verified to his satisfaction, left many of his projects and ideas only in note form and not published. Some subjects he disliked he purposefully suppressed.

Teaching, a part of his duties, was distasteful to Gauss. He was impatient with lesser minds in his field, although he could unbend and inspire those few he felt worth his while.

Concerned only with contemplation and with the

intensity of work within his circle, Gauss disliked controversy and avoided the usual formalities and social graces. Even to his admirers Gauss was considered cold and uncommunicative. Although possessed of a good classical education, and being a wide reader, his expressed cultural attainments were bourgeois. But none of these personal characteristics in any way dimmed the transcendency of his creativity and reputation in the fields in which he far surpassed his peers.

Gauss's range of subjects was prodigious. Mathematics, including arithmetic, number theory algebra, statistics and probability occupied his earliest efforts and continued throughout his life. Astronomy dominated much of his career. Geodesy, geomagnetics, mechanics, dioptrics, electricity and magnetism were fields in which he contributed during his middle and later years.

In the last 15 years of his life the intensity of Gauss's output gradually decreased, but he was by no means inactive. He completed his work on the Hannover geodetic survey, and continued to report on a variety of mathematical problems. His publications became extensions and variations of his earlier projects. He took on some teaching and followed the political, economic and technological events of the world. He added Russian to his long list of languages.

Although afflicted with various ailments, Gauss took good care of himself in a self-treated and diligent health regimen. Until the end he avoided the physicians.

**The 50th YEAR** annivarsary celebration of his doctorate, held in 1849, brought him formal honors from around the world, In the last five years of his life a developing heart disease slowed the pace of Gauss's activities and increased his isolation. He finally was approachable only by those in his personal circle; to others he was fast becoming a legendary figure.

Gauss died in his sleep in February 1855. Although he had been considered a dreamer by some of his early townspeople, and was called a "star-gazer" by his father, Gauss in death revealed his shrewd, worldly and practical side. He left an estate of 200 times his annual salary.

**The GAUSS** - in 1894, the American Institute of Electrical Engineers, meeting in Chicago, proposed *gauss* as the name for the CGS electromagnetic unit of *magnetic flux density.* It is equal to one emu line of magnetic force (one maxwell) per square centimeter, and is the magnetic flux density that will induce an electromotive force of one hundred-millionth of a volt in each linear centimeter of wire moving at one centimeter per second at right angles to the magnetic flux.

In the MKSA system of units one weber/meter$^2$ is equal to $10^4$ gauss.

**GAUSS'S NAME** lives on in the terminology of the many "gaussian" methods, effects, concepts and laws he innovated. In World War II, the word "degaussing" entered the common vernacular in expressing the measures taken by naval forces to neutralize the external magnetic field of a vessel to avoid being blown up by triggering of magnetic mines.

In geomagnetics, on which Gauss spent so much of his efforts, he left us, on the earth's surface, immersed in a magnetic field which is measured in his name, and which has a value of about ½ gauss.

---

J.C. MAXWELL, in his *A Treatise on Electricity and Magnetism,* commented on Gauss's work:

*"It is hardly necessary to enlarge on the beneficial results of magnetic research on navigation, and the importance of a knowledge of the ture direction of the compass, and of the effect of the iron in a ship. But the labours of those who have endeavoured to render navigation more secure by means of magnetic observations have at the same time greatly advanced the progress of pure science.*

*Gauss, as a member of the German Magnetic Union, brought his powerful intellect to bear on the theory of magnetism, and on the methods of observing it, and he not only added greatly to our knowledge of the theory of attractions, but reconstructed the whole of magnetic science as regards the instruments used, the methods of observation, and the calculation of the results, so that his memoirs of Terrestrial Magnetism may be taken as models of physical research by all those who are engaged in measurement of the forces of nature."*

**THE YEAR 1831** witnessed two births that would have colossal impact on world history. They would reshape the whole progress of science and technology. One birth was that of the electric power industry, germinating from Michael Faraday's discovery, on 28 October 1831, of electromagnetic generation of electricity. The other birth, on 13 July 1831, was that of James Clerk Maxwell, the Scottish genius, whose career, greatly influenced by Faraday's ideas, climaxed with presentation of the laws of electrodynamics. Building on the goundwork of Faraday's experimental researches in electricity, Maxwell created a mathematical formulation for four famous equations whose physical meanings are the foundations of the electromagnetic theory. And, from these equations he drew the astounding conclusion that energy could be transmitted by electromagnetic waves at the speed of light - from which emerged a new world of communication.

**FARADAY,** in trying to account for results of his experiments, had put forth a speculation, daring for its time, that electrical and magnetic action took place in fields of force. In his mind's eye he saw these fields as *lines of force.* Curving in space from their sources in electricity and magnetism, the lines of force existed in a state of tension and compression. They put the intervening medium in a state of strain. To Faraday the very substance of matter, and even the origin of light waves, might be found in the lateral vibrations of these strains.

Unable to express mathematically the laws which his theories should obey, Faraday's hypotheses were cooly received by his orthodox contemporaries in science. Fortunately, however, to the young mind of Maxwell, the ideas of Faraday had the appeal of novelty and belief. They began to appear to him as concepts that could be given precision, and that held the promise of leading to new and exciting discoveries. Finally he accepted them whole-heartedly and under the magic of his mathematical symbols Faraday's lines of force evolved into reality. In the course of time they could be measured, and they have been the great unifying structure for the subsequent advances in electrical theory.

**IT WOULD BE DIFFICULT** to pick two scientific geniuses whose differences in origin and in procedure were so contrasted. Faraday was the son of a blacksmith; Maxwell came from a line of distinguished ancestors. Faraday grew up in poverty; Maxwell's family had ample means. Faraday had only a few years in common day school; Maxwell attended the finest institutions in Scotland and England. Faraday became one of the world's greatest experimenters in science. Maxwell became one of the world's most famous theoretical physicists. Yet both were complementary, and both made indispensable contributions to the fields of electricity and magnetism.

In the early 1860s, as Faraday's own life-time work came to a close, Maxwell had progressed to fully developed ideas on electrical forces, based on Faraday's laws of induction, as presented in his memoir *On the Dynamical Theory of the Electromagnetic Field.* When Faraday died in 1867, Maxwell had begun to lay the foundations for his *A Treatise on Electricity and Magnetism,* one of the world's most celebrated compendiums in the advancement of the electrical science.

a unit of magnetic flux is named for
the giant of 19th century physics, J.C. MAXWELL,
whose discovery of
the electromagnetic theory of light, and
the laws of electrodynamics
are two of the most significant events
in the history of electrical science

JAMES CLERK MAXWELL (1831-1879) from a portrait by R.H. Campbell, 1929, hanging in the Institution of Electrical Engineers, London. The apparatus on the table, designed by Lord Kelvin, was used by Maxwell in determining the value of the ohm. It consisted of a coil revolving around a small suspended magnet needle. Current induced in the coil by the earth's magnetism deviated the needle for the computation of the ohm in electromagnetic measurement. Photograph courtesy of the Institution of Electrical Engineers.

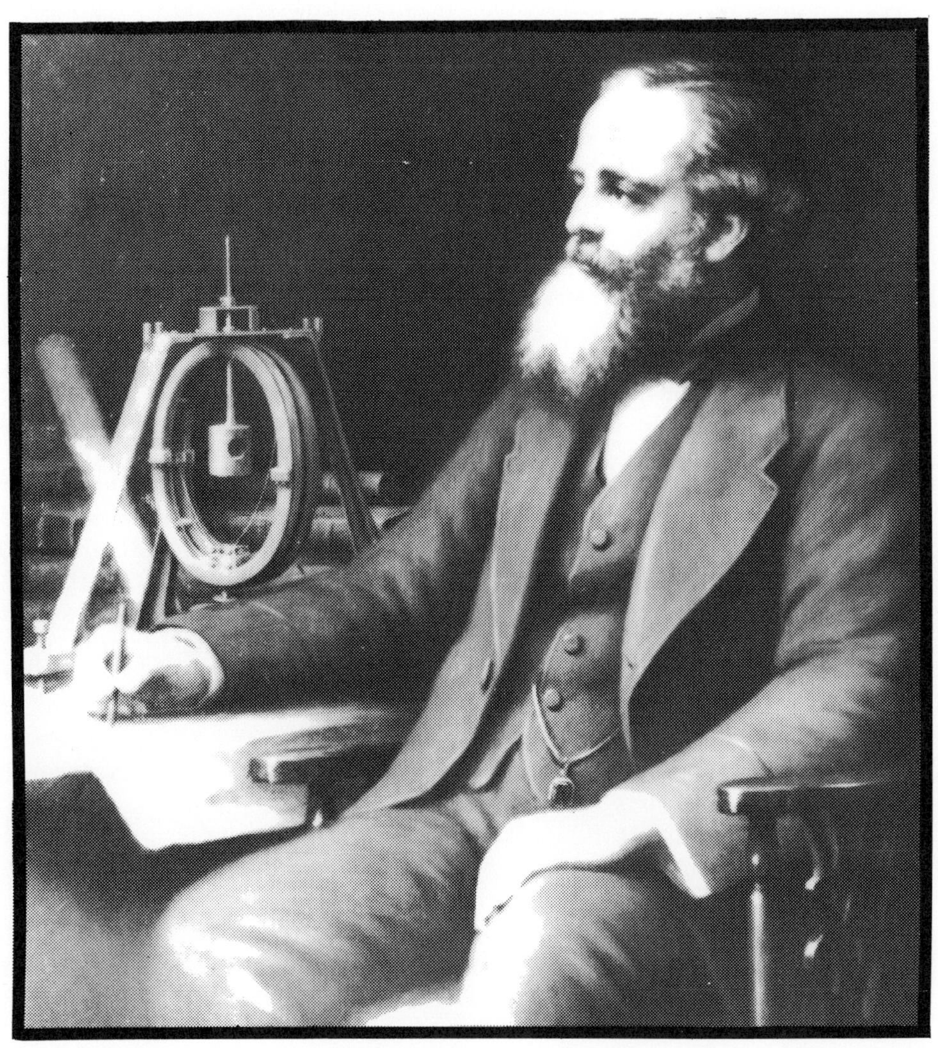

**JAMES CLERK MAXWELL** was descended from the Clerks of Penicuik in Midlothian, near Edinburgh. They were a well-known Scottish family, whose history had been traced back to the 16th century, and whose members included eminent judges, politicians, merchants and men and women of art and music. James father, John, was a brother to Sir George Clerk, M.P. of Penicuik, but he had adopted the surname of Maxwell on succeeding to an estate of 1500 acres of farmland in Galloway on the southwestern border of Scotland that had descended to him through a previous family marriage to the Maxwells.

John Clerk Maxwell was trained in the family tradition of law, which he practiced for some years, but his interests were elsewhere. He had a strong practical ability in mechanical pursuits for which he felt the farmland estate offered more opportunity for expression. Relinquishing the bar in 1826, he and his wife moved to Galloway where he built a home named "Glenlair," located in an area of fine woodland scenery. There he proceeded to develop and enlarge the farm, took part in local county affairs and enjoyed the quiet life of the country. James's mother, the former Frances Cay, whose antecedents came from Northumberland in northern England, was described as " a lady of strong, good sense and resolute character."

James Maxwell was born on 13 July 1831, in Edinburgh, where the mother had gone for medical treatment. He was an only son. Soon after her recovery the family returned to the Glenlair home. Glenlair, in Galloway, was then in a remote and undeveloped area of Scotland, a two hard days journey from Edinburgh. Separated from England by the Solvay Firth, the old castles in the Galloway countryside gave evidence of ancient disputes between countries and between clans for the rights of the area. In these the Maxwells had played an important and hereditary part,

In this rustic and scenic corner of Scotland, James Maxwell spent the first ten years of his childhood. Here in relative isolation, his education came under the loving attention of his parents, with their Edinburgh culture, and frequently augmented by artistic influences of visiting relatives. James had as his playmates the children of the cottagers on the estate, and from them he picked up a strong local accent that marked his speech for the rest of his life.

Young Maxwell was described as a nearsighted and affectionate boy. Endowed with a wonderful memory, the pride of his mother, he could quote, at age eight, long passages from Milton and from the Bible. Like his father, he was persistently fascinated with mechanical contrivances. Apparently he was curious as a little dog's nose about things and how they worked, continually questioning his father on "what is the go of it?" He showed a knack in making drawings and diagrams, and was clever with his hands in assemblying objects. His first creation was a pictured toy wheel giving an illusion of continuous motion. This early occupation was a prelude to his disposition and ability in later life to design dynamical models for displaying complex motions and physical processes.

When James was ten years old his mother died of cancer, an affliction that forty years later was to kill him also. The tragedy was a shock to the boy and it brought father and son even closer together. With the loss of the mother a tutor was hired to carry on with James's education. This did not work out well, however, and it was decided to send him to Edinburgh Academy, which had been founded in 1824 to provide Scottish youth with a classical education along English lines. James would spend most of the next eight years in Edinburgh, living with his two aunts and shuttling back and forth to Glenlair.

James's first day at the Academy created a sensation. His clothes, which were designed by his logically but not fashionably minded father, included square-toed "hygenic" shoes and a laced tunic. This, plus his Galloway accent brought derision and finally a pummeling by his schoolfellows, and a nickname of "Dafty." But Maxwell was stubborn, resourceful and brave, and could give as well as take. He soon assumed a place at school which his friendly disposition and mental powers entitled him. He formed associations which were to be life long. One was with Lewis Campbell, his future biographer, and another was with P.G. Tait, who would become a celebrated physicist and mathematician.

**AT EDINBURGH ACADEMY** young Maxwell experienced a gradual awakening of interest in mathematics and geometry. He constructed models of polygons, and corresponded with his father about them. Then suddenly, at age 14, he prepared his first scientific paper, on a method of tracing perfect oval curves with pins and thread. This method had been well known for constructing ellipses, but Maxwell's arrangement for producing ovals was new. John Maxwell proudly showed his son's work to J.D. Forbes, professor of natural philosophy at Edinburgh University, who saw its originality and merit. He decided it should be communicated to the Royal Society of Edinburgh. It was not thought proper for an author so young to appear on the rostrom, so Professor Forbes agreed to present the paper presonally.

At age 16 Maxwell entered the University of Edinburgh. Here he came under two inspirational men who were to have a lasting influence on his future. One was Professor Forbes, whom Maxwell already knew, and the other was Sir William Hamilton, chairman of logic, and an erudite metaphysician.

Both men, though personal enemies, took a special interest in Maxwell's progress. Forbes, who was an experimental physicist, let Maxwell use his physical apparatus, and encouraged him to develop an experimental technique. This type of skill was rare in theoretical physics at the time, but was to be a valuable support for Maxwell in his future work. Maxwell studied mathematics and some physics and chemistry. In addition, it was traditional in Scottish universities for those in science to take courses in philosophy. This he did under Hamilton who awakened in Maxwell a keen interest in logic and morals, and stimulated a love of speculation to which his mind was already prone.

Maxwell distinguished himself while at the university with two additional papers read to the Royal Society - one on *The Theory of Rolling Curves,* in 1849, and another *On the Equilibrium of Elastic Solids,* in 1850. These from a student of nineteen, showed a gift for originality and an ability for concentration and sustained exertion. Maxwell read voraciously, and occupied much of his time with mathematical speculation, and with a variety of experiments in optics, magnetics and chemistry. He was a talented adolescent who was developing a roving enthusiasm for many fields of knowledge, sscientific and philosophical, and a mind ripening for the work of his future years.

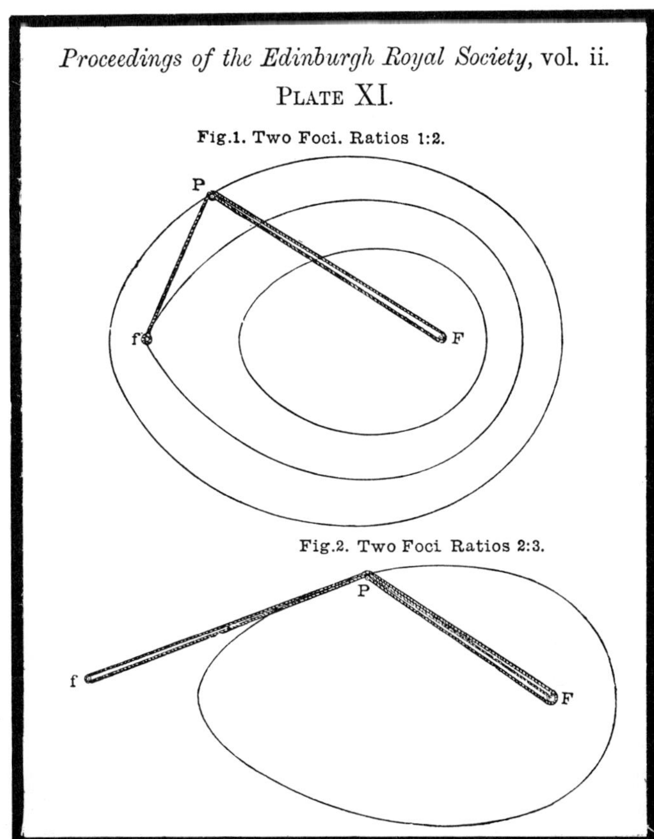


MAXWELL'S FIRST PUBLICATION - AT AGE 14

Fig. 12.1 - Drawing oval curves by wrapping string around pins, in a manner modified from that used to make an ellipse, was the subject of Maxwell's first paper, written at age 14, It was printed in the Proceedings of the Royal Society in Edinburgh in 1845. Because of Maxwell's youth, the paper was presented to the Society by Prof. J.D. Forbes head of Natural Philosophy at the University of Edinburgh.

**YOUNG MAXWELL** was originally intended to follow his father in the profession of law, but this was abandoned when it was obvious that his talents were so gifted in the direction of science. To carry on with his mathematics and other aspects of an education he went on, in 1850, to Cambridge University. There, at Trinity College, he became one of the private pupils of William Hopkins, a geophysicist and highly regarded mathematics coach, whose students included some of the foremost minds of that generation in science and mathematics. Hopkins immediately recognized Maxwell's talents. He saw in the young student one whose grasp of physical subjects was phenomenal, and he

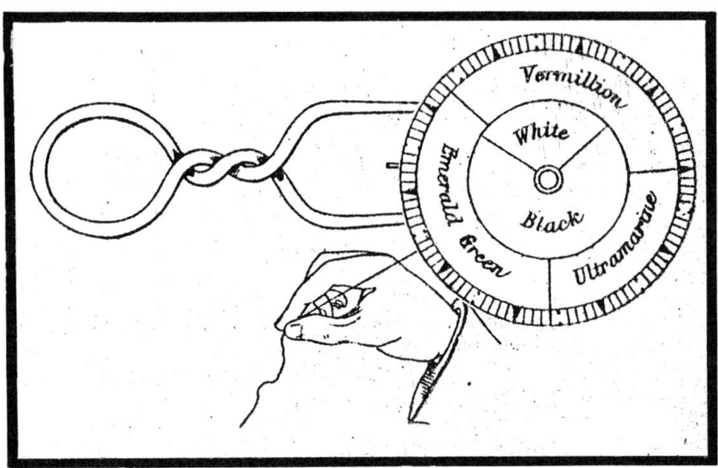


MAXWELL PIONEERED THE PHYSICS OF COLOR

Fig. 12.2 - Maxwell's earliest experiments in color analysis used the color top, a spinning disk to which sectors of colored paper were attached. He found the primary colors to be vermillion, ultramarine blue and emerald green.

later described Maxwell as "the most extraordinary man I have ever met."

Maxwell worked hard on his subjects, and found time also to carry on investigations outside of his courses. But he did not neglect social accomplishments, and was continually broadening his interests outside the sphere of science. Absorbed in philosophy and in literature, and with a liberal turn of mind, Mawell was soon specially recognized in the student body. He was made an associate of the "Apostles," a famous and select coterie of twelve members. This intellectual group of unusual students engaged in cultivated discussion and speculative essays on subjects beyond the regular courses at the university.

Maxwell was described, by one of the men in the group, as "the most genial and amusing of companions, the propounder of many a strange theory and composer of many a poetic *jeu d'esprit.* Maxwell was disposed to carry on bizarre investigations. One study had to do with how a cat manages always to land on its feet. He demonstrated that a cat could right itself when dropped upside down from only a few inches. He could also be a joker. At a tea-party he exhibited a brass top which he used for experiments in dynamics. His friends left it spinning, and the next morning Maxwell, spying one of them coming across the street, lept out of bed, started the top, and retired between the sheets. The startled morning visitor greeted the still spinning top with amazement and new respect.

**In JANUARY 1854,** Maxwell's undergraduate career closed. He shared Tripos Examinations as Second Wrangler (at that time Cambridge University's designation of the highest class of honors in mathematics) and was tied in the Smith's Prize with E.J. Routh, later a celebrated mathematician. Maxwell stayed on for two years at Cambridge. He was elected a Fellow of Trinity, and placed on the staff of Lecturers.

In addition to tutoring and lecturing, Maxwell immediately plunged into experimental work, the most noteworthy of which was in color optics. He had begun color analysis in 1849 in Professor Forbe's laboratory in Edinburgh. The apparatus he used was the simplest possible - a spinning "color-top" fitted with a disk on its spindle to which sectors of various colored paper could be fastened, Fig. 12.2. When the top was whirled rapidly the color sensation of the disk blended to produce various shades.

Forbes had tried to get "white" shade using the traditional artist's primary colors: red, bue and yellow, but could get only an assortment of grays. Maxwell soon found that yellow and blue did not make green, but only a pale pink. Then Maxwell, who was early exhibiting the flash of originality that could break tradition, decided the primary colors should be red, blue and *green.* They worked to produce white, and the colors of the spectrum.

Maxwell resumed work on color after his graduation at Cambridge, and improved on color analysis by devising the "color-box" in which desired colors could be matched and blended more accurately by light projection. For his work in color sensation and on the causes of color blindness, Maxwell was later awarded the Rumford Medal of the Royal Society. Also, later, in the Royal Institution, he projected the first trichromatic picture; he may thus be considered the pioneer of color photography.

**MAXWELL'S MOST SIGNIFICANT ACTIVITY** in his two post-graduate years at Trinity was his entrance into the study of electricity, a field which ultimately led to the discoveries for which he is famous. As a youth Maxwell had been taken by his father to an exhibition of electromagnetism in Edinburgh put on by a Mr. Davidson of Aberdeen. The exhibition showed how electricity from a battery could provide magnetism to lift heavy iron weights, and to produce rotation by "Electromagnetic Engines" of the type that had been developed by Thomas Davenport in America. The novelty of this new type of power doubtless greatly impressed young Maxwell, and was a stimulus that eventually set him on the course of resolving the mystery of the new, wonderful force, an effort that was to occupy a large part of his life's work.

In approaching the subject of electricity while at Trinity, Maxwell decided that before making any analyses of his own he would first sit at the feet of wisdom in the form of the greatest master of the time in revealing the powers of electricity and magnetism - Michael Faraday - who was winding up a lifetime of brilliant discovery.

Maxwell resolved first of all to read through Faraday's *Experimental Researches in Electricity.* The result of that reading was a paper, published in 1856, *On Faraday's Lines of Force.* This was the first of the remarkable series of works on the fundamentals of electricity and magnetism which continued to pour out of Maxwell's pen until his death in 1879.

Maxwell's stay at Cambridge ended when he became a candidate for a vacant position, the chair of Natural Philosophy at Marischal College, Aberdeen. The move was in deference to his father, who was in failing health, and wanted his son near him. He obtained the professorship in early 1856. Unhappily, the kind intentions that prompted him to apply were frustrated by his father's death in April 1856. It was a tragic loss for Maxwell; father and son were as close as it is possible to be.

Although his teaching load at Aberdeen was not heavy, and he took the work seriously, Maxwell was not entirely in his element at the job of drilling young and raw pupils in the principle of mechanics and physics. He found it difficult to pace himself to the capabilities of the average student. His lectures often involved an excursion on a idea that flashed into his mind; discussion that would be inspirational to a few bright members of the class, but beyond the comprehension of most. Maxwell missed the type at Trinity who could appreciate his originality, and in whose instruction he himself could derive an advantage.

**SATURN'S RINGS** - Maxwell's studies in electricity and magnetism in Aberdeen were interrupted by a project so demanding that it occupied him for two years. He entered the competition for the Adam's Prize Essay of Cambridge University, the topic of which was *The Stability of Saturn's Rings.* These thin disks circling the equator of the second largest planet in our solar system, had intrigued physicists and astronomers ever since they were first seen in Galileo's primitive telescope in 1610. Were the rings rigid? Were they fluid or airy? Did they consist of matter not coherent, i.e., what was the structure and what accounted for their motion and permanence?

Maxwell put an immense amount of labor into his 69-page prize-winning essay, which Sir George Airy, Astronomer Royal, described as one of the most remarkable applications of mathematics to physics he had ever seen. After stability analyses for a number of ring models, Maxwell concluded that the only ring system that could work must consist of an assembly of countless tiny, independent satellites revolving with different velocities in accordance with their respective distances. Maxwell's Saturn essay, and the dynamical principles it used, established him as a foremost mathematical physicist of the day.

**STATISTICAL PHYSICS** - The Saturn conclusion that an uncountable number of individual tiny masses with random motion and velocities could compose a stable system, excited Maxwell's interest in a field involving similar principles: the *kinetic theory of gases.* This theory which had been proposed by Rudolf Clausius in Germany, Daniel Bernoulli in Switzerland and James Joules in England, and others preceeding Maxwell in the field, was successful in explaining pressure, temperature and density on the basis that gases were composed of innumerable particles in exceedingly swift random motion. Early proponents of the theory, to simplify the mathematics, assumed that all particles moved at the same speed. Maxwell, on common sense reasoning, and based on probability, held that this assumption was untenable because random collisions among the molecules would necessarily give them a wide range of velocities.

Maxwell then proceeded to treat the problem of molecules colliding at random by introducing a statistical method for solving the dynamics. He found it necessary to use a probability distribution

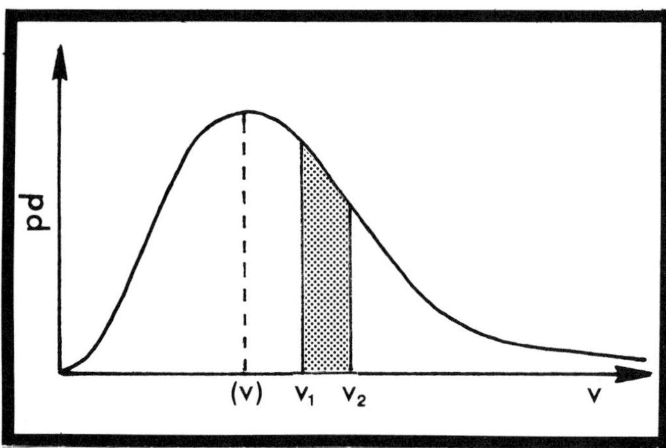


Fig. 12.3 - MAXWELL'S DISTRIBUTION CURVE for the molecular velocities in a gas. Each point on the curve indicates the probable density (pd) of the molecules moving at that speed, v. The molecules may have any value of velocity, but are expected to be near average value (v). Molecular density with the velocities between the points $v_1$ and $v_2$ is shown by the shaded area under the curve.

in which the molecules could have any speed, but some speeds were more likely than others. As the gas molecules strike each other they are pushed around in all directions. Each speed of a molecule depends only on where it was after the last push, and the new speed will depend on the next hit. The resulting jiggling pattern of the molecular path is called a "random walk," in which each step is not related to the previous step nor to the next step. Maxwell decided that this velocity description had the same pattern as Gauss had developed for the normal distribution of errors, represented by a bell-shaped probability curve, Fig. 12.3.

On the basis that ideas of probability are essential to the description of molecular happenings, Maxwell, using the statistical method, arrived at a formula for gas pressure. He then went on to apply the statistical method to other properties of gases. He explained viscosity in terms of molecular momenta, which were in turn dependent on the mean free path, or the average distance of molecular travel between successive collisions. Gases with long mean free paths would have high viscosity. Maxwell found that the viscosity of a gas at a given temperature was independent of the gas pressure.

The application of statistical methods to the study of gases by Maxwell was a scientific achievement of a high order. His several memoirs on the *Dynamical Theory of Gases* inspired other investigators to enlarge and refine the mathematics to the point where explanation of the remaining discrepancies from a complete representation had to await development of the quantum theory.

**IN FEBRUARY 1858** Maxwell wrote Miss Cay, his aunt: " This comes to tell you that I am going to have a wife . . . dont be afraid, she is not mathematical, but there are other things besides that, and she certainly will not stop mathematics." His bride was Katherine Mary Dewar, daughter of the principal of Marischal College, and seven years older than he. It was reported that Mrs. Maxwell was of neurotic disposition. But judging from the tone of the many letters between them, recorded in Campbell's biography of Maxwell, the marriage union was close and affectionate, and Maxwell was devoted to her. She was a helper in some of the experiments on color vision and the kinetic theory of gases performed at his home. They had no children.

In 1860 Marischal College and King's College, the other one in Aberdeen, were joined into a new institution - the University of Aberdeen. The new chair of Natural Philosophy thus created was filled by David Thomson, who held that post at King's College, and was Maxwell's senior. In the retrenchment that followed, Maxwell was dismissed. There was some popular resentment at this in the town, and charges of favoritism in the selection of Thomson. But Maxwell was secure in his reputation and soon received an appointment to the chair of Natural Philosophy at King's College, London, where he would remain until 1865.

**LONDON**, Maxwell's new home, now gave him the opportunity of occasionally consulting with Faraday, with whom he had previously had only correspondence. Both exchanged ideas on the electrical work in which they were so intimately connected, and both held each other in the highest respect. Living in London also gave Maxwell social and professional contacts with men in his own departments of science. He was active in the British Association for the Advancement of Science, which at that time was setting up standards for electrical measurements. Despite his arduous teaching duties and social distractions, Maxwell's five London years were distinguished by the production of his most important papers.

He resumed work on color, and carried on with experiments in the large garret of his home in Kensington to obtain data that would verify and add to his determinations in quantitative color-imetry. He also worked in the garret, with the help of his wife, to make additional measurements of gas viscosity in relation to its temperature and pressure.

But of greatest importance, Maxwell returned to his work in the electrical science, from which he had been diverted in Aberdeen by the Saturn essay. He was now ready to attack on a broad scale, and to gather into a unified theory, the disparate elements of electricity and magnetism.

**ELECTRICITY AND MAGNETISM,** at the time Maxwell began his investigations, had become the central focus of physical science. This was largely as a result of Faraday's amazing and wide-ranging experimental discoveries. Faraday's achievements crowned a century of research originated by his predecessors: Coulomb, Volta, Oersted, Ampère and others. Their efforts had produced a body of facts on electric charges and currents and their links with magnetism and chemistry.

Electric charges had been proved to be related to each other with a force law like gravitation, that is, as the product of their charges and inversely as the square of their distance. Electric currents were shown to be encircled by magnetism, and hence currents themselves could react electro-dynamically. Faraday had demonstrated electro-magnetic motor action, and had revealed that current was induced in a wire when it was swept by a field of magnetism. He had shown that magnetism could rotate the plane of polarized light, had analyzed electrochemical reactions, and had even hinted at a connection between electro-magnetism and light.

Each discovery introduced problems demanding explanation. What were the relationships between electricity and magnetism, and how did they influ-ence each other through space? What was a field? Could the elements of electric and magnetic action be brought into a unified theory? Did light waves have an electromagnetic character? The answers to these and other questions of electromagnetic cor-relation occupied the best minds ofthe time.

To many theorists of the mid-nineteenth century the manifestations of electricity and magnetism through space were explainable by Newtonian physics. The framework of electrostatics and magnetics worked on by the Continental physicists: Laplace, Poisson, Gauss, Weber and others were employing the mathematics of instant action-at-a-distance. In analogy with the laws of gravitation, the forces of electricity and magnetism traversed instantaneously from their origination in one body to the next. The Newtonian physicists directed attention primarily at the places where electric and magnetic forces happened to be located. The condition of the intervening space was not consid-ered in the interaction between the force centers.

The Newtonian physics worked all right for charges and currents, but when Faraday and Henry discovered electromagnetic induction, involving a time element for action, it became necessary to consider what was going on in the space in which the dynamic forces were operating.

**FARADAY** had a theory that he thought was a truer picture of nature. From his earliest experi-ments he had developed the conception that the forces of electricity and magnetism were evidenced by physical strains. These strains were a property of matter and of space and they were transmitted contiguously with a specific velocity in the medium between the interacting electrical and magnetic bodies.

Thus, to Faraday, the scene of action was in the medium. The strains represented a dynamical condition of space in which tension, compression and progressive motion were the fundamental properties. This condition of space he visualized in terms of "force fields" made up of "lines of force" which acted progressively and could curve to span and fill the space between centers of forces. The direction and density of the lines gave a picture of force potential, and of quantity.

Maxwell had thoroughly read Faraday's great *Experimental Researches in Electricity.* Maxwell was struck with Faraday's concept of fields of force. He highlighted Faraday's view and that of proponents of action-at-a-distance:

"Faraday, in his mind's eye saw lines of force traversing all space, where the mathematicians saw centres of force attracting at a distance.

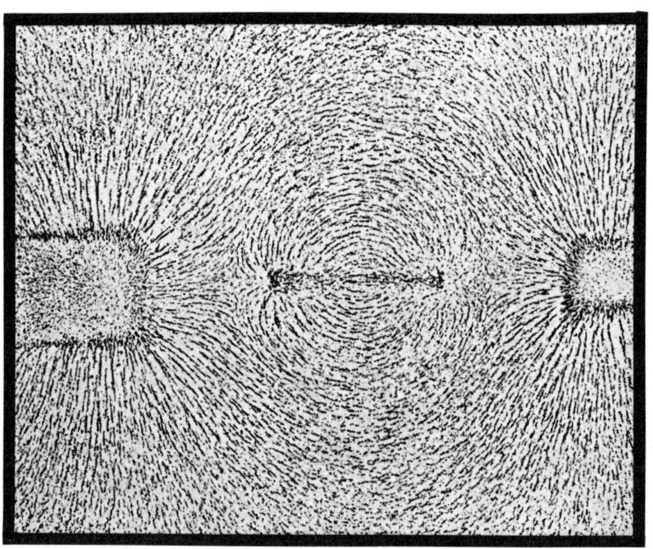

MAXWELL UTILIZED FARADAY'S LINES OF FORCE

Fig. 12.4 - In his paper "On Faraday's Lines of Force," Maxwell developed a hydrodynamical model for magnetic action based on streamlines of an incompressible fluid having forces equivalent to the tension and compression attributed by Faraday for the lines of force as pictured above. This iron filings representation of a magnetic field between two bar magnets was prepared by Faraday.

Faraday saw a medium where they saw nothing but distance; Faraday sought the seat of the phenomena in real actions going on in the medium, they were satisfied that they had found it in a power of action at a distance impressed with electric fluids."

Maxwell, who was gifted with a physical imagination similar to that of Faraday, was profoundly impressed with the field of force concept, and he proceeded to develop it in his paper, *On Faraday's Lines of Force,* written while he was at Trinity.

Maxwell's motivation and method were due partly to the influence and example of a life-long friend and confidant, William Thomson, later to become Lord Kelvin (1824-1907), a fellow Scotsman, also with a Cambridge degree. Thomson was a brilliant mathematician and physicist who had been the first to provide an analysis of transient electric currents. Thomson had also been interested in Faraday's experiments and had tried to explain electromagnetic induction by a model in which propagation of electric and magnetic forces were analogous to changes in the displacement transmitted through a perfectly elastic solid.

Thomson, in common with other scientists of the Cambridge school, placed great importance on the science of *dynamics,* that is the science of force and motion. Dynamics, based on nature's most widespread activity, motion and its communication was placed, with mathematics, at the head of all the sciences. Thus, mechanical models, providing demonstrable properties, were essential for analysis and formulation. Thomson declared that he could not understand anything unless he could make a model of it. This could be intricate in the case of unobservable motions such as electric current and the electromagnetic field.

Maxwell, believing also that physical action was founded on dynamics, and impressed with Thomson's attempts at physical analogy, was inspired to proceed with his own description of electric and magnetic action by use of mechanical models. He began with an examination of Faraday's lines of force.

**FARADAY,** in attempting an explanation of electromagnetic induction, had invented what he called the *electrotonic state.* This he conceived as the special state developed in a wire when in the presence of a magnetic field. Any change of the electrotonic state, such as by movement of the wire in the lines of force of the field, or a change of the lines of force relative to the wire, always resulted in an electromotive force, and if the circuit was complete, in an electric current.

Faraday's electrotonic state was a starting point for Maxwell. As he had written in 1855 at Trinity: "By a careful study of the laws of elastic solids, and of the motions of viscous fluids I hope to discover a method of forming a mechanical conception of the electrotonic state adapted to general reasoning."

Maxwell proposed a hydrodynamical model for his purpose. In the model each line of magnetic force was represented by a fine tube, of variable section, carrying a constant flow of an incompressible fluid. The tubes of fluid force, side by side, filling the magnetic space, were vectors indicating the velocity and direction of flow. By analogy, this was the direction of flow of the magnetic circuit in that location, and the magnetic intensity at any point was represented by the velocity of the fluid.

The streamlines of flow were applicable likewise to electric lines of force. The magnitude of velocity of flow represented the amount and direction of electrical potential. Difference of flow represented difference of electrical pressure. Maxwell used the fluid analogy also for dielectrics and for diamagnetics. Materials like these, interposing a change of medium to the flow, he assumed to be equivalent to a resistance, which varied in proportion to the velocity of flow. Thus a change in the medium tended to divert the electric and magnetic forces.

The model enabled Maxwell to establish a distinction in two vector functions of electric and magnetic action. One function represented the intensity or "force" of the electricity or magnetism; the other function represented the resulting electric or magnetic density, which he called the "flux."

Maxwell proceeded with the model to determine the geometrical significance of the electrotonic state. In his analogy he now envisioned the tubes of flow as being elastic, expanding or contracting with any variation of the magnetic intensity. A changing flow of magnetism would thus exert a dimensional change equivalent to an electric pressure. Pressure variation was transmitted from tube to tube through the elastic surfaces and this was a change in the electrotonic state with a resulting production of electromotive force.

Maxwell concluded that "the electromotive force on any element of a conductor is measured by the instantaneous rate of change of the electrotonic intensity on that element, whether in its magnitude or direction."

The paper on Faraday's lines of force was concluded with a mathematical analysis of some examples of electromagnetic induction.

**FARADAY** was delighted with Maxwell's interpretation of his researches and wrote his appreciation: "I was at first almost frightened when I saw such mathematical force made to bear on the subject and then wondered to see how the subject held up so well."

In addition to the commendation, and speaking of mathematics, the sixty year old Faraday had a suggestion for the author who was then only twenty six years old:

"There is one thing I would be glad to ask you. When a mathematician engaged in investigation of physical actions and results has arrived at his conclusions, may they not be expressed in the common language as fully, clearly and definitely as in the mathematical formulae?

. . . If this be possible, would it not be a good thing if mathematicians working on these subjects, were to give us the results in the popular, useful, working state, as well as in that which is their own and proper to them?"

**MAXWELL'S** paper *On Faraday's Lines of Force* was the first installment of his search in the field of electromagnetic action. Six years elapsed before he could get on with the job of expanding on Faraday's work. He was now at King's College. Here, in 1861, he resumed formulation of the physical nature of Faraday's electrotonic state, proposing to "examine magnetic phenomena from the mechanical point of view." In pursuit of this he invented an ingenious thought model. With it he hoped to explain the operation of changing fields.

Faraday had given a hint for the nature of a model when he said of magnetic lines of force that " . . . there is a tension along the lines and a pressure between them." The best method of representing this type of stress distribution, Maxwell decided. was to assume magnetism was a rotary phenomenon that could be pictured by "molecular vortices" whirling in an incompressible fluid.

Consider the array of parallel molecular vortices shown end on by the hexagons of Fig. 12.5, an illustration that appears in the Maxwell paper: *On Physical Lines of Force,* published in 1862. Each vortex can be thought of as a tube rotating axially around a line of force in order to produce the magnetic intensity. The speed of the vortices at the surface is identified with the amount of the intensity. To produce a given magnetic field the vortices were presumed to rotate at constant speed in the same direction.

But how could the vortices in contact with each other all rotate in the same direction? "I have found great difficulty," said Maxwell, "in conceiving of the existence of vortices in a medium side by side, revolving in the same direction about parallel axes. The continguous portions of the consecutive vortices must be moving in opposite directions."

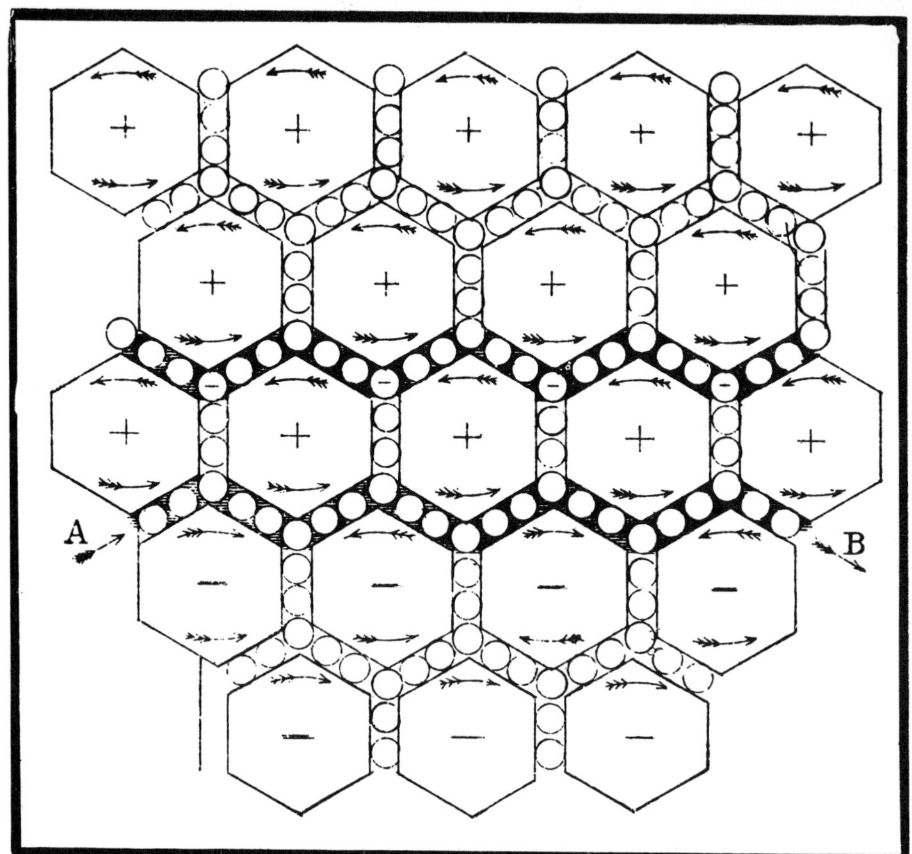


MAXWELL'S MODEL ILLUSTRATING ELECTROMAGNETIC ACTION

Fig. 12. 5 - Mechanical model which appears in Maxwell's paper: "On Lines of Physical Force." He used it to explain the actions of an electromagnetic field by analogy with the dynamics of elastic solids. The rotating parallel "magnetic vortices," which he made hexagonal to accomodate the geometry, are seen end-on. Between are spherical "electrical particles" that in translatory motion or in deformation simulated a current. A sudden flow of spheres from A to B rotates the magnetic vortices to move adjacent spheres, thus illustrating the principle of electromagnetic induction.

Maxwell's solution to the rotation puzzle was ingenious: he interposed a layer of small spheres, "idle wheels" he called them, so as to counter-rotate between the vortices. Now, linked to each other by the spheres, the vortices could all rotate in the same direction.

With the new arrangement, the relative motions of the vortices and spheres could be used to explain electromagnetic phenomena. Maxwell thought of the spheres as being "particles of "electricity." Revolving in place they were simply stationary electric charges. But when rolling between the vortices in a translating motion, the spheres constituted an electric current. Hence the model could combine electricity and magnetism based on magnetism as a phenomenon of rotation of vortices, and electric current as a phenomenon of sphere translation.

Turning at a uniform speed in the same direction the vortices represented a steady-state magnetic field. In this condition the revolving spheres did not move out of position, hence there was no potential to produce a current.

Suppose, however, a sudden increase in the magnetic intensity so that the rotary velocity of some molecular vortices are accelerated. The relationship of these speeded-up vortices to their neighbors is now altered and this affects the intermediate spheres. They are subject to a differential velocity and can no longer revolve in a stationary position.

The tangential action on the spheres causes them to circulate or "curl" around the accelerated vortices. This tangential action during a change in magnetic intensity is equivalent to the induction of an electromotive force, and the subsequent motion of the spheres constitutes an electric current. Thus the model verifies Faraday's law that a magnetic field in action can induce an electromotive force.

Maxwell then let his model explain electromagnetic induction when a conductor was in motion across a stationary magnetic field. Assume a wire moving through the vortices at right angles to their axes, and at right angles to the axis of the wire. The moving wire will stretch the vortices in front of it, speeding them up. This differential speeding exerts a tangential force on the enclosed spherical particles, causing them to move along the axis of the conductor as an electric current. Thus, as Faraday had shown, a conductor cutting across a magnetic field has an electromotive force induced in it.

Thus far, operation of the model was successful. It accounted for situations where the spheres were free to move relative to the vortices, thus simulating a continuous conduction current. But how would it explain action in materials that did not conduct? Faraday's experiments with insulators provided the next challenge to Maxwell's model.

In his electrostatic condenser experiments, Faraday had found that various insulators, when electrified, were put under a strain, in the direction of the electric force. He called this strain effect *polarization.* Polarization was an internal displacement of the dielectric medium due to the electric pressure. It implied acceptance (and return) of an electric charge in an amount depending on the specific inductive capacity of the insulator.

**MAXWELL** said of conductors and insulators: "If a body is a conductor, an actual passage of electricity takes place . . . bodies which do not permit a flow of electricity through them, are called insulators . . . but though electricity does not flow through them, the electrical effects are propagated through them, and the amount of these effects differs according to the nature of the body, so that equally good insulators may act differently as dielectrics.

A conducting body may be compared to a porous membrane which opposes more or less resistance to the passage of a fluid, while a dielectric is like an elastic membrane which may be impervious to the fluid, but transmits the pressure of the fluid on one side to that on the other.

The effect of this action on the whole dielectric mass is to produce a general displacement of the electricity in a certain direction. This displacement does not amount to a current, because when it has attained a certain value it remains constant, but it is the commencement of current, and its variations constitute currents in the positive or negative direction, according as the displacement is increasing or decreasing."

How could a mechanical model of vortices and spheres explain displacement current? Maxwell's answer: by giving the spherical electrical particles elastic walls. Now the spheres, although confined and stationary in the dielectric, still had freedom of elastic yielding in response to a force. They could produce linear displacement by altering their shape. With an applied electromotive force they would distort in one direction. Distortion represented a movement of charge, and although small it was the beginning of a current. On release of force the distorted spheres would snap back into original shape, and this movement also produced the beginning of a current.

Thus, in addition to the well-known current occurring in conductors, Maxwell inferred that there must be added another current which he called *displacement current.* In the case of a condenser discharging through a connecting wire, for example, the total current consisted of the conduction current in the wire, and the displacement current in the dielectric of the condenser.

Recalling from Fig. 12.5, that the spherical particles are enmeshed in the vortices, the model now revealed another important fact. As the spheres are distorted, and then returned to normal shape with application and removal of electromotive force, there is a corresponding displacement movement of elongation and then contraction. This movement, by a tangential contact, is transmitted to the vortices on either side, thus setting them into rotation. This is the condition for producing magnetism. Thus Maxwell came to the highly significant conclusion: *The current of displacement creates a magnetic field exactly as does a conduction current.*

**MAXWELL'S MODEL** had achieved its purpose. It interpreted the function of a dielectric in a circuit in terms of displacement current with accompanying magnetic field. Now an important corollary followed, and Maxwell pursued it to an outstanding conclusion. If the spheres are set into elastic motion by an oscillating, wave-like electric field, there should be a corresponding wave-like displacement current accompanied by transverse waves of magnetism. These waves should then propagate.

But waves require space for propagation. Faraday confined his dielectrics to ponderable materials. However, in Maxwell's model, dielectric space had no limitations. The space envisioned by Maxwell was not confined to a material like a wire, but included that space surrounding it - whether it be air or a vacuum.

Waves also require time for propagation, and this implied a velocity. What would this be? Maxwell set out to determine it on the basis of his model. To do so he went back to the older physics of elastic solids. Previous studies had shown that viscosity and elasticity were related to each other through time. Would the propagation of electric and magnetic forces be analogous to the transmission of displacement through an elastic solid?

Maxwell applied to the matter of the magnetic vorticies, a *density,* $\rho$, and to the transverse elasticity of the spherical particles a *rigidity coefficient, m.* It was known that in an elastic medium which involved density and rigidity, vibrations would be propagated with a velocity equal to:

$$V = \sqrt{m/\rho} \qquad (1)$$

To equate the density and rigidity of the model with the electromagnetic properties, Maxwell defined a coefficient of magnetic permeability, $\mu$, proportional to the density, $\rho$, so that:

$$\mu = \pi\rho \qquad (2)$$

Then, from (1) $\pi m = V^2\mu$ (3)

Now the rigidity ($\pi m$) was equated to a quantity $E$, such that:

$$E^2 = \pi m \qquad (4)$$

Maxwell had previously determined, by the relationship of Coulomb force between unit charge and the equivalent force between unit current, that the quantity, $E$, was a number by which the electrodynamic measure of any quantity of electricity must be multiplied to obtain its electrostatic measure.

Substituting (4) into (3):

$$E = V\sqrt{\mu} \qquad (5)$$

Thus, in Maxwell's model the velocity, $V$, was shown to be the same as the velocity ratio, $E$, between electrostatic and electromagnetic units.

This ratio had already been determined by Weber and Kohlrausch in their celebrated experiment reported in the previous chapter. Adapting those results to the equation, Maxwell found that for a medium for which the permeability, $\mu$, was equal to unity, equation (5) now became simply:

$$V = E$$

The Weber-Kohlrausch experiment had determined the velocity ratio as $3.1 \times 10^{10}$ centimeters per second, or 193,088 miles per second. Fizeau had measured the velocity of light at 195,647 miles per second.

The velocity coincidence was astounding, and Maxwell gave it powerful significance in the conclusion which he placed upon it:

"The velocity of transverse undulations in our hypothetical medium, calculated from the electromagnetic experiments of MM. Weber and Kohlrausch, agrees so exactly with the velocity of light calculated from the optical experiments of M. Fizeau, that we can scarcely avoid the inference that *light consists in the transverse undulations of the same medium which is the cause of electric and magnetic phenomena.*"

**MAXWELL'S EXCITEMENT** was evident in the italics with which he ended the statement.This was the first step in his forthcoming electromagnetic theory of light!

But so novel a concept as displacement current, its part in forming a closed current circuit, and the ability to undulate in free space, was understood with difficulty, and accepted with reluctance by many of Maxwell's contemporaries.

Maxwell made his own determination of the electrostatic-electromagnetic ratio. Using the apparatus of Fig. 12.6, he balanced the electromagnetic repulsion between two parallel coils with the electrostatic attraction of two electrified disks to each of which one coil was attached. The ratio was measured in terms of a resistance of over one million ohms connected between the disks. The average of several readings gave a velocity equivalent of $2.8798 \times 10^{10}$ centimeters per second, which was in general agreement with the Weber-Kohlrausch figure. Tests made by others using various apparatus and test methods were also in agreement.

**THE MECHANICS** of Maxwell's model had now delivered two portentous discoveries that were highlights of his paper *On Physical Lines of Force.* One was a new kind of current called *displacement.* The other was a hypothetical *undulating medium* that related light and electromagnetic phenomena by a common velocity. With this achievement Maxwell proceeded to build in a more developed form a unified theory of electricity and magnetism. This he displayed in mathematical formulation in *A Dynamical Theory of the Electromagnetic Field,* presented to the Royal Society in 1864.

"The theory I propose," wrote Maxwell in the preface, "may be called a theory of the *electromagnetic* field because it has to do with the space in the neighborhood of the electric or magnetic bodies, and it may be called *dynamical* theory because it assumes that in space there is matter in motion, by which the observed electromagnetic phenomena are produced.

"The electromagnetic field is that part of space which contains and surrounds bodies in electric or magnetic condition . . . It may be filled with any kind of matter, or we may endeavour to render it empty of all gross matter . . . as in the case of vacua. There is always, however, enough matter left to receive and transmit undulations of light and heat . . . the undulations are those of an aethereal substance . . . a pervading medium of small but real density, capable of being set in motion, and of transmitting motion from one point to another with great, but not infinite velocity."

In building his new framework on the behavior of electricity and magnetism, Maxwell now set aside his model. The whirling vortices and the rolling spheres were discarded. But his fundamental concept of the electromagnetic field as the mechanical state of a dielectric media remained. He substituted two supports: the field and the undulating medium. "The energy of electrification," wrote Maxwell, "resides in the dielectric medium whether that be solid, liquid, gaseous, dense or rare, or even what is called a vacuum, provided it still be capable of transmitting electrical action."

Thus, to Maxwell, the electromagnetic field was not an independent dynamical entity. It was but one of the mechanical states of a material sub-

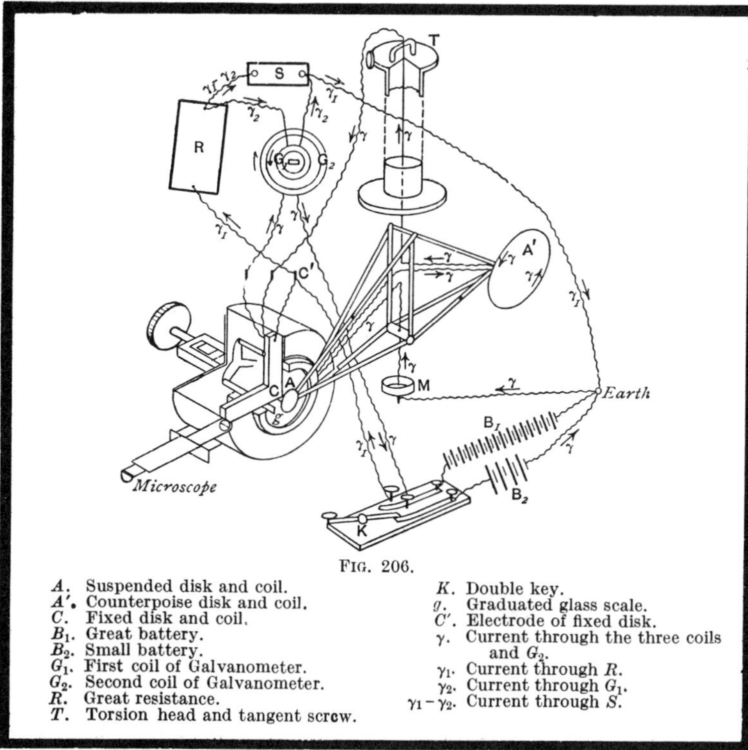


Fig. 12.6 - MAXWELL'S APPARATUS for determining the velocity ratio of electrostatic and electromagnetic units, in terms of a great resistance. The value obtained was 2.8798 $\times 10^{10}$ centimeters per second, about equal to the velocity of light. Illustration from Andrew Gray, "Absolute Measurements in Electricity and Magnetism, Dover Publications.

stance which he described as " of a more subtle kind than visible bodies, supposed to exist in those parts of space which are apparently empty" . . . in other words the *ether.* This hypothetical substance existed not only in free space, but in the supposed vacuous interstices of a ponderable matter itself. An elastic, quivering, universal medium was essential as the seat of storage and as the vehicle of transmission for the electrical and magnetic energy.

**EQUIPPED** with fields of ethereal matter, Maxwell now applied himself to the task of setting up in a single theory the known facts of electricity and magnetism in the form of *General Equations of the Electromagnetic Field.* Called one of the great triumphs of 19th century science, the equations formulated the discoveries of his predecessors and added the perceptions of his own genius to

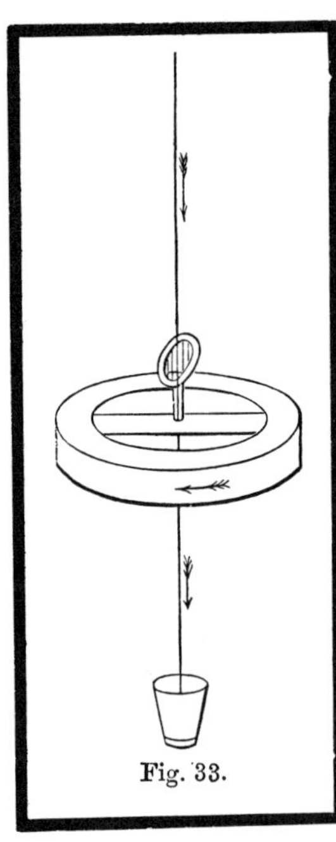

Fig. 33.

IS ELECTRICITY A FLUID?

Fig. 12.7 - If it was like water flowing along a wire, then at the moment of starting and stopping a force would be exerted in changing its momentum. To test this Maxwell proposed a coil with many turns suspended by a wire so that it could rotate about its vertical axis, and its motion indicated by a mirror. Current to be fed through the suspension wire to one below dipping into a mercury cup. To prevent action on the coil from the earth's magnetism, the earth's field was to be exactly neutralized by fixed bar magnets.

Now, if electric current is a material substance, its angular momentum on its application to the coil should produce a reaction, indicated by the mirror. But, wrote Maxwell, "no phenomenon of this kind has as yet been observed."

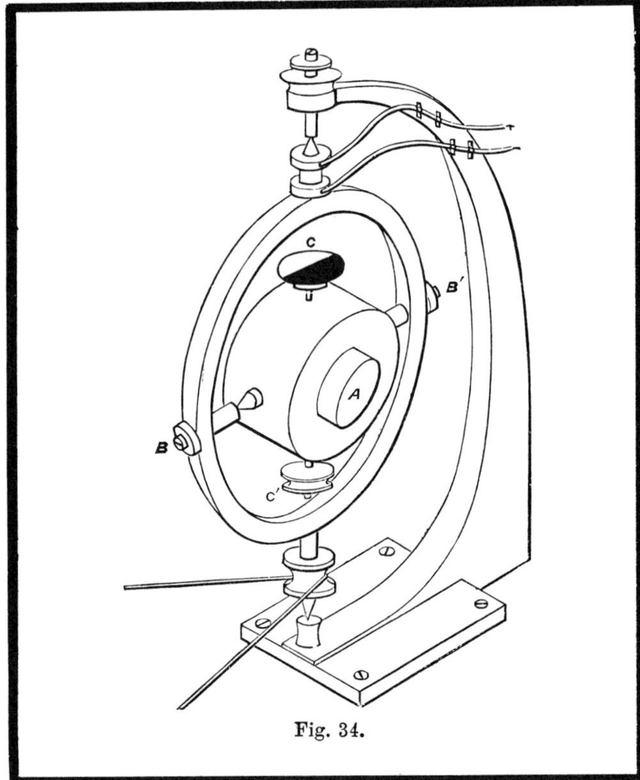


Fig. 34.

Fig. 12. 8 - If electricity were a material fluid, a rotating electromagnet should exert a gyromagnetic effect. To detect this Maxwell devised the apparatus above for rapidly rotating a magnet, A. No evidence of the action was found.

provide in compact mathematical language the body of electromagnetic action.

As listed in his *Dynamical Theory of the Electromagnetic Field,* Maxwell's eight general equations expressed:

*(A) The relation between electric displacement, true conduction, and total current compounded of both.*

*(B) The relation between the lines of magnetic force and the inductive coefficients of a circuit, as already deduced from the laws of induction.*

*(C) The relation between the strength of a current and its magnetic effects, according to the electromagnetic system of measurement.*

*(D) The value of the electromotive force in a body, as arising from the motion of the body in the field, the alteration of the field itself, and the variation of electric potential from one part of the field to another.*

*(E) The relation between an electric current and the electromotive force which produces it.*

*(F) The relation between electric displacement, and the electromotive force which produces it.*

*(G) The relation between the amount of free electricity at any point, and the electric displacements in the neighbourhood.*

*(H) The relation between the increase or diminution of free electricity and the electric currents in the neighbourhood.*

**THE 8 EQUATIONS** covering the relations above were somewhat revised and reordered in Maxwell's later *Treatise on Electricity and Magnetism.* On this foundation, and as usually written, the equations became four in number, traditionally called "Maxwell's Equations," which are one of the pivotal generalizations of physics.

They set down all modes of behavior of electromagnetism, in terms of the electric field, $E$, the magnetic field, $B$, and the currents. In his original work Maxwell used partial differential equations for his formulas. Later he adopted a form of vector calculus, and used the terms "divergence" (div), "curl" and "gradient" (grad) to express the mathematics more compactly, and to indicate somewhat the physical idea of the mathematical operation.

Of the four Maxwell equations, the first two indicate the nature of the continuity of the electric and magnetic forces, and the second two tell how their field forces affect each other when changing in time in any way.

The **first** equation is:

$$\text{div}\,\mathbf{E} = 4\pi\rho$$

Derived from Coulomb's and Gauss's laws, the equation states that, in the presence of a charge density, $\rho$, the number of lines of electric force, **E**, that diverge from a closed surface, flowing either outwardly (positive) or inwardly (negative), is a function of the electrostatic charge within. For a unit charge there is a divergence of $4\pi$ lines of electric force.

In empty space, with no charges present:

$$\text{div}\,\mathbf{E} = 0$$

The **second** equation is:

$$\text{div}\,\mathbf{B} = 0$$

is also derived from Gauss's law. It is a formal way of saying that all magnetic fields invariably consist of closed lines; no isolated magnetic poles have ever been found. As Ampère had deduced, there is no source of magnetic fields except from electric currents, hence there is no magnetic divergence from points in space.

The **third** equation is:

$$\text{curl}\,\mathbf{E} = -\frac{1}{c}\frac{\partial \mathbf{B}}{\partial t}$$

is a law for the electrical effect of changing magnetic fields, discovered by Faraday, and put into differential form by Maxwell. The production of an electromotive force was demonstrated by Faraday in his two crucial experiments: by plunging a bar magnet into a coil of wire, in which case the magnetic field was in motion, and by rotating a copper disk in the field of a magnet, in which case the conductor was in motion.

The time rate of change of the magnetic field, **B**, is denoted by $\partial \mathbf{B} / \partial t$, and the resulting effect is a circulation of an electric field, curl **E**, about an area of the magnetic lines of force which the curl encloses. The changing magnetic field, **B**, produces the electric field, **E**, which "curls" in closed loops about the magnetism. The minus sign indicates that the electric field acts in a sense to oppose what causes it.

If an electric charge was tracked around the path of the curl in Fig. 12.9, it would indicate a work potential or electromotive force with a value dependent on the ratio of the path of circulation to the area enclosed. In the mathematics of the curl, the circulation contour, and with it the enclosed area is allowed to shrink to a point. The curl then represents the electromotive force at this point, and, from the equation, is equal to the rate of change of the magnetic field at that point times the

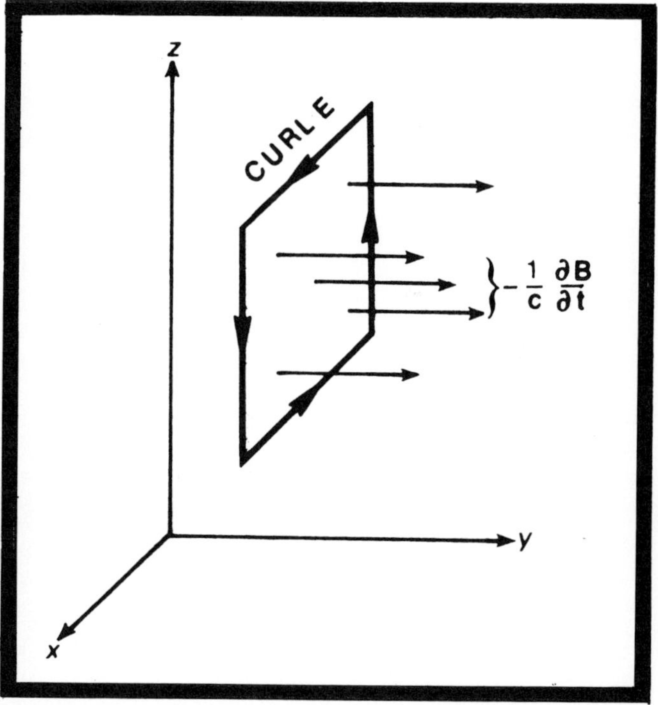


Fig. 12.9 - CURL OF MAXWELL'S EQUATION for a changing magnetic field is shown by this Cartesian coordinate diagram. The meaning of the curl is that the work required to take a unit charge once around the closed curve against the generated electric field is equal to the rate of diminution of the number of lines of magnetic force passing through the enclosure of the curve.

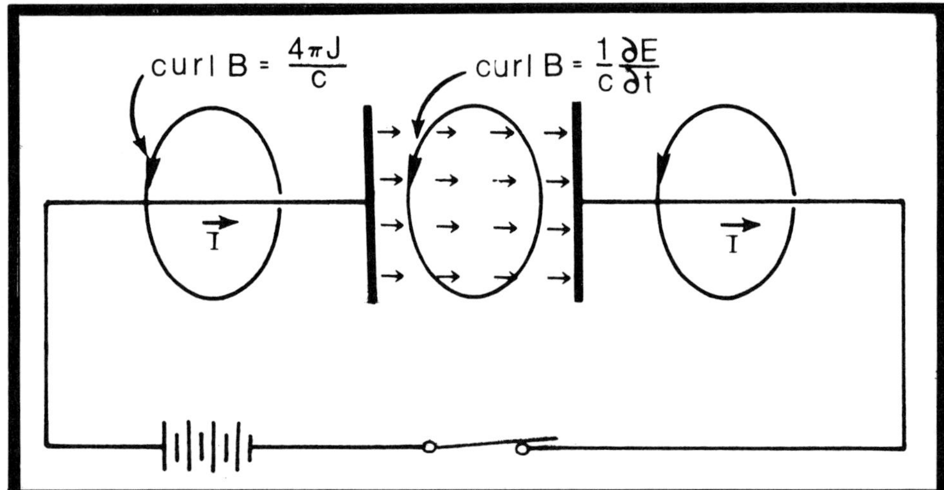


ELECTROMAGNETISM IN SPACE

Fig.12.10- Showing displacement current in a circuit with an air-gap condenser. When the switch is closed to charge the plates, a magnetic field, B, is produced by the conduction current, I. Is there also a magnetic field in the air-gap? Maxwell assumed there was. It was produced by a displacement current caused by change of the electric field $\partial E/\partial t$. The diagram is, modified, from E.P.Ney, "Electromagnetism and Relativity," Harper & Row.

fraction $1/c$. The factor, $c$, as Maxwell had shown, was necessary to relate the electrostatic units of **E** to the electromagnetic units of **B**, and was in fact a velocity of the same magnitude as light.

A wire loop in the path of curl **E**, in Fig. 12.9, would show experimentally that an electromotive force was induced by the changing magnetic field. But even without a conductor there would be an electromotive force in the space, and its presence and role in the next equation was Maxwell's brilliant contribution to expansion in the usefulness of electromagnetism.

The **fourth** equation is:

$$\text{curl } \mathbf{B} = \frac{4\pi\mathbf{J}}{c} + \frac{1}{c}\frac{\partial \mathbf{E}}{\partial t}$$

The roles of **E** and **B** have been interchanged. Now the magnetic field, **B**, curls as a result of the two components on the right side of the equation. These components represent two historical highlights of electromagnetic formulation - one due to Ampère, the other to Maxwell.

The first component, $4\pi\mathbf{J}/c$, represents the "conduction current," generally considered to flow in wires. By Ampère's law the circulation of the magnetic field, **B**, about the wire, is equal to the current density, **J**, in the wire, multiplied by $1/c$ to adjust the electrostatic and electromagnetic units. Prior to Maxwell this was the only member of the equation, because the only electromagnetic fields known were due to conduction currents.

But there were "open" circuits also, such as in Fig. 12.10, where an air-gap condenser interrupts the conduction current. On closing the switch a momentary conduction current, accompanied by its magnetic field, flows in the circuit to charge the condenser plates. Does a momentary current and a magnetic field also exist in the space between the plates? Maxwell said yes. Using his model, as we have seen, he had extended his inquiry beyond the conduction current to derive the behavior in an insulator. In a speculative discovery of the first rank he hypothesized that an electric pressure on a dielectric of any kind produced a displacement current, with an accompanying magnetic field that traversed the dielectric space. It was evident that the space had to be an active part of the electric and magnetic circuit.

Thus, as seemed reasonable to Maxwell, a continuity of current was necessary and could be predicted. The true total current must include the displacement current, which could exist even in free space. To complete the equation for true currents he therefore added the second component $\partial\mathbf{E}/\partial t$. Now the insulating medium, which could be empty space, or its near equivalent, air, became the conductor, and was, under some conditions, an equal or predominantly active part of the system.

Radiating into the vastness of empty space, free of conduction currents, what would be the nature of Maxwell's two time-varying equations for electric and magnetic fields?

They became simply:

$$\text{curl } \mathbf{E} = -\frac{1}{c}\frac{\partial \mathbf{B}}{\partial t}, \text{ and curl } \mathbf{B} = \frac{1}{c}\frac{\partial \mathbf{E}}{\partial t}$$

These symmetrical equations state that a magnetic field changing with time produces a changing electric field, and an electric field changing with time is accompanied by a magnetic field also changing with time.

**THE PROPERTIES** of the equations in free space immediately predicted to Maxwell a self-propelled electromagnetic wave motion. Any disturbance causing a changing electric field must produce a changing magnetic field which must produce a changing electric field, and so on. Energy is alternately exchanged between the two fields, and the waves advance outward in the process. Transported by an undulating ether, the vibrating electric and magnetic fields take off and wing their way through space.

What was the nature of this flight of waves through space? Maxwell was on the track of one of the great unifications in physics - the identity of light and electromagnetic waves. The correlation between the two depended on their being the same type of wave, of equal velocity, and transmitted through the same elastic ether.

Light had long been held, by experimental evidence, to be polarized, that is to consist of transverse waves. These vibrated at right angles to the direction of the beam of light. Light was assumed to be transmitted by a medium called the "luminiferous ether." The velocity of light had been derived by Fizeau and others, with optical measurement, to be about $3 \times 10^{10}$ centimeters per second.

Maxwell now used his equations to obtain similar characteristics for the magnetic and electric fields. By differentiating the two time-varying equations and substituting from one equation to the other, the vectors for the electric and magnetic field were shown to satisfy the traveling wave equation. Furthermore the **E** and **B** fields were shown to be perpendicular to each other, representing a polarized front for the wave, as shown in Fig. 12.11.

Solutions of the equations also resulted in a velocity, $v$. The value of $v$ entered into the equations as the equivalent of the constant, $c$, which was the ratio of unit size between the electrostatic and the electromagnetic systems. The equivalency was remarkable. The constant "c" was a universal velocity of electromagnetic waves, equal in amount to the ratio of the units, whose velocty equivalent had been derived by the ratio measurements of Weber and Kohlrausch, and confirmed by Maxwell in his own determinations.

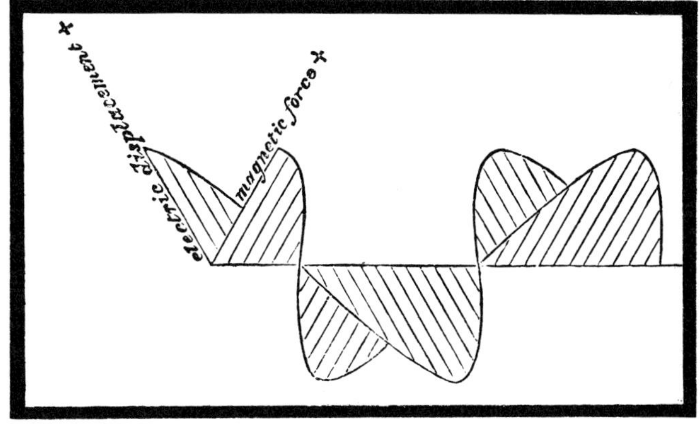


IN A MUTUAL EMBRACE THE ELECTRIC AND MAGNETIC WAVES FLY THROUGH SPACE

Fig. 12.11 - Maxwell pictured the propagation of electromagnetic waves with this diagram. Magnetic force is in a plane perpendicular to that of electric displacement, and both are transverse to propagation direction.

In his *Dynamical Theory of the Electromagnetic Field* Maxwell stated: *This velocity is so nearly that of light that it seems we have strong reason to conclude that light itself (including radiant heat, and other radiations, if any) is an electromagnetic disturbance in the form of waves propagated through the electromagnetic field according to electromagnetic laws.*

Before Maxwell there was only light; now optics was one phenomenon in the family of electric and magnetic waves.

**INCLUSION** of light into electromagnetics, with a definite ether velocity, made the Newtonian instantaneous action-at-a-distance theory less tenable. Time was implicit in wave propagation. The perception of an electromagnetic disturbance at a receiving point in space implied from Maxwell's field theory, a time-delayed response to a field disturbance from another point in space.

The wave theory of light, formally deduced from electromagnetic field forces, provided, in Maxwell's view, the principles for testing the reality of propagating fields. But it was not apparent at the time how experimental proof could be carried out, and Maxwell did not offer any specific method. There was no conceivable way of producing electrically the enormous frequency of oscillations that light involved, or the means of measuring the associated fields. As a result of the lack of experimental verification, Maxwell's theory of electromagnetic waves was not universally accepted by his scientific contemporaries. They were split into opposing points of view; one was a rejection of the field theory and skepticism concerning the electromagnetic character of light; the other was acceptance of Maxwell's conclusions but with modifying and alternate interpretation of the role of displacement current and the conditions of wave propagation. There were several investigators attempting experiments that would test aspects of Maxwell's theory.

Light was a visible evidence for an electromagnetic wave theory. But it was recognized that there were other sources of disturbance to the ether that could produce waves of non-visible frequencies. Sparks from electrical machines, for example, had been observed to radiate enough energy to magnetize steel needles at a distance, implying a propagation at frequencies other than light. However, during Maxwell's time no technique for transmitting or detecting electromagnetic waves of any kind had been developed.

**HEINRICH HERTZ** in Germany in 1879, took up the investigation of electrical oscillations, and ten years later was the first to succeed in showing indisputably the existence and properties of waves.

In a brilliant series of experiments he produced waves of frequencies equivalent to present-day radio that could be manipulated by his laboratory apparatus. As described in a succeeding chapter, he generated, detected and measured the waves, and found that they exhibited the same characteristics as light, that is, interference, reflection and refraction.

A new range of self-propagating waves had been added to the electromagnetic spectrum - and light now became but one element in the wave domain. Hertz's discovery of the time dependency and transmission characteristics of electromagnetic waves was a defeat for Newtonian instantaneous action theory. It was also a victory for the field theory - electricity and magnetism were continuous functions of time and position.

**FIVE YEARS** of teaching and high research productivity had affected Maxwell's health, and he looked forward to relief from the pressures of life in London. Resigning the professorship at King's College in 1865, he returned to his country seat. For the next few years he led a secluded life at Glenlair, varied by annual visits to London for work on electrical standards under sponsorship of the British Association for the Advancement of Science, to Cambridge as an Examiner on the Mathematical Tripos, and to Italy for a tour in 1867.

Maxwell, at Glenlair, soon resumed a high level of activity in his "retirement." He kept up a heavy schedule of social and scientific correspondence and worked on a number of topics in physics. He enlarged his house, and enjoyed walking and riding and the neighborliness afforded by country life.

But his chief employment during the Glenlair years was the preparation of his celebrated *Treatise on Electricity and Magnetism.* Inspiration for this comprehensive work was enhanced by the swiftly growing importance of its subject. As expressed by Maxwell: "It appears to me that the study of electromagnetics in all its extent has now become of the first importance as a means of promoting the advancement of science."

In prefacing the book Maxwell proposed . . . "to describe the most important of these electrical and magnetic phenomena, to shew how they may be subjected to measurement, and to trace the mathematical connexions of the quantities measured. The most important aspect of any phenomenon from a mathematical point of view is that of a measureable quantity. I shall therefore consider electrical phenomena chiefly with a view to their measurement, describing methods of measurement, and defining the standards on which they depend."

**THE TREATISE** is partly based on Maxwell's earlier papers, expanded greatly in subject matter and in mathematical treatment, and formulated largely around Faraday's intuitive ideas of field force. Covering almost every branch of electrical and magnetic theory, the *Treatise* is divided into four parallel parts: Electrostatics, Electrokinetics, Magnetism and Electromagnetism. Each part begins with an interesting historical background review. For example, Ampère's investigation of the mutual action of electric currents, is covered in detail with illustrations of the experimental arrangements used by Ampère. Instruments, measurements and the experimental methods related to the various topics form a substantial part of each discussion.

Maxwell introduced the use of the three fundamental physical quantities: length [L], mass [M] and time [T] to express the dimensions of the electric and magnetic units. He also lists the practical units, three of which, *volt, farad,* and *ohm,* had already been named for the eminent scientists. Maxwell used the vector operations: gradient, divergence, convergence (non-divergence) and curl, coining the names for the latter two.

A chapter in final pages of the *Treatise* covers the "Electromagnetic Theory of Light," derived from his earlier paper *A Dynamical Theory of the Electromagnetic Field,* in which electricity, magnetism and light are described as a uniform system.

Maxwell extended the wave theory to cover the velocity of light in materials other than air - the general idea being that the velocity was dependent on the "specific inductive capacity" or "dielectric constant" of the material. Maxwell also examined the relation of the refractive coefficient of transparent media to the dielectric constant, and the relation between the reflection coefficient and conductivity of metals. These relations could only be treated approximately by physics of the time.

Energy considerations enabled Maxwell to predict that light falling on a surface would exert a radiation pressure. Confirmation of this extremely minute force had to await development of detection devices of the 20th century.

The first edition of the *Treatise* was published in 1873. Prompted by the rapid pace of theoretical and practical advance in the field Maxwell began an extensive revision in 1877, but did not live to finish it. A second edition appeared in 1881, edited by W.D. Nevin of Trinity College, Cambridge. The present edition, printed in 1891, and reprinted by Dover Publications Inc., N.Y., was edited by J.J. Thomson of the Cavendish Laboratory, Cambridge.

**THE TEACHING OF SCIENCE** as related to its application to industry declined in England in the 1860s compared to the uses related to technology

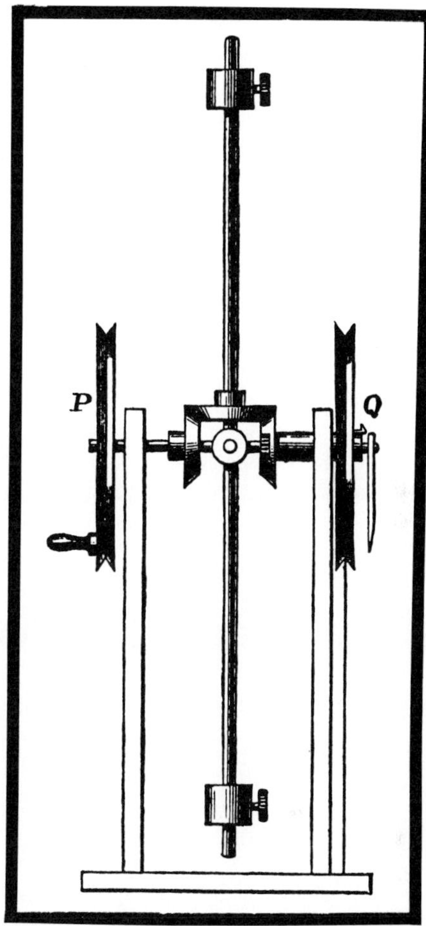


Fig. 12.12 - MAXWELL'S MODEL for illustrating the laws of inductive electrical circuits. P and Q are disks connected by a differential gearing. P represents the primary current and Q the secondary current. Inertia of the intermediate rotary arms can be varied by moving the weights inward or outward. Friction of a string on Q, kept tight by an elastic, represents the resistance of the secondary circuit.

If the disk P is set in rotation (simulating a current started in the primary) the disk Q will turn in the opposite direction (inverse current when the primary is started). When the rotation of P becomes uniform, Q is at rest (no current in the secondary when the primary current is constant); if the disk P is stopped, Q starts to rotate in the direction in which P was previously moving (direct current in the secondary on breaking the circuit). Effect of an iron core in increasing the induction can be illustrated by increasing the moment of inertia of the flywheel. Diagram is from Maxwell's *Treatise on Electricity and Magnetism.*

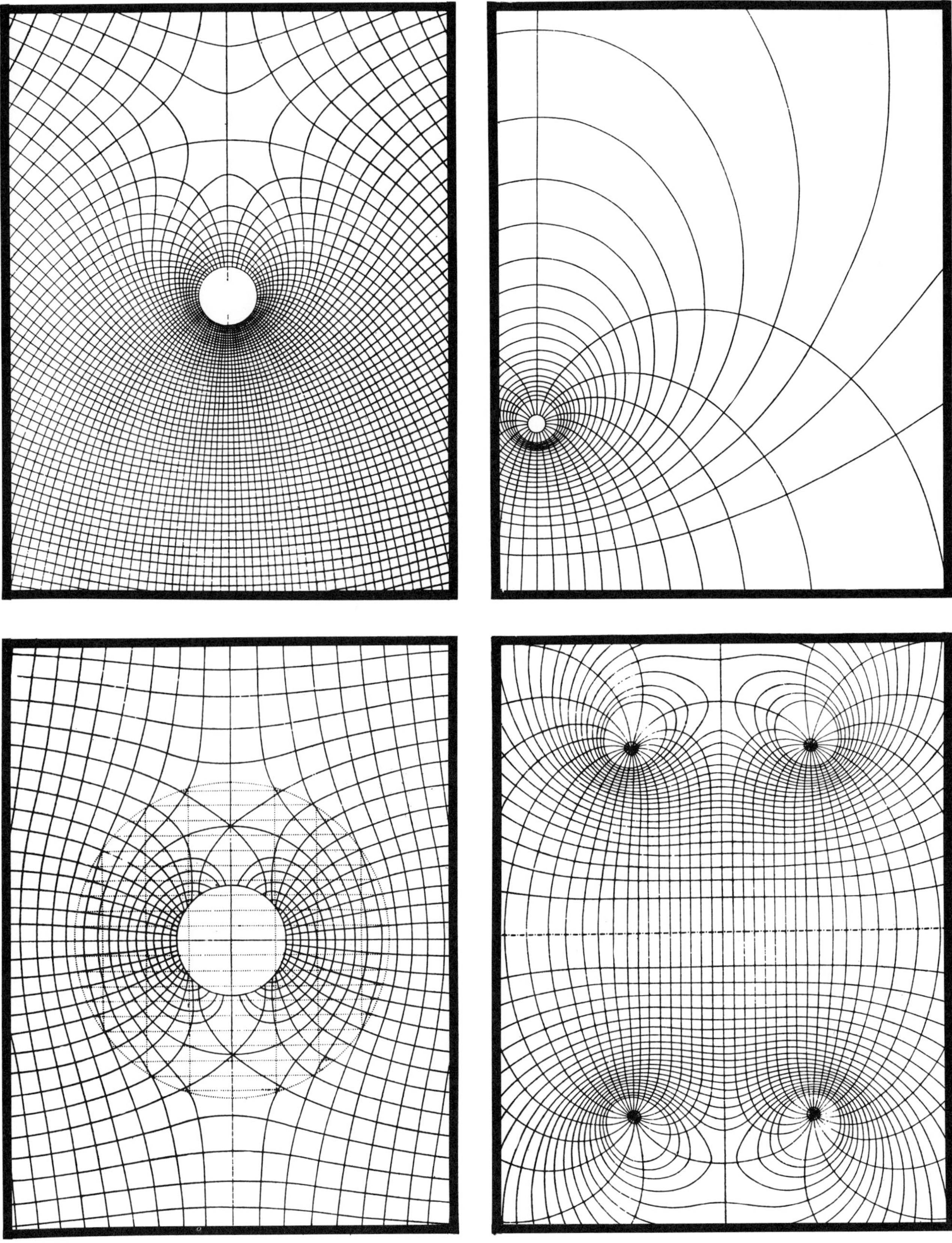

Fig. 12.13 - SOME OF THE MOST BEAUTIFUL FLUX maps ever produced are these lines of force diagrams appended to Maxwell's *Treatise.* They picture electromagnetic fields in the modern point of view as representing a dynamic state of strain in space. Upper left - electric conductor in a magnetic field; lower left - transverse magnetized cylinder in a uniform magnetic field; upper right - magnetic field in a portion of a circular current; lower right a uniform magnetic field produced by two current coils.

being fostered by institutions on the Continent. This began to reflect on the higher centers of learning. Cambridge, which was traditionally aloof to the commercial values and technical skills, found itself out of touch with the scientific movement around her. Complaints that the school was out of step, and cries for teaching of such subjects as heat, electricity and magnetism, reached the ears of the university's chancellor, William Cavendish, Seventh Duke of Devonshire. A man of means and sympathetic to the new knowledge, he soon took action by initiating the founding of a professorship in physical science at Cambridge, and implementing it with a grant, generous at the time, of £6300 for the building and apparatus.

The choice to plan and operate the new laboratory fell on Maxwell. He was induced, in 1871, to quit his Glenlair retreat to become its first professor of experimental physics. He assumed a major role in the design and construction of the building itself, and took a particular interest in the arrangements for the laboratories. Among the special designs were allowance of ample clearance for those experiments that required large horizontal and vertical space, devising vibration-free tables for sensitive instruments and provision of an iron-free room for magnetic measurements. Maxwell selected the finest equipment, often with improvements he found desirable, to make the institution the most advanced in instrumentation and in facilities.

The laboratory was formally opened in 1874, and given the name of Cavendish for the family which had counted among its members the celebrated 18th century scientist, Henry Cavendish. Maxwell was now an administrator, and a teacher of courses in heat, electricity and magnetism. He directed the character of the new studies, and in addition took an active part in the general business of the University. Maxwell set the tone of the Cavendish in his opening lecture: "it is natural to expect that the knowledge of physical science obtained by the combined use of mathematical analysis and experimental research will be more solid, available and enduring than that possessed by the mere mathematician or the mere experimenter." Maxwell brought to the laboratory his high skills in experimental procedure.

**FASCINATED** by the obscure career of the man whose name designated the new laboratory, Maxwell, as a sideline to his scheduled activities, spent spare hours for five years examining and editing and providing notes on the unpublished papers of Henry Cavendish (1731-1810), a forebear of the chancellor, and one of the most reclusive of all English scientific investigators. Cavendish, a Cambridge graduate, led a strange retired life as an experimenter, confining himself to his library and laboratory. In a lifetime of careful research he had made discoveries that anticipated Coulomb in the inverse square law of electrostatic charges, Faraday in the measurement of specific inductive capacity, and Ohm in his famous law.

Cavendish failed to publish his important works, and apparently cared little about priority of discovery. His findings went unnoticed for a century, and the credit for discoveries he had made went to others who later unknowingly repeated his work.

Maxwell's two-volume book: *Unpublished Experimental Researches of the Hon. Henry Cavendish,* published in 1879, finally established in the scientific world the reputation of this experimental pioneer.

**RAPID EXTENSION** of telegraph lines had brought to the forefront in the electrical community the need for determination and standardization of the quantitative units. The unit of electrical resistance and establishment of the standard ohm were first given attention. The existing standards were incoherent, varying in size from country to country. To remedy this situation the British Association for the Advancement of Scicnce appointed a committee in 1861 to determine what would be the most convenient unit of resistance and the best form and material for the physical standard representing the unit. A dozen distinguished British scientists made up the Committee, including Maxwell and his friend William Thomson, later Lord Kelvin.

In 1851 Wilhelm Weber, in Germany, had proposed a system of electric and magnetic measurement in which resistance would be expressed as a velocity. Adopting the Weber system of absolute units, the Committee recommended a velocity of $10^9$ cm/sec. With this value of a standard decided upon, the next determination was the physical form of the ohm and how it should be calibrated.

The first physical determination of the standard ohm in electromagnetic units was made by Maxwell and his associates Balfour Stewart and Fleeming Jenkins, using a revolving coil apparatus designed by William Thomson, as shown on the title page of this chapter. The subsequent history of the standard ohm involved the selection of the material and form for the practical type that could be distributed to and duplicated in other physical laboratories. This effort is described in the chapter on Ohm, the following chapter on the Siemens brothers, and in the final chapter of this book. Maxwell was active on the units committee for the selection and nomenclature of electrical quantities until 1870.

Although in this sketch first importance has been given to the work of Maxwell in electricity and magnetism, his activities in other scientific subjects ranged widely. In addition to his major papers he contributed articles to the celebrated ninth edition of the *Encyclopedia Britannica* on "Faraday," "Atom," "Ether" and subjects of a similar nature. He encouraged the popularization of science with simple and lucid presentations. Among his books for the layman was one on dynamics titled *Matter and Motion.* He gave frequent addresses to the scientific institutions of London, and was a regular visiting professor at Cambridge.

He began work on an *Elementary Treatise of Electricity,* but did not live to finish it. His health gave way in 1879 and he returned to Glenlair with hopes of improvement in the country air. Brought back to Cambridge for ministrations by his favorite doctor, he died on 5 November 1879 of abdominal cancer, and is buried at Glenlair.

**The MAXWELL** - In honor of Maxwell, the 5th International Electrical Congress, meeting in Paris in 1900, suggested the *maxwell* for the unit of magnetic flux in the CGS electromagnetic system of units. This was confirmed at the Stockholm meeting of the Congress in 1930. The maxwell is equal to one unit of magnetic flux per square centimeter in the CGS system, or $10^{-8}$ webers in the MKSA system. The maxwell is the magnetic flux which will produce a tension in its own direction of $1/8\pi$ dynes when distributed uniformly over one sq. cm. of cross-section.

Max Planck, winner of the 1918 Nobel prize in physics for his formulation of the quantum theory, wrote of Maxwell's work:

*" . . . it was his task to build and complete the classical theory of electromagnetism, and in doing so he achieved greatness unequalled. His name stands magnificiently over the portal of classical physics, and we can say this of him: by his birth James Clerk Maxwell belongs to Edinburgh, by his personality he belongs to Cambridge, by his work he belongs to the whole world."*

* * * * * * * * * * * * * * *

**The ETHER** - If electromagnetics in the form of light is propagated by waves, they are waves of what? Light travels throughout interstellar space and through the most perfect vacuum that can be obtained. If there are light waves they must be in some medium which extends throughout space as far as the most distant body from which we receive light. Maxwell, and most scientists of his time, could not conceive of electric and magnetic fields as entities propagating through space without a supporting medium. Maxwell expressed his view:

"If something is transmitted from one particle to another at a distance, what is its condition after it has left the one particle and before it has reached the other? If this something is the potential energy of the two particles, as in Neumann's theory, how are we to conceive this energy as existing in a point of space, coinciding neither with one particle nor with the other? In fact, whenever energy is transmitted from one body to another in time, there must be a medium or substance in which energy exists after it leaves one body and before it reaches the other . . . we ought to endeavour to construct a mental representation of all details of its action."

**Maxwell wrote poetry both serious and humorous:**

VALENTINE BY A TELEGRAPH CLERK (Male)
TO A TELEGRAPH CLERK (Female)

*"The tendrils of my soul are twined*
*With thine, though many a mile apart,*
*And thine in close-coiled circuits wind*
*Around the needle of my heart.*

*"Constant as Daniell, strong as Grove,*
*Ebbulient through its depths as Smee,*
*My heart pours forth its tide of love,*
*And all its circuits close in thee.*

*"O tell me, when along the line*
*From my full heart the message flows,*
*What currents are induced in thine?*
*One click from thee will end my woes."*

*Through many an Ohm the Weber flew,*
*And clicked this answer back to me,—*
*"I am thy Farad, staunch and true,*
*Charged to a Volt with love for thee."*

LECTURES TO WOMEN IN PHYSICAL SCIENCE

PLACE — A small cove with dark curtains.
The Class consists of one member.
SUBJECT — Thomson's Mirror Galvanometer.

*The lamp-light falls on blackened walls,*
*And streams through narrow perforations,*
*The long beam trails o'er pasteboard scales,*
*With slow-decaying oscillations.*
*Flow, current, flow, set the quick light-spot flying,*
*Flow current, answer light-spot, flashing, quivering, dying.*

*O look! how queer! how thin and clear,*
*And thinner, clearer, sharper growing*
*The gliding fire! with central wire,*
*The fine degrees distinctly showing.*
*Swing magnet, swing, advancing and receding,*
*Swing magnet! Answer dearest, What's your final reading?*

*O love! you fail to read the scale*
*Correct to tenths of a division,*
*To mirror heaven those eyes were given,*
*And not for methods of precision.*
*Break contact, break, set the free light-spot flying;*
*Break contact, rest thee, magnet, swinging, creeping, dying.*

---

**THE NOTION** of a universal medium goes back to the Greeks, who had a word for it - the *ether.* To them it filled all space of the starry heavens, and in it the planets moved. With the investigations of optics in the 17th century, the idea of an ether was taken over as a way of explaining the transport of light. With the introduction, in 1864, of Maxwell's equations demonstrating that light was an electromagnetic phenomenon, the ether became more essential than ever. Exceedingly varied theories were put forward to explain this mysterious medium. To carry light implied that it must be vastly different from any known substance.

Ether was of such importance to science of the 19th century that it took up four pages in the ninth edition of the *Encyclopedia Britannica,* prepared by Maxwell in 1879 just before his death. In it Maxwell presented the ether as a necessary transport mechanism for light.

To prove the existence of the ether Maxwell had suggested an optical experiment that would measure the "ether wind" as the earth orbited through it. No instrument of the sensitivity required was available in Maxwell's time, but his suggestion was picked up by an American optical scientist, A.A. Michelson (1852-1931). He, together with E.W. Morley (1838-1923), performed in Cleveland in 1887, one of most remarkable experiments of the 19th century.

Using an interferometer, in which a single beam of light was split with one beam in the direction of the earth's orbit and the other at right angles, they found no difference in the light speeds, hence no indication of the earth's motion through an ether. The null result was of great importance in leading to Einstein's special theory of relativity. Einstein derived two postulates: one, the impossibility of detecting absolute motion of a body, and two, that the speed of light in empty space is a universal constant, not influenced in any way by the motion of the light source or the observer. These postulates made the ether superfluous.

Today there is no understanding of a universal medium for the electromagnetic wave phenomena. But the ether theory, in its time, played an important part in the development of modern physics.

SIR WILLIAM SIEMENS (1821-1883)

ERNST WERNER SIEMENS (1816-1892)

# SIEMENS. . . the unit of conductance is named for two brothers who pioneered in electrical science, engineering and manufacture,

## WERNER and WILLIAM SIEMENS

"**TIME WILL PROBABLY REVEAL** to us the most effectual means of carrying power to great distances, But I cannot refrain from alluding to one which is, in my opinion, worthy of consideration, namely the electrical conductor. Suppose water power to be employed to give motion to a dynamo-electric machine. A powerful electric current is the result. This may be carried to a great distance through a large metallic conductor, and there be made to impart motion to electromagnetic engines, to ignite the carbon points of the electric lamp, and separate metals from their combinations.

"A copper rod of 3 inches diameter would be capable of transmitting 1,000 horsepower a distance of, say, thirty miles, an amount sufficient to supply one quarter of a million candle power, which would suffice to illuminate a moderately sized town.

" The use of electrical power has sometimes been suggested as a substitute for steam power, but it should be borne in mind that so long as the electrical power depends upon a galvanic battery it must be much more costly than steam power, inasmuch as the combustible consumed in the battery is zinc, which is a substance necessarily nuch more expensive than coal. But the question assumes a different aspect if in the production of the electric current a natural force is used, which could not otherwise be made available."

Thus spoke Dr.C. William Siemens, the new president, in an inaugural address to the London Iron and Steel Institute, as reported in the May 1877 issue of "Scientific American." Dr. Siemens was a notable example of a new class of men entering the electrical field. In addition to being a scientist-inventor, he was also an engineer-promoter, with a practical vision of electricity as a source of power for the expanding industrial age. In hands such as his, electricity and magnetism were to come out of the laboratory. Electromagnetic forces would be put into the bustling world to drive the wheels of industry, and thus reshape the fortunes of mankind in the coming age of technology.

Highlights of the previous chapters have covered the work of eminent scientists who pioneered in establishing the fundamentals of electricity and magnetism. Those individuals in general, were content with the discoveries they had made, the laws of nature they had defined and the pathways they had opened to further investigation. Having thus achieved, they went on to new scientific inquiry, and left to others to follow up and reap the rewards of putting the new knowledge to work in practical applications.

**THE ERA** of practical application began in the last half of the 19th century. An accumulating fund of scientific knowledge quickened the pace of industrial technology, and with it generated a new array of gifted men in the field of electricity. These were individuals mainly concerned with putting basic principles to work for a profit. They were the inventors, engineers and entrepreneurs who were devoting their talents to the ways and means of applying electrical power to usable devices. In doing so they provided and promoted historical advances in processes and apparatus for widespread and beneficial use of electromagnetics.

**NOW,** in the last half of the 19th century, *the world was waiting for a practical dynamo.* Batteries as Dr. Siemens stated, were far too costly for large-scale power. Mechanically generated electricity was essential. Economical power would promote the beneficial and profitable use of electrical apparatus and encourage developments for wider application.

Essential for progress also was development of a standard system of electrical units and the instruments to measure them. Methods of calculation had to be formulated for prediction of machine performance. Manufacturing needed a type of organization suited to the design and construction of the new electrical technology. Practical electricians needed training to operate and maintain the new electrical equipment.

One of the outstanding European groups associated with the advance of electrical science and the application of electrical products was the Siemens family of Germany. Four brothers, Ernst Werner and Karl Wilhelm (later Sir William), in collaboration with Karl and Freidrich, were pioneers in the development of the electrical and steel industries. All four played a part in the Siemens manufacturing and engineering operations which eventually became international in scope.

**WERNER von SIEMENS,** the eldest brother, and the oldest in a family of ten surviving children, was born on 13 December 1816 in Lenthe, near Hannover, Germany. The father, Christian Ferdinand, was a well-to-do farmer and estate manager, prominent in local affairs. Displaying an early interest in science and engineering, Werner, at age 17, received an appointment as officer candidate with the Prussian artillery. For three years he studied mathematics, physics and chemistry at the Berlin Artillery and Engineering School.

Attracted to the rapidly opening field of electricity, Werner secured a transfer, in 1841, to the artillery workshops in Berlin. Here he was given the opportunity of getting acquainted with electrical apparatus and of studying developments in the applications of electricity, the most popular and promising of which was telegraphy.

The early moving-arm type of semaphore was being replaced by the electromagnetic telegraph. The new method held such attraction for quick, long-distance communication that many were at work devising various types of apparatus for signalling and receiving messages. The electric telegraph aroused the interest particularly of the railroad and military authorities.

One of the most successful inventors was Charles Wheatstone (1802-1875) in England. Siemens was fascinated by the operation of the indicator-type Wheatstone telegraph instrument, Fig. 13.1. Realizing the potential of the new method of signalling for exchange of information in the military, Werner decided to specialize in telegraphy, and he proceeded to make modifications of existing systems to improve the speed, range and reliability of transmission.

**IMPRESSED** with the progress being made by Siemens, his army superiors gave him a commission in 1847 to lay an underground telegraph for use of the services. It was essential for the success of this venture to provide a covering for the wires that would keep out the ground mositure. Siemen's innovation for this was a coating of the recently discovered "gutta percha." This durable, rubber-like material from the latex of certain trees in Malaysia was a natural plastic that provided an excellent seamless, waterproof insulation for the conductors. Gutta percha was later to become an indispensable material for insulation of submarine cables. The success of the Siemens telegraph brought other military communications assignments.

To construct the apparatus for the army projects Siemens persuaded a young maker of scientific instruments at the artillery school, Johann Georg Halske, to establish, in partnership with him, a plant for building telegraph equipment. The demand for telegaph installations increased rapidly, and the business of the firm prospered.

A telegraph network for northern Germany was built under contract for the new Prussian state telegraph department, and Siemens was asked to become its manager. But he decided to make a new independent career for himself in the telegraph

Fig. 13.1 - WHEATSTONE NEEDLE INDICATOR of the type that first interested Werner von Siemens, and led to his career in land and underwater telegraphy. From W. James King, *The Telegraph and the Telephone.* U.S. National Museum Bulletin 228, Smithsonian Institution.

manufacturing business. Resigning from the military, he began to extend the sale of his telegraph system beyond Germany. A line from St. Petersburg to the Crimea on the Black Sea was built for the government of Russia. It was used in the 1854-56 Crimean War, the first application of the electric telegraph in military combat.

The Morse telegraph, which had been developed in the United States, eventually replaced others and became internationally used. Siemens adopted the system and proceeded to improve its operation by introducing a polarized principle for the relay used to provide the local contact in long-distance transmission. By supplying a magnetic bias the sensitivity of the relay was increased. Also it was made double-acting because the current in the receiving coils sent in one direction attracted the armature of the relay, and sent in the other direction, repelled it.

Siemens carried on with investigations for the improvement of conductor insulation for the telegraph cable being laid underground and under water. This would lead eventually to the involvement of the company in vast operations of producing and laying undersea cables for trans-oceanic communication. He also patented a process for electroplating gold and silver that was to be an entering wedge for the company's products in England.

**AS THE 19th CENTURY** moved on, growth in use of electricity was limited by the high cost and bulk of battery installations. The heavy current demands of electroplating and arc lighting spurred the need for more effective and economical production of electric power. This could only be met by development of suitable electric generating equipment driven by steam engines and water-wheels. Responding to the opportunity for expansion, Siemens & Halske branched out from telegraphy into engine-driven rotating generators. Werner soon began to apply the skills he had shown in telegraphy to the new and wider field.

In his early examination of the design of existing generators Siemens found that the disk-type armatures then in use, Fig. 13.2A, were bulky and inefficient. The bobbins of wire around the armature disk used the flux of the field magnets poorly and incompletely. How could the armature winding be made more effective? Siemens answer was a simpler armature shaped something like a weaver's shuttle, invented in 1856.

The **SHUTTLE-TYPE** armature, Fig. 13.2B, used an iron rotor with a cross-section like a double-T, into the slots of which the coil was wound and its ends brought out to a two-segment commutator. The armature revolved in a circular opening between the field poles so that the full magnetic flux density was delivered to the total axial surface of the winding. The shuttle armature allowed higher speed operation, a smaller air-gap for more efficient performance, and a greater rigidity and compactness than the previous types of construction.

The electric generator had progressed in the mid-1800s from the "magneto-electric" machine as in Fig. 13.3A, using a bulky and costly cluster of permanent horseshoe magnets for field excitation, to the more compact and efficient machine using electromagnetic field excitation. The coils of the field winding were originally excited by a small separate magneto, Fig. 13.3B, driven from the power shaft or from the shaft of the generator. It occurred to Siemens that the separate exciter could be dispensed with by using current from the generator itself. In 1867 he announced to the Berlin Academy of Sciences his discovery of the principle of "self-excitation," a superior and a self-contained method of energizing the field winding, Fig. 13.3C.

In the self-excited machine the field winding is connected in the circuit of the generator to divert some of the current generated in the armature to energize the field magnets. The principle works automatically because the iron core of the field poles, once energized, retains some magnetism. Thus whenever the generator is started up the small residual magnetism is enough to begin inducing current in the armature and this cumulatively builds up the field excitation for full output of the machine.

Siemens coined the name *dynamo-electric machine,* later shortened to *dynamo,* for his new, self-excited generator. Siemens was not alone in this discovery, however. The self-excitation idea had occurred also, at about the same time, to Sir Charles Wheatstone and S. Alfred Varley, in England, and to others. But it was Siemens who foresaw the advantages of the dynamo for heavy-current applications, and who rapidly pushed its development and use in providing power for industry and for electric traction.

While the Siemens shuttle-type armature was an improvement over previous types, its single coil construction presented problems with sparking at the commutator, and with heating of the armature. Zènobe T. Gramme patented an armature with a toroidal core around which were wound many individual coils connected to form one continuous closed coil, Fig. 13.4A. This armature provided better commutation and easier cooling, and the machines were widely applied. However the ring armature suffered in efficiency because the wire on the inside of the toroidal coils was shielded by that on the outside, hence was not productive, and only added to the resistance of the coils.

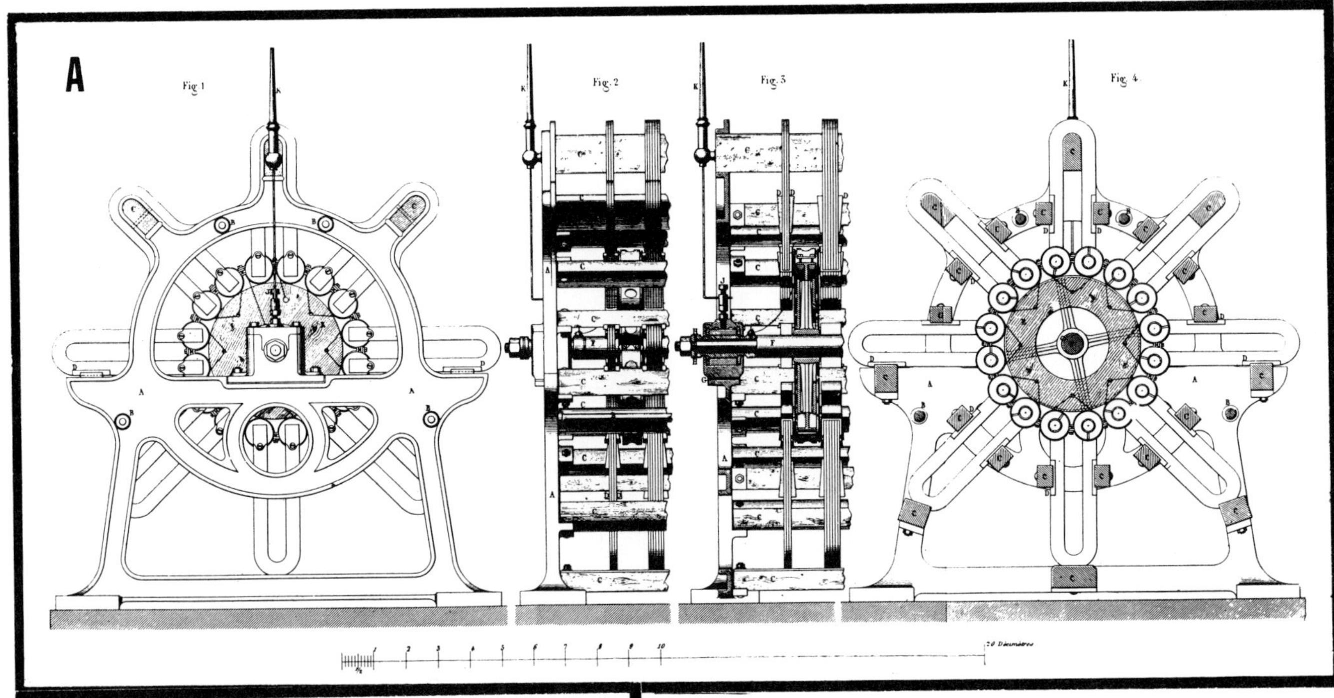


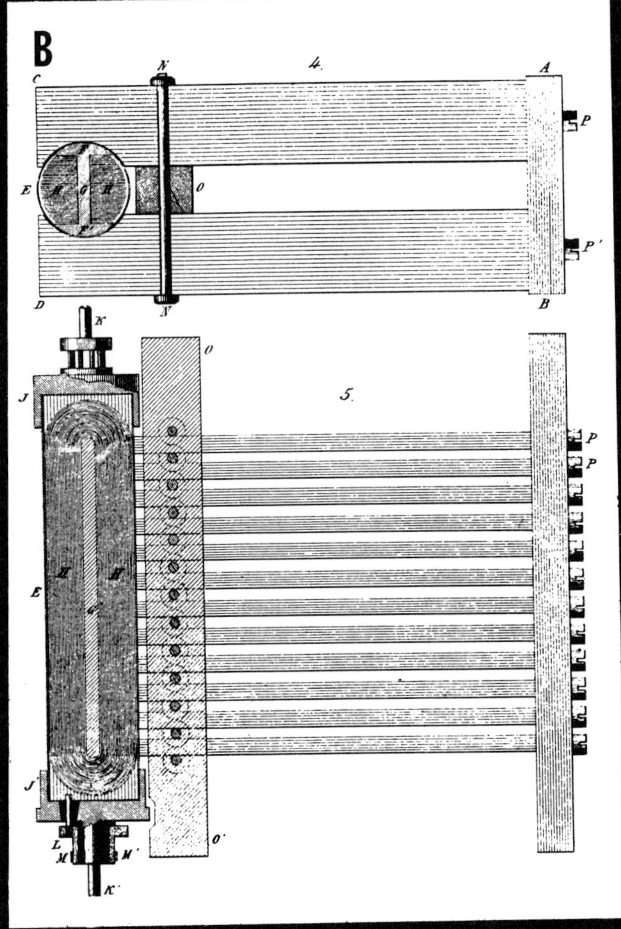


FROM BOBBIN TO SHUTTLE

Fig. 13.2 - (A) the "disk-type armature used in the magneto-electric generator above, with bobbin coils on its perimeter, was bulky and made poor use of the magnetism supplied by the horseshoe field magnets.

(B) Siemen's "shuttle-type" armature, left, made possible higher speed and greater compactness. It was the first armature construction improvement for greater efficiency and for space and weight saving. Illustrations from W. James King, ***The Early Arc Light and Generator.*** United States National Museum Bulletin 228. Smithsonian Institution, Washington, D.C.

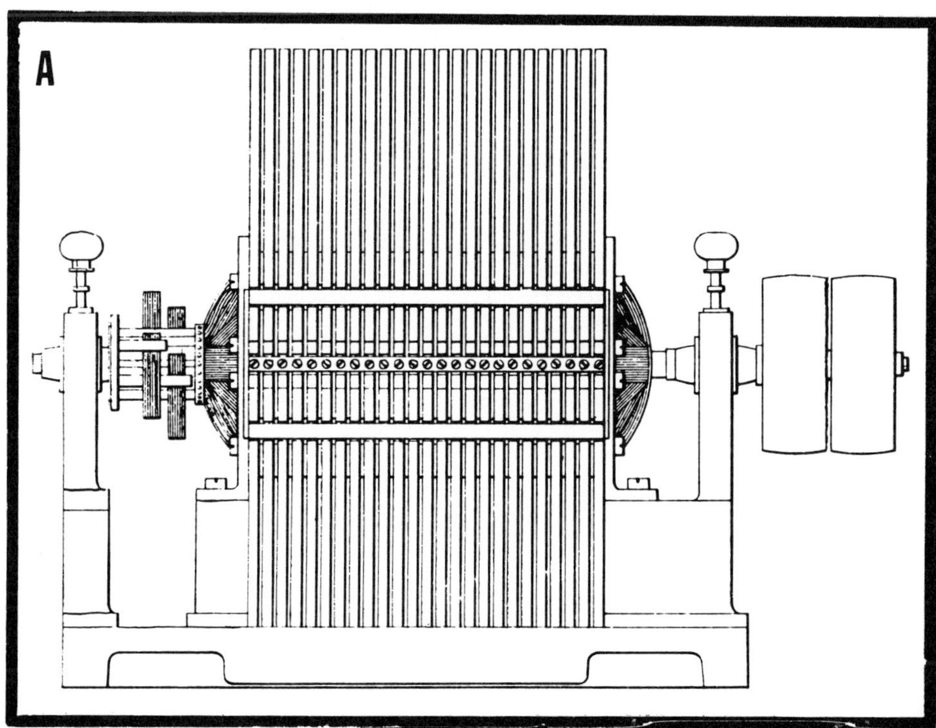


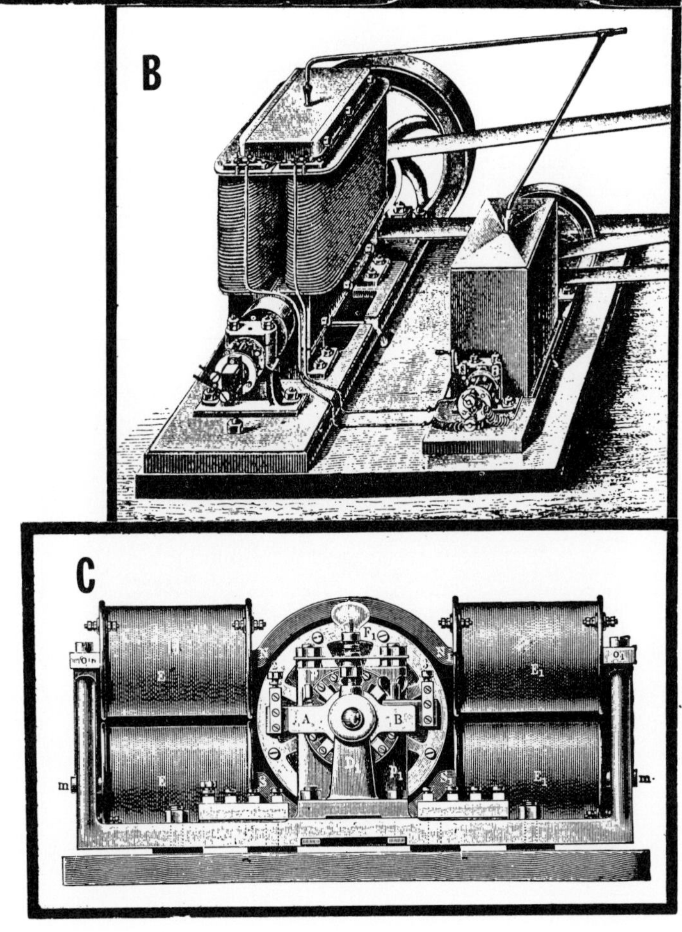


FROM PERMANENT MAGNET TO ELECTROMAGNET

Fig. 13.3 - (A) The magneto-electric generator, above, used clusters of permanent horseshoe magnets for the field excitation. The magnets were extremely bulky and delivered a relatively weak field strength.

(B) Substitution of electromagnets, right, provided a much stronger and controllable field excitation in a radically smaller space. The field of the early generators was energized by an external magneto-electric machine driven separately from the power shaft.

(C) The principle of "self-excitation," below, using current from the generator itself to excite the field winding, was discovered by Werner von Siemens in 1867. It eliminated the need for a separate magneto and provided a compact, self-contained machine, which he called "dynamo-electric," later shortened to "dynamo."

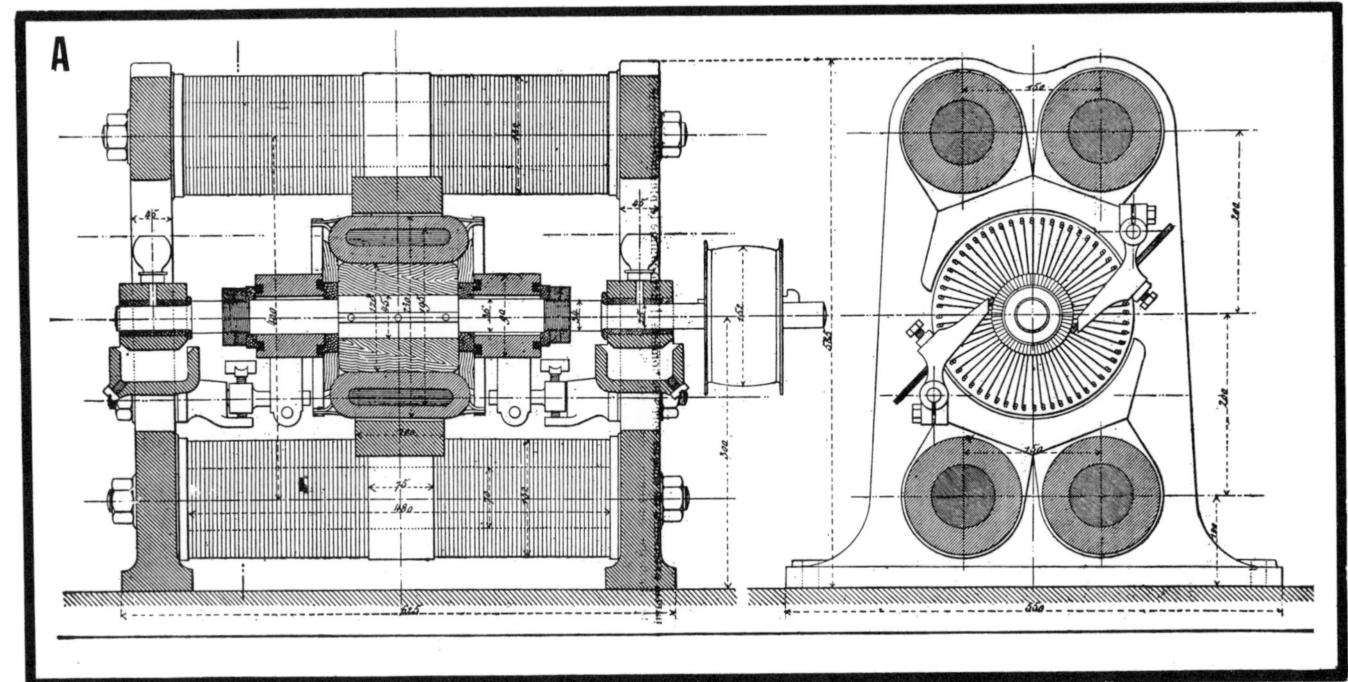

SIEMENS DEVELOPS THE MODERN ARMATURE

Fig. 13.4 - (A) The Gramme generator above used a "ring-type" armature in which only the outer part of the winding was effective. (B) The ring armature was replaced by the more efficient "drum-type" armature winding developed in 1872 at the Siemens & Halske works, and used on the generator at the left. The drum-type winding, or its equivalent, is used on all present-day machines. Illustration courtesy the Deutsches Museum, Munich.

To overcome disadvantages of the ring armature, and to obtain the highest efficiency, Siemens and his chief engineer at Siemens & Halske, Friederich von Hefner Alteneck, devised the "drum-type" armature in 1872. Instead of winding the coils around a torus, they were wound around a drum.

The **DRUM-TYPE** armature, Fig. 13.4, used a cylindrical iron rotor whose surface was slotted axially at regular intervals around the perimeter of the rotor to receive the windings. Except for the end-turns of the windings, the entire length of the conductors were effective in cutting the magnetic flux from the field. The drum-type armature was the most efficient and practical ever invented for a rotating winding. It gave the most effective distribution and protection of the coils. It is basic to the rotating machines of today.

The firm of Siemens & Halske became very active in the manufacture of dynamos and motors, using their continuing improvements in design. In 1882 they introduced the compound winding for the field coils, in which a winding in series with the armature was added to the usual field winding shunted across the armature. The series winding, responding to the load current, acted to strengthen the field excitation to compensate for the drop in voltage as the load on the dynamo was increased. By the late 1880s dynamos in both the horizontal and vertical constructions were being built in capacities over a thousand kilowatts. In 1879 Siemens & Halske developed the world's first practical electric railway, which initially operated on a demonstration line at the Berlin Trade Exhibition. In 1880 the first electric elevator was exhibited by Siemens & Halske at the Mannheim Exhibition.

Although an inventor-engineer, Werner von Siemens considered himself primarily a scientist and experimental physicist. In his love for science he promoted a two-way relationship - using science to develop technology - then extracting from technology the advances for science. He was instrumental in convincing the government of the crucial necessity of scientific excellence as a base for advancement of technology. He promoted and helped fund the Physikalisch-Technische Reichsanstalt, in 1887, at Charlottenberg, the first national scientific and technological institution.

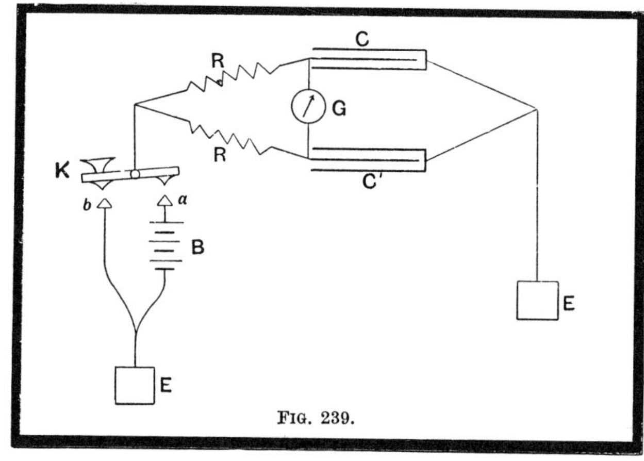


Fig. 13.5 - SIEMENS'S arrangement for testing dielectrics and determining their capacities. The rocker key, K, permits rapid charging and discharging of capacitors C and C', through resistances, R, from the battery, B, through the grounds, E. The capacities, and hence the specific inductive capacity of the dielectrics could be determined as a function of the voltage, resistance and frequency of reversals of current. Siemens found that the capacity of a given dielectric was independent of voltage used for the test

**WERNER von SIEMENS** was a close follower of the work of Michael Faraday. Siemens's extensive concern and experience with insulation of cables led to a special interest in dielectrics, and in Faraday's views on those materials. Faraday held that the capacity exhibited by various insulators was not from any penetration of current through them, but due to a polarization produced by the electric tension that gave each molecule of the material a displacement charge.

Siemens developed the arrangement shown in Fig. 13.5 for testing dielectrics and determining capacities. He also confirmed Faraday's discovery that the amount of charge in a solid dielectric was a function of a factor called the "specific inductive capacity," which varied with the nature of the material.

Siemens developed a current-measuring instrument which needed no iron or other magnetic for its operation. Called an *electrodynamometer,* the measurement depended on the mutual reaction of the magnetic fields of two coils one stationary and the other movable, when the current was

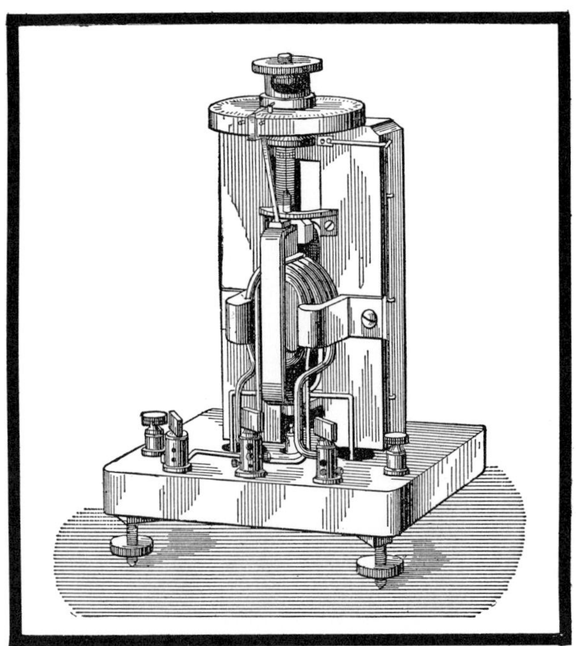

IT COULD MEASURE ALTERNATING CURRENT

Fig. 13.6 - SIEMENS ELECTRODYNAMOMETER - The current passes through the fixed coil and is led into the suspended movable coil through two mercury cups into which its ends dip. The magnetic action of the coils upon each other causes the suspended coil to turn against the torsion of the spring which suspends it. Deflection of the instrument was unaffected by reversal of current, hence it could be used on alternating current.

passed through them in series. Because deflection of the instrument was unaffected by the direction of the current, the electrodynamometer could be used for measuring alternating current, Fig. 13.6.

Werner, and his brother, William, were early contributors to the activities of the Committee on Electrical Standards of the British Association for the Advancement of Science. This committee was organized in 1861 to establish practical standards for the electrical quantities. Werner was a member of the 1862 group to set up the physical standards for electrical resistance. He developed the first internationally accepted measure for the unit *ohm,* namely the resistance of a one meter column of mercury one square millimeter in section at the temperature of melting ice.

As a notable inventor, engineer, and bestower of progress in electrical science and technology, Siemens was showered with honors by scientific organizations and by a grateful government. He was made a member of the Prussian Chamber of Deputies, and was raised to the rank of nobility. He died in Charlottenberg, near Berlin, in 1892.

It was Werner's invention in electroplating of gold and silver that launched his younger brother, Wilhelm, on his career.

**WILHELM SIEMENS,** later Sir William, was born in Lenthe, Germany, in 1823. He was intended by his parents for a business and banking career. But his elder brother, Werner, decided that engineering had a better future, and Wilhelm was sent to a technical school, then to the University of Gottingen where he studied chemistry, physics and mathematics. After an apprenticeship in a steam engine factory near Magdeburg, Wilhelm decided to strike out on his own as an engineer and as a salesman for his brother's invention of the electroplating process.

After a modest successin Germany, Wilhelm was commissioned by his brother to try to establish a market for the patent in England, so he traveled to London in 1843. He was to call on the Elkingtions in Birmingham, a firm that had pioneered in electroplating. Though less than twenty years of age, and with the handicap of being a foreigner, Wilhelm's solid engineering ability, his talent for persuasion, his ambition, and the merits of his product, brought success at the Elkington works. The sales price was 1600 pounds, a substantial sum at the time, of which Wilhelm received one half of the profit - and this was the means of continuing the English enterprise.

Finding the marketing prospects in England favorable for inventions, William decided to adopt England as his future home. A water meter he had invented in 1851 caught on and he soon was receiving large royalties, so that he was able to afford an office. He purchased a home in the fashionable Kensington district of the city. Here he lived with his younger brothers, Karl and Friedrich until marriage in 1859 to Anne Gordon, the sister of an engineering professor at the University of Glasgow. He then became a British citizen.

William and his brother Friedrich, in 1856, began to emerge as major contributors to industrial innovation. Friedrich had patented a regenerative principle in which the waste heat of the flue gases from industrial operations using large-scale combustion was recovered for pre-heating the incoming air, thus improving the efficiency of the process.

England had gone from a former wood-burning economy, and had turned to the use of domestic coal for fuel. The invention of coal gasification provided an ample, cheap and convenient heat that was taking over for industrial operations like the glass industry. The first application of the Siemens regenerative furnace was to a coal-gas-fired glassworks in Birmingham in 1861. There its merits in providing continuous production, high fuel efficiency and control of quality were such as to promote its eventual world-wide adoption. The Siemens regenerative principle was soon applied to the steel industry with equal success.

**The IRON and STEEL INDUSTRY,** in the last half of the 19th century, was revolutionized by the introduction of two methods of making cheap steel. In this development the Siemens brothers took a vital part.

The first method of quickly converting pig iron from the smelters into steel was announced by Sir Henry Bessemer in 1856. The Bessemer process uses a tiltable, egg-shaped, open ended furnace to hold a batch of liquid iron. Air under pressure is introduced at the bottom of the upright furnace and this burns the carbon in the iron. The furnace is then tilted to empty the batch. The first American steel plants used the Bessemer process.

In 1860 the Siemens brothers applied their regenerative principle to steel making with the introduction of the revolutionary "open-hearth" method. In this process coal-gas-fired flames are passed over the liquid pig iron in a shallow, covered hearth to burn out carbon and other impurities. The hot combustion gases are then diverted to a brick-work heat-absorbing regenerator below the hearth. The stored heat is then used to pre-heat the incoming air for the flames. The open-hearth was a continuous process, and its advantages were such that by the end of the century it accounted for most of the steel production. In recognition of his genius as a major source of innovation for British industry, William Siemens was elected to the Royal Society in 1862.

William was also to follow in the family electrical tradition. Beginning in 1850 he had acted as the British agent for the Siemens & Halske Berlin firm. Later he organized a separate British company under the same name. In this he was to make another reputation and fortune in Siemens telegraphy. He soon became a recognized authority on submaries cables, and he published papers on the subject. Siemens was appointed as scientific consultant to the British government on under-water telegraphy.

Established as Siements Brothers in 1858, a cable factory was set up near London. Gutta-percha was found to make an excellent under-water electrical insulation, and Siemens used it in fabricating thousands of mile of undersea conductors. The firm achieved a world-wide reputation for laying submarine cable. For that purpose the company designed a special ship, named the "Faraday." With that new facility they completed, in 1875, a trans-Atlantic link with the United States, and in the next ten years laid five more cables across the Atlantic. An even greater achievement was the construction by Siemens Brothers of the Indo-European cable from London to Calcutta, via Berlin, Odessa and Teheran, completed in 1870.

In addition to attending to his business enterprises, William Siemens was active in professional societies, and was a frequent lecturer on the science and applications of electricity. For his contributions Siemens was honored by the many scientific societies, by universities and professional institutions. He was president of the British Association for the Advancement of Science in 1862. He was knighted by Queen Victoria in 1885, the year of his death.

CONSUMPTION of ELECTRICITY expanded rapidly in the late 19th century. Arc lighting and electroplating were large users. Electric motors were increasingly installed in factories, and began to be applied to electric traction. The first arc furnaces for metal processing were put into operation in 1879. Direct-current generators, sometimes backed up by banks of batteries, were the predominant source of power.

Alternating current, the other form of electric power, was waiting in the wings for applications suitable for its service. The first of these appeared in 1876 with the introduction of the Jablochkoff arc candle.

The ordinary high-intensity arc light used a

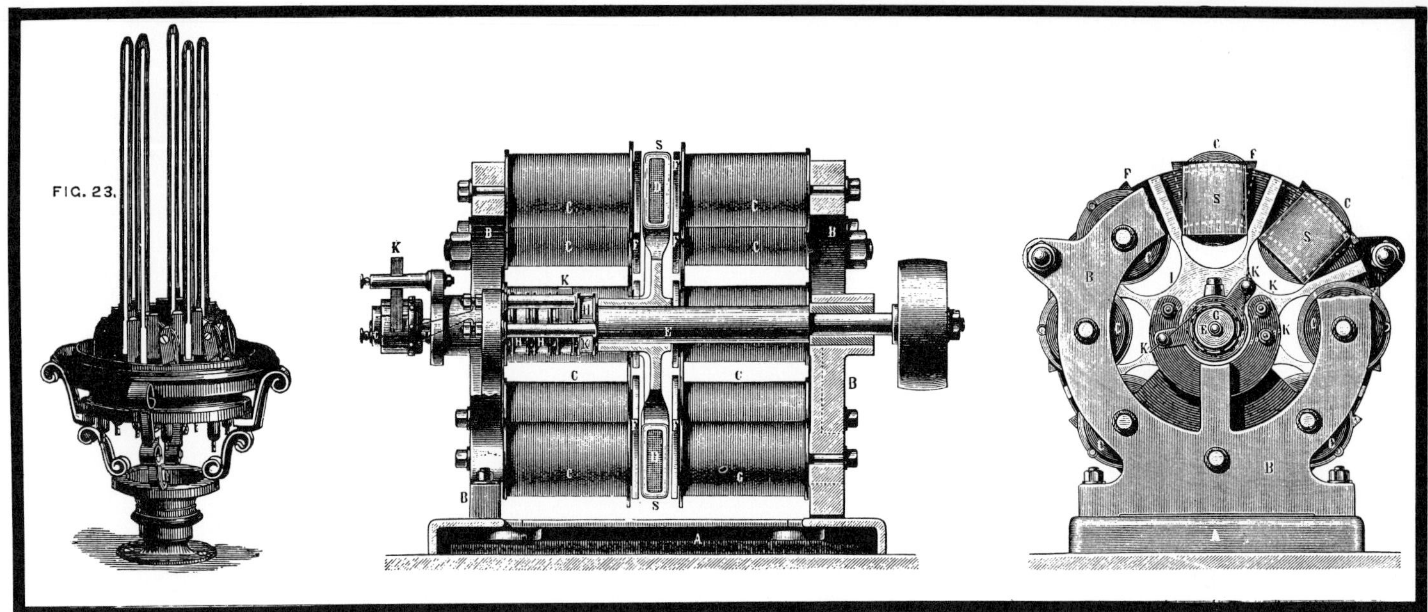


JABLOCHKOFF CANDLE AND ITS A-C GENERATOR

Fig. 13.7 - To burn evenly, the Jablochkoff Candle (left) required alternating current, and Siemens & Halske developed a self-excited alternator for this purpose (above). Illustrations from W. James King *The Early Arc Light and the Generator.* Smithsonian Institution Bulletin No. 228.

complicated mechanism for regulating the feed of the carbons to maintain constant light. It was bulky, did not lend itself to subdivision of lights, and was suited mostly to outdoor use. The situation was greatly altered when Paul Jablochkoff, a Russian engineer, found a way of producing a continuous arc without need of a regulating mechanism, and of reducing intensity of the light to permit the convenient use of multiple units.

Called the "Jablochkoff Candle," Fig. 13.7, the arc was produced at the ends of two carbon rods separated by a fusible material such as kaolin. Enclosed in an onyx glass globe, the arc candle lasted one to two hours, producing a pleasing light. Direct current was not suitable for the candles because the carbons burned unevenly, hence it was necessary to employ alternating current. There was an immediate demand for alternators suitable for powering groups of candles which might be at different locations.

A line of a-c machines for this purpose was quickly developed by Siemens & Halske, one form of which is shown in Fig. 13.7 in side and end views. It used a disk armature rotating between two stationary circles of electromagnets which were excited by an auxiliary winding on the armature through a commutator on the shaft.

With the advent of incandescent lighting in the 1880s the arc candles soon went out of use. But the Jablochkoff system provided the first impetus for the development of large alternators, and it initiated the use of a central power station feeding multiple distribution circuits, a forerunner of later central station systems.

**TO PROMOTE** orderly, effective progress in the generation and utilization of electricity, William, and his brother Werner, saw the need for consensus on the measurements used for electricity and magnetism. Rapid advance in electrotechnology was making it imperative that the various and unrelated electrical units in use at the time be standardized.

In 1861 the British Association for the Advancement of Science selected a committee of leading scientists to begin a program for reference standards for "absolute" electrical quantities. William and Werner Siemens took an active part in this effort.

First needed standard was for resistance, which had become a critical factor in the operation of telegraph lines. Early telegraph engineers had to make use of extemporized units and convert between them. To resolve the situation Werner Siemens, taking mercury as a suitable material, proposed as a practical standard, the resistance of a column of mercury one meter long and on square

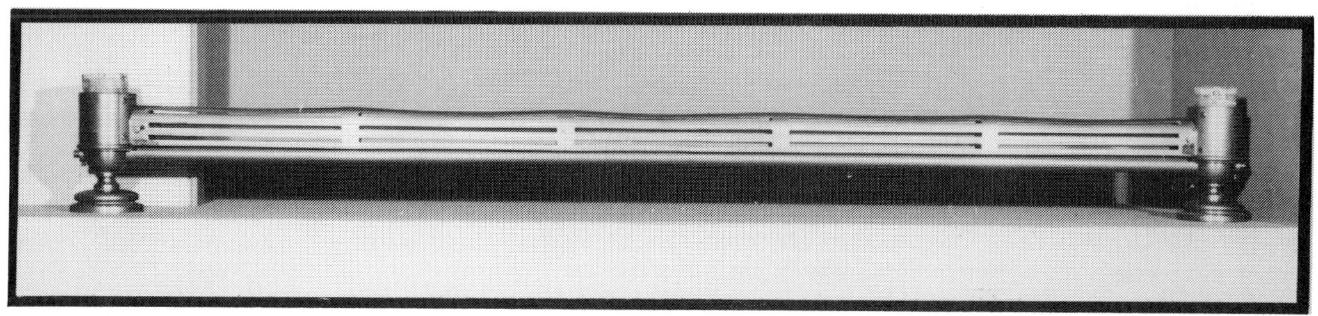

millimeter in cross-section at zero degrees C. This gained acceptance and was used as the "siemens unit" of resistance.

Fig. 13.8 - SIEMENS MERCURY COLUMN standard ohm, prepared in 1881. Photograph- Deutsches Museum, Munich.

The year 1881 saw the first of a series of international electrical congresses serving as forums for discussion, approval and promulgation of the practical electrical units. These Congresses, in 1893 and 1908, modified the specifications of the Siemens unit, which in 1867 had been named the *ohm.* This modification was necessary to conform to the value of the ohm in absolute units which had been determined by the British Association Committee of Electrical Standards for Resistance. In 1908 the International Ohm was defined as the resistance to an unvarying current of a column of mercury of mass 14.4521 grams, 106.300 centimeters in length and of constant cross-section at temperature of melting ice.

The Siemens mercury standard continued in use until 1948. Since that time the legal standard ohm has been determined in terms of the reactance to an alternating current by an inductor or capacitor, as described in the chapter on Ohm.

**The "SIEMENS"** - The contribution of the Siemens brothers to electrical science was recognized in 1935 by the Commission on Electric and Magnetic Magnitudes and Units, by naming the unit of electrical conductance the *siemens.* This was made official in the International System of Units (SI) by the 14th General Conference of Weights and Measures in 1971.

The *siemens* (formerly called the reciprocal ohm, or *mho*) is the conductance of a conductor such that a constant voltage of 1 volt between its ends produces a current of 1 ampere it it.

In the calculation of alternating current parallel circuits, the real part of the total admittance, $Y$, of the circuit is $R/(R^2 + X^2)$, and is the conductance.

**IN NAMING** the siemens, the Electrical Commission left unspecified which of the two Siemens brothers was honored by the unit. Dr. Forest K. Harris, Consultant, Electricity Division, National Bureau of Standards, has offered this opinion:

*"I have not found any definite statement as to which of the Siemens brothers was to be honored by having his name attached to the SI unit of conductance.*

*"However, it was Werner who proposed as a unit of resistance - a mercury column 1 metre in length and 1 square millimetre in section (Pogg. Ann. - CX, p. 1) - that was used for a time on the continent under the name "siemens einheit," and it was Werner who was largely responsible for establishing the first great national laboratories, the Physikalisch Technische Reichsanstalt, in Berlin.*

*"It seems to me quite possible that the British delegate, in voting for the "seimens" thought he was honoring William, and the German delegate thought he was honoring Werner. It I were to choose on the basis of contribution I would without question pick Werner. Actually these brothers worked so closely together throughout their careers, they should perhaps both be honored with the name of the unit of conductance."*

The Siemens brothers were always in the closest laison - in business, in engineering and in preserving family ties. The close harmony was celebrated every five years by a reunion in the Hartz mountains of Germany. Both the rich and the poor came, and help for the less fortunate members was provided from a special fund. The business of the Siemens firm continues to this day.

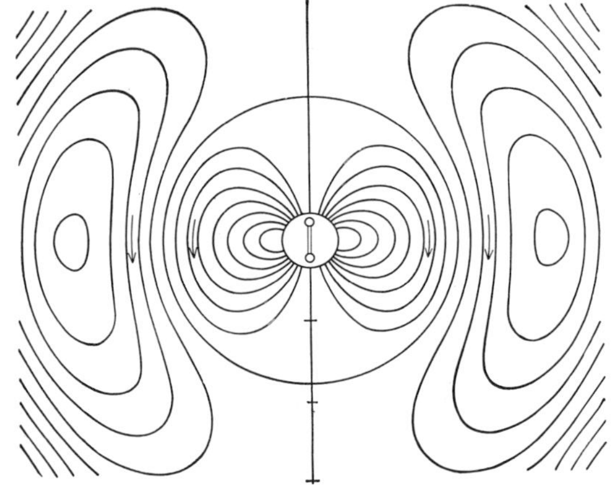

# HERTZ...

## his discoveries opened the age of radio

**TELEVISION** antennas, sprouting from rooftops everywhere, are visible reminders of the invisible waves of electricity that crowd the air and bring mass communication to us from all around the world. Television, radio, radar, and all the other marvels of high-frequency electromagnetic radiation are the legacies of the German physicist Heinrich Hertz who in his short life made discoveries that were turning points in world history.

Building on the prophecy of Maxwell that electric oscillations should produce waves traveling with the speed of light. Hertz devised a remarkable series of investigations. With a mind of crystal clarity and unerring direction he soon unlocked the secrets of the electric wave - and gave the world the basics for communication that would girdle the globe. Using the simplest of experimental apparatus Hertz proved the existence and transmission of electric oscillations, demonstrated the connection between electromagnetism and light, and had a glimmering of electrons in being the first to note the photo-elelctric effect.

**HEINRICH RUDOLF HERTZ** was born on the 22nd of February 1857 into a prosperous and cultured family in the great merchant port of Hamburg, the oldest of three brothers and a sister.

## the unit of frequency, one cycle per second, is named for the German theorist and experimenter

# HEINRICH HERTZ

## whose astonishing researches were a turning point in world history

His father, Gustav F. Hertz, was a man of substantial position as a lawyer and senator of the city.

During Hertz's youth Germany was undergoing an intense period of consolidation. A domain of two dozen independent principalities was being forged into a single nation under the iron hand of the empire builder Count von Bismarck, a Prussian prince. France, fearful of a united Germany, tried to thwart Bismarck's ambitions, and thus brought on the Franco-Prussian War of 1870. France was easily defeated and lost the provinces of Alsace and Lorraine. The new German empire was established, and to advance its economy the new government promoted expansion in science and technology. In this atmosphere Hertz matured. He showed an uncommon gift for languages, a trait shared with his father, and an interest in science stimulated by his grandfather who had pursued natural science as a hobby.

Hertz attended a private school until he was fifteen and there gave the intimations of his future brilliance by being first in his class. He early exhibited a practical bent, setting up a woodworking shop in his home basement and fitting out a laboratory where he carried on experiments in chemistry and physics. Enrolled in the Johanneum Gymnasium in Hamburg, he finished his classical studies, adding Greek and Arabic to his list of languages, and graduating first in his class.

Hertz liked mechanical work, and in line with the tone existing in his native Hamburg that students should direct their efforts toward something practical, had decided on an engineering career. For practical experience he spent a year with an engineering firm in Frankfort before entering the School of Polytechnics at Dresden. At age 19 he was called upon to put in his prescribed year of military service which he did in the railway department in Berlin.

On completion of his year in the army, Hertz moved to Munich with the intention of continuing his engineering education in the Technical High School there. However, since his Gymnasium days, Hertz had also been pursuing the study of mathematics and natural science on the side because he was attracted to these subjects and increasingly impelled toward consideration of an academic future. Hertz remained in doubt for two years but as he grew in knowledge he grew in conviction that he could find an enduring satisfaction only in scientific work.

Weighing a practical life in engineering, as compared to a scientific career, he made a momentous decision for the latter. With his father's approval and promise of financial help he dropped his plans for the technical school and matriculated at the University of Munich. The university, he felt, better fitted his idealistic ambition for a success in fundamental study and research.

To insure the parallel growth of science and technology in the new German empire, those pursuing these fields were expected to be equally

Fig. 14.1 - HERMANN von HELMHOLTZ (1821-1894), a 19th century German scientist, was a deep and versatile investigator. Trained as a physician, he made important contributions to physics, physiology and the theory of knowledge. His interest in Maxwell's electromagnetics led directly to Hertz's involvement with finding its truths and the electric waves. Illustration The Burndy Library.

proficient in mathematics and in laboratory technique. This dual role was in line with Hertz's desires and his capabilities. Under the mentorship of P.G. von Jolly, a pioneer is setting up several great university laboratories in Germany, Hertz was advised to read the treatises of Lagrange, Laplace, Poisson and others who were creating the mathematics of science. He also obtained a grounding in experimental practice in Jolly's laboratory at the university, and found the experience very satisfying. During all his later life Hertz made it a practice to alternate between the book and the bench - between intense theoretical study and experiment in the laboratory.

It was customary in Germany for a student to progress from one university to another, in order to gain in stature and promotion. Hertz made the decision in 1878, after considering alternates, to leave Munich and to enter the Berlin Physical Institute which was then Germany's greatest scientific center. The choice was a momentous step in establishing the direction of Hertz's future road to scientific fame. At the Institute Hertz soon caught the eye of Hermann von Helmholtz, one of Germany's greatest 19th century men of science. Helmholtz had, in 1871, become the first professor of physics at the Institute where he had a well-equipped laboratory, and students to help carry out research.

**HERMANN von HELMHOLTZ** was the first to propose, in 1847, on a purely theoretical basis from the conservation of energy, that the spark discharge of a Leyden jar was not a simple one-way flow of electricity, but a series of oscillations at a very high frequency. It had been frequently observed that these bursts of oscillating energy from the jar produced radiation effects, such as magnetizing steel needles, over a considerable distance from the source. But efforts to explain the radiant action were unsuccessful.

Then, in the 1860s, came Maxwell's paper predicting that oscillating disturbances resulted in the propagation of electric waves. The significance of this theory was not lost on Helmholtz. He was the first scientist on the Continent to study Maxwell's works and to appreciate their importance in the future development of physics.

A deep and versatile investigator, Helmholtz proceeded to find out what was verifiable and what might be wrong with the predictions of Maxwell's equations. In this effort he found himself faced with rival and conflicting electromagnetic theories, mainly those of Weber and Neumann in Germany, who had developed mathematical expressions on the interaction of electric and magnetic forces.

"Weber," said Helmholtz, "sought to trace back

electric and magnetic phenomena to a modification of Newton's assumption of direct forces acting at distance and in a straight line . . . he assumed that the magnitude of the force between the two quantities of electricity might be affected by the velocity with which the two quantities approached towards or receded from each other, and also by the acceleration of such velocity . . . there existed a number of others, all of which had this in common - that they regarded the magnitude of force expressed by Coulomb's law as being modified by the influence of some component of the velocity of the electrical quantities in motion.

"This plentiful crop of hypotheses had become very unmanageable, and in dealing with them it was necessary to go through complicated calculations, resolutions of forces into their components in various directions, and so on. So at that time the domain of electromagnetics had become a pathless wilderness. With the object of clearing up this confusion I had set myself the task of surveying the region of electromagnetics, and of working out the distinctive consequences of the various theories, in order, wherever that was possible, to decide between them by suitable experiments."

**TO RESOLVE** the contradictions, Helmholtz published, in 1870, a paper in which he had constructed a general electrodynamic theory the equations of which could test those of Maxwell, Weber and Neumann as special cases. Helmholtz found that all three agreed in the prediction for electromagnetic action through continuous and closed metallic circuits, for which the laws were well known and had been carefully investigated. But in the case of unclosed conductors between which insulating materials were interposed, the situation was different. Surgings of electricity at the ends of unclosed circuits caused surgings of charge accumulating in the insulator, an understanding of which could be obtained only by a thorough investigation of the processes that were occurring. Weber had attempted to provide for this in his hypothesis by suggesting that electricity possessed a certain degree of inertia. Weber's and Neumann's equations did not, however, meet the requirements of his general theory. Helmholtz was then left with Maxwell's theory of open circuits.

In this theory, variation in the polarization of dielectrics induces displacement currents which cause electromagnetic effects in the dielectric. The ultimate test of the various theories was experimental verification, and Helmholtz proceeded to set up problem topics that might lead to productive research in the various areas. His better students were invited to enter in competition for the best solution to a problem for which an academic prize was awarded.

Recognizing Hertz's exceptional talents, he called his attention to a prize being offered by the Berlin Philosophical Faculty for the solution to an experimental investigation he had originated to test the correctness of Weber's idea that particles of electricity had mechanical inertia. Hertz, eager to get on in research, enlisted for the prize under the sponsorship of Helmholtz and was given laboratory space in the physics department.

**HERTZ** was now in his chosen field and in an environment conducive to challenging his inquiring mind and ingenuity in experimental procedure. In entering for the prize he was responding to a conviction, as he wrote his parents: " that his greatest satisfaction would be in exploring for new truths in nature."

Hertz won the Philosophical Faculty prize for his paper "Experiment to Determine an Upper Limit to the Kinetic Energy of an Electric Current," and was awarded a gold medal. Hertz's paper essentially negated the Weber theory of electrical inertia as a factor in electrical induction. His experiments, on the other hand, impressed him with the exceeding mobility of electricity, which was to be a factor in his future studies.

In 1879 Helmholtz proposed another prize problem to the Berlin Academy of Sciences. It was for a demonstration of the relationship between dielectric polarization and electromagnetism to test this critical assumption of Maxwell's theory. Helmholtz wanted Hertz to try for this much more valuable prize. But, feeling that the time required and the probability that the effects sought might be beyond the limits of detection, Hertz declined.

Hertz received his doctorate "magna cum laude" from the Berlin Institute in 1880 for his paper on "Induction in Rotating Spheres." He then began an academic career as salaried assistant to Helmholtz in the practical work of the Berlin Physical Institute.

For the next three years Hertz occupied himself in a diverse series of investigations. An interest in mechanics, reflective of his earlier studies in engineering, resulted in research on the hardness of materials, the elasticity of solids and evaporation of liquids.

Hertz's studies on the compression of elastic solids and the stress on elastic spheres during impact, were of value to later generations of engineers who were involved in the design of ball-bearings and had to determine the maximum loading they could withstand. He was one of the first to interest himself in the hardness of materials, and he proposed a hardness test based on the load required to produce the first crack in the material. Indicative of his early development of manual dexterity, Hertz invented a new hygrometer, built a battery of 1000 small galvanic cells, and blew glass tubes for his gas discharge experiments.

But his foremost interest was in electrical phenomena. In this field he covered such subjects as electromagnetic induction, residual charges in dielectrics, and the constitution of cathode rays which had been identified in 1870 by Sir William Crookes in England. Papers on these investigations were read at meetings of the Berlin Physical Society where he was in the companyof some of the notable scientists in Germany.

As an assistant to Helmholtz, Hertz was able to have close relations with the great man whom he considered Germany's foremost scientist. He often dined with Helmholtz and his family. But Hertz was ambitious to step up on the academic ladder, the first rung of which was becoming a "Privatdozent." There were already too many in this hierarchy in Berlin at that time for him to consider staying there. Hertz's chance to move along came when the University of Kiel requested a privatdozent for mathematical physics and Hertz was recommended. He went there in 1883.

**KIEL** had a poorly equipped laboratory so Hertz set up one in his home in the effort to continue with his work. But he did not carry on his usual experimental activity in the two years at Kiel. He published three theoretical papers, the most important of which was an in-depth study of Maxwell's electrodynamics. Hertz was deeply impressed with Maxwell's work and became an adherent to Maxwell's theories and electromagnetic equations. This was at a time when few in Germany, because of Maxwell's new concepts and the difficulties of the mathematics, had made the effort to understand them. Hertz doubtless had considered, as had others, the possibilities of a direct experimental verification of Maxwell's theory of the electromagnetic nature of light. In a few years he would do so and would become Maxwell's greatest champion.

The years at Kiel convinced Hertz that he did not want to become a purely theoretical physicist. He declined a promotion offered at Kiel, and accepted instead a professorship in physics at the Technische Hochschule at Karlsruhe, a beautiful city on the Rhine. It was there shortly later that Hertz met Elizabeth Doll, the daughter of a professor at the school. After a courtship of three months they were married in July 1886. Hertz now was entering upon a very happy and productive period of his life. In November of that year he began, in the well-equipped laboratory at Karlsruhe, the experiments on electric waves that would prove a turning point in world communication history, and that would make his name immortal.

When Hertz began his studies in electromagnetics in the late 1870s the subject was in an unsettled state. There were contending theories differing in their formulation and their views on the nature of of electricity itself. In Germany, the physicists Wilhelm Weber and F.E. Neumann, extending the Newtonian theory into electricity, explained electrodynamics in terms of the motion of electrical particles acting at a distance. In Britain the thinking was different. Action-at-a-distance had been sidetracked in favor of the theories begun by Faraday and developed by Maxwell that the electrodynamic action was seated in fields of force from points of origin. These fields, when changing, proceeded from their sources contiguously and with finite velocity.

**MAXWELL** had recognized quite clearly the importance of the electrical condition in the dielectric media in accounting for electrical phenomena in a circuit. Any alteration of the electrical condition of dielectric material of any kind, including as a special case the ether, was accompanied by a displacement current which produced magnetic and electric fields. He had derived this from his hydrodynamical analog studies of the electric and magnetic circuits, and had arrived at the concept that oscillating electrical disturbances proceeding in free space should be in the form of electrical and magnetic waves traveling with a finite velocity equal to the speed of light. Maxwell did not offer any method by which the propagation of light waves, or electromagnetic waves of longer longer length which he thought might exist, could be proved experimentally. His short life left to his successors the problem of demonstrating directly the existence of the electromagnetic waves and their identity to light.

The solution to the problem of producing and detecting the electric waves predicted by Maxwell lay in a leap of thinking from the existing state of electrodynamics in which electrical energy was exchanged from one body to another in a system having lumped elements, to a state, represented by optics, in which energy fields radiated freely through space unattached to any body.

**THAT SPARK DISCHARGES OSCILLATED,** and could emanate energy through space, had been known for a long time. In 1750 Luigi Galvani in Italy observed the twitching of a frog's leg when it was near sparks from an electrical machine.

William Woolaston, in England in 1801, noted that when a Leyden jar was discharged through water between two wires, hydrogen and oxygen appeared on both wires, as though the current went both ways through the water.

Felix Savary in Germany found in 1827 that steel needles in the vicinity of spark discharges were magnetized in different directions in alternate layers of the metal.

Joseph Henry in America in 1842 also noticed reversals of magnetization of steel needles from spark discharges and reasoned that the electricity oscillated back and forth from one coating of the Leyden jar to the other. He also reported that a single spark could magnetize needles at a distance of 30 feet in a cellar with two floors intervening.

William Thomson, who later was distinguished by being created Lord Kelvin, unaware of the discovery by Henry, showed in 1855, by the principle of conservation of energy, that the discharge of a spark must oscillate until all the original energy of the charge was expended in sound, heat, light and radiation, and that when resistance of the discharge circuit was very small the period of oscillation, $P$, was given by:

$$P = 2\pi\sqrt{LC}$$

where $L$ is the coefficient of self-induction, and $C$ is the capacity of the jar. The number of oscillations could be on the order of ten million per second. Visual confirmation of Kelvin's formula was first made in 1857 by Berend Fedderson in Germany. Using a rapidly rotating mirror, he was able to photograph the spark discharge waves.

Fig. 14.2 - FEDDERSON spark discharge photograph.

Thomas A. Edison, in 1875, reported that the sparks produced by the operation of a telegraph key were repeated when a wire attached to a metal plate in the vicinity of the key was placed near another metal object. The sparks were also noted when two carbon points, separated very slightly, were put in the vicinity of sparks from the key.

G.F. FitzGerald (1851-1901), a professor of philosophy at Trinity College, Dublin, was the first to predict, in 1883, with the aid of his theory, the possibility of producing measurable electromagnetic waves. He reasoned it to be highly probable that part of the energy of oscillating currents was radiated into space, and the energy of radiation would be proportional to the fourth power of the frequency. How, he questioned, could an electromagnetic means be used to produce the electromagnetic waves? His answer was that it should be possible to generate radiant energy using purely electrical means if the current was alternated with frequency high enough for detection.

To get frequency high enough for detection he suggested the oscillatory discharge of a condenser such as the Leyden jar. For the radiator he proposed a small circular loop whose terminals were connected to the jar. FitzGerald, however, knew of no way to detect the waves when produced. And his apparatus would not have been effective because the electrostatic energy, which would have been most easily detected, was confined to the glass dielectric of the jar, while radiation from the loop was largely magnetic, derived from the discharge current. Hertz, in his experiments four years later, altered the radiation apparatus to make the effect due to the electrostatic charges the predominant effect.

In England, five years before FitzGerald had made his prediction, there was a discovery, which if pursued might have anticipated many of the later developments in electric waves. David Edward Hughes (1830-1900), a telephone engineer in London, had invented a carbon microphone. In experimenting with it in a telphone receiver circuit he noted, in 1879, that there were occasional unaccounted for crackling sounds. He traced their origin to sparks occurring at the terminals of a nearby energized coil whenever the circuit to it was interrupted. Curious as to effects of other spark discharges, he obtained the same results when separated a considerable distance from the microphone. Hughes recognized that he was dealing with a form of radiation. But his discovery and procedures were empirical, not motivated by any detailed knowledge of the theories of Maxwell, and apparently he did not perceive them as electromagnetic waves.

Fascinated by the effects and deciding that they were due to some new kind of aerial conduction, he went on with experiments until he was able to perceive the signals 500 yards distant. Seeking professional confirmation of what he considered an important phenomenon, Hughes demonstrated the reception to three distinguished British scientists in 1880. They, however, did not connect the results to the wave predictions of Maxwell and FitzGerald, and gave a negative opinion of the possibility of a new phenomenon. They concluded that the aerial transmission was due to the already well-known effect of electromagnetic induction. Discouraged, Hughes discontinued experiments and did not publish his observations. It was not until 1899, when an inquiry was made in connection with the preparation of a history of wireless telegraphy, that Hughes agreed to describe the results of his work. Hughes had produced and detected electromagnetic waves, but his path to discovery led to no glory.

**HERTZ** was not aware of Edison's observations nor of FitzGerald's predictions, nor did he know of Hughes's discoveries, when in 1886 he returned to electrical experiments at Karlsruhe. He was now in his professional capacity in physics, with a suitable laboratory and a keen interest in Maxwell's works and in electrical radiation. Although he had in 1879 declined the problem put to him by Helmholtz for testing Maxwell's theory, Hertz had continued to be alert to the possibility of a solution. A stumbling block to progress was lack of a means of detection. In attempting to observe or measure oscillations it would have been hopeless to use a a galvanometer or any other physical means available at that time. The period of oscillation involving a fraction of a millionth of a second could not displace bodies even as light as the needle of a galvanometer. Thus no progress could be made until some fresh knowledge arrived. This came not from theory, but from a chance observation by an acute and prepared mind, that at last put him on the path of experimental demonstration. Hertz describes the situation in his book *Electric Waves.*

" . . . in spite of having abandoned the solution (to Helmholtz's problem proposal in Berlin) at that time, I still felt ambitious to discover it by some other method, and my interest in everything connected with electric oscillations became keener. It was scarcely possible that I should overlook any new form of such oscillations, in case a happy chance should bring such within my notice. Such a chance occurred to me in the spring of 1886, and brought with it the special inducement to take up the following researches.

"In the collection of physical instruments at the Technical High School at Karlsruhe (where these researches were carried out), I had found and used for lecture purposes a pair of so-called Reiss or Knochenhauer spirals. I had been surprised to find that it was not necessary to discharge large batteries through one of these spirals in order to obtain sparks in the other; that small Leyden jars amply sufficed for this purpose, and that even the discharge of a small induction-coil would do, provided it had to spring across a spark-gap. In altering the conditions I came upon the phenomenon of side-sparks which formed the starting point of the following research.

"At first I thought the electrical disturbances would be too turbulent and irregular to be of any use; but when I had discovered the existence of a neutral point in the middle of a side-conductor, and therefore of a clear and orderly phenomenon, I felt convinced that the problem of the Berlin Academy was now capable of solution."

**HERTZ** had found a means for proceeding with experiments on the nature of the response to electrical oscillations. The detector was simplicity itself - a circle or rectangle of wire interrupted in the middle by a small, adjustable air-gap. Placed where he wished to detect the force, the oscillations set the electricity of the detector in motion and caused flashes at the gap. Electric sparks across the minute gap could become visible even when the potential causing them arose for only a millionth of a second. The intensity and length of the sparks was a measure of the force of the oscillations. Although he could not have foreseen it, within three years, with the use of this detector, he would verify Maxwell's theory of wave propagation and the electromagnetic theory of light.

Hertz first used the arrangement of Fig. 14.3A. An induction coil, A, produced sparks across the gap at B which Hertz estimated had a frequency of about one hundred million oscillations per second. Connected to this discharge circuit, B, was a side-circuit, M, with a micrometer spark gap, 1, 2. The sparks at B produced vigorous side-sparks in the air-gap at M, showing that the disturbance at B was affecting the whole circuit. But, more important, Hertz reasoned that the sparking at M indicated a potential difference reaching the spark knob, 1, in a shorter time than at knob, 2, thus indicating to him that there was a finite velocity for the electrical oscillations.

Hertz now modified the arrangement as shown in Fig. 14.3B. The connection from the induction spark coil at, e, split the side-circuit, M, into equal parts. Sparking at 1,2, now disappeared. Sparking resumed, however, if the connection at, e, was moved toward, c or d. Lack of sparking when, e, was at the mid-point Hertz interpreted as due to every variation arriving at, 1,2, at the same phase. Arriving simultaneously, there was no difference of potential between them. There were only standing waves in the wires. Only with alteration of the balance, such as by attaching a length of wire at, 1, or, 2, did the sparking resume, which Hertz attributed to a potential difference due to reflected waves from the extra wire.

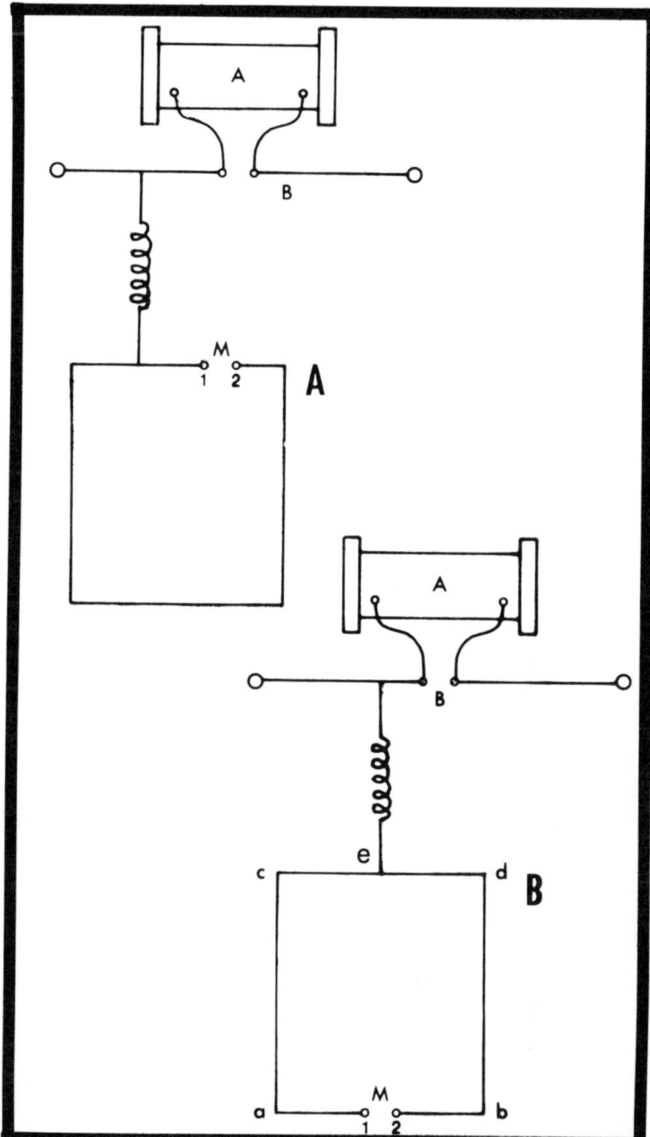


HERTZ FINDS THE BASICS OF RADIO

Fig. 14.3 - Hertz's four initial experiments are illustrated in figures A to D, above and on the following page. Using an induction spark coil and a spark-gap detector he found, in four steps, the essential elements for transmission and detection of electromagnetic waves; The drawings appear in Hertz's book *Electric Waves,* translated by D.E. Jones, 1893; Reprint edition, 1962, Dover Publications, N.Y.

Hertz again modified the arrangement as shown in Fig. 14.3C. This experiment was significant in proving that an open circuit could provide a source of oscillating radiation, and that this could be detected in a neighboring conductor without any

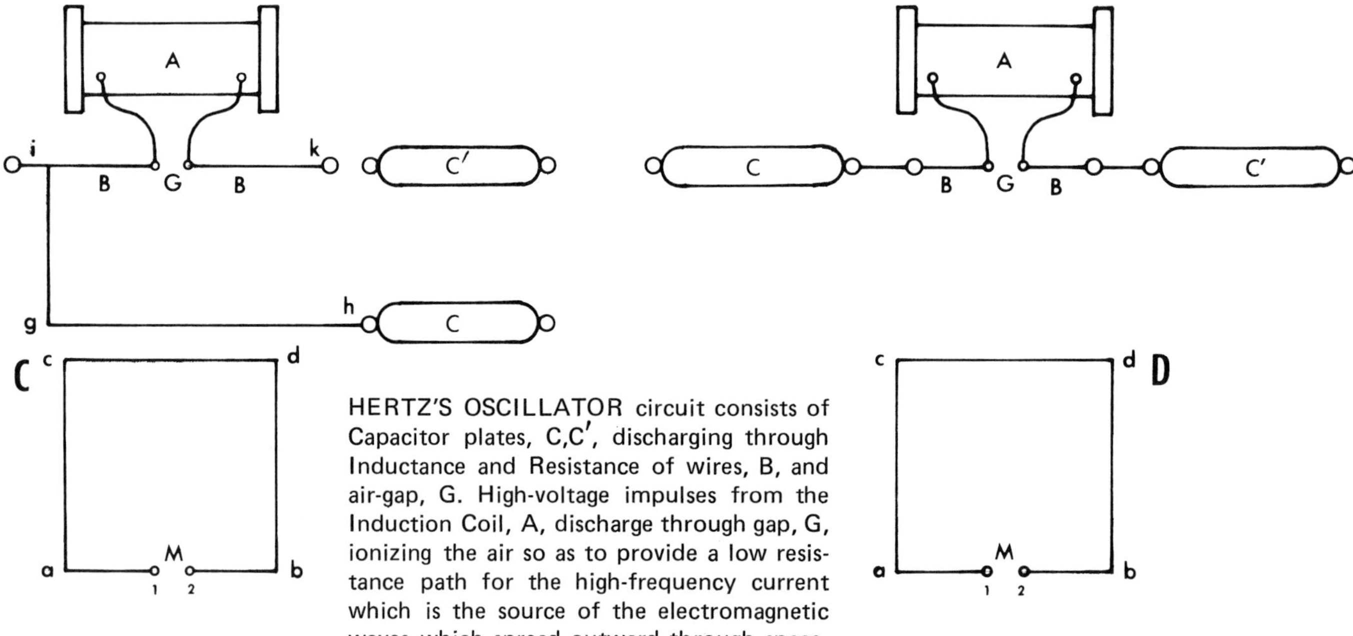


HERTZ'S OSCILLATOR circuit consists of Capacitor plates, C,C′, discharging through Inductance and Resistance of wires, B, and air-gap, G. High-voltage impulses from the Induction Coil, A, discharge through gap, G, ionizing the air so as to provide a low resistance path for the high-frequency current which is the source of the electromagnetic waves which spread outward through space.

direct connection to the primary circuit. The side-circuit, k,i,g,h, was now open-ended, and, a,b,c,d, served as a detection circuit. As long as the end, h, of the wire, gh, was free, only weak sparks were seen at the micrometer spark-gap, M. But when a large conductor, C, was added at, h, to increase the capacity of the circuit, much larger sparks appeared. When another conductor, C′, was added at, k, there was further increase of sparking at M. The adding of capacity to the spark circuit was thus shown to increase the radiation.

In the next experiment, Fig. 14.3D, Hertz introduced a well-known wave phenomenon, *resonance,* to give proof of the existence of regular oscillations. He made the following comment:

"We may now regard it as having been experimentally proved that currents of rapidly varying intensity, capable of producing powerful induction effects, are present in conductors which are connected with the discharge circuit. The existence of regular oscillations, however, was only assumed for the purpose of explaining a comparatively small number of phenomena, which might be accounted for otherwise.

"But it seemed to me that the existence of such oscillations might be proved by showing if possible, symphonic relations between the mutually reacting circuits. According to the principle of resonance, a regularly alternating current must (other things being similar) act with much stronger inductive effect upon a circuit having the same period of oscillation than upon one of only slightly different period. If, therefore, we allow two circuits, which may be assumed to have approximately the same period of vibration, to react on one another, and if we vary continuously the capacity or coefficient of self-induction of one of them, the resonance should show that for certain values of these quantities the induction is perceptibly stronger than for neighboring values on either side."

In other words, as tuning forks vibrate in sympathy when of the same frequency, the detector should respond most vigorously when in resonance with the main sparking circuit.

Hertz modified the arrangement to put the capacity terminals, C, and, C′, in opposite directions and in a straight line as in Fig. 14.3D. By altering the length of the wires leading to, C, and C′, the inductance of the oscillating circuit could be varied; by modifying the size of, C, and, C′, the capacity of the circuit could be varied. By adjusting these two factors Hertz achieved a desired period of oscillation to which he could match the detector.

To bring the detector into resonance with the oscillator, Hertz progressively altered the size of the detector. Sides, a,b, and, c,d, were kept the same length, while, a,c, and, b,d, were varied from

10 cm to 250 cm. The length of spark for variation in detector wire length was plotted to give the resonance curves shown in Fig. 14.4. By similar experiments Hertz proved conclusively that the most effective action of the system was not only due to the qualities of the oscillating circuits, but also to the proper harmony, or resonance, between the two.

**A NOVEL EFFECT** during the course of his experiments attracted Hertz's attention. The first indication of this he stated as follows:

"In all the experiments described the apparatus was set up in such a way that the spark of the induction-coil was visible from the place where the spark from the micrometer took place. When this is not the case the phenomena are qualitatively the same, but the spark lengths appeared to be diminished."

The spark from the detector micrometer was sometimes not very luminous, and so to improve the visibility Hertz occasionally enclosed it in a dark case. In doing this he noticed that the spark length was reduced, and when the box was removed the sparking resumed normal length. Curious about this effect and sensing it worth investigation, he carried on a remarkable series of experiments to track down the cause. Interposing a large variety of materials between the spark and the detector he found that the effect on the length of the micrometer spark was due to some portion of the light given off by the main spark discharge. By spectrum analysis he isolated the effect as solely due to the ultra-violet portion of the light. Hertz concluded on this discovery as follows:

"According to the results of our experiments, ultra-violet light has the property of increasing the sparking distance of the discharge of an induction-coil, and of other discharges. The conditions under which it exerts its effect upon such discharges are certainly very complicated, and it is desirable that the action should be studied under simpler conditions, and especially without using an induction-coil. In endeavouring to make progress in this direction I have met with difficulties (that is in finding a simpler means). Hence I confine myself at present to communicating the results obtained, without attempting any theory respecting the manner in which the observed phenomena are brought about."

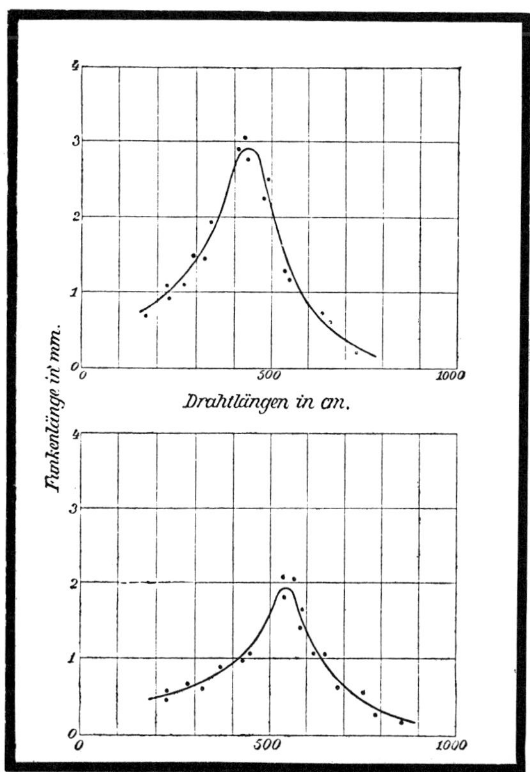


Fig. 14.4 - EFFECT OF RESONANCE on the strength of sparking at the detector for various size and proportions of the detector wire rectangle are shown by the curves above. Abscissae are the total length of the detector circuit, and ordinates the corresponding length of spark. Upper curve is for a straight wire, and lower for a spiraled wire.

Thus the mechanism of the effect was obscure at the time. Later it was shown that Hertz had discovered the "photo-electric effect." Ultra-violet radiation from the spark discharge was releasing electrons from the metal of the detector.

**HERTZ REPORTED** on all the foregoing work to date in two papers: *On Very Rapid Electric Oscillations* and *On An Effect of Ultra-Violet Light Upon the Electric Discharge,* both published in 1887. These papers and others issued subsequently on his experiments with electric oscillation were collected in Hertz's book *Electric Waves,* translated into English by D.E. Jones and published in 1893 by Macmillan & Company, London. A reprint of the Jones edition was issued in 1962 by Dover Publications, Inc., New York.

Hertz, in 1887, with the completion of his first work on rapid oscillations, was now in a position to enter for the Berlin Academy Prize Problems of 1879, which he had originally turned down, and which were still waiting for solutions. These problems had been set up by Helmholtz to confirm or disprove Maxwell's equations, based on three assumptions:

(1) that the variation of dielectric polarization of a non-conductor produced a magnetic field which was equivalent to that of a conduction current,
(2) that magnetism sweeping through a dielectric could produce polarization, and,
(3) that in respect to these two assumptions, air and empty space behave like other dielectrics.

Proving all these assumptions appeared later to be unreasonable, so the Academy required for the prize a confirmation of only one of the first two.

**HERTZ** decided to probe the first of the Maxwell equation assumptions, namely that the polarization and depolarization of an insulator should produce in its neighborhood the same electromagnetic effects as a current in a conductor.

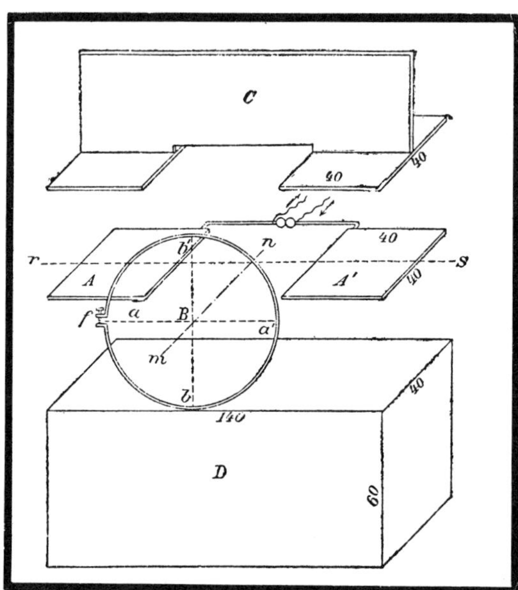


Fig. 14.5 - MAGNETIC FIELD in an insulating block, D, was shown to be equivalent to that in a conductor, C, when subjected to oscillations from A and A'. This was a proof of Maxwell's theory of displacement current in a dielectric.

His apparatus, Fig. 14.5, consisted of the two essentials of his previous experiments, the oscillator, A,A', and the detector, B. His procedure was straightforward and ingenious: to see if the effect of a conductor, C, on the primary circuit was duplicated by a block of insulation, D. The detector, B, a circle of copper wire 35 cm in radius, with a micrometer gap, was of such dimensions as to be resonant with A,A', and thus be most sensitive to oscillations from that source. The detector could be rotated about the axis m-n. With C and D absent, and the detector in the position a,a', a neutral point in the inductive balance, there was no sparking at f.

Hertz now introduced conductor, C, and due to its inductive action the balance of A,A' was shifted, indicated by sparking at, f. When f was turned toward b,b', sparking was again eliminated, giving evidence of the effect of C. Moving C up and down, Hertz found a position of resonance at which the effects were at a maximum, giving further indication that the oscillations had wave characteristics and a finite propagation rate.

Removing C, Hertz now placed plates A,A' just above the block of insulation, D, for which he piled up books to a half cubic meter mass. The expectations of Maxwell's theory were confirmed - block D acted like a conductor, C. Sparks jumped at f where there were none before, and f had to be turned downward to make them disappear. Thus encouraged, Hertz proceeded to investigate with blocks of other materials - asphalt, pitch, wood, sandstone, sulfur, paraffin, and even a tank of petroleum. All gave similar results in varying degrees. Hertz published his findings in November 1887 in a paper: *On Electromagnetic Effects Produced by Electrical Disturbances in Insulators.*

The first Maxwell assumption was proved correct and Hertz now considered attacking the second. But he had a better idea:

"I felt," he said, "that in the third hypothesis is contained the gist and the special significance of Faraday's and therefore of Maxwell's view, and that it would thus be a more worth goal for me to aim at."

He would go directly to testing the third assump-

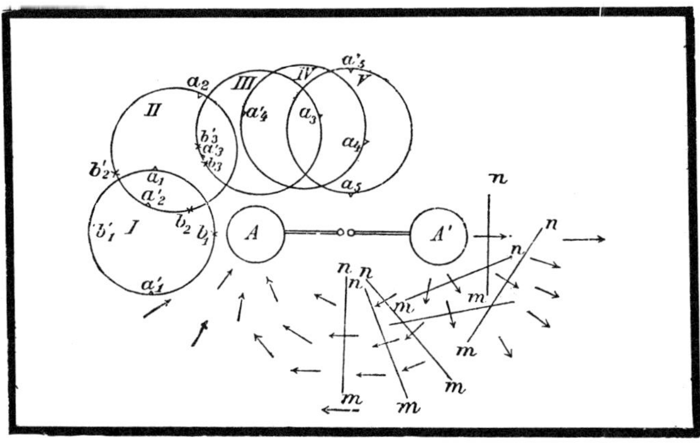

ELECTRIC AND MAGNETIC FIELDS CIRCLE A SPARK

Fig. 14.6 - Fields around the oscillator terminals were plotted by Hertz from determinations made by placing the circular detector in various positions shown by I, II, etc. Arrows depict directions of the electrostatic field at one instant of polarity, simulating that which would be due to two oppositely charged spheres A and A'. Magnetic field around the osillator terminals is shown by lines m-n.

tion: that air and empty space acted like other dielectrics, and that as Maxwell claimed, waves traveled in empty space with a finite velocity.

Hertz had been able to map the oscillating field as shown in Fig. 14.6. And in the course of his experiments had been able to detect signals up to 12½ meters. According to his calculations there should be several wave-lengths in that distance. How could they be detected? He had an idea for an experiment:

"When variable electric forces act within insulators whose dielectric constants differ appreciably from unity, the polarizations which correspond to those forces exert electromagnetic effects. But it is quite another question whether variable electric forces in air are also accompanied by polarizations capable of exerting electromagnetic effects.

"While I was vainly casting about for experiments which would give a direct answer to the question raised, it occurred to me that it might be possible to test the conclusion, even if the velocity under consideration was considerably greater than that of light. The investigation was arranged according to the following plan:- In the first place, regular progressive waves were to be produced in a straight, stretched wire by means of corresponding rapid oscillations of a primary conductor propagated through the air; and thus both actions were to be made to interfere. Finally, such interferences were to be produced at different distances from the primary circuit, so as to find out whether oscillations of the electric forces at greater distances would or would not exhibit a retardation of phase, as compared with the oscillations in the neighborhood of the primary circuit."

**HERTZ'S APPARATUS**, Fig. 14.7, was designed to send the oscillations from the induction coil along two paths, one through the air and the other along a wire, and to observe with detectors the nature of the interference between them. The wire and the detectors were designed to be in resonance with each other.

It was known that, due to reflections from the end of the wire, standing waves would be set up in a length of wire connected to an oscillator. By noting the wax and wane of the sparking as he moved the detector along the wire Hertz measured the length of the half-wave - about 2.8 meters. By calculating the frequency of the primary oscillator from its dimensions he could determine the velocity of the waves in the wire.

If the wave velocity in air was the same for the wire, the waves should travel together, adding to or canceling each other as the case might be. Hertz found, however, that the pattern was not regular. The length of wave interference became greater with an increase in distance from the origin of the oscillations. Perplexed, Hertz first concluded that the air wave velocity must be infinite. Further consideration led him to decide on a finite velocity but that waves in the wire must be retarded in relation to those in air.

It was shown later that Hertz's wire wave velocity was in error because calculation of the frequency of the primary oscillator was not correct and that the velocity in both air and wire should be the same - the velocity of light. Another factor making the results somewhat illusory, but not recognized at the time, was the effect implied by Maxwell's theory - that electromagnetic waves would be reflected by any conductive surfaces. In retrospect Hertz noted that there was an iron stove in the vicinity of the primary oscillator and several iron posts around the room that could have badly

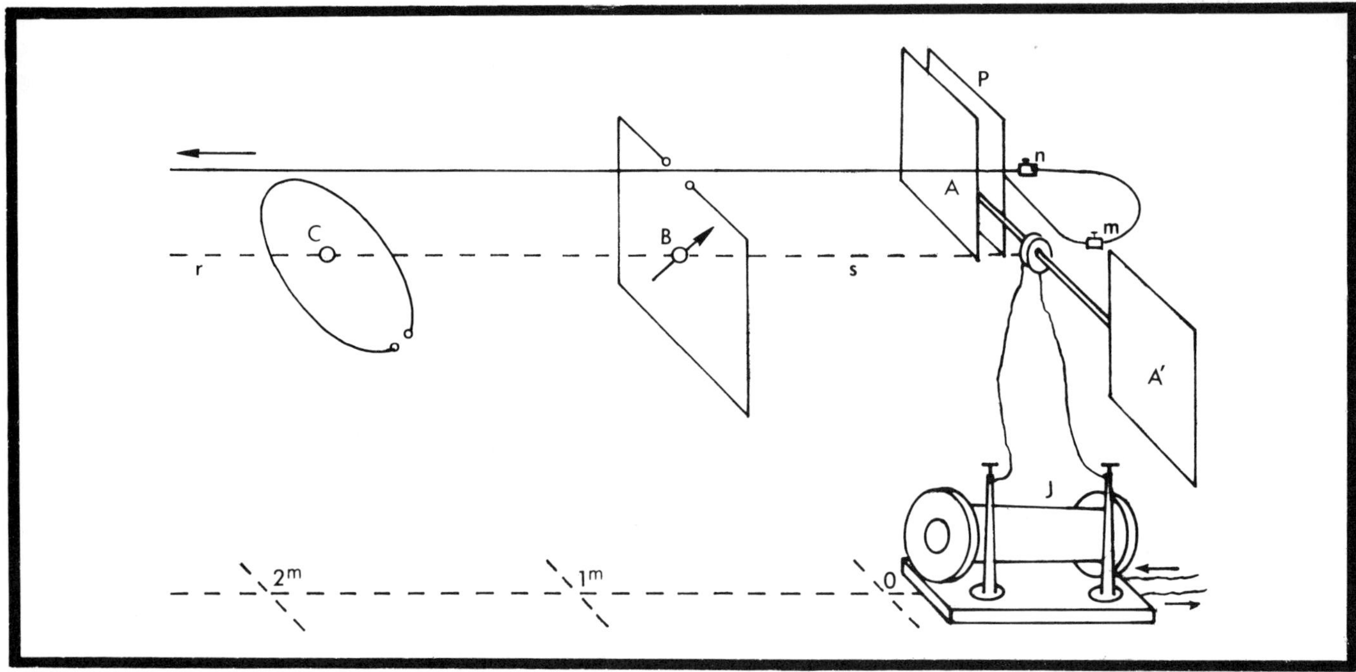


Fig. 14.7 - HERTZ'S APPARATUS for determining the relative velocities of electrical propagation in air and in a wire (solid line). Induction coil, J, provided powerful oscillations to the spark-gap, A,A′, and by induction to plate, P, connected to the wire. By manipulating detectors, C and B, Hertz found the wave length to be about 2.8 meters, from which he calculated a velocity of about 280,000 km/sec in air, and a retarded velocity in the wire. Later tests showed that the velocity should be the same in both air and wire.

affected the results. Later Hertz repeated the experiment taking into account the disturbing reflections, and the outcome was what he had hoped for originally, and in line with Maxwell's theory.

**HERTZ** had been aware of unusual effects when working with the oscillations at a distance from the origin. Something like shadows were formed behind the conducting masses, and there was reinforcement of action in front of these shadow-forming masses and in the vicinity of walls. But though he was conversant with Maxwell's theory, the idea of reflections appeared inadmissable at the time. However, after he had established the existence of waves, the principle of reflection seemed a likely explanation of the observations, and a phenomenon for further investigation.

In approaching the subject, Hertz wrote:

"I observed that in most positions of the secondary conductor the sparks became appreciably stronger when I approached a solid wall, but again disappeared almost suddenly close to the wall. It seemed to me that the simplest way of explaining this was to assume that the electromagnetic action, spreading outwards in the form of waves, was reflected from the walls, and that the reflected waves reinforced the advancing waves at certain distances, and weakened them at other distances, stationary waves in air being produced by the interference of the two systems. As I made the conditions more and more favorable for reflection the phenomenon appeared more and more distinct, and the explanation of it given above more probable."

Hertz now proceeded to determine whether standing waves could be set up in the air in the way they had been developed in a wire, that is, by reflection. In his physics lecture room, 15 meters in length, he placed the primary oscillator at one end. On the wall at the other end he fastened a sheet of zinc 4 meters high and 2 meters broad to get a good conduction reflecting surface. The detector, a circle of 35 cm radius, was mounted in a wooden frame for manipulation by hand; it was

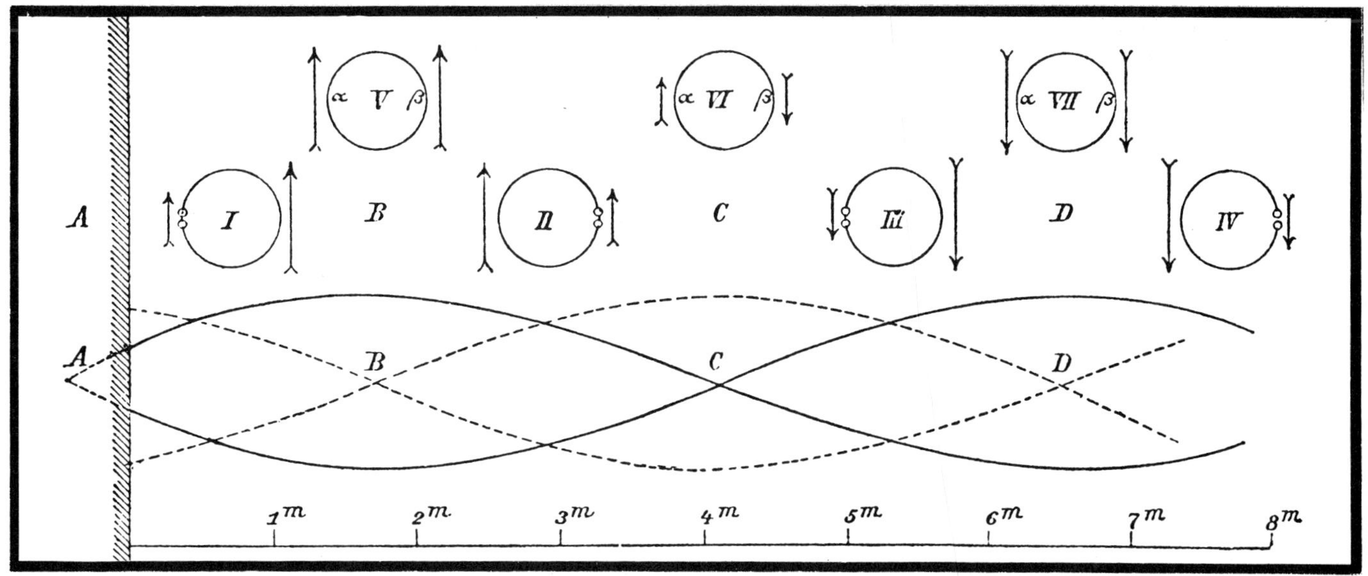


Fig. 14.8 - HERTZ placed an oscillator about 12 meters from a zinc reflecting plate to create standing waves of about 4 meters length, the intensity and polarity of which could be measured. By use of the detector ring he traced the amplitude and reversals of the electrical component, A-C, and the magnetic component, B-D, of the oscillator waves, thus demonstrating their periodic relationship.

also arranged for remote control since the body of the observer had an influence on the sparks.

The sequence of Hertz's detector tests is shown in Fig. 14.8. In the use of the detector, the magnitude of the field in any given position was determined by its effect on the part of the detector ring opposite to its gap.

Starting with the detector near the wall in position I, Fig. 14.8, Hertz determined the strength of the spark with the detector spark-gap pointed toward the wall, then with the gap pointed away from thewall. Arrows in the two positions show the strength and direction of the field force that was indicated in the two orientations. The amplitude rose in progressing from the wall.

Moving to position II the amplitude showed a declining trend. At position III the amplitude again increased, and by orienting the detector Hertz confirmed that the rise was in the opposite direction. Position IV showed another declining amplitude. By putting the detector in positions V,VI and VII there was further confirmation of the standing wave character of the electromagnetic energy, and a further defining of the node and change of sign at C. The node at A appeared to be behind the wall, which Hertz attributed to the wall not being perfectly conducting.

In addition to the experimentally determined electric wave with nodes at A-C (solid lines), Hertz theorized that it must be accompanied by another wave with nodes at B-D (dotted lines). "We are quite certain," he wrote, "that we have recognized nodes of the electric wave at A and C, and antinodes at B and D. We might, however, in another sense call B and D nodes, for these points are nodes of a stationary wave of magnetic force which, according to theory accompanies the electric wave and is displaced a quarter wave-length to it."

**HERTZ'S EXPERIMENT** was reported in his paper *On Electromagnetic Waves in Air and Their Reflection,* which was published in 1888. He had obtained remarkable evidence of the periodicity of the emission from an oscillator. He concluded the paper with a tribute to Maxwell: "it is clear that the experiments amount to so many reasons in favor of that theory of electromagnetic phenomena which was first developed by Maxwell from Faraday's views. It also appears to me that the hypothesis as to the nature of light which is connected with that theory now forces itself upon the mind with still stronger reason than heretofore."

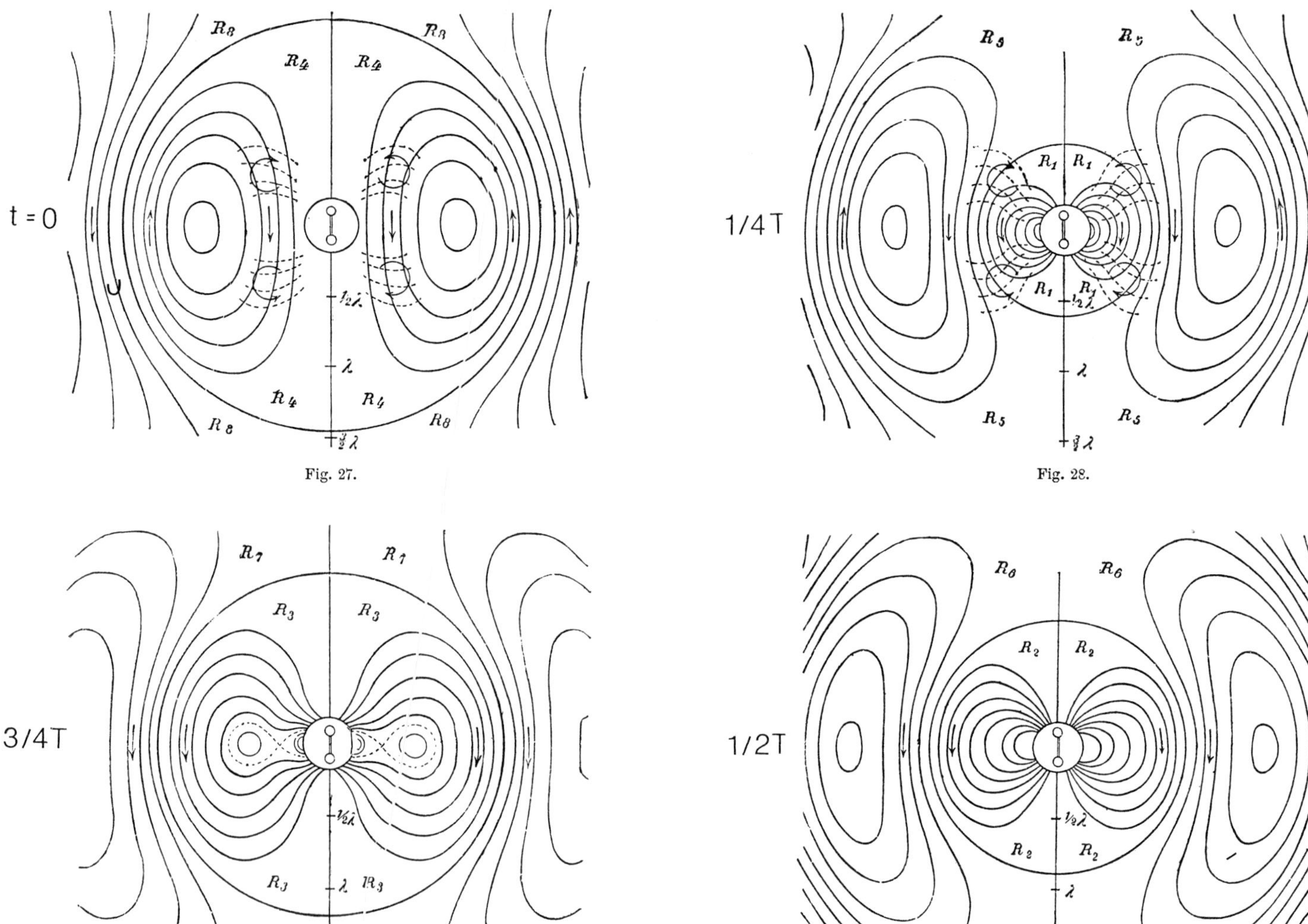


Fig. 14.9 - HERTZ'S theoretical plots show the expansion and radiation of the electromagnetic energy through space from an oscillator for 4 times periods from original charge.

Having completed the experiments just described Hertz returned to his theoretical analysis, and in 1889 published a paper on *The Forces of Electric Oscillations Treated According to Maxwell's Theory.* That theory, Hertz concluded in the paper, had been found to account most satisfactorily for the majority of the phenomena revealed by his experiments. And that, Hertz acknowledged, "was no mean performance."

One of the illuminating parts of the paper is the construction shown in Fig. 14.9. The four illustrations depict the action taking place near the spark-gap of the oscillator at time intervals during a wave-length. The figures exhibit the distribution of force at the times: T = 0, ¼T, ½T and ¾T.

The action proceeds as follows: At t = 0, the current is at its maximum strength, but the poles of the oscillator are not charged with electricity - no lines of force converge toward them. But from time t = 0 onwards, such lines of force begin to shoot out from the poles and are comprised within a sphere which rapidly enlarges. At t = ¼T the lines which have their ends on the poles, fill the space $R_1$. At t = ½T the lines have expanded symmetrically to fill the space $R_2$. At this time the electrostatic charge of the poles in at its greatest development and the lines cease to issue outward.

As time progresses no fresh lines of force proceed from the poles, but the existing ones begin to retreat toward the oscillator. Now, at t = ¾T comes the action for radiation. A portion of the inner lines contract in such a way that their upper and lower parts nearest the oscillator touch each other. The outer lines act differently however - they form loops and detach themselves as self-enclosed lines of force, advancing independently into space. Thus part of the energy of oscillation leaves the source and speeds outward with the velocity of light.

Before the close of 1888 Hertz began to sharpen the course of his experiments. What had started as an investigation of dielectric polarization in insulators, and the action of unclosed circuits was taking a new track. He had already, on a general scale, shown that the action of an electric oscillation spreads out as a wave in space, and that the waves could be reflected. He was now zeroing in on the properties of the oscillations themselves to prove that their behavior was equivalent to that of light.

**TO PURSUE** his investigations most effectively Hertz gave further evidence of his experimental inventiveness. He designed new apparatus to permit beaming of the waves, and to provide more precise measurement. He shortened the radiator, A, in Fig. 14.10 to give much higher frequency of oscillation, hence a much shorter wave-length that could better be handled within the space limits of the laboratory. The detector, B, Fig. 14.10, consisted of two straight, opposite-pointing wires, the ends of which were connected to the micrometer spark-gap. The gap was placed so that observations could be made from the rear to prevent interfering with the incoming waves. The oscillator was then mounted on the focal line of a parabolic concave mirror or reflector consisting of sheet zinc fastened to a wooden frame. The detector was mounted on an identical concave mirror.

Preliminary testing of the oscillator showed that there was no action behind the mirror or on either side of it. The waves were in the direction of the optical axis of the mirror and could be distinguished at distances up to 9 or 10 meters. Repeating previous tests with reflected waves, using a circular detector, he was able to find the nodal points that indicated a half-wave of 33 centimeters. Using the detector in the concave mirror he found

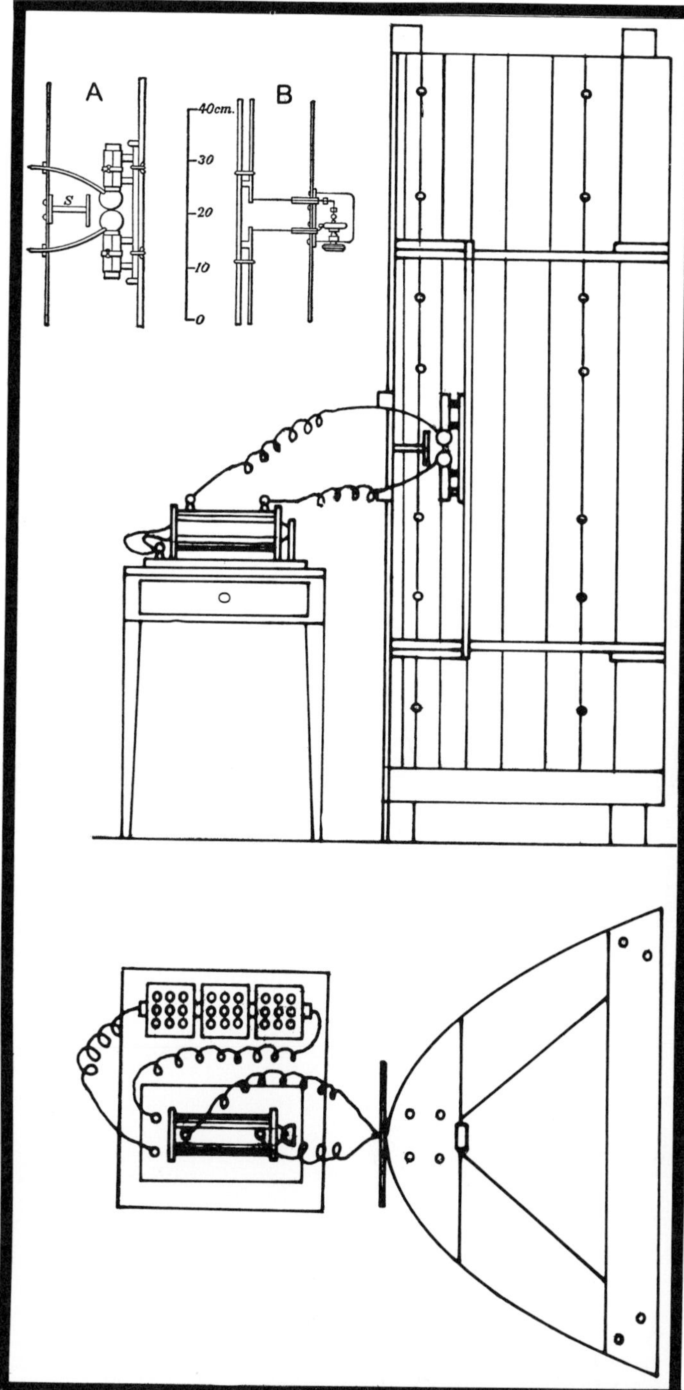


Fig. 14.10 - HERTZ'S APPARATUS for focusing a beam of electric waves, and for receiving them. Height of the mirror was 2 meters, width 1.2 meters, depth 0.7 meters. Hertz was fortunate that this apparatus produced waves of such short length, about 25 centimeters, now called short radio waves, that he could show, within the room space that the waves had properties identical to those of light.

that his expectancies for it had not been mistaken. He was easily able to perceive the ray at the full length of the laboratory - about 16 meters. For the experiments to be made, he found a distance of 6 - 10 meters most suitable in providing clearly distinguishable spark lengths at the detector.

Having produced distinct rays of electric force, Hertz used them to carry out four experiments associated with the characteristics of light and radiant heat.

**RECTILINEAR PROPAGATION** - On placing a zinc sheet 2 meters high and 1 meter broad on the straight line joining both mirrors and at right angles to the direction of the ray, Hertz found that the detector sparks disappeared completely. Screens of tinfoil and gold-paper also blocked the ray. Insulating substances such as wood were, however, transparent to the rays. Setting metal sheets horizontally on each side of the ray did not affect it as long as the opening was not narrower than the breadth of the mirror. When the oscillator was turned to the right or left about its optical axis the ray was weakened, and disappeared at a rotation of 15 degrees. The ray was thus shown to propagate rectilinearly.

**POLARIZATION** - In view of the method of producing the oscillating waves Hertz concluded they must vibrate transversely as they propagated, and in a preferred plane corresponding to the focal line of the concave reflector. To prove this he rotated the detector from its vertical position about the axis of the ray; the sparks became more feeble and disappeared when the detector was in a horizontal position with respect to the focal line of the oscillator. The waves were shown to be *polarized* in the vertical direction, the oscillator serving as the polarizer.

To further test polarization Hertz used an octagonal frame 2 meters high and 2 meters broad, across which were stretched copper wires parallel to each other and 3 cm apart. This served as a differential screen. When the screen was interposed between oscillator and detector with the wires in a perpendicular direction to the focal line of the mirrors there was no interference with the detector sparks. But turning the screen 90 degrees to make the wires parallel to the focal line stopped the sparks completely, implying the waves were polarized in that direction. The screen was acting toward the rays like a tourmaline crystal plate was known to act toward a plane-polarized beam of light.

With the detector in the horizontal position with respect to the oscillator there was no sparking. If now the screen was inserted in the path of the ray in such a position that the wires were inclined 45 degrees to the horizontal on either side, detector sparks were immediately seen. In this position the screen resolved the advancing oscillations into two right-angle components, and transmitted only that component perpendicular to the direction of its wires.

Hertz summed up his polarization results:

"With regard to the polarization it may be further observed that, with the means employed we are only able to recognize the electric force. When the primary oscillator is in a vertical position the oscillations of this force undoubtedly take place in the vertical plane through the ray, and are absent in the horizontal plane. But the results of experiments with slowly alternating currents leave no doubt that the electric oscillations are accompanied by oscillations of magnetic force which take place in the horizontal plane through the ray and are zero in the vertical plane. Hence the polarization of the ray does not so much consist in the occurrence of oscillations in the vertical plane, but rather in the fact that the oscillations in the vertical plane are of an electrical nature, while those in the horizontal plane are of a magnetic nature."

**REFLECTION** had already been proved by Hertz when an advancing wave impinging on a flat conducting plate caused standing waves that must have resulted from interference by a wave reflected from the plate. To further analyze the reflective properties he separated the wave systems to permit reflection at a right angle.

Hertz explained:

"I allowed the ray to pass parallel to the wall of the room in which there was a doorway. In the neighbouring room to which the door led I set up the receiving mirror, so that its optic axis passed centrally through the door and intersected the the direction of the ray at right angles. If the plane conducting surface was now set up vertically at the point of intersection, and adjusted so as to make

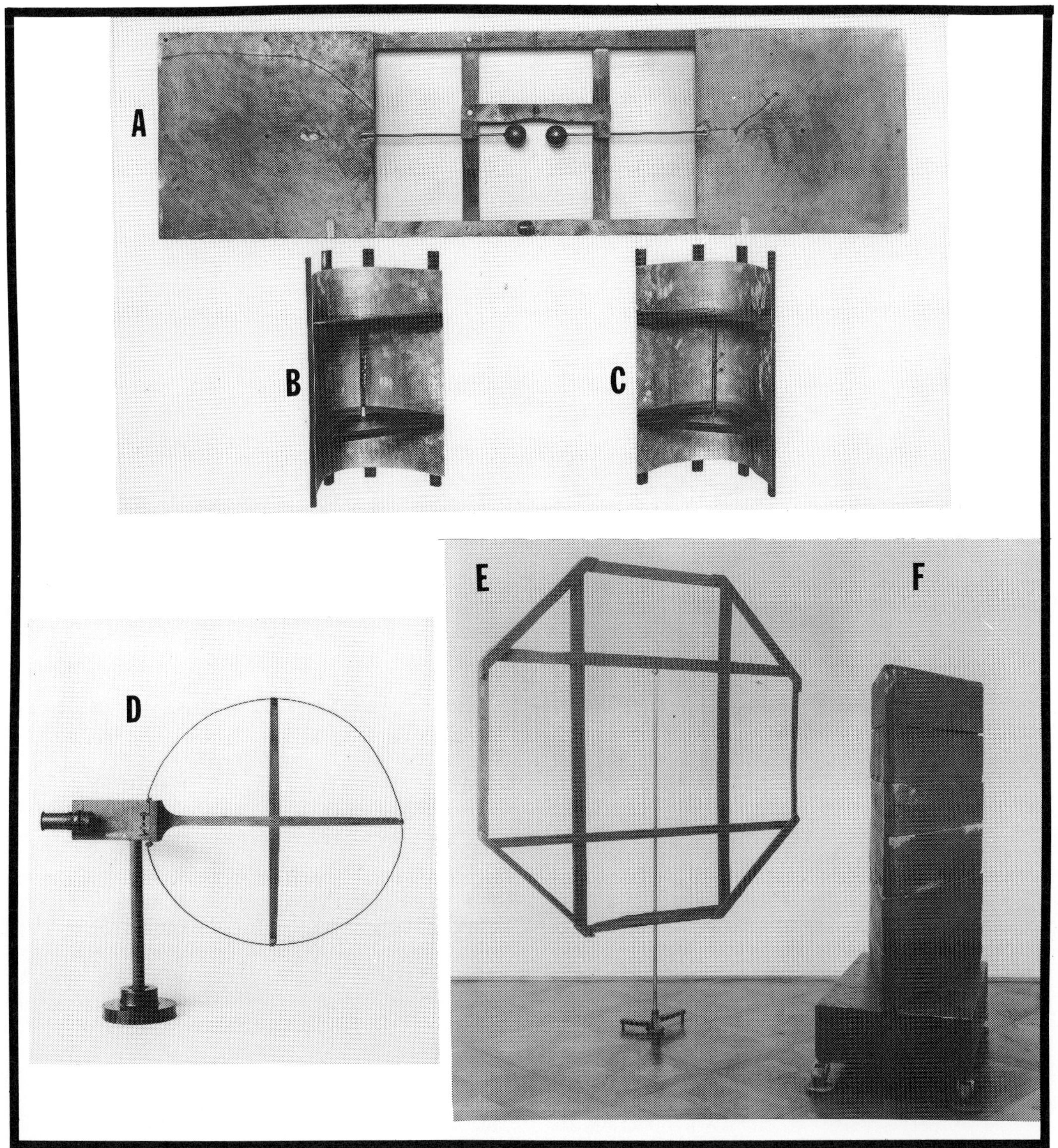


Fig. 14.11 - APPARATUS USED BY HERTZ for the study of the characteristics of electromagnetic waves. (A) Oscillator plates and spark-gap operated by the induction coil; (B) Oscillator with parabolic reflector for beaming the radiation; (C) Detector with parabolic receptor; (D) Circular detector wire with telescope for observing the air-gap; (E) Octagonal screen for polarization tests; (F) Prism for refraction tests. Photographs - Deutsches Museum, Munich.

angles of 45 degrees with the ray and also with the axis of the receiving mirror, there appeared in the secondary conductor a stream of sparks which was not interrupted by closing the door. When I turned the reflecting surface about 10 degrees out of the correct position the sparks disappeared. Thus the reflection is regular, and the angles of incidence and reflection are equal."

This was further evidence that the waves, though imperceptible to the eye, had properties equivalent to light. But surprisingly, as he stated, his waves, longer by far than those of light, were not stopped even by a wooden door. Conductors blocked them but insulators let them pass.

**REFRACTION,** or the bending of a light beam as it enters one medium from another, was a well-known optical phenomena, easily observed when light changes from a path in air to one in water. Would refraction occur to the electric rays from the oscillator with a change from air to another medium? To determine this with the relatively long wavelengths produced by the oscillator, Hertz needed to use a large insulating body. He poured pitch to make a prism, Fig. 14.11F. Its base had the form of an isosceles triangle and was 1.2 meters on a side, with a refracting angle of 30 degrees. The height was 1.5 meters and the weight 1200 pounds - so heavy that it had to be cast in several sections.

Hertz beamed the oscillator ray to the refracting face at an angle of incidence of 25 degrees. On the side of the emerging ray he marked a circle of 2.5 meters radius on the floor, around which the detector mirror was moved with its aperature always directed to the center of the circle. Starting with the detector directed at the same angle as the incident ray there was no sparking. Moving the detector around the circle, sparks were first observed at 11 degrees from the first position, reaching a maximum at 22 degrees, and being extinguished at a deviation of 34 degrees. The angular relationship of the incident and emerging rays corresponded to a refractive index of 1.69 compared to an index of 1.5-1.6 for light with pitch-like materials. This was the approximate value that had been predicted by Maxwell. Hertz's results showed that refraction was a common characteristic of all electromagnetic radiation.

Summing up the experiments, Hertz stated in his paper: *On Electric Radiation,* which appeared in December 1888:

"We have applied the term rays of electric force to the phenomena which we have investigated. We may perhaps further designate them as rays of light of very great wavelength. The experiments described appear to me, at any rate, eminently adapted to remove any doubt as to the identity of light, radiant heat and electromagnetic wave-motion. I believe that from now on we shall have greater confidence in making use of the advantages which this identity enables us to derive both in the study of optics and electricity."

And thus Hertz had concluded a series of experiments among the most notable in the history of electrical science. They were a victory for the field-force theory of Faraday and Maxwell, and a denial of the Newtonian action-at-a-distance idea. They were proof that electromagnetic forces were propagated in air with the speed of light.

Hertz's discoveries were also considered a triumph in finally defining the *ether* as a mechanical substance pervading all space, and responsible for electromagnetic action. Maxwell had conceived the electromagnetic field as a special state of matter, and not as an entity itself. In Hertz's time it was inconceivable that electricity, magnetism or light could be propagated without an intervening medium - a material substance which by strain, compression or undulation was capable of transmitting the transverse waves.

Thus the evidence of electromagnetic waves traveling through space with finite velocity in effect credited Hertz with establishing the existence of ether as a demonstrated fact. To many 19th century scientists this was the main achievement of Hertz's experiments. To Hertz himself also, a fuller understanding of the ether was, in the future, to be a prime factor in determining the fundamental nature of electricity.

**HERTZ'S WAVE EXPERIMENTS** brought him international renown, and led to some tempting offers from other institutions. In September 1888 the University of Geissen tried to hire Hertz away from Karlsruhe, and the Berlin Academy asked him to return. There he would again be associated with Helmholtz at the new Physical-Technical Institute which he headed. Clark University at Worcester,

Massachusetts, also wanted him to head up their new physical institute, which had a fine laboratory. These proposals he turned down. Then, in December 1888, the Prussian Ministry of Culture asked Hertz to take the professorship of physics at the University of Bonn. The position had been left vacant by the death of Rudolf Clausius (1822-1888) the famous thermodynamics physicist. The offer was a signal honor for so young a man, and Hertz accepted it. He was attracted by the quiet, beautiful setting of Bonn on the Rhine. The Hertz family moved there in January 1889, occupying the house of the famous predecessor.

The laboratory facilities at Bonn were run down and Hertz had to spend considerable time to put them in order. But the new position had the advantages of requiring less teaching and providing more time for research. And, as in his early student days he had assisted Helmholtz in Berlin, Hertz now could call on his capable students to assist in working on electromagnetic investigations.

Hertz resumed the theoretical studies on Maxwell's electrodynamics he had begun at Karlsruhe, and in 1890 published a paper: *The Fundamental Equations of Electromagnetics,* which is the concluding chapter of his book *Electric Waves.* The paper dealt with the equations of electromagnetics for bodies at rest and in motion, and was an elucidation and extension of the system of ideas and formulas by which Maxwell presented electromagnetic phenomena.

Of Maxwell's work, Hertz wrote:

"The system of ideas and formulae by which Maxwell represented electromagnetic phenomena is in its possible development richer and more comprehensive than any other of the systems which have been devised for the same purpose. It is certainly desirable that a system which is so perfect, as far as its contents are concerned, should also be perfected as far as possible in regards to its form."

Convinced of the correctness of the physical content of Maxwell's theory, he set about to interpret the logic of structure of Maxwell's *Treatise,* and to simplify the formalism involved in Maxwell's equations. He, and also the British electrical engineer, Oliver Heaviside (1850-1925), were the first to bring Maxwell's equations into the form that is now presented in textbooks.

**"ELECTRIC WAVES,"** the Hertz book comprised of his papers on electricity, is a model of lucidity and thoroughness in showing a problem-oriented scientist at work. The logic of his thought processes, the skill and energy in his experimental maneuvering, and the successful resolutions despite some faulty steps, are a delight to follow. His laboratory work is an example of the fact that physical questions can often only be settled by experiment, and in his case this involved much prosaic and patient, but always purposeful, labor. Hertz's character is revealed in the credit he gives to others whose investigations predated or paralled his, and in the frank assessment of his accomplishments in terms of those who made them possible.

**HONORS** poured in on Hertz from all over Europe. In 1888 he was awarded the Matteucci Medal of the Italian Scientific Society, in 1889 the La Caze Prize of the Paris Academy of Sciences and the Baumgartner Prize of the Imperial Academy of Vienna, and in 1891 the Bressa Prize of the Turin Royal Academy. In 1890 he received the Rumford Medal of the Royal Society in London where he was feted by Crookes, Fitz-Gerald, Lodge, Kelvin and other eminent British scientists. The Prussian Government awarded him the Order of the Crown. Hertz moved on equal ground with Helmholtz, Kohlrausch, Siemens and other leading German men of science.

He was invited to make a major address on the relations between light and electricity at the 1888 meeting of the society of German Natural Scientist and Physicians, in Heidelberg. In this presentation he reviewed his experiments on electric waves.

In the last few years of his life Hertz skirted the possibility of discovering the electron. In early experiments at Karlsruhe he had found that ultraviolet light was an agent in touching off sparks at a detector ring spark-gap. Identified later as the *photo-electric effect,* the high-frequency portion of the light was initiating spark discharges by knocking electron charges off the gap metal. In 1833 at Berlin Hertz had studied cathode rays and in 1890 at Bonn resumed work on them. Hertz sought to find out whether the cathode rays were beams of particles or were electromagnetic waves. If particles, the beam should be deflected by a strong electrostatic field. He did not find a deflection. Unfortunately the vacuum of his tube was not high enough to minimize the ionization of the residual gases which shielded the electric field. With a much higher vacuum J.J. Thomson (1856-1940) at Cambridge in 1897 identified the rays as a stream of charged particles he called "corpuscles," thus achieving the identification Hertz had sought. The corpuscles were named *electrons* by the Irish physicist G. Johnstone Stoney (1826-1911).

Hertz's cathode ray experiments were to be his last. His health was now a serious problem, and he turned to his final theoretical work - a book on mechanics. In Karlsruhe he had complained of toothaches and in 1889 had all his teeth removed. Then, in Bonn, he developed nose and throat pains so severe he was forced to stop work. He sought to free himself from what he thought might be hay-fever by a stay in a spa but got no relief. A bone malignancy was discovered and he was operated on in 1892. Several head operations gave only brief respite from pain. Resuming work in the fall of 1893 he carried on until December when he gave his last lecture. He died on New Year's Day of 1894 at age 37.

In their two short lives, Maxwell achieved a purely theoretical prediction that electromagnetic energy could propagate as waves with the velocity of light, and Hertz had shown experimentally the truth of that prediction. Hertz demonstrated how to produce and to manipulate the high-frequency displacement currents that were formalized by Maxwell's electromagnetic theory of light. The waves that Hertz employed differed from light rays only in that they had much longer wavelengths.

**FROM HERTZ TO RADIO** - Hertz's work had opened a new electromagnetic spectrum, and its use for practical ends was not long in forthcoming. Hertz apparently saw no immediate employment for his waves. But his demonstrations that they could leap space and be detected soon revealed to others the possibilities for wireless communication.

Sir William Crookes (1832-1919), the noted English chemist, who was familiar with the work of Hertz, made in 1892 an accurate assessment of the potential of electromagnetic waves for signalling using the Morse code then employed for wire telegraphy. He had assisted in the type of experiments in which Hughes, whose investigations have previously been described, had succeeded in receiving signals over considerable distances.

Crookes prescribed what needed to be discovered to make wireless communication possible: "firstly simpler and more certain means of generating electric waves of any desired wavelength, from the shortest, say of a few feet in length, which will easily pass through buildings and fogs, to those long waves whose length are measured by tens, hundreds and thousands of miles; secondly, more delicate receivers which will respond to wavelengths between certain defined limits and be silent to all others; thirdly, means of darting the sheaf of rays in any desired direction.

As is often the case, the path between discovery of a secret of nature and its practical application is strewn with the names of those who had the possibilities with their grasp but missed the final accomplishment and the popular acclaim. This was true of radio.

The dashed hopes of Hughes have already been related. Also, as previously noted, in 1876 Edison had witnessed signalling at a distance but did not follow it up. "What has always puzzled me," he is quoted as saying many years later, "is that I did not think of using the results . . . If I had made use of my work I should have had long-distance wireless telegraphy." Elihu Thomson (1853-1936), an American engineer and inventor, Amos E. Dolbear (1837-1910), a professor of physics at Tufts College, Boston, S. P. Thompson (1851-1916), an engineering educator in London, and others, had all worked with the essence of wireless communication but did not go on the exploit it.

**THE FIRST PATENT** for wireless telegraphy in the United States was issued on July 20, 1872, to Mahlon Loomis, fifteen years before Hertz had completed his work. The Patent No. 129,971 was for an "Improvement in Telegraphying," and covered "aerial telegraphy by employing an 'aerial' used to radiate or receive pulsations caused by producing a disturbance in the electrical equilibrium of the atmosphere." That this patent and Lommis's successful experiments entitled him to be recognized as the inventor of radio, was presented by Otis B. Young in an article, *The Real Beginning of Radio - The Neglected Story of Mahlom Loomis,* appearing in the Saturday Review March 7, 1964.

*Sketch from Mahlon Loomis's application for a U.S. patent on radio.*

Fig. 14.12 - MAHLON LOOMIS demonstrated in 1866 that by aerial transmission of electrical disturbances he could signal between peaks fourteen miles apart in the Blue Ridge Mountains of Virginia. This sketch and the excerpts below are copyright 1964 by Saturday Review. All rights reserved. Reprinted with permission.

Mahlon Loomis (1826-1886), a dentist, inventor and experimenter in Washington, D.C., wrote in 1864 . . . "I have been for years trying to study out a process by which telegraph communications may be made across the ocean without wires, and also from point to point on the earth, dispensing with wires."

In October 1866, in the presence of U.S.Senators from Kansas and Ohio, Loomis demonstrated his system of wireless signalling, Fig. 14.12. From the summits of two peaks about twenty miles apart in the Blue Ridge Mountains of Virginia he sent up kites. Each had a piece of wire gauze about fifteen inches square attached to the under side, and was held aloft by 600 feet of wire connected at its support end to the binding post of a galvanometer. The other post of the galvanometer was connected to a plate buried in the earth. The kite arrangement on the two peaks was identical and either served as transmitter or receiver.

With a prearranged time schedule, the signals were sent by making and breaking the aerial connection of one galvanometer and noting the response of the galvanometer on the other peak. The operation of each station was then reversed. The signalling was a success, and in 1869 a bill was introduced in Congress to appropriate $50,000 for the furtherance of Loomis's telegraph. Even though the bill was signed by President Grant in 1893, the funds never materialized. A group of Boston promoters promised to support Loomis's wireless scheme but were ruined by the financial panic of 1869. Loomis died in 1886 without seeing development of his idea, and without recognition.

Loomis's work appeared to be an early instance of radio communication. Whether it was done with Hertzian waves is uncertain. But Loomis's simple antenna-galvanometer-ground arrangement is said to have been duplicated by the military engineers and its workability substantiated.

**THE SUCCESS** of wireless telegraphy was dependent on discovery of a reliable and sensitive detector of electromagnetic waves. This came in the form of the "coherer," so named in 1894 by Sir Oliver Lodge (1851-1940). The coherence phenomenon had been noted by several investigators who found that finely powdered metal in its normal loose state was highly resistant to the passage of current, but when subjected to high electrostatic voltage or discharge the metal filings were excited into a condition of conductivity. When the filings were shaken they returned to their high resistance condition.

The coherer was first made into a workable device by Edouard Branly (1846-1940) a professor at the Institut Catholique in Paris. Branly, in experimenting with a glass tube filled with metal filings, found a change of resistance from several million ohms to a few hundred ohms in the vicinity of electric discharges. He further

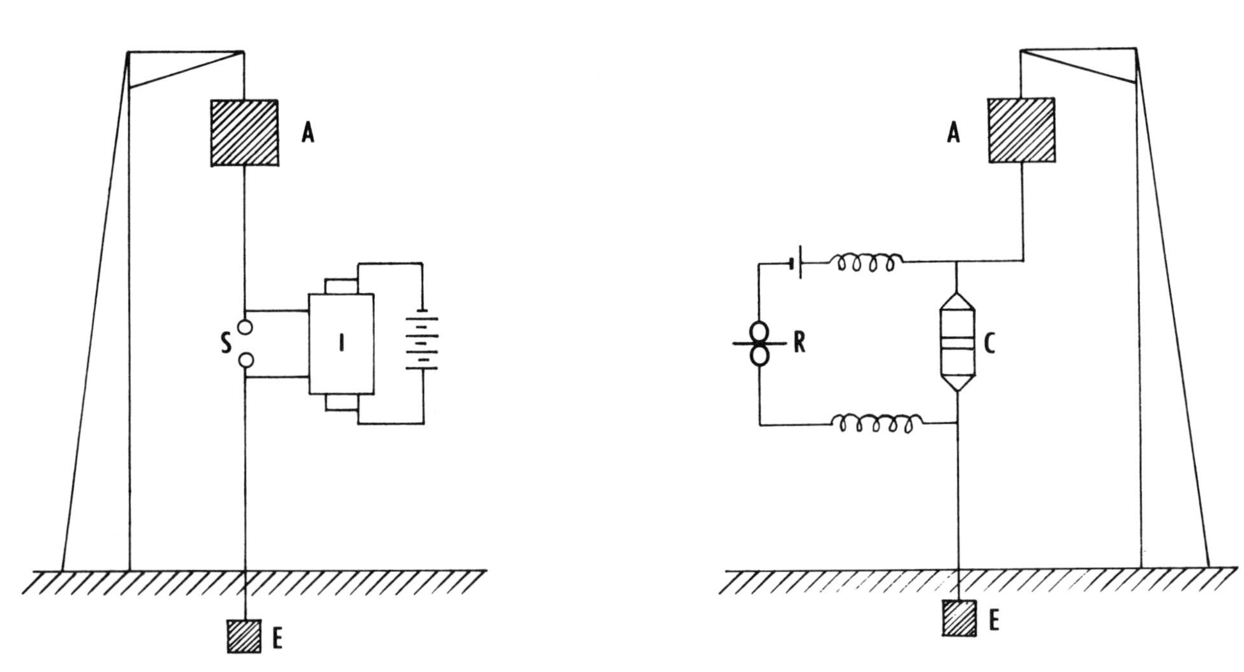


MARCONI'S EARLY SYSTEMS OF WIRELESS TELEGRAPHY
(Sketches from his Nobel Prize paper of 1909)

Fig. 14.13 (Above) Marconi's first use of Hertzian waves for wireless telegraphy, employing antenna, A, and earth, E, to increase transmission distance. I - Induction Oscillator; S - Spark Gap; C - Coherer R - Recorder. Coherer used silver-nickel filings.

Fig. 14.14 (Below) Marconi's first transatlantic signal of code letter "s" from Poldhu, Cornwall to St. Johns on the coast of Newfoudland on December 1901 was accomplished with circuits using resonant coupling with inductance and capacitance

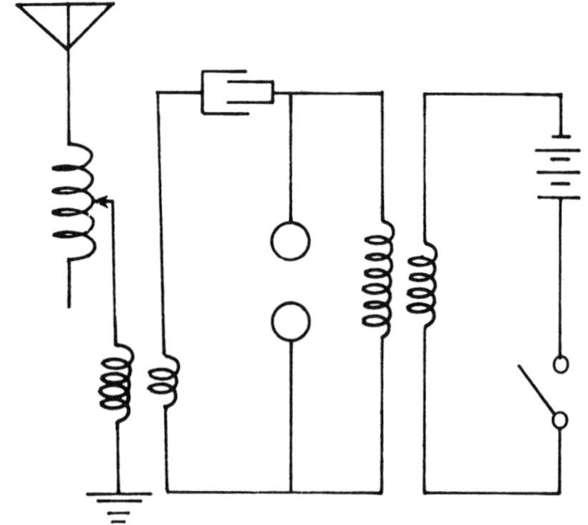

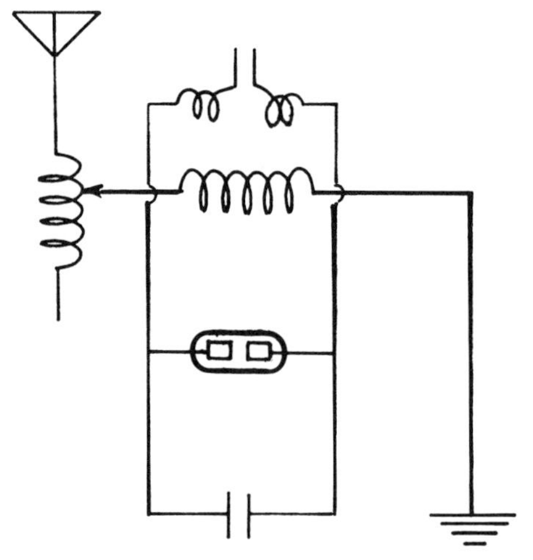

found that the original resistance could be restored by merely tapping the glass tube. However, in reporting his findings in 1891, Branly made no mention of using the coherer for the detection of electromagnetic waves.

Lodge, at the University of Liverpool, who was working on the principle of electrostatic cohesion of particles for the purpose of precipitating them, and for lightning protection, devised the first automatic "decoherer." It consisted of an electric bell in the coherer circuit which when energized by the coherer action tapped the coherer tube to restore the filings to their original condition. Thus each spark in the presence of the coherer would be signalled by a stroke of the bell. Lodge did not carry through this idea for telegraphic use. He did show that by "tuning" the circuit into resonance with inductance and capacitance the sensitivity of the system was increased.

Among those who read of Hertz's experiments and those of Branly and Lodge was the young Guglielmo Marconi (1874-1937) of Bologna, Italy, son of an Italian father and an Irish mother. He was fascinated with the results of Hertz's work and immediately grasped the communication possibilities of electromagnetic waves.

"I commenced early in 1895 to carry out tests and experiments with the object of determining whether it would be possible by means of Hertzian waves to transmit across a distance telegraphic signs and symbols without the aid of connecting wires," Marconi stated in receiving the Nobel Prize in 1909 for wireless telegraphy.

"After a few preliminary experiments with Herztian waves, I became convinced that if these waves could be reliably transmitted and received over considerable distances, a new system of communication would become available, possessing enormous advantages over flashlights and optical methods, which are so dependent for their success on the clearness of the atmosphere."

When barely out of his teens, Marconi set up an apparatus using Hertz's oscillator and a coherer receiver to see if a system of telegraphy could be achieved. In 1895, on the grounds of his father's estate, he connected his transmitter and receiver to an antenna and ground, Fig. 14.13, and using a telegraph key to produce dots and dashes, sent the first wireless messages. He was able to transmit up to a mile and a half. Packing up his equipment Marconi went to London, the center of world commerce and shipping, to make his fortune as an inventor and promoter of wireless telegraphy.

**MARCONI** received his first radio-telegraphy patent in 1897, and British capital was subscribed to form an organization that became the Marconi Wireless Telegraph Company, Ltd. The value of wireless for ship to shore and ship to ship communication was soon demonstrated, and in 1899 England and France were linked by wireless.

On 12 December 1901, Marconi achieved his greatest triumph and world-wide acclaim by receiving the first wireless signal across the Atlantic: the code letter "s" ( ... ) which was being broadcast from Poldhu on the coast of Cornwall to a kite-supported antenna on the coast of St. Johns, Newfoundland.

The dawn of the radio age had arrived! Hertzian waves would soon circle the globe, and in the course of time would span the vastness of outer space.

**TWO DISCOVERIES** in Hertz's lifetime would have tremendous impact on the future of Hertzian waves, and of electrical technology.

Thomas A. Edison, in 1883, found that an electric current could flow in the space, from filament to plate, in an incandescent light bulb. Later called the *Edison effect,* the phenomenon led to invention of the electron tubes, and to the age of electronics.

Karl Ferdinand Braun (1850-1918), a German physicist, in 1874, discovered rectification in the contact of a semiconductor and a metal. This figured in the invention of the *transistor* in 1948, opening the age of solid-state technology.

# TESLA . . .

## he pioneered the ideas that are fundamental to the art of radio today . . .

LIGHTNING MADE TO ORDER BY TESLA IN HIS COLORADO SPRINGS LABORATORY

# the unit of magnetic flux density is named for NIKOLA TESLA, a lone genius, poet, visionary and practical engineer. . . his polyphase system and induction motor are the basics of a-c power

**THE GREAT ROOM WAS ABLAZE** with electric fire . . . purple flashes zigzagged and roared . . . the air reeked with the acrid fumes of ozone. Crashing with millions of volts the streamers shot out, some leaping a hundred feet or more from the top of a giant coil in the center of the building. Here, for the first time by the hand of man, lightning was being made to order. Nikola Tesla, seated in his grand experimental station on a 2000 meter high plateau near Colorado Springs, was hurling thunderbolts rivaling those of the thunderstorms that often lighted up Pike's Peak and other nearby Rocky Mountain tops.

Tesla was pursuing a dream. He would send the power of electricity around the world - without wires!

In 1898, in his New York City laboratory, he had conceived the idea of radiating electrical energy through the air by using special booster coils he had developed. The new coils stepped up ordinary power to unheard of millions of volts and millions of vibrations per second to electrify the air with lightning-like flashes.

Envsioning his laboratory coils enlarged on a mighty scale, he predicted they could generate tremendous bursts of electromagnetic radiation hurtling through space. These could be picked up at remote places to provide signals and to deliver electric power. Gone would be the poles, the wires and other trappings of electric power lines. Instead, a few large towers strategically located would radiate the electrical energy, and small receiving antenna on homes and factories would tap off the power as required.

Now, in 1899, with the advice of friends, and with power services to be provided by Colorado Springs Electric Company, Tesla had set up a gigantic experimental laboratory to test his ideas.

**TESLA** suspected that the earth and the atmosphere would be found alive with its own electrical vibrations, and these could be reinforced to serve his purposes. His laboratory location was ideal for observing the vibrations because the nearby mountains rumbled with frequent and violent lightning storms. His detection apparatus could follow and measure them for hundreds of kilometers coming and going. Tesla wanted to study the properties of lightning discharges as they related to his own plans for electromagnetic waves of great power.

NIKOLA TESLA at the age of 36

On one occasion he recorded twelve thousand discharges in two hours. He noticed that several times his instruments were affected strongly by discharges taking place as the storm was passing away. This puzzled him.

As recorded in his journal, an answer to this activity came in July 1899. A violent storm broke loose, and after expending most of its fury in the mountains was driven away at high speed eastward over the plains. With the storm's retreat the indications on his instruments grew faint. Then, as he eagerly watched, the unexpected happened - the intensity grew stronger and stronger, and after passing through a maximum, decreased and ceased once more. The same strange waxing and waning repeated with undiminished force when the storm was calculated to be three hundred kilometers away.

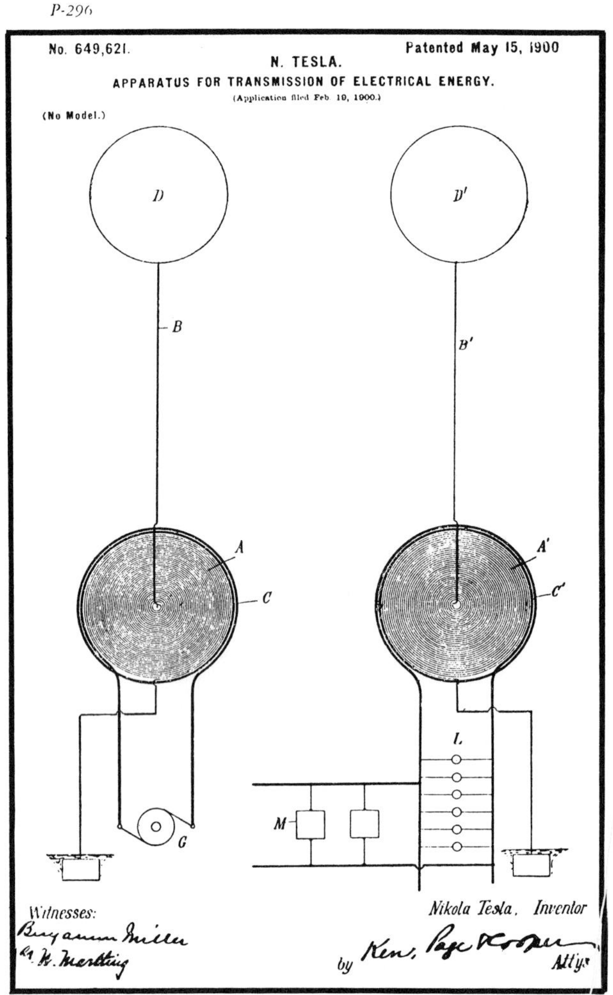


TESLA'S SCHEME FOR WIRELESS WORLD POWER

Fig. 15.1 - Global power was based on his belief that at a suitable elevation the rarified air would be a true conducting medium for electromagnetic oscillations, which would then return through the ground. Power was supplied at high frequency and exceedingly high voltage to the elevated transmitter, D, through wire, B, from step-up transformer, C-A. A similar system would receive and step down the oscillations to power lights and motors at a desired frequency and voltage.

What was the phenomenon?

To Tesla there was no doubt: "I was," he wrote, "observing stationary waves."

His receiving circuit was recording their successive nodes and loops, a fact which took on tremendous significance in Tesla's dreams for transmission of energy by his system.

"Not only," he predicted, "would it be practicable to send telegraphic messages to any distance without wires, as I recognized long ago, but also to impress upon the entire globe the faint modulations of the human voice, far more still, to transmit power, in unlimited amounts, to any terrestrial distance and almost without any loss."

**THE MAIN APPARATUS** at Colorado Springs was a giant version of what is now called the "Tesla Coil," a device occasionally seen in physics laboratories. It is an air-core transformer in which the discharge of a condenser through the primary provides a discharge in the secondary of exceedingly high voltage and high frequency, capable of spectacular effects.

The 75-foot diameter, 3-turn primary coil circled the perimeter of the room. Within it in the center of the room, tuned to resonance, was a 10-foot diameter, 100-turn secondary, one end connected to the ground, the other to a 200-foot mast with a 3-foot diameter copper ball that could be raised and lowered. When the switch was thrown the canonading of millions of volts flashing from the ball could be heard by the gold miners in Cripple Creek fifteen miles away.

Tesla hoped that the tremendous bursts of elevated high-frequency energy would excite electrical oscillations radiating far and wide. Stationary waves, spreading from his coils, the kind he had observed with lightning, would transmit instant communications and power around the globe. Actually he was able to light 200 incandescent lamps at a distance of 30 kilometers, and he produced detectable wireless signals at 1000 km.

News reporters, awe-struck at the electric fire streaming from the ball on the mountain top, spread bizarre stories of Tesla's man-made lightning. Spectacular electric effects like this were novel and bewildering, attracting great attention.

**ON TESLA'S RETURN** to New York in early 1900 he was hailed as an electrical wizard. Among other marvels, he was promising a future world of instant communication. "I have no doubt," he stated, "that it will prove very efficient in enlightening the masses . . . it involves employment of a number of plants, all of which are capable of transmitting individual signals to the uttermost confines of the earth. Each of them will be preferably located near some important center of civilization and news it receives will be flashed to all points of the globe. A cheap and simple device, which might be carried in one's pocket, will record the world's news or such special messages as may be intended for it."

A fantastic vision of future broadcast!

The prophecy aroused great public interest. But although not announced, Tesla was at the same time scheming to promote his main goal: transmitting electric power without wires.

In a remarkable article published in the Century Magazine for June 1900, entitled *The Problem of Increasing Human Energy,* Tesla enlarged on his plans for a wireless "World Telegraph."

" . . . we have every reason to anticipate that in a time not very distant most telegraphic messages across the oceans will be transmitted without cables . . .communications without wires ought to have many times the working capacity of a cable, while it will involve incomparably less expense."

Tesla's article and his grand design for a world wireless telegraph aroused the interest of the financier J.P. Morgan. His House of Morgan was banker for American Telephone & Telegraph Company. Sensing an opportunity for further control of world communication, he agreed in November 1900

Fig. 15.2 - TESLA'S TOWER for worldwide telegraphy and transmission of electric power without wires, erected in 1900 at Wardenclyffe, Long Island. The 180 foot tower, with 30 foot half-dome, was destroyed in 1917 to prevent its use by enemy agents. The generator station at the foot of the tower, designed by Stanford White, is still standing.

to advance $150,000 for a transatlantic wireless station. A site was selected on Long Island about 100 kilometers east of Brooklyn, and Tesla named it "Wardenclyffe." Here the station consisted of an administration building and laboratory designed by the noted architect Stanford White, behind which was the transmission tower with generators and transformers for power supply. The wooden beam lattice tower stood 57 meters high, surmounted by a 30-meter diameter metal half-dome weighing 60 tons. A station to receive the messages from this transmitter was planeed for a coastal site in Cornwall, England.

**LATE 1901** was the time set for energizing Wardenclyffe. But delays in delivery of the very special electrical equipment postponed completion. Then a notable event occurred. On December 12, 1901, Marconi, using Hertzian waves, transmitted the code letter "s" from Cornwall to Newfoundland - opening the age of wireless telegraphy.

Marconi accomplished this feat with a low power for the transmission, and a simple coherer-type receiving equipment. The Marconi success cast doubt on the future of Tesla's huge, complex and expensive project, raising questions among his financial backers. These were heightened when it was revealed that Wardenclyffe was intended not only for telegraphy but also for transmitting energy without wires.

To complete Wardenclyffe required far more funds than Tesla had estimated.His pleas for more financing began to fall on deaf ears. Finally, in the midst of the finanacial panic of 1903, he appealed to Morgan, revealing his true ambition of transmitting power. But he was refused - Morgan had lost faith in Tesla's telegraph venture, and the idea of wireless power radiation had little appeal. Fast accumulating debts finally put an end to Tesla's grand plans. After a few preliminary experiments in 1906 the station was abandoned.

**IT WAS A FAILURE** to everyone except Tesla. He maintained that the massive bursts of electromagnetic power from his station could be collected at points all over the globe for a revolution in the use of electricity.

Writing in his journal he vented his frustrations: ". . . it is a simple feat of scientific electrical engineering, only expensive - oh blind faint-hearted doubting world."

And thus ended Tesla's grandiose and expensive plans for world telegraphy and power radiation. His exploits became a butt for ridicule by some of his contemporaries. Had this been Tesla's only contribution in the field of electricity his name might be remembered mostly for that suited to science fiction; a spectacular showman and visionary who failed of accomplishment.

**BUT FORTUNATELY FOR MANKIND,** Tesla, earlier in his career, had another dream - not that of an eccentric - but of a thoroughly practical engineer. It was a vision that became a reality and that, despite great opposition, revolutionized the world's system of electric power. Nowadays, whenever we press a switch to light a lamp or start a motor we put to work the alternating currents that Tesla's creative genius established as the most effective system of electrical power.

**IN THE 1880s** there were two types of electricity supply in the United States. One was the *direct-current* system developed and championed by Thomas A. Edison, which was rapidly going into use for incandescent lighting in commercial city areas.

Edison's d-c system hadone critical disadvantage however. The direct current was generated and distributed at lamp voltage, usually not higher than 250 volts. Since electric power is the product of the voltage and the current, and the voltage was fixed at the lamp value, every increase in lamp load meant a corresponding increase in current. Serving high density lamp areas thus required very large currents using huge and expensive copper conductors. Furthermore, with large currents and long distribution distances there would be undesirable heating losses in the conductors which went higher with the square of the current. There would also be a drop in voltage due to the resistance of the conductors that could dim lamp illumination. As a result the distances of distribution from a single power station had to be limited and localized with a powerhouse needed about each square mile.

The second system of electricity supply was an *alternating current,* promoted by George Westinghouse (1846-1914), the founding genius of the Westinghouse Electric Company. Alternating current had the unique advantage that the voltage of a power line could be raised or lowered by means of a *transformer* - a simple stationary device having adjacent primary and secondary coils, separated electrically, but joined magnetically, when energized by their respective magnetic fields. By altering the ratio of the turns of wire in the secondary coil with relation to the number of turns on the primary coil, the voltage output of the secondary could very efficiently be raised or lowered.

Michael Faraday, in England, and Joseph Henry, in America, in the 1830s, had independently discovered the principle of voltage transformation. The first practical transformer for use in stepping the voltage up or down on an alternating-current power distribution line was patented in 1881 in Europe by Lucien Gaulard and John D. Gibbs.

Westinghouse, in 1885, bought the Gaulard and Gibbs patents. He foresaw that the use of transformers would give alternating current power systems a formidable advantage over direct-current systems which could not use the transformer principle. Electricity at the generating plant could be transformed up to thousands of volts, with corresponding reduction in the current, and hence smaller wire size, for long-distance transmission. Then, at the point of use, the voltage could be transformed down to the required value. One generating plant, at a most suitable location, such as at a waterpower site, could serve a greatly expanded area of utilization.

By 1886 the newly formed Westinghouse Electric Company had demonstrated an a-c supply at Great Barrington, Massachusetts, in which the a-c power was stepped up to 500 volts, transmitted 4000 feet to the center of town, and there stepped down for lights in stores and offices in 20 buildings. This transmission system and several others like it proved successful. But while alternating current promised to overcome the distance handicap of direct currennt, it had the disadvantage of lacking a suitable general-purpose motor.

**THAT WAS THE ELECTRICAL SITUATION** when Nikola Tesla, a tall, lanky, dark-haired young immigrant from Croatia stepped off the boat in New York City in 1884. Three years earlier, in a flash of inspiration, he had conceived an idea for a rotating magnetic field that would produce a practical alternating-current motor. Failing to find support for his concept in Europe he had come to America to seek acceptance and financial help.

His head was full of ideas for an alternating-current system of which his motor would be a vital part. In the new world Tesla was initially employed as an assistant in the Edison works at Menlo Park, New Jersey. But in his preoccupation with alternating current he was soon at odds with the great inventor who was championing the use of direct current. Tesla would soon enter the arena of the desparately fought "battle of the currents," - ac vs. dc. In the ultimate outcome of this battle his induction motor and his system of alternating current would play a deciding part.

**NIKOLA TESLA** was born on 10 July, 1856, of Serbian parents in Smiljan, Lika, Croatia, then a province of Austria-Hungary, and now Yugoslavia. His father was a clergyman in the Serbian Orthodox Church. His mother, although unlettered, was remembered as of poetical disposition, with an excellent memory and an inventiveness in home and farm labor-saving devices. Tesla, in his later life, credited his mother's qualities as the source of his own exceptional memory and unusual innovative skills.

Young Tesla showed early signs of inventive genius. He made his own toys, and once built a waterwheel. He also made a tiny windmill powered by the fluttering wings of June bugs glued to its arms. At the age of seven he showed quick perception when a new fire engine being demonstrated failed to pump. An intuitive flash prompted him to dive into the river to remove a kink obstructing the suction hose. The event made him town hero for the day.

In his eleven years of study at primary and secondary schools Tesla displayed an agile and scholarly mind. He was quick in mathematics and had an incredible memory. He loved to recite poetry and early began the study of languages. From his childhood Tesla had been intended for the clergy by his parents, but the young boy looked forward to this future with dread. He was far more interested in the field of physics, mathematics and in the workings of nature.

The matter was resolved when Tesla, taken so critically ill with cholera that his life was despaired of, was promised by his father that on recovery he could take up the study of his choice. Thus in 1877 Tesla entered the Polytechnic Institute of Graz, Austria, one of the oldest technical schools in Europe. Here he became fascinated with electricity and resolved to follow that as a career. But in addition to physical studies Tesla read poetry and literature and carried on with his love of languages.

In his second year at Graz, a laboratory incident played a crucial part in shaping his future. The class was working with a Gramme machine from Paris, a direct-current type which could operate either as a motor or a generator. The brushes pressing on the commutator of the machine sparked and flashed when it was in operation. Professor Poeschl, who headed the physics department, was demonstrating the machine as a motor, and there was very lively sparking at the brushes. Tesla took particular note of this and ventured to remark to the professor that it might be possible to devise a machine in which the sparking parts were eliminated.

Tesla related what followed:

"**PROFESSOR POESCHL** said that it was quite impossible and likened my proposal to a perpetual motion scheme, which amused my fellow students and embarrassed me greatly. For a time I hesitated, impressed by his authority, but in my conviction which grew stronger I decided to work out a solution. At that time my resolve meant more to me than the most solemn vow.

"I undertook the task with all the fire and boundless confidence of youth. To my mind it was simply a test of will-power. I knew nothing of the technical difficulties. All my remaining term in Graz was passed in intense but fruitless effort, and I almost convinced myself that the problem was unsolvable. Indeed, I thought, was it possible to transform the steady pull of gravitation into a whirling force? The answer was an emphatic no. And was this not true of magnetic attraction? The two propositions appeared very much the same."

**THE PROBLEM** of eliminating commutator and brushes became an obsession with Tesla. But it was not until four years later that the basics of the new motor and generator came to him, instinctively and fully formed. In the meanwhile he had gone to Prague, Bohemia, carrying out his father's wish to complete his education at a university. Here he felt the atmosphere of the old and interesting city was favorable for inventive thinking, and the solution to the brushless machine appeared forthcoming. In a year, however, his father died, and realizing that his mother would be making too great a sacrifice for continuing his education, he decided to relieve her of the burden. On learning that the American telephone had reached Europe and that the system was to be installed in Budapest he seized the opportunity for employment with the company. Starting as a draftsman he advanced, at the age of 26, to chief electrician.

Tesla's mind continued to turn on the subject that absorbed him - a motor and a generator that would eliminate the problems of d-c machines. The solution involved the use of alternating current, in itself not new, but utilized in a way that would create the rotational torque such as was accomplished by the commutator of the d-c machine.

Something of Tesla's mysticism and poetical nature is evident in the manner of his first grasp of an answer to the problem. He was walking one day, in 1882, in Budapest's City Park with his friend Mr. Szigeti, enjoying the summer air and reciting poetry of which he was passionately fond. "At that age," he said, "I knew entire books by heart and could recite from memory word by word. One of these poems was *Faust.* It was late afternoon and the sun was setting, and I was reminded of the passage:

*The glow retreats, done is the day of toil,*
*It yonder hastes, new fields of life exploring;*
*Ah, that no wing can lift me from the soil,*
*Upon its track to follow, follow soaring!*

*Alas! the wings that lift the mind no aid*
*Of wings to lift the body can bequeath me.*

"As I spoke the last words, plunged in thought, and marvelling at the power of the poet, the idea came in a lightning flash. In an instant I saw it all, and I drew with a stick on the sand the diagrams, which Mr. Szigeti understood perfectly."

Tesla had conceived the principle of the spatially shifting electromagnetic field that produces the torque in an induction motor. He reasoned that by using two or more out-of-phase alternating currents flowing through geometrically displaced coils it would be possible to develop a progressively rotating magnetic field that he had long sought as the basis for a new motor. He had hit upon the rotating idea, but it was to be some time before he could exploit his invention commercially.

**IN LATE 1882** Tesla moved to Paris to join the Continental Edison Company. He was engaged as an engineer on incandescent lighting plants being installed in Europe under the Edison patents. In this capacity he met several American engineers to whom he explained his motor invention. They suggested the formation of a company to market it. Before Tesla could act on this he was transferred to Strasbourg in Alsace. There he had the first opportunity to construct a model incorporating the new motor principles. As he related later:

"I had brought some material from Paris and a disk of iron with bearings was made for me in a mechanical shop close to the railroad station in which I was installing the electric light and power plant. It was a crude apparatus, but afforded me the supreme satisfaction of seeing for the first time rotation affected by alternating current without a commutator. I repeated the experiment with my assistant twice in the summer of 1883.

"My intercourse with the Americans had directed my attention to the practical introduction and I endeavored to secure capital, but was unsuccessful in this attempt and returned to Paris early in 1884. Here, too, I made several ineffectual efforts."

Tesla's connection with the Paris branch of Edison had opened his eyes to the progress being made in America in developing electric power systems. He decided that his best chances for introducing his motor commercially would be across the Atlantic. To secure immediate employment Tesla arranged through the Paris branch for a job at the Edison Machine Works in New Jersey, where he was to work on the design of dynamos. There, in late 1884, he was introduced to his new employer, the great American inventor.

"I met Edison, and the effect that Edison produced upon me was rather extraordinary. When I saw this wonderful man, who had had no training at all, no advantages, and did it all himself, and see the great results by virtue of his industry and application. I felt mortified that I had squandered my life - you see I had studied a dozen languages, I had delved in literature, and had spent the best years of my life in ruminating through libraries and reading all sorts of stuff that fell into my hands, I thought to myself, what a terrible thing it was to have wasted my life on those useless things, and if I had only come to America right then and there and devoted all of my brain power and inventiveness to my work, what could I have not done? In later life I realized I might not have produced anything without the scientific training I got, and it is a question as to whether my theory as to my possible accomplishments was correct."

Edison's lighting plants were being widely installed, and the apparatus was undergoing an improvement in design, a task to which Tesla was assigned. Edison found Tesla to be a capable and incredibly hard worker. His hours on the job were far more than required by regular employment. But Tesla's main thoughts were elsewhere. He was getting more and more anxious about his invention. He tried to interest Edison.

**EDISON** would hear nothing of Tesla's plans for his induction motor and alternating current system. Edison was totally entrenched with direct current and unalterably opposed to the use of alternating current for power transmission. It is not surprising, therefore, that Tesla, given to equal obstinacy in his dreams for alternating current, found the association unsatisfactory. He left the Edison works in 1885 to participate in a company organized by some friends of his to develop an industrial arc light.

Tesla's preoccupation with his alternating current ideas soon put him at odds with the arc light promoters. "I told those friends of mine," said Tesla, "that I had a great invention relating to alternating current transmission, and they said, 'No, we want the arc lamp. We dont want this alternating current.' " Tesla did finally perfect an arc lamp, but the experience was unrewarding and he left the company in 1886. His resources were then so meagre that he eked out a living digging ditches and by other odd jobs. But he was soon to enter the most productive period of his life with, in his words, "the invention that swept the world."

**WITHIN A COUPLE OF YEARS** Tesla was back on his feet. He had obtained two patents on an arc lamp, and these attracted enough financing to enable him to establish a laboratory in New York City. Here he could enjoy a free rein for his research, and pursue the vision of converting his schemes for alternating current into a commercial reality.

**STEP BY STEP** Tesla developed his motor system. The current to the motor was supplied by a generator, Fig. 15.4, with two separate windings displaced by 90 electrical degrees, to give a two-phase current. These windings generated separate waves of current that followed each other in rotational sequence by 90 electrical degrees. When fed into a motor having similarly disposed stationary windings a revolving magnetic field was established, Fig. 15.3. A rotor winding placed within this field is magnetically energized by induction, and the interaction of the two magnetic fields causes the rotor to revolve in unsion with the rotating magnetic field of the stator, thus providing an *induction motor.*

Tesla worked with two types of rotors for his motors. One had only iron poles without windings. This rotor started with very feeble torque but when up to speed locked into synchronism with the rotating field of the stator, thus operating as a *synchronous motor.* The other rotor was wound with coils closed on themselves. This rotor started with high torque and operated at a speed slightly below that of synchronism. The slip in speed was just enough to induce in the rotor winding, by transformer action, the magnetic torque to drive the motor and its load, thus operating as an *induction motor.* This was Tesla's dream motor - a motor without commutator. The invention was covered by patents Nos. 381968, 382279, and 390721, granted in 1888.

**THE INDUCTION MOTOR** was only a part of Tesla's overall conception. In a series of history-making patents he demonstrated a polyphase alternating-current system, consisting of a generator, the transformers, the transmission layout, and the motor and lights, see Fig. 15.4. From the power source to the power user it provided the basic elements for electrical production and utilization.

But, as Tesla had found in the past, invention was one thing and commercial development was another; Teslas's ideas were in advance of the time so that exceptional vision was required to finance and promote a new and more complicated system in the face of Edison's direct-current system which was providing reliable and satisfactory service.

Tesla was a practically unknown inventor. But his succession of patents attracted attention in the electrical profession, and the significance of his

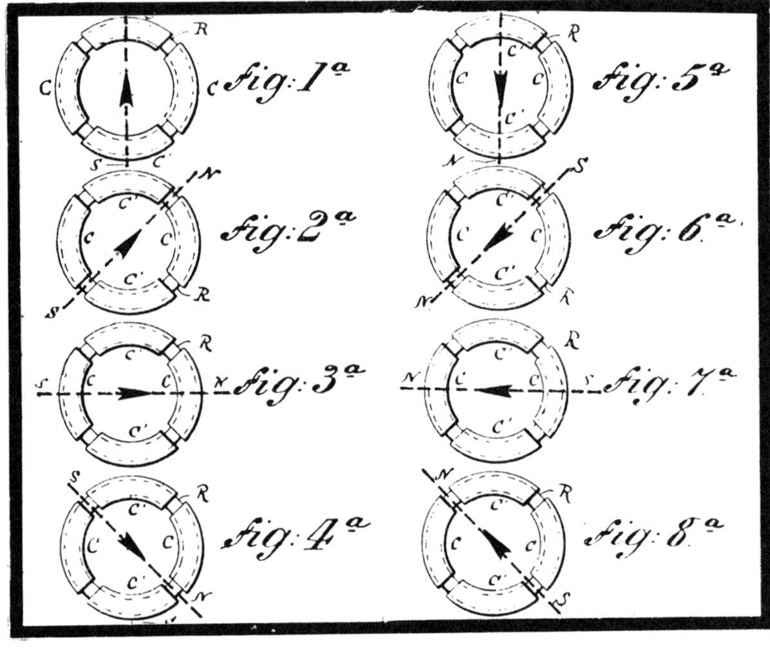


Fig. 15.3 - ROTATING MAGNETIC FIELD as pictured in Tesla's Patent No. 381968 for an Electromagnetic Motor. A two-phase current from a generator (Fig. 15.4, opposite page), with waves separated by 90 electrical degrees, is supplied to the pairs of coils on the stationary winding of the motor shown schematically at the left. This produces a rotating magnetic field in the stator indicated by the succession of arrows, Figs. 1 to 8. The rotating field in the stator induces magnetic fields in the winding of a rotor within it, and the interacting magnetic fields cause the rotor to revolve in unison with the stator field. This is *induction motor* action.

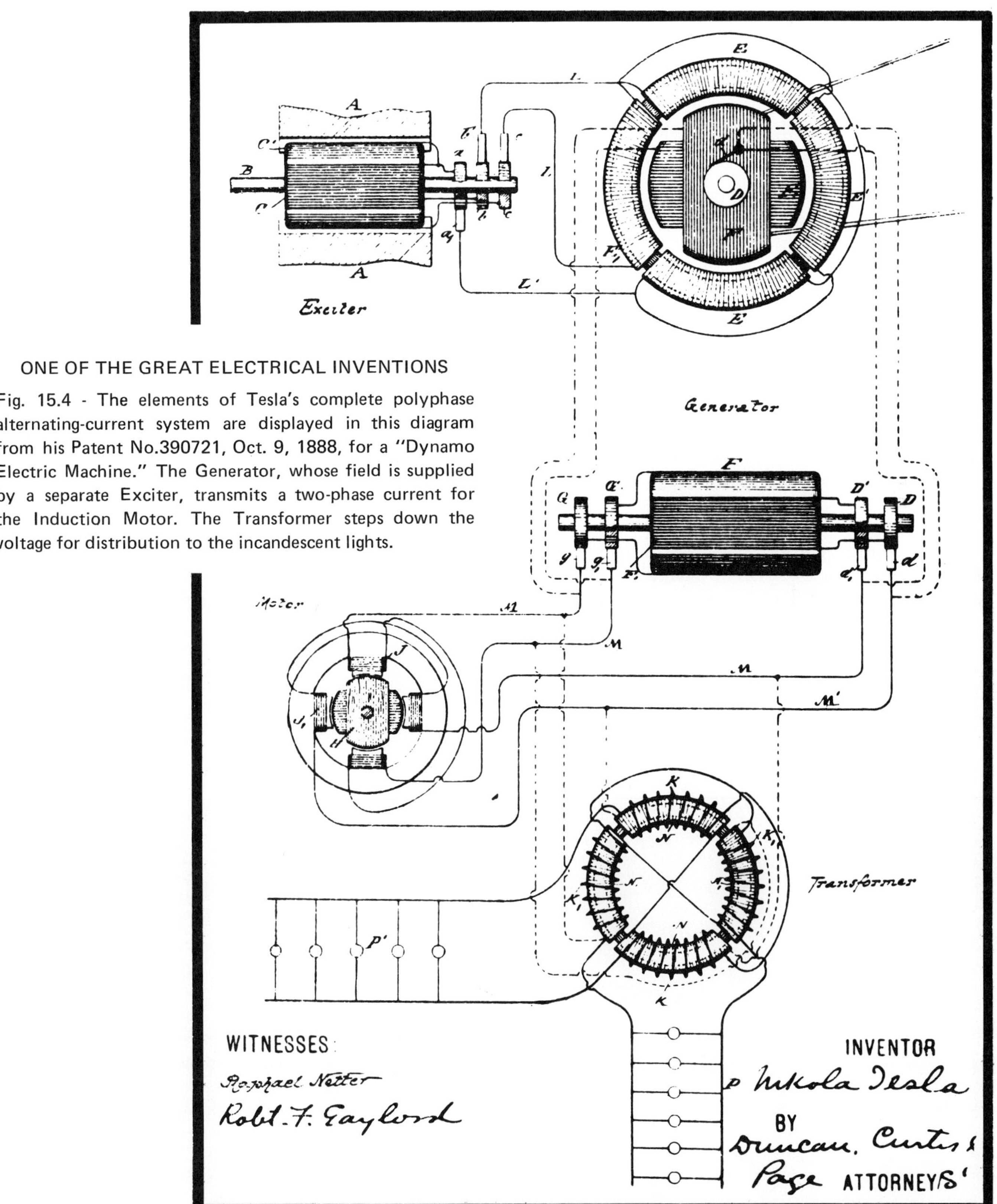


ONE OF THE GREAT ELECTRICAL INVENTIONS

Fig. 15.4 - The elements of Tesla's complete polyphase alternating-current system are displayed in this diagram from his Patent No.390721, Oct. 9, 1888, for a "Dynamo Electric Machine." The Generator, whose field is supplied by a separate Exciter, transmits a two-phase current for the Induction Motor. The Transformer steps down the voltage for distribution to the incandescent lights.

epoch-making discoveries was quickly recognized. Thus, shortly after he had obtained his basic patents, Tesla was invited to demonstrate his system to the New York Chapter of the American Institute of Electrical Engineers. Tesla accepted the invitation with alacrity and put his heart into a thorough and illuminating lecture covering the theory, application and advantages of the tremendous new concept for electric power.

His paper on *A New System of Alternate-Current Motors and Transformers* was a classic in the field of electrical engineering. With it Tesla's stature as an inventor was established for all time, and his invention laid the foundation on which the electrical power systems of the world are based.

One of those who heard of Tesla's lecture was the industrialist, George Westinghouse, who had been interested in the possibilities of alternating-current distribution. He saw a future in Tesla's ideas and he was in a position to promote them.

**WESTINGHOUSE** was already well-known for his invention of the railroad air-brake, and for a system of distributing natural gas. He had also entered the electrical field with his Union Switch & Signal Co. which made signal equipment for the railways. Having organized an electrical design group for that business he then branched out into the manufacture of direct-current machinery and had established a factory at East Pittsburgh, Pennsylvania.

Westinghouse had become very much aware of the transmission problem with direct current. The 110-volt distribution for incandescent lamps entailed large currents that required massive and expensive conductors. Higher transmission voltage, with corresponding reduction in current and a smaller size of conductor, was clearly needed. And, furthermore, high-voltage transmission would permit longer lines, allowing the generation station to be located in the most desirable place independent of the load.

When in 1885 Westinghouse learned of the 2000-volt a-c system devised by John Gibbs in England and Lucien Gaulard in France which used transformers to perform the function of increasing the voltage for transmission and decreasing the voltage at the load, he immediately took an option on the patents. He realized they were an essential link in the expansion of alternating current for commercial service. Several of the European devices, then called "inductoriums" or "secondary generators," were shipped to East Pittsburgh. Here they were redesigned for quantity manufacture, and, most importantly, modified for use in the

THE MOTOR THAT MADE ALTERNATING CURRENT PRACTICAL

Fig. 15.5 - Tesla's induction motor as demonstrated in his historical lecture before the American Institute of Electrical Engineers, at Columbia University, 16 May 1888. The motor developed about ½ hp. Photo courtesy Nikola Tesla Museum, Belgrade, Yugoslavia.

parallel connection instead of the series connection which was in use in Europe.

The pioneering demonstration of a system employing transformers at Great Barrington, Mass. has previously been described. The 500 volts from the generator was stepped up for transmission, and stepped down to 110 volts for load distribution. The practicability of alternating current was proved by this installation, and it was followed by stations at Greensburg, Pa. and at Buffalo, N.Y. where the power was transmitted at 1000 volts. Buffalo's alternating-current system was the first to operate commercially. Other a-c plants followed.

**BUT** alternating current lacked a general-purpose motor, whereas direct-current motors, which were developed out of the experience with dynamo design, were in successful operation giving reliable and efficient service. Thus when his representatives sent in favorable reports on Tesla's demonstration to the AIEE at New York, George Westinghouse immediately sensed the competitive commercial possibilities of the new induction motor. Together with his transformer system of distribution, the motor would provide another vital element for successful alternating-current service.

On July 7, 1888, only 68 days after Tesla's presentation, he had purchased the exclusive patent rights for one million dollars, plus a royalty on each motor sold. Westinghouse also secured the services of Tesla himself. Before the year was over Tesla arrived with a working model of the motor and went to work redesigning it to fit the existing commercial applications.

**TESLA'S INDUCTION MOTOR** could not simply be connected to the existing alternating current services. They were generally single-phase type, providing one wave of alternating current, used mainly for supplying incandescent lamps and arc lights. Tesla's patents, on the other hand, covered a motor of the polyphase type, and was part of a polyphase system. At this point the employment of Tesla's original induction motor principle required the availability of polyphase service on a broad scale. This was not yet developed, and had yet to overcome obstacles arising in the path to achieving commercial preference over the prevailing direct current.

To carry on in the meanwhile with single phase, Tesla invented a split-phase motor by an ingenious adaptation that made a single phase do the work of two phases. This motor, although it eventually achieved wide use, presented initial difficulties because it would not work well on the 133$^1/_3$ cycles per second current then available. A-c frequencies in use at the time ranged from 16$^2/_3$ to 133$^1/_3$ cycles, each frequency having its ardent proponents. One of Tesla's lasting contributions was to force a decision on the standard frequency of 60 cycles per second, which he calculated as right for both lighting and for his motor, and which is now standard in North America and other parts of the world.

**AVAILABILITY** of an a-c motor, and growing recognition of the advantages of high voltage and long-distance transmission made it appear that alternating current was in a position to move ahead. But Edison and his backers who had consolidated the patent situation on the incandescent lamp against infringers, and were now profitably promoting and installing d-c systems to replace gas lights, reacted vigorously to the threat of a competitor in their field. It was apparent to Edison that there was no room for the two currents side by side, and he determined that his direct-current system would have to prevail.

Edison pulled out all stops in a struggle for dominance in what was popularly called "the battle of the currents." As a great and popular inventor he had a great deal of previous experience in publicity, and used it constantly to promote the commercial success of his developments. Edison waged a ruthless and well-financed campaign against acceptance of Tesla's system. He sought to inflame public opinion against alternating current by newspaper advertisements that imputed and dramatized dangers to human life with the use of high voltage. The propaganda climaxed with the spectacle of electrocuting a horse with high-voltage alternating current.

"My personal desire," said Edison, "would be to prohibit entirely the use of alternating currents. They are unnecessary as they are dangerous. I can therefore see no justification for introduction of a system which has no element of permanency and every element of danger to life and property."

**TESLA'S SYSTEM** had made some progress, but dramatic increase in its acceptance was not to be had without convincing evidence of its superiority.

**THE FIRST OPPORTUNITY** for grand public display and for showmanship to counter Edison's efforts, came at the great Chicago Exposition celebrating the 200th anniversary of the discovery of America. The management of the Columbian Exposition of 1893 asked for bids to electric light the Fair Grounds. Westinghouse seized the chance to publicize the Tesla system and he put in a low bid to obtain the contract. He and Tesla then proceeded, against seeming insuperable time limits, not only to light the Fair but also to demonstrate the advantages and practicability of a complete system of polyphase a-c generation, transmission and distribution.

Because there were no large polyphase generators at that time Tesla improvised twelve 1000 hp two-phase generators by the use of 24 single-phase, 500 hp engine-driven machines mechanically connected 90 degrees out of phase, as shown on the opposite page. The 2300-volt, 60-cycle polyphase power output was transmitted and transformed in voltage to supply a whole gamut of equipment designed to display the capabilities of alternating current. This included induction motors, synchronous motors and split-phase motors. To show the complete adaptability of the system Tesla developed a rotary converter for changing polyphase alternating current into direct current which was then used to run railway motors, to operate arc lights and the electrolytic tanks for electroplating.

**WORLD'S FAIRS** always offered the opportunity for spectacular lighting effects. They were a great magnet that drew the crowds. The Columbian Exposition was no exception. But it was not apparent how the Fair could be lighted by lamps made by Westinghouse without infringing the single-piece sealed bulb Edison incandescent lamp patents. The crisis was met with a novel two-piece "stopper" lamp - a glass globe "corked" with a ground-glass stopper upon which the carbon filament and lead-in wires were mounted.

In a flurry of activity involving new methods of glass blowing, of grinding the stoppers and evacuating the bulbs, 100,000 stopper lamps were produced. They bathed the Fair grounds and the Exhibition Halls in a brilliant blaze that astounded the Fair visitors, and showed electricity coming of age. Also, to amaze the public, Tesla had developed an oscillator coil producing millions of volts, but at such high frequency that its electrical sparks did not harm the body. Tesla boldly immersed himself in the awesome electrical fire, a feat of showmanship aimed at countering the public image of high voltage.

The displays at the Electricity Hall, pictured on the opposite page, were a dramatic step in tilting the balance in the "battle of currents" in favor or alternating current.

There was soon new, important evidence that the dominance of the direct-current system was to be further threatened. It came from Niagara Falls where the local power company, financially backed by the House of Morgan, had obtained a charter for using part of the enormous energy of Niagara's 164 feet of plunging water. The project presented an engineering challenge of unprecedented magnitude. How could the 120,000 horsepower be most effectively harnessed and transported?

**NIAGARA'S** original planning called for an intake canal 2½ miles in length along the river front above the falls, supplying several hundred individual turbine pits each powering a nearby factory. This was found impractical and its costs excessive. It became clear that the most economical solution was to locate the waterwheels at one point and to distribute the large amount of power over long distance to industrial locations, such as Buffalo, where it could be profitably utilized. The existing systems of direct current and single-phase alternating current appeared inadequate for disposing of the great amount of power and for spanning the great distances of transmission contemplated from the Falls. To get technical advice an International Niagara Commission was set up in 1890, headed by the renowned Lord Kelvin of Great Britain. The Commission included Dr. Coleman Sellers, Prof. H.A. Rowland of Johns Hopkins University and Prof. George Forbes of England. The Commission inaugurated a world-wide quest for methods and apparatus, and they invited plans and offered a $22,000 prize.

The seventeen projects submitted by 20 representatives from six countries revealed the state of the art in 1890. Several proposed hydraulic pressure systems through a pipe network, others employed compressed air, and one system called for transmitting the power with manilla and wire ropes. Of the six electrical proposals, four used direct current, one of them comprising a system

EXHIBIT OF THE TESLA POLYPHASE ALTERNATING-CURRENT SYSTEM AT THE 1893 COLUMBIAN EXPOSITION

The power system used twelve 1000 hp, 2300 volt, 60 cycle 2 phase generators, the largest in America. Nearly a half million feet of conductor was used for the machinery and for a total of 92,632 incandescent lamps. The electrical exhibit featured a model kitchen with electrical appliances such as heated saucepan, chafing dish, coffee pot and grill.

Photographs courtesy Westinghouse Electric Corporation

of ten d-c generators at 1000 volts connected in series for transmission at 10,000 volts to drive a similar arrangement of motor-generator sets at Buffalo for local power distribution, One plan proposed a single-phase a-c system but the details were not fully given.

The remaining plan, by Prof. George Forbes, one of the Commission members, advocated a polyphase alternating current installation.

Forbes reported that: "It will be somewhat startling to many, as I confess it was at first to myself, to find as the result of a thorough and impartial examination of the problem, that the only practical solution for Buffalo, and the best solution for the new industrial city which it is proposed to build near Niagara, lies in the adoption of alternating-current generators and motors."

Forbes further stated: "The only non-synchronizing motor which has been developed in practical form is the Tesla motor manufactured by the Westinghouse Electric Company and which I have myself put through a series of tests at their works at Pittsburgh . . . The torque on starting is considerable . . . the largest I have tested was five horsepower . . . they have no commutator or even brushes or other collectors."

Professor Forbes was prophetic but flawed in lack of imagination in the size of the units which he proposed. Limiting himself to the size of equipment then available he recommended 500 hp generators and 100 hp transformers in a station being constructed to provide 50,000 horsepower!

**THE COMMISSION** looked favorably on the electrical proposals but found none of the specific plans acceptable. They were not yet convinced that alternating current was practical. Moreover, Lord Kelvin was adamantly opposed to using alternating current. Meanwhile construction of a tunnel for 100,000 horsepower was underway.

Alternating-current generators had already been in commercial use in Europe for about a decade. E.A. Adams, president of Cataract Construction Company, and his chief engineer, Dr. Coleman Sellers, inspected installations on the continent where transmission of large amounts of a-c power were making headway. Most notable was the three-phase, 110-mile transmission at 15,000 volts, using bare wire, from a generating plant at Lauffen to the 1891 Electrotechnical Exhibition at Frankfurt, Germany. Carrying a load of about 250 kilowatts the line supplied power, transformed down in voltage, for lights, and for a 100 hp, 3-phase induction motor designed by the German firm of AEG. It was acknowledged by the Exhibition authorities that: "The three-phase current as applied at Frankfurt is due to the labors of Mr. Tesla." This long-distance transmission was considered a sensational achievement in engineering circles, and improved the atmosphere for use of alternating current in other parts of the world.

Adams and Sellers were also familiar with the achievements at the Columbian Exposition. The development of rotary converters was particularly impressive to them because these machines complemented the advantages of high-voltage a-c transmission with a means for conversion to direct current which was needed for industrial plants using electrochemical reduction. Thus the polyphase system could comprise a complete system providing the requirements for both alternating and direct current usage. At this point the simultaneous availability of the Tesla system and the on-going plans for Niagara proved fortuitous.

**CATARACT** Construction Company cast their vote on May 6, 1893 for polyphase alternating current, and invited proposals from American companies. Two manufacturers submitted bids, the newly formed General Electric Company, and the Westinghouse Electric & Mfg. Company. The construction was awarded in October 1893 to Westinghouse for three 5000 hp Tesla polyphase generators operating at 2250 volts.

A controversy arose over the alternating-current frequency to be generated. A low frequency was considered desirable for operation of the rotary converters that were to provide current for industrial and electrochemical uses. Prof Forbes wanted 2000 alternations per minute ($16^2/_3$ cycles per second). The Westinghouse engineers wanted 4000 alternations per minute ($33^1/_6$ cycles per second). Finally a compromise was reached for generators of 12 poles rotating at 250 rpm to give 25 cycles per second.

The generators were far larger than any built, and involved much new design. The first machine was installed in 1895 and power was first delivered in August of that year to the first customer, the Pittsburgh Reduction Company, now Aluminum

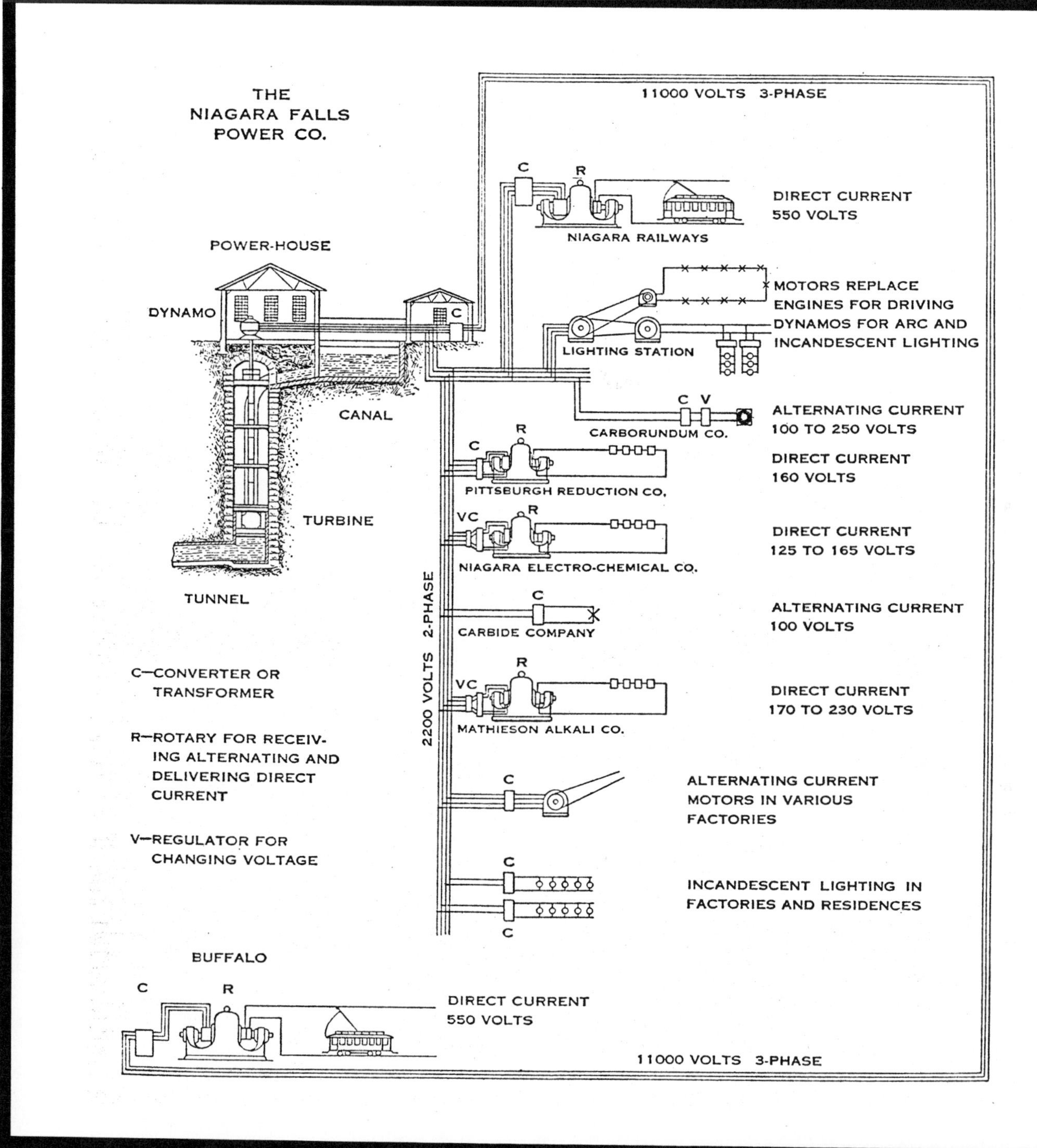


Fig. 15.6 - THE NIAGARA FALLS POWER SYSTEM of polyphase alternating current provided the advantages of high voltage, long-distance transmission, the simplicity of induction motor drives, and the use of rotary converters for direct current where that was needed. It was the first commercial demonstration of a universal system of electric power. The Niagara system went into operation in 1896. Courtesy Niagara Mohawk Power Corporation,Buffalo,N.Y.

Company of America, for producing aluminum by the Hall electrolytic process. By November all three generators were in operation and on November 16, 1896, circuits were closed that inaugurated electric power service to Buffalo 22 miles away. Eventually seven additional 5000 hp vertical generators were installed in the hydroelectric plant.

**BUFFALO**, which in 1887 had offered $100,000 as a prize for the most effective method of power transmission from Niagara Falls was now a city electrified. It soon became the largest electrochemcal center in the world. The miracle of transmitting electric power 22 miles was made possible with Tesla's current. Niagara was demonstrating the versatility of alternating current for all purposes, as shown in Fig. 15.6.

The development of large generators for Niagara led immediately to a big installation of steam-driven units in New York City. The reciprocating steam engines drove 5000 and 7500 kilowatt alternators for electrifying the elevated and street railways, for subways, for steam railroad electrifica= tion, and the Edison lighting systems, either by converting ac to dc, or by a changeover to alternating-current service.

The Tesla system was on the verge of revolutionizing the method of generating, transmitting and utilizing electric power. It portended the future development of huge utilities, with monstrous generators sending electricity at thousand of volts to cities and countryside to turn motors, make illumination, deliver heat, transmit voice, music, pictures, and to power the myriad other working gadgets of the nation.

**TESLA** had reached one of the peaks of his career. In knowledgeable circles his name shared equal recognition with that of Edison. The polyphase system of electric power had been successfully launched, attested by the fact that General Electric Company, which had originated on the basis of Edison's direct-current service, had obtained license rights under Tesla's polyphase patents. These patents, which had been subject to long and constly litigation, had been made invincible by a sweeping decision in Tesla's favor by the U.S. District Court of Connecticut. Westinghouse and General Electric settled 300 patent infringement suits between them by a cross-licensing agreement that would promote a-c power.

## TESLA and HIGH-FREQUENCY ELECTRICITY

Tesla's long and creative scientific and engineering career can be related in two great time periods. The are marked by the frequency of the electricity with which he worked. The first, spanning the early 40 years of his life, was the low-frequency period. This ended with Tesla's triumph for his polyphase power system at Niagara Falls. The second, beginning in 1890, and covering the rest of his life, was the high-frequency period. In it Tesla's vision took off from the confines of 60-cycle earthly power, and soared, with frequencies of millions of cycles, into the vast realm of electromagnetic space.

Tesla severed his connection as a consultant with Westinghouse in 1889 and returned to New York. He was essentially a "loner" who was unable to work with others on a production team. With a characteristic indifference to commercial pursuit of his ideas, he left to others the development of his motor and polyphase system. He was now well off financially from the Westinghouse purchase of his 40 patents on the polyphase system (however he would cancel his motor royalty agreement in an impulsive action to help save the Westinghouse Company from threatened bankruptcy in the financial panic of 1893).

Back in the "Tesla Laboratories" in New York City he was embarked on conquering the new fields inspired by his fervent scientific vision. He was experimenting in a field of electricity that was to occupy most of his active remaining years - the phenomena of high-frequency alternating currents.

During the early years of the 1890s Tesla startled and challenged the scientific world with his public demonstrations of electrical discharges and their transmission through space. Producing voltages and frequencies far above the customary limits, he was working with radiant energy, a field little understood at the time, but significant for future science Hertz's brilliant investigations, recounted in the previous chapter, had opened to the world the actuality of an electromagnetic radiation, and Tesla was revealing with his spectacular experiments the further manifestations of this exciting phenomena.

With a high flair for the dramatic, Tesla began to thrill scientific and lay audiences in a series

Fig. 15.7 - MULTI-POLAR high-speed alternator built by Tesla to produce high-frequency currents. Ouputs up to 30,000 cycles per second were obtained with this machine.

of lectures in which he featured brilliant displays of electric glow in low-pressure gas tubes. He had earlier been intrigued with the curious color of Geissler tubes and their possibility for illumination.

The incandescent lamp was in its infancy, using carbon filaments which gave a yellow light of poor luminosity, and it had a short irregular life. Tesla was planning to provide something better.

**TO GENERATE** high-frequency currents Tesla developed several types of multi-polar alternators with large diameter armatures rotating at very high speed, Fig. 15.7. With these machines he obtained frequencies up to 30,000 cycles per second.

But for lighting and for the other special effects he was envisioning Tesla needed to use enormously higher frequency and voltage than he could obtain from mechanical generators. To produce the outputs needed Tesla invented the famous "Tesla Coil," which he called an "electric oscillator."

For this apparatus, one form of which is shown in Fig. 15.8, he invoked for the first time the principle of "tuning," to achieve a resonancy with

Fig. 15.8 - "TESLA COIL" - The voltage from generator, A, is stepped up about a hundred times by an iron core transformer, B. Sparks from the rotary gap, G, set the resonant circuit, C,E, into high-frequency oscillation. Coil, E, of few turns, is air coupled to coil, F, having many hundreds of turns, thus producing a frequency and voltage output in the millions. The current, however, has been stepped down in the ratio that the voltage has been stepped up, so that discharge at G, is harmless.

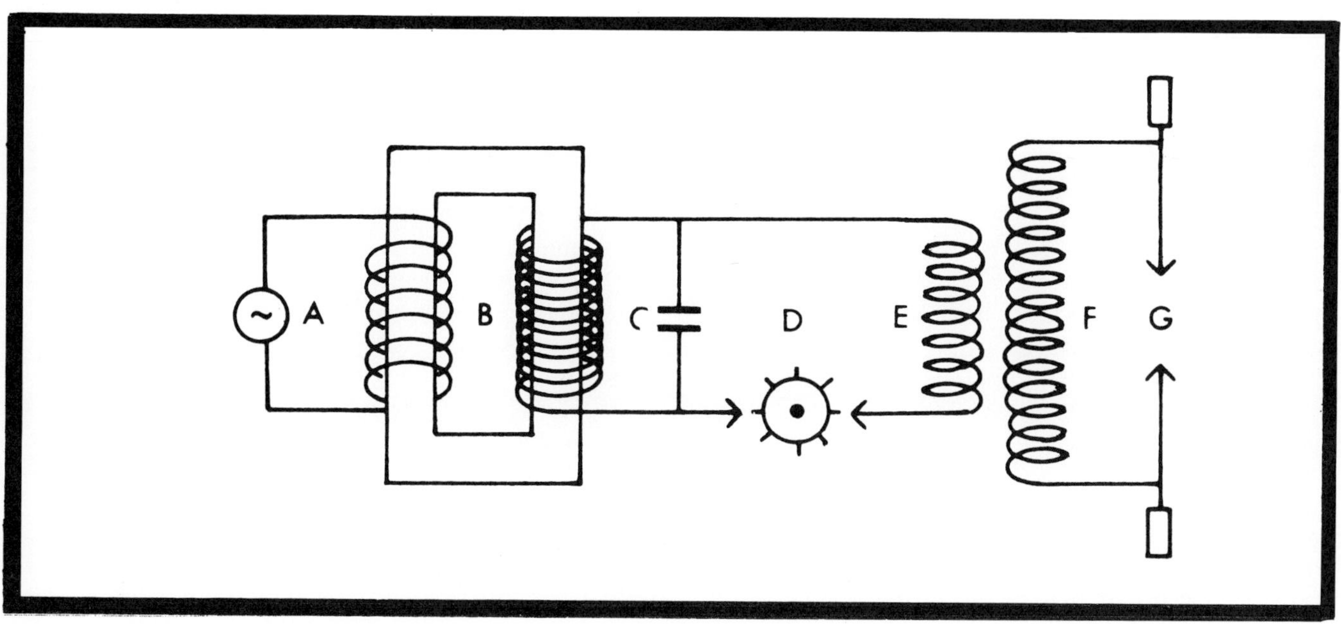

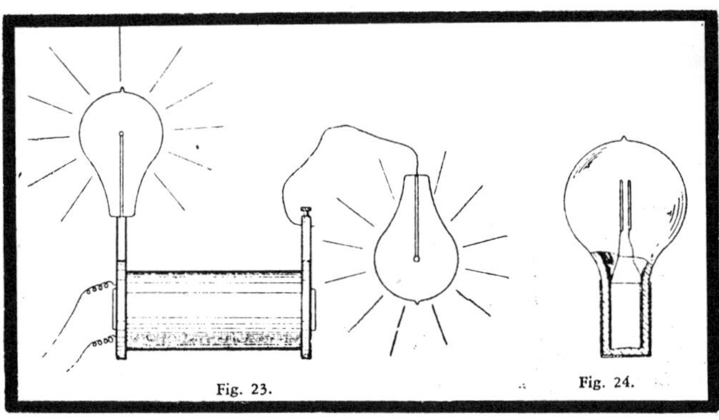


Fig. 15.8 - SINGLE-ELECTRODE LAMP, left, provides illumination by bombardment with the high-frequency radiation reflected from the interior surface of the bulb. Double electrode lamp at the right becomes incandescent by a capacitative bombardment between the electrodes.

the source frequency. Thus an alternating current circuit could be set in a state of maximum oscillation. This was provided by combining the proper values of capacitance and inductance that would make the circuit resonate. Tesla's years of experience with alternating currents had given him an expert understanding of the part played by the capacity and inductive elements in determining the behavior of oscillating circuits. The Tesla Coil represented the first application of what is now known as a "coupled oscillatory circuit." This type of circuit was be an essential element in future radio equipment.

**MAGNIFICENT SHOWERS** of electricity lept from Tesla's machines. The fiery display was something brand new, awesome and wonderful to behold. Tesla delighted in showing it off. He often staged elaborate dinners for friends at the Waldorf-Astoria, serving rare foods and exquisite wines. After the meal he would invite the guests to his South Fifth Avenue laboratory and hold them spellbound with the man-made lightning and the kaleidoscopic burst from the strange coils. He approached the spraying electricity with tubes and globes held in his hand and they became alive with weird and beautiful colors.

The tremendously high frequency enabled him to perform his feats without harm. A natural for news reporters, Tesla's performances were sensationalized. He was pictured as a modern-day wizard whose eyes flashed and hair floated as he played magic in the midst of electrical fire. Tesla was especially eager to talk about his plans for producing light by electrical vibrations. He claimed that by making the vibrations fast enough their light would rival that of the sun. Referring to Edison's incandescent lamps with some disdain, he promised that his method would give a much better and more efficient light.

Myriads of ideas to exploit the possibilities of high-frequency electricity were being hatched in his laboratory. And whereas Hertz and his followers were satisfied with relatively small outputs for

TESLA'S "IDEAL" METHOD OF LIGHTING A ROOM

Fig. 15.10 - The tubes, without electrodes or filaments, are made brilliant in an alternating electrostatic field.

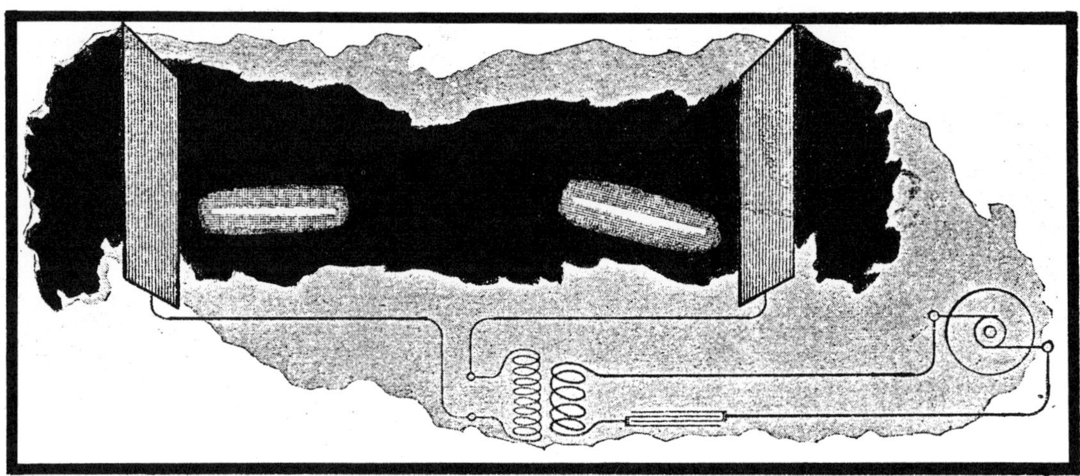

experiments in electromagnetic radiation, Tesla had grander designs for an immense increase in power that could bridge space to activate lighting and other effects he was dreaming up. Tesla was ready to present his devices to his peers. It was done in a lecture on *Experiments on Alternate Currents of Very High Frequency and Their Application to Methods of Artificial Illumination,* given before the American Institute of Electrical Engineers at Columbia University on 20 May 1891.

In a long evening filled with fascinating displays, Tesla began with an elementary show of the light that could be produced by simply using high-voltage discharges in free air. He then turned to low-pressure, gas-filled tubes and bulbs, some with various electrode arrangements, two of which are shown in Fig. 15.9.

He associated the generation of light with the molecular vibrations induced by the rapidly alternating fields of electromagnetism and showed that any bottled gas could give illumination when excited with the rapidly oscillating field. His underlying theme was that this could yield more efficient light than from a hot filament incandescent lamp. Evacuated bulbs glowed when connected to only one terminal of his machine. Bulbs provided with electrodes and buttons of various materials became brilliant from the bombarding effect of the strong electrostatic field reflected from the inner surface of the glass.

His ultimate scheme, shown in Fig. 15.10, was the use of glass tubes with low gas pressure and without an electrical connection. These were energized to room-filling luminescence between the plates of a high-frequency electrostatic field.

Tesla's Columbia University presentation was an historic event. Attended by some of the biggest names in electrical engineering, the lecture left an indelible impression on many in the audience, and was an inspiration to some in their later work.

**IN THE YEAR 1892** Tesla went abroad. His revolutionary achievements with polyphase apparatus, and the brilliancy of his performances with high-frequency oscillations, had not only put him in demand for lectures in America, but also in Europe. Foreign papers echoed the praise for his accomplishments that were regularly appearing in publications in the United States. Tesla made a noteworthy appearance before the Institution of Electrical Engineers in London in February 1892. Tesla's spectacular demonstrations on high-potential, high-frequency currents to an audience of some of the most notable names in British science brought the same acclaim as in America. He also gave this same lecture to the Physical Society of Paris.

Fig. 15.11- PHOSPHORESCENT BULB is the source of light for time-exposure photograph of Tesla taken in 1894.

As part of his European trip Tesla had planned to visit his mother in Croatia. On receiving a telegram that she was gravely ill he took the train to her home in Gospic, reaching the town only a few hours before her death. Tesla's relationship with his mother had always been close and the loss was very distressing to a man whose life included few women other than those in his family. The shock of his mother's passing and the fatigue from his lecture exertions exhausted Tesla and lengthened his stay in Europe until late in the year.

On his return to the United States Tesla found that the engineering on his polyphase system was progressing. But the most striking news was that alternating current had leaped one barrier to acceptance by being chosen for the 1893 Columbian Exposition in Chicago. The opportunity to light a World's Fair for the first time with incandescent lamps, and to exhibit the Tesla polyphase system, had been seized by the Westinghouse Electric & Mfg. Company. Awarded the contract in a bitterly fought battle of price, and of displaying the merits of ac vs. dc, they were busy with the plans for manufacturing and installing the machinery for the great Exhibition. Tesla was immediately engaged as a consultant, and he threw his whole energy and expertise into the project.

The Chicago World's Fair was a triumph for electric lighting and for alternating current. Tesla's spectacular demonstrations of electric fire with his oscillation transformer were a highlight of the exhibits in the Electrical Hall.

**TESLA'S AMBITIONS** for the future of high-frequency electricity grew rapidly in scope and boldness as he enlarged the power of his oscillation apparatus. Always a believer in electrical vibrations as the prime form of energy, he was planning unheard of feats for the waves from his electric oscillators. His visions were first revealed in lectures before the Franklin Institute at Philadephia in February 1893, and before the National Electric Light Association in St. Louis the following month. In addition to producing luminous effects he was now predicting the possibilities of using his powerful oscillations to transmit energy without wires. Tesla's imagination was taking on global dimensions, as shown by his words at the Franklin Institute lecture:

"The plan I have suggested is to disturb by powerful machinery the electricity of the earth, thus setting it in vibration. Proper appliances will be constructed to take up the energy transmitted by these vibrations, to transform them into a suitable form of power, to be made available for the practical wants of life."

This was to be accomplished by induction and capacity devices set into powerful action by resonance. Suitably adjusted and operating on a vast scale. Tesla believed that his coils could thunder out oscillating energy that would activate the great standing waves of electricity that he believed were available in the interior of the earth.

**1895 WAS A DISASTROUS YEAR** for Tesla. On the 13th of March his evening meal was interrupted by apalling news. A fire starting in the basement of his laboratory had raced upward gutting the building and collapsing the upper floors. The loss was total, including not only equipment and historical documents, but also plans and a nearly completed model of a wireless guiding system. Tesla carried no insurance, and rebuilding the laboratory took all his resources plus the financial help of friends. By mid-1896 the plant was back in operation, and because he was possessed of a remarkably photographic memory, Tesla was able to reproduce the details of several inventions that were underway.

Expensive litigation on his past patents had made Tesla wary of disclosing details of his current plans. But to newspaper reporters eager for tips on his latest exploits, Tesla was giving broad hints of grand designs for methods of guiding movable objects by using remote control. Headlines soon proclaimed that Tesla was prophesying a revolution in warfare by a means to start, stop, guide and explode projectiles and torpedoes by wireless signals. There was a great deal of skepticism in the more conservative engineering publications, however. They were critical of Tesla for seeking self-advertising for something they considered obviously impossible. But Tesla was soon to silence his critics.

Having secured Patent No. 613809 Tesla put on, in 1898, a series of the most memorable and controversial scientific exhibitions that Madison Square Garden had seen. He was demonstrating his *Method of and Apparatus for Controlling Mechanism of Moving Vessels or Vehicles.* Floating in a huge tank in the auditorium was a metal boat several feet long equipped with electric lights. Tesla, who was in evening dress that gave him the appearance of a magician, stood by an assortment of controls. He announced that the boat would be piloted by a "teleautomatic brain" that would respond to his will or that of anyone present.

**TO THE AMAZEMENT** of the packed crowds he then maneuvered the boat, and invited some of the spectators to try their hand at starting, guiding and stopping the obedient vessel. It was all done, he explained, by the mysterious powers of wireless electricity. Furthermore, he hinted, that they were witnessing only a few of the shipboard operations that would be responding in the future to invisible wireless commands.

Tesla had invented the first radio remote-control system, and had laid the basis for telemechanics. The boat operated by a succession of radio impulse commands impinging on the boat's antenna to energize, through relays, the battery-operated propeller motor and rudder mechanism, Fig.15.12. At the will of the operator, the sequence and duration of the impulses were such as to determine speed of the boat and the direction and amount of swing of the rudder.

The United States involvement in the Spanish-American War, heightened the significance of Tesla's invention, particularly when he revealed the possibility of a crewless torpedo boat with submarine destroyers that could be directed at enemy vessels without their knowledge. Some were skeptical of Tesla's visions for a future automated warfare, others recoiled and were critical of the greater destruction in the next conflict.Tesla, however, stressed that his inventions would help to prevent war.

**IN THE LAST DECADE** of the 19th century Tesla had obtained nearly twenty five patents dealing with high-frequency oscillations and their transmission through what he called "natural media."

In a scientific world busy with investigations of electromagnetic radiation, Tesla was treading a singular path. His fundamental aim in the patents was to implement his vision of a transmission of intelligence and power with electric oscillations. But to achieve these ends he disavowed the use of Hertzian waves as wasteful of energy for the purposes intended. For his oscillations Tesla planned to use what he considered superior paths: the stratosphere as a natural conductor and the earth as a sympathetic vibrator, a combination which he claimed would give better results than with actual wires.

His patent No. 649621 on *Apparatus For Transmission of Electrical Energy* stated:

"It is to be noted that the phenomenon here involved in transmission of electrical energy is the one of true conduction and not to be confounded

Fig. 15.12 - TESLA'S "AUTOMATON" It was operated by a ship-robot powered by storage batteries and piloted by a "teleautomatic" brain responding to "wireless electricity." This was the first use of remote control by radio. The model below is in the Nikola Tesla Museum, Belgrade. Tesla's Madison Square Garden demonstration vessel was completely enclosed in a steel hull to resemble a ship.

with the phenomena of electrical radiation which have heretofore been observed and which from the very nature and mode of propagation would render practically impossible the transmission of an appreciable amount of energy to such distances as are of practical importance."

His research, carried on for a number of years, led Tesla to recognize three important necessities for further action: first to develop a transmitter of great power embodying the principles of his electrical oscillator; second, to perfect a means for isolating and individualizing the energy transmitted and, third, to ascertain the laws of propagation of currents through the earth and the atmosphere.

Fig. 15.13 - TESLA'S EXPERIMENTAL LABORATORY on a plateau in the foothills of the Rocky Mountains near Colorado Springs, erected in the summer of 1899.

To carry out these ambitious plans his New York laboratory was far too small and the surroundings unsuitable. So Tesla cast around for a new site. For the new laboratory location Tesla moved two thousand miles west and 2000 meters up, to the top of a plateau at the foot of Pike's Peak near Colorado Springs. The selection was made largely with the advice and the financial help proferred by his friend Leonard E. Curtis, who had gone west for his health. Curtis had arranged for the availability of power from the Colorado Springs Electric Company.

The purity of the air, the unequalled beauty of the mountains and the restfulness of the surroundings greatly exhilarated Tesla. He proceeded with high spirits in building his new laboratory and in carrying on with his new experiments.

Erected in the summer of 1899, Tesla's wooden building housed the largest electrical oscillator ever made, producing the longest point to point electric discharge ever achieved by man - over 200 feet. In his year and a half at Colorado Springs Tesla carried out innumerable experiments on the most effective design and operation of his resonant apparatus for transmitting wireless power and wireless communication.

**TESLA'S RESONANT TRANSFORMER** had an input of 200 kW, and worked at output wavelengths ranging from 2 to 10 kilometers. Tesla described in his diary how he calculated the genertive, inductive, condensive, switching and other elements to produce oscillations at this great power. He also described the numerous problems he had to overcome in pioneering in the radio field.

Some of the problems successfully tackled by Tesla at Colorado Springs, as revealed in his diaries, were stated in the 1976 "Tesla Symposium" paper by Prof. Vojin Popovic of Belgrade University:

"Tesla wrote in his diary of an investigation for definition of optimum correlation between inductance and capacitance in the circuit to achieve the highest overvoltages and effects, selection of most suitable wavelengths with regard to the frequency of switching in primary circuit and frequency of free oscillations in the secondary circuit, adoption of the best form of antenna with either spherical or other final capacitance body at the end of the oscillatory coil, by which the most effective discharge is achieved, namely, effect of radiation in space. Tesla's experiments for determination of the effects of the height of the ball on its capacitance to the ground, as well as in finding out, by means of measurements, of electrostatic and electrodynamic capacitance of complex forms of construction which served for the various antenna systems, are particularly original.

"Further in Tesla's notes are descriptions of joint applications of new rotary and vibrating detectors and relays for switching of the electric circuits of bigger power, aimed at investigation in various ways of adjustment of coupled resonant coils and determination of so-called resonant jump, was performed by Tesla by special methods and sensitive indicators of resonance, and many pages of his manuscript are devoted to it. Tesla also described in detail his investigations into effects of oscillator for determination of nodal points in the earth as a conductor of waves."

**THE EXPLORATIONS** and the high-frequency constructions developed at Colorado Springs were preliminary to an even more powerful radio and power broadcast station he planned to put up in the east. The original purpose of a new station would be its incorporation, along with many other similar stations, into a planned system of his world communications. That system, according to his documents, should have served for transmission of radiotelegraphic signs and facsimilies and signals of exact time.

Back in New York City Tesla was at work constructing his famous "World Telegraph" transmitter at Wardenclyffe, Long Island. The funds had been provided by the J.P. Morgan interests who had also financed the Niagara power development. While the tower and station were being erected, Tesla wrote in the March 5, 1904 issue of *Electrical World and Engineer,* this vision of the future:

"When it is fully recognized that this planet, with all its appalling immensity, is to electric currents virtually no more than a small metal ball and that by this fact many possibilities, each baffling the imagination and of incalculable consequence, are rendered absolutely sure of accomplishment; when the first plant is inaugurated and it is shown that a telegraphic message, almost as secret and non-interferable as a thought, can be transmitted to any terrestrial distance, the sound of the human voice, with all its intonations and inflections, faithfully and instantly reproduced at any other point of the globe, the energy of a waterfall made available for supplying light, heat or motive power, anywhere - on sea, or land, or high in the air - humanity will be like an ant heap stirred up with a stick: See the excitement coming!"

**ALAS**, as has been related previously, Tesla failed to reach his goals. His magnificent visions, although appealing to the electrical outlook of the times, were being circumscribed by the forward march of science. Tesla's projects, based primarily on transmitting power rather than signals, could not compete with progress in wireless telegraphy using far less massive equipment and less power.

END OF A DREAM . . .

Fig. 15.14 - Tesla's life had elements of monumental triumph and spectacular failure. The toppling tower at Wardenclyffe, blasted during World War I to foil enemy agents who might use it for communications, ended Tesla's grand dreams for transmitting world telegraphy and power;

The infant commercial electric power industry was beset with machinery and expansion problems, and had no time or money for untried and little understood projects of global wireless power.

The turn of the century saw a series of extraordinary discoveries: the *electron* by Thomson in England, the *x-ray* by Roentgen in Germany, and *radioactivity* by Becquerel in France. These opened fields for investigative pursuit far more promising and exciting than Tesla's high-voltage, high-frequency currents. Thus Tesla's global visions and fundamental work on the radiation of oscillatory circuits were eclipsed for the time being by the dramatic shift to new areas of developments in science and technology. Tesla's ideas for world power and telegraphy moved into an unproductive direction and he was never able to translate them into reality. Wardenclyffe was never operative as planned, and Tesla abandoned its laboratory in 1906.

Although nearly bankrupted by the heavy expenses of Wardenclyffe, and unable to get more financial help, Tesla never admitted defeat for his dreams of transmitting power world wide. To him there was simply the bitter realization that he was years ahead of the times and would have to wait out an interim that would bring acceptance and he could take up his uncompleted plans. The reverses by no means signaled a slowdown in his inventive activities.

Tesla continued to work on improvements of the polyphase system. He also turned his attention to a new area which was growing in importance for motive power - the steam turbine. Developed by C.A. Parsons (1854-1931) in England, the turbine used the axial flow of steam at high pressure to spin a fan-bladed rotor. Providing simplicity, high speed, compactness and economy, the turbine was becoming popular for driving electrical generators.

**UNCONVENTIONAL,** as usual, Tesla theorized that the properties of steam could be used to provide better performance. Instead of jetting the steam flat at the turbine blades he proposed to admit it radially between smooth, unbladed rotor disks, claiming that the natural adhesion and viscosity of the vapor would provide greater drive efficiency. He was able to interest one manufacture in developing his idea, but within a few months progress was stalemated. Tesla accused the engineers of lack of imagination in converting his theory to practice. They, on the other hand, claimed that it was not possible to work from designs existing only in Tesla's head.

Tesla continued to work on several other projects which because of lack of funding did not go farther than entries in his notebooks. With advancing years his schemes became more and more esoteric. Their sensational nature and humanitarian implications brought exaggerated press releases that stretched the credulity of editors and readers alike.

**TESLA'S REPUTATION,** bold imagination and the vivid projection of his ideas into the future made his every utterance and proposal newsworthy and fascinating to his followers. He produced a mechanicsl oscillator which by sympathetic vibration he claimed could topple buildings or even split the earth; he proposed concentrated high-voltage beams, called "Death Rays," aimed at enemy troops, to annihilate them; he considered it feasible to use electrical waves for controlling the atmospheric moisture. Setting his sights on the universe, he proclaimed his conviction that life existed on other planets and urged communication with the inhabitants of Mars. People shook their heads and some decided they were listening to a carckpot.

**RUMORS APPEARED** in 1914 that America's two great inventors, Edison and Tesla, were to be jointly awarded the Nobel Prize in physics. Their selection was announced as a fact with great satisfaction in the November 13, 1915 issue of the trade journal *Electrical World.* But the news proved premature and incorrect. A Reuters dispatch the next day reported the prize had been bestowed on Sir Henry Bragg of the University of Leeds and on his son William Lawrence Bragg of Cambridge for their joint study of crystal structure by x-rays.

There was consternation among those who expected the Americans to be honored. Why the switch in awards had been made nobody knew, but it sparked accusations that dislike between Edison and Tesla, and unwillingness of one to be associated with the other, had affected the decision. It was also rumored that Tesla had turned down the honor out of pique because six years previously Marconi had been given the Nobel Prize for wireless telegraphy in contradiction to Tesla's claim for priority in discovering the principles of wireless transmission. Tesla, despite having established the essential elements of radio communication, had been denied an injunction against Marconi for his claim to basic patents. None of these speculations for failure in the Nobel award could be proved, and were later largely discounted.

The secrecy surrounding award selection in Stockholm left no clue as to how seriously Tesla and Edison had been considered in the first place nor why they had not been chosen. But the dismay among Tesla's admirers at his being passed over did bring immediate and keen awareness of inexcusable neglect in recognizing Tesla's many contributions to the electrical field.

As if to make amends for a long delay in honoring Tesla for his accomplishments, he was presented with the Edison Medal in a notable program at the Annual Meeting of the American Institute of Electrical Engineers in New York City on May 5, 1917. The medal originated with a group of Edison's friends who set up a fund in 1904 for an annual award, under sponsorship of the AIEE, to a living electrician for "meritorious achievement in electrical science and art." Tesla's award was particularly for his polyphase system and his work in high-frequency currents.

**GREAT EFFORT** was needed to get Tesla to accept the award. It was only with persuasion by his friend B.A. Behrend, who had developed the circle diagram technique for illustrating motor performance and who had been one of the first electrical engineers to grasp the significance of Tesla's discoveries, that he agreed to take part.

Tesla felt keenly the neglect of recognition due him in a lapse of nearly thirty years since his announcement of the rotating magnetic field and the polyphase system. A new and young generation of electrical engineers, some of whom were in attendance at the award meeting, had entered the field unaware of Tesla's original discoveries. Text books they had used in classes hardly mentioned Tesla's name.

In responding to the award Tesla admitted that from his early efforts . . . "a gigantic revolution has been wrought in the transmission and transformation of energy. While we are pleased with the results achieved," he said, "we are pressing on, inspired with the hope and conviction that this is just the beginning, a forerunner of further and still greater accomplishments." Tesla followed with a fascinating account of his life and some revelations of a personal and intimate character bearing on his work as an inventor.

**FROM THE 1920s ON** Tesla's fortunes began to decline. Losses from the ill-fated Wardenclyffe, and an accumulation of older debts, left him practically broke. Although he had enjoyed elegant living when he was well off, Tesla seemed not to be motivated only by a desire for money. Despite his mastery of science for beneficial ends he appeared to lack interest in the commercialization of his abilities for the betterment of his income. He always believed that his work was primarily for the improvement of mankind's existence - for an easier and better life. But for himself Tesla seemed content with an austere living in the New York hotel room that his limited resources allowed. Having few close friends, Tesla became more and more a recluse, brooding on his schemes for seeking new sources of natural power and radiation that had been the mainstream of his life's work.

But although Tesla had withdrawn, he had not been forgotten. His 75th birthday, in 1931, occasioned an international reassessment of Tesla's genius and a proclamation of his many contributions to electrical science and to mankind. The anniversary was instigated by Kenneth M. Swezey, a science journalist, with cosponsorship of Professor Charles F. Scott, who had been associated with Tesla at Westinghouse, and of Dr. B.A. Behrend, an electrical engineer and close friend of Tesla's.

Swezey had followd Tesla's career with a passionate interest and deplored the neglect by a world that had gained so much by the discoveries and inventions that Tesla had showered on it. Largely through the campaign he sparked, letters of greetings and congratulations to Tesla poured in from the former presidents and the prominent members of the American Institute of Electrical Engineers, and from engineers and scientists in America and abroad. Many acknowledged the inspiration from Tesla's work that had furthered their careers.

**TIME** magazine, in its July 20, 1931 issue, featured Tesla on the cover and made an occasion of his 75th birthday by interviewing the fabled and secluded inventor. Reporting on the meeting the

Fig. 15.15 - TESLA at 75. Vibrant, alert and dreaming ahead of his time of beaming energy into the cosmos.

interviewers remarked that they wished they might have seen him as he used to be in the Colorado Springs laboratory a generation ago: "strolling or sitting like a calm Mephistopheles amid the blazing, thundering cascades of sparks 30 ft. long, Tesla currents alternating at such prodigious frequency that they would not harm a kitten."

Instead they found him, not without some difficulty, in seclusion on the 20th floor of Manhattan's Hotel Governor Clinton. He looked "pale but healthy, thin to ghostliness but strong and alert as ever . . . his hair is slate gray, overhanging eyebrows almost black. His eyes are blue. Only their sparkle and the shrillness of his voice indicate his psychic tension."

**HE HAD RECEIVED** congratulatory birthday greetings from Sir Oliver Lodge, E.F.W. Alexanderson, Robert A. Millikan, Lee DeForest and other notables in science and engineering. Some commented with gratitude on how they had been influenced by Tesla's spectacular experiments and lectures and the impetus these had given their own work. Their greeting conveyed hope that Tesla's genius would spawn yet another astounding device for mankind. Tesla, in the interview, was again willing to expand on his thinking. He was working to develop a new source that would transmit energy, thousands of horsepower, from one planet to another regardless of distance. He explained:

"I think that nothing can be more important than interplanetary communication. It will certainly come some day, and the certitude that there are other human beings in the universe, working , suffering, struggling, like ourselves, will produce a magic effect on mankind and will form the foundation of a universal brotherhood that will last as long as humanity itself."

Asked when this might come, he said:

"I have been living a secluded life, one of continuous, concentrated thought and deep meditation. Naturally enough I have accumulated a great number of ideas. The question is whether my physical powers will be adequate to working them out and giving them to the world."

**IN HIS ADVANCING AGE** Tesla still followed the lure of new challenges, and his projects became more grandiose. One of them was his "death ray," which was to be a beam of concentrated energy capable of devastating destruction. The idea was not exactly new. A British scientist in 1924 had claimed discovery of a diabolical ray of electric current that could wipe out airplanes. The Germans were reported to have death rays capable of spreading certain death. The military on both sides of the Atlantic were interested and concerned.

Tesla claimed priority for the death ray, citing the experiments at Colorado Springs as originating the first principles. On Tesla's 78th birthday newspapers reported he was producing a ray based on a new principle of physics. The beam was to be a tiny one but with such a wallop that it could destroy airplanes at 250 miles. A ring of beam projectors around the United States could, he claimed, defend the country from all invaders. The device would use millions of volts, amplified and concentrated so it could penetrate interstellar space.

**TESLA'S DEATH RAY** died with him on January 7, 1943, its secret unrevealed, and considered by many as the mere rambling of a senile mind. But this was Tesla the visionary, a man beyond his time, predicting as he had done in the past, the waves of the future. Could he have foreseen the laser and other electronic and nuclear beams that loom so large today? His death preceded by only two years the atomic bomb blast that revealed the awful destruction of which man is capable.

A grateful Yugoslav government established in Belgrade in 1936 the Tesla Institute. Its opening celebration marking Tesla's 80th birthday was attended by seventy delegates representing seventeen countries. Tesla was given an annual honorarium of $7200. With it he managed comfortably the requirements of the bare subsistence to which he had become accustomed. In his final wishes Tesla asked that his entire possessions, papers and personal belongings be brought to Belgrade; in this way he felt he was paying a debt to the Institute which bore his name.

The collection is now preserved in the Nikola Tesla Museum, which was opened in Belgrade in 1955. It contains over a hundred thousand documents pertaining to Tesla's lifetime of stupendous activity. His correspondence alone involved over 7000 individuals. The Museum also houses an exhibition of working models of Tesla's most important inventions.

**The 100th ANNIVERSARY, 1956,** of Tesla's birth, brought forth, under the sponsorship of the Nikola Tesla Museum, a year of commemorative programs devoted to Tesla's work. Many distinguished scientists gathered to pay homage to Tesla's creative genius and to testify to the universality of the contributions Tesla made to mankind.

The fall meeting of the American Institute of Electrical Engineers in Chicago, October 1-5, 1956, was dedicated to Tesla. Chicago was a logical place for the gathering because it was here 63 years earlier that the Tesla polyphase current system was demonstrated at the Columbian Exposition. Here, in Chicago, Tesla had also exhibited his first radio-guided missile. The meeting was also an anniversary of the founding, in 1893, of the Chicago chapter of the AIEE, the oldest in the nation.

The mayor of Chicago dedicated October 1 as Nikola Tesla day, and a special message from President Tito of Yugoslavia was displayed.

**The "TESLA"** - As part of the 100th anniversary celebration, the International Electrotechnical Commission, meeting in Munich, June 29-July 7, 1956, formally agreed to adopt the name *tesla* for the unit of magnetic flux density. Tesla thereby joined the company of the other immortals to whom this book has been devoted.

**A TESLA SYMPOSIUM,** in honor of the great Yugoslav-American inventor, was held in New York City, January 30-31, 1976. The Symposium was sponsored by the Power Engineering Society of the Institute of Electrical and Electronics Engineers, and was an event of the American Bicentennial. The affair included formal papers, informal discussion, and presentation of the Nikola Tesla Award which was established by the IEEE in 1975 to honor an individual who had made an outstanding contribution in the field of electric power generation or utilization. Symposium speakers from Yugoslavia's Academy of Science and Arts, the Nikola Tesla Museum and the University of Belgrade presented tributes to their countryman. A feature of the Symposium papers was an updating of contemporary developments related to Tesla's many innovations in power. lighting, and in energy transmission.

**NIKOLA TESLA** has been popularly called an "eccentric" - and most geniuses are that to some extent. However, Tesla had a life-style that easily brought on the eccentricity label. He never married and while he idealized women in an objective way, he excluded them from his life. This choice he justified by his plans for a life devoted to his work, a life in which he would allow no non-participating persons to distract him.

Tesla had idiosyncracies that were an essential part of his character. He had an extreme germ phobia, possibly as a result of his boyhood bout with cholera. He had his own private washroom in his office, so that he could clean his hands at the slightest pretext. Except for his elaborate dinners during his years of wealth, he always ate by himself, cleaning each dish and piece of silverware with napkins. He avoided coffee, tea and cocoa, but savored whiskey, maintaining that it was beneficial in prolonging life.

A loner, he was not an easy person to work with. He had a brilliant mind and an exceptional memory which could not easily accomodate associates with lesser talents or those unable to visualize detail. He was not an organization man, preferring to work independently. Tesla had very few personal friends, but with these the association was warm and lasting.

Although most people think of Tesla as an inventor, he considered himself a discoverer, one who revealed new principles which the inventor then converted into useful applications. Tesla's aloofness to commercialization of his discoveries was a factor in his growing lack of recognition by the newer generation of engineers. He became disinterested in the ventures which would have kept him in the public eye, and with advancing age he was largely forgotten. Secretive about his plans, he never shared with others the goals that might have been carried on by those after him to bring attention to his name'

**A LIST** of Tesla's patents is a veritable storehouse of the state of electrical developments at the turn of the twentieth century.

They include 36 for motors and generators, 9 for transmission of electric power, 6 for electric arc lighting, 17 for high-frequency apparatus and circuit controllers, 13 for radio, 7 for steam turbines and similar machinery, and 11 for various other apparatus.

Tesla's contributions to electrical science and their influence in establishing technology that followed, can be conveniently classed by whether they used high or low frequency.

**TESLA'S LOW-FREQUENCY** polyphase generation system, installed at Niagara in 1896, was one of the great events in engineering history. By permitting for the first time economical distribution of power over long distances by alternating current it created and fulfilled a tremendous demand for electric lighting. Tesla completed the advantage and practical use of alternating current in 1888 by using the rotating field of the polyphase system in his invention of the induction motor - a motor that is one of man's most versatile servants, with a utility that spread far beyond Tesla's fondest dreams.

The basics of these magnificent concepts have not changed materially since Tesla's time. The fundamental processes are at work in all the enormous power generation and utilization systems today, but made vastly more applicable and efficient by continuous improvement over the years in design and manufacture.

Leaving his polyphase system for others to commercialize, Tesla, in 1889, stepped into the new horizons of high frequency. He was soon dazzling the scientific world with his brilliant experiments in which the known limits of voltage and frequency were being far exceeded. Using the Tesla oscillator, he was exploring high voltage and frequency and their combination for their novel effects in lighting and radiation.

**TESLA'S HIGH-FREQUENCY** experiments led him to conclude he could use these currents for more efficient electrical lighting than was being provided by the carbon filament lamp. Using low pressure gas tubes he got luminous discharges that can be regarded as giving impetus to modern methods of gas-display lighting. Tesla has been called the "father" of the neon tube.

Tesla described four new concepts of electric light using electrically activated gas-filled tubes:

1 - Tubes in which a solid body is rendered incandescent.

2 - Tubes in which phosphrescent and fluorescent materials are caused to luminesce.

3 - Tubes in which rarified gases become luminous.

4 - Tubes in which luminosity is produced in gases at ordinary pressures.

The fluorescent lamp, which now furnishes a major part of our commercial illumination, owes its development to the early work of Tesla.

Another of Tesla's lighting discoveries used a carbon button in a spherical sealed glass globe, connected by a single wire to a high frequency current. The carbon button electrostatically repelled the gas molecules around it toward the inner surface of the glass globe. They were then repelled backward to heat the button to incandescence. This kind of illumination is now being reinvestigated.

Tesla, in one of his visionary projections proposed an artificial display of auroral lights to transform the globe into a giant lamp. Activating the high conductivity layers of the stratosphere, he would bathe the earth in the glow of a thousand moons.

**THE BEAUTIFUL EXPERIMENTS** of Hertz in demonstrating electromagnetic radiation acted as an immense stimulus to succeeding research. Tesla soon saw the transmission possibilities of electric waves for energy and signalling, and in the early 1890s turned his exceptional talents in this direction.

Tesla quickly disavowed a connection with Hertz's relatively weak spark radiations, citing the old-fashioned equipment used by Hertz and the lack of tuning. Tesla's aim was an apparatus developed on an engineering scale that would give large power effects over long distances using the natural oscillating properties of the earth. Tesla envisioned the earth as a vast reservoir of electricity whose charges could be disturbed effectively and released by a properly designed and powerful electrical machine. The earth's properties would be assisted by the conducting qualities of the atmosphere.

To produce the radiating power Tesla invented a novel air-core transformer in which the oscillating discharges of a condenser were magnified in a resonant coupled circuit that delivered unheard of high voltage and high frequency in the millions. Built with enough oscillating power this magnifying transmitter could, he claimed, enhance the natural vibrations of the earth to radiate unlimited electromagnetic energy. For transmission Tesla proposed that the high-frequency power be led to an elevated antenna plate, and connected by wire to a similar metal plate buried in the earth. For reception there would be an identical arrangement. This was the first known disclosure of the antenna-earth combination for electromagnetic radiation, thus establishing one of the basics for wireless communication.

**TESLA HAD A DOZEN PATENTS** for radio apparatus, but none of his schemes for signalling became commercially operative. Nor were his procedures and devices taken up immediately by others in the developments that followed. One reason for this was probably Tesla's emphasis on transmission of large-scale power rather than exploitation of the signalling potential.

"This seems somewhat unfortunate," wrote L.P. Wheeler in *Electrical Engineering* for August 1943, "for a perusal of his pertinent patent specifications would appear to indicate that, although not exploited effectively for communication purposes, there are at least three matters of prime importance to the radio art today on which Tesla's ideas were clearer than those of his contemporaries and on which he is entitled to either distinct priority or independent discovery. These are:

1 - The idea of inductive coupling between the driving and the working circuits.

2 - The importance of tuning both circuits, that is, the idea of an "oscillation transformer."

3 - The idea of a capacitance loaded open secondary circuit.

"It seems incontestable that all three of these fundamental ideas are clearly revealed in Tesla's patent specifications and lectures prior to 1894, although their application to communication purposes, while mentioned, is made incidental to the power-transmission objective. As none of these ideas appear in the specific literature of the radio art prior to the patent specifications of Marconi, Lodge and F. Braun of the years 1897-1900, it would seem that Tesla's name is worthy of perpetuation as a pioneer of these ideas which have been so basic in the radio art down to the present . . . "

"In addition to this major pioneering activity, Tesla made at least two contributions specifically in the communications field that are not generally known. The first is that embodied in the patent specification entitled: *Method of and Apparatus for Controlling Mechanism of Moving Vessels or Vehicles.* This method of remote control operates on a succession of radio impulses whose incidence on a receiving antenna energizes through relays the battery-powered steering and propelling motors of the moving vessel . . . it would seem that this invention deserves to be mentioned as the earliest radio remote-control system with which he is acquainted.

"The second of these less known contributions is contained in a patent entitled: *Apparatus for the Utilization of Radiant Energy.* These describe a scheme which, in so far as it would be actually operative, depends on the changes in the charge on a capacitor produced by the incidence of the radiation (light) on an elevated-capacitance plate antenna connected to one of the capacitor terminals. It is thus seen to embody an application of the photoelectric effect discovered by Hertz in 1887, although Tesla seems to have been ignorant of that fact."

Mr. Wheeler states that: "After studying his pattent specifications and the record of his public lectures, Tesla and his work in the high-frequency field would seem to justify a place in the history of radio engineering not so very far below that due to his accomplishments in the field of low-frequency alternating currents."

**"ELF" COMMUNICATION** - One aspect of Tesla's long-ago experiments at Colorado Springs has come alivc. He there envisioned sending power pulses using the earth as one conductor and the stratosphere as the other for signaling. This method is in use by the U.S. Navy for communicating with submarines when deeply submerged. Power signals of the Tesla type, called "ELF" (extremely low frequency - 30 to 300 Hz, giving wavelengths of 1000 to 10,000 kilometers), penetrate hundreds of meters into the ocean. Sent from electric grids in various key locations to submarines at great depths, ELF provides jam-proof communication.

K.W. Swezey, in an article appearing in *Electrical Engineering* for September 1956, entitled: *Nikola Tesla, Pathfinder of the Electrical Age,* reported that after Tesla's death, the British Institution of Electrical Engineers arranged a memorial lecture at which Dr. A.P.M. Fleming recreated some of Tesla's classic demonstrations. At the end of the lecture W.H. Eccles, a noted electrical pioneer, said the he "had been driven to the conclusion that Tesla was the greatest electrical inventor we have on our role of membership; in fact we might go so far as to say that he was the greatest inventor in the realm of electrical engineering."

**THAT MARCONI** invented the radio is the common conception. Thus Tesla has never received the popular recognition for his pioneering work in that field. Marconi's claim for radio patents was tested before the U.S. Supreme Court in 1942-43. In a suit: *Marconi Wireless Telegraph Company of America vs the United States,* Marconi claimed patent infringement by the U.S. Government in producing radio equipment during World War I. In a decision of 21 June 1943, the year of Tesla's death, the Court denied Marconi's claim, ruling that Tesla's patents covering the basic principles of radio communication had been filed seven years before Marconi filed his.

**EVERY SCHOOLBOY** knows about Edison and his electric light. He is popularly considered a founder of electric power, a fact of which we are reminded by the inclusion of his name in that of the big power companies. But Edison's direct-current system died out long ago. Its demise was the work of the young immigrant who once assisted Edison, and who parted with him to carry out his vision of a new age of electricity.

His name is not on the popular tongue. But when put on the measuring scales of contribution to the realization of universal electric power, Tesla's name has an equal weight with that of Edison. Tesla's polyphase alternating-current system rules the world of electric power today.

## STANDARDIZING THE A-C FREQUENCIES FOR LIGHT AND POWER DISTRIBUTION

**WHEN TESLA** invented the induction motor in 1888 there were no established standards for alternating-current frequency. Each a-c generating undertaking selected its own frequency depending on the judgement of the engineers and the type of apparatus available. As a result a wide range of frequencies was in use. In the U.S. there were about ten different frequencies from $16^{1}/_{3}$ to $133^{1}/_{3}$ cycles per second. In Great Britain there were more than twenty in a similar range, and the same differences existed in Europe.

The early load on a-c power systems was largely from incandescent lamps and arc lights. These could accept a wide range of frequencies since the illumination was not sensitive to frequency unless so low as to cause objectionable light flicker.

With the advent of induction motors, however, the situation changed. Speed of the motor was strictly related to frequency by the equation: *rpm = frequency in cycles per second* X *60* X *number of pairs of poles in the motor.* The higher the frequency the higher the motor speed for a given number of pairs of poles. While the induction motor might be designed electrically for proper operation at any reasonable frequency, the peripheral speed at higher frequencies might be greater than desirable for mechanical reasons. Thus the selection of a suitable frequency standard had to accomodate requirements of the induction motor.

No one was in a better position to make the intial choice of a frequency standard than Tesla, who in 1888 had been engaged by Westinghouse to develop the new induction motor commercially. Out of a patchwork of existing frequencies Tesla urged the use of 60 cycles as the best all-around frequency. Westinghouse engineers, who were preparing the huge electrical machinery to light the 1893 Chicago Columbian Exposition, accepted Tesla's choice. The Columbian Exhibition became the world's biggest showcase for electric lighting and the new polyphase a-c system, all at 60 cycles.

General Electric Company, which had acquired rights to build induction motors under the Tesla patents, was also faced with a choice of a commercial frequency. The decision fell into the hands of their engineering wizard, Charles P. Steinmetz, who from extensive motor design considerations had also chosen 60 cycles.

Thus, when the American Institute of Electrical Engineers met in Chicago in 1894 to adopt a frequency standard for lighting and power distribution in the U.S. they selected 60 cycles, based largely on the recommendations of the two big American electrical manufacturers. Provinces in Canada which were operating at 50 cycles converted eventually to 60 cycles for the advantages of interconnection with the U.S 60 Hz is now standard in North America, in several South American countries, and elsewhere as shown on the map.

**In GREAT BRITAIN and EUROPE** the early electrical supply situation was complicated by the large number of local installations monopolizing particular districts, each operating on its own frequency and voltage, with little incentive for standardization. In 1924 there were still 17 different frequencies in use in Britain. The main industrial areas had adopted 50 cycles. The Manchester area was operating on 40 cycles. Large parts of the Midlands and South Wales were using 25 cycles, which was also a dominant frequency in Scotland.

There had been several government proposals in the early 20th century for interconnection of power systems for a common frequency and voltages. They all failed for fear of government takeover, and for financial reasons.

Finally, the cost benefits of interconnection and standardization to the electric power industries and to their customers alike became so urgent that Parliament in 1926 passed the Electricity Supply Act creating the Central Electricity Board. Eventually the frequency was standardized at 50 Hz.

The adoption of 50 Hz in Europe appears to have followed a similar pattern of national programs of interconnection and standardization. Introduction of the British-developed Parson steam-turbine, direct-connected a-c generators, operating at 3000 rpm, into several countries on the Continent, was a factor in the standardization of 50 Hz there.

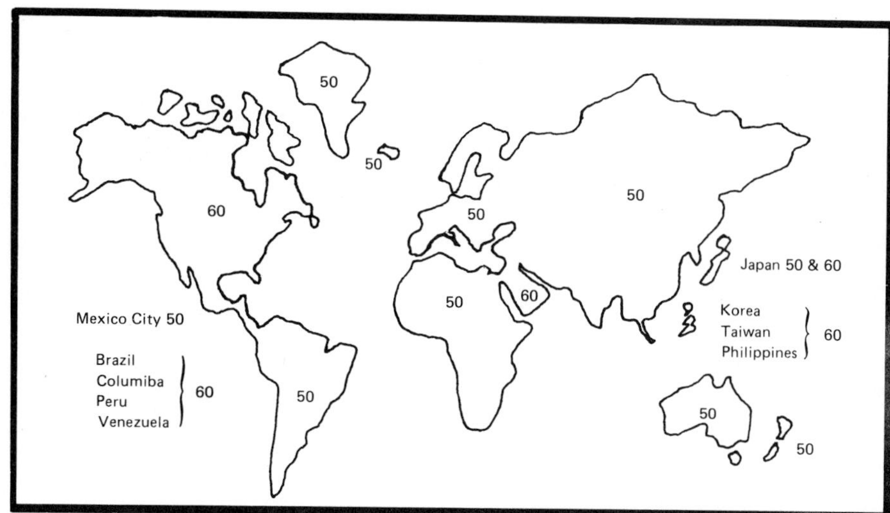


World standards of a-c power frequencies

# EPILOGUE

THIS SAGA of sixteen heroes has been one of glorious adventure and struggle along the pathways of enlarging the human understanding for the human betterment. With mighty strokes of their scientific swords each of these pioneers has cleared the way for his followers. They all have left their *"footprints in the sands of time"* as guides for the next steps into the great unknowns. Their deeds in scientific discovery are a continuing testament to the insatiable curiosity of the human mind, and its joy in unraveling the works of nature.

The table on the opposite page shows the chronological progress in naming units of electricity and magnetism for the heroes of this history. It covers nearly a century - from 1862 to 1956, ending with Tesla, whose long life spanned almost the same period of time. He was born when the buffalo still roamed the western plains, and died in the exploding age of electronics technology.

Our history ends with Tesla as the last electrical pioneer to be remembered with a unit. Will new electrical science produce new units . . new names? The only answer is that the inexhaustible richness of the field of electricity will surely bring future innovations requiring new names as they have in the past, and new heroes to be immortalized by them. As we began in the Foreword of this book: *THE WONDERS OF ELECTRICITY never cease!*

To you alone, true philosophers, ingenuous minds, who not only look in books, but in things themselves for knowledge . . . . .

CHRONOLOGICAL ORDER OF THE NAMING OF THE ELECTRIC AND MAGNETIC UNITS

| DATE | NAME | UNIT | SYMBOL | NAMING BODY |
|---|---|---|---|---|
| 1862 | OHM | *electrical resistance* | Ω | British Association Committee (London) |
| 1867 | FARAD | *electrical capacitance* | F | British Association Committee (London) |
| 1881 | VOLT | *electric potential* | V | 1st International Electrical Congress (Paris) |
| 1881 | COULOMB | *electric charge* | C | 1st International Electrical Congress (Paris) |
| 1881 | AMPERE | *electric current* | A | 1st International Electrical Congress (Paris) |
| 1889 | WATT | *power, radiant flux* | W | 2nd International Electrical Congress (Paris) |
| 1891 | GAUSS* | *magnetic field intensity* | | 3rd International Electrical Congress (Frankfort) |
| 1891 | WEBER* | *magnetic flux* | | 3rd International Electrical Congress (Frankfort) |
| 1893 | HENRY | *inductance* | H | 4th International Electrical Congress (Chicago) |
| 1894 | GILBERT* | *magnetomotive force* | $\mathscr{F}$ | AIEE Standards Committee (Chicago) |
| 1894 | OERSTED* | *magnetic reluctance* | $\mathcal{R}$ | AIEE Standards Committee (Chicago) |
| 1900 | MAXWELL* | *magnetic flux* | Φ | 5th International Electrical Congress (Paris) |
| 1930 | OERSTED* | *magnetic field strength* | Oe | IEC Technical Committee (Paris) |
| 1932 | HERTZ | *frequency* | Hz | IEC Technical Committee (Paris) |
| 1932 | WEBER | *magnetic flux* | Wb | IEC Technical Committee (Paris) |
| 1932 | SIEMENS | *electrical conductance* | S | IEC Technical Committee (Paris) |
| 1956 | TESLA | *magnetic flux density* | T | IEC Technical Committee (Munich) |

* CGS units

in the discovery of secret things and hidden causes, stronger reasons are obtained from sure experiments - Gilbert, *De Magnete*

# CGS to MKSA to SI

## THE RISE OF THE UNITS. . . .

**WHEN**, in the early 19th century, Georg S. Ohm was carrying on the experiments with electric circuits that led to discovery of his famous law: $V = IR$, he worked with the simplest of instruments. Measurements of current intensity were made by how much a compass needle, placed over the wire being tested, was deflected in opposition to the attraction on the needle by the earth's magnetic field. The deflection provided a relative measure of the current in the wire. The amount of deflection would be affected somewhat by local changes in the strength of the earth's magnetism. Electromotive force was crudely measured by the amount of deflection of the leaves of an electroscope.

Terminology was diverse and indefinite, and there was vagueness in the specification of the physical quantities to be determined. No coherent system of units for measuring current, resistance, electromotive force or capacity had yet been developed. Electrical characteristics were variously described by such terms as "tension," "intensity," "quantity," "voltaic excitation," and "electroscopic force."

Measurements made by one experimenter were expressed in arbitrary figures depending on the apparatus he employed. His results were thus not generally usable by other investigators. For example, resistance might be reported as that of a certain length and size of copper wire. This kind of information could not be conveniently used by others.

Ohm initially described the resistance of his test wires in terms of the "reduced length" of the particular circuit he was using. He knew nothing about volts or amperes (or even about ohms).

By the mid-19th century the need for standard units and generally approved terminology for required electrical quantities had become acute. The first answer to this need came in 1862, only eight years after Ohm's death. In that year, largely due to the initiative and efforts of the British Association for the Advancement of Science, the first electrical unit was adopted in England - that for resistance - and was named the "ohm."

**ACTION** for a standard of electrical resistance was prompted by the first practical use for electromagnetics - the telegraph. Put into commercial operation in the 1840s, the telegraph lines were soon spanning hundreds of land miles in both Europe and America, and by 1858 an Atlantic telegraph cable had joined the two continents.

Resistance measurement was important in the electric telegraph to insure that the metal of the wires was suitable for the distances to be covered. And when a fault occurred in the line its position could be located by the measurement of resistance. Submarine cables were tremendously expensive, requiring constant monitoring during manufacture, laying and in use to insure proper operation. This involved resistance measurements that could be understood and duplicated by others.

"In the present state of electrical science," Maxwell wrote in his *Treatise on Electricity and Magnetism,* "determination of the electrical resistance may be considered as the cardinal operation in electricity, in the same sense that the determination of weight is the cardinal operation in chemistry."

The commercial measurement of resistance in the various countries had resulted in locally used units based on lengths of telegraph wire. In England one unit of resistance was a mile of No.16 copper wire, in Germany a mile of No. 8 iron wire, in France a kilometer of iron wire 4 millimeters in diameter. Other units were proposed, such as

LORD KELVIN, William Thomson (1824-1907) – British physicist and mathematician, and professor of natural philosophy at the University of Glasgow from 1846 to 1899. He is famous for his work on heat and electricity. He introduced the Kelvin, or absolute scale of temperature, and discovered the Thomson effect in thermo-electricity. He was a leading authority in the field of submarine telegraphy, and the inventor of several instruments that led to successful operation of the transoceanic cables. Kelvin pioneered the work of the British Association for the Advancement of Science in establishing standards for electrical units, and was a member of the British Association Electrical Standards Committee for over thirty years. Photo courtesy the Burndy Library, Norwalk Connecticut.

that of Werner Siemens in Germany. His unit of resistance was a column of mercury one meter long, one square millimeter in section, measured at zero degrees Centigrade.

There was no general acceptance of these more or less arbitrary standards, and the situation drifted to the point when, in 1861, two telegraph engineers, Latimer Clark and Charles Bright, presented a paper at the Manchester meeting of the British Association for the Advancement of Science calling attention to a need for scientific units, standards and measurements for telegraph engineering.

**THE BRITISH ASSOCIATION** for the Advancement of Science was founded in response to the need for a society that would promote the interests and accelerate the progress of science. The first meeting was held in York beginning 26 September 1831. The British Association soon established its position as a forum for the discussion of scientific discoveries and events. It provided for regular meetings in the various cities of Britain. Its membership included the leading scientists, engineers and educators, plus interested laymen. They were active in promoting science as a profession and in providing opportunities for the exchange of information and promulgation of guide posts for scientific progress.

One of the active and far-seeing members was Professor William Thomson (later Lord Kelvin). Acting on the need for the electrical units and standards as presented by Clark and Bright, he arranged for the appointment of a committee of members of the B.A. to initiate a project for determining the best standard of electrical resistance. The members of the Committee were eminent in basic and applied science, and included Prof. James Clerk Maxwell, who was to be one of the great contributors to the successful outcome of the endeavor to standardize the electrical units.

The Committee of Electrical Standards of Resistance began its work in 1861. The path for its progress had been partially paved by four previous scientific events that were pertinent to the efforts of the Committee, namely:

1 - by the mechanical system of Newton as contained in his three laws of motion, in which the fundamental measurable quantities are length, mass and time, as expressed, for example, in his second law defining the force as mass times acceleration, or, $f = ma$. The force factors in this equation could be denoted in appropriate units such as centimeters and grams, which had been provided

2 - by the adoption by an International Commission meeting in Paris in 1875 of a metric decimal system of weights and measures. The system, conceived in 1791, during the French Revolution. was

founded on a base unit of length, the *meter,* equal to one ten-millionth of the length of the quadrant of the earth meridian through Paris. From this length the *kilogram* was derived as a cylinder of platinum of mass equivalent to a cubic decimeter, or one liter, of distilled water at 4 degrees Centigrade; and,

3 - by the publishing, in 1827, of Ohm's law relating voltage, current and resistance, and his discovery that resistance is a property of a conductor; and,

4 - by the work in Germany of C.F. Gauss, the mathematician, and his collaborator, the physicist W.E. Weber, as has been described in a previous chapter. In his famous paper of 1833: *Determination of the Earth's Magnetic Field in Absolute Measure,* Gauss showed how magnetic forces could be expressed in absolute units, and thus be connected with the dynamical units based on chosen fundamental units of length, mass and time. Weber, in 1851, extended Gauss's metric methods to the measurement of electrical quantities based on dynamic units.

**WEBER** had determined that electrical resistance is basically related to and could be measured by a *velocity.* He made the first distinct proposal, in 1851, of a definite system of electrical measurements according to which resistance would be measured in terms of an absolute velocity using fundamental units of length and time. This system he called "absolute electromagnetic measure" in analogy with Gauss's nomenclature of "absolute magnetic measure" as used in his original determination of terrestrial magnetism in absolute units.

The word "absolute" was used as opposed to the word "relative," and implied that the measurement instead of being a simple comparison with an arbitrary quantity of the same kind as that measured, is made by reference to certain fundamental units of another kind as postulates - in the case of resistance - a unit of length per unit of time. The absolute system had the advantage that all units formed part of a coherent system, avoiding useless coefficients in passing from one kind of measurement to another.

Weber proposed that the unit of resistance correspond to a velocity of one millimeter/second. He also suggested a revolving coil arrangement by which resistance of the coil could be determined in electromagnetic measure. Weber's absolute unit had no physical existence, however, and was of no practical value until converted into a readily usable material standard that could be duplicated.

**THE PRINCIPAL OBJECTS** of the B.A. Committee of 1861 were, first, to determine what should be the most convenient *unit* of resistance, and second, what would be the best form for the *standard* representing that unit. The first B.A. Committee report of 1862 recommended that the unit should provide the following desirable qualities:

*1 - The unit should be of convenient size for the more usual electrical measurements.*

*2 - The unit should form part of a complete system of electrical measurements - bear a definite relation to units of electrical quantity, current and electromotive force.*

*3 - The unit of resistance in common with other units of the system, should bear a definite relation to the unit of work.*

*4 - The unit should be prefectly definite, and should not be liable to require alteration or correction from time to time.*

*5 - The unit should be reproducible (a) in case the original standard was impaired, or (b) in case copies were not obtainable they could be independently manufactured.*

To meet these qualities, the B.A. Committee in 1862 stated the rationale of their recommendations for procedure as follows:

"a meter of mercury or some other arbitrary

THE FIRST ELECTRICAL STANDARD - A VELOCITY

Excerpts and basic equations on the opposite page show the formal development of electrical resistance as an absolute velocity. They were written by James Clerk Maxwell, and appear in the 1862 report of the Committee on Electrical Standards appointed by the British Association for the Advancement of Science.

## RESISTANCE HAS THE DIMENSION OF A VELOCITY . . .

A current ($C$) in a straight conductor of length ($L$) crossing the lines of force of a magnetic field of the intensity ($S$) at right angles will experience the same force ($f$) as if all the points of the conductor were at the unit distance from a pole of the strength ($S$). The force in this case exerted on the magnet is, by equation (5), equal to $SLC$, and, conversely, an equal force is exerted by the magnet on the current. Hence we have equation (7), expressing the value of the force ($f$) exerted on a current crossing a magnetic field at right angles,

$$f = SLC. \quad \text{.........................(7)}$$

Let us imagine this straight conductor to have its two ends resting on two conducting-rails of large section in connexion with the earth, and let the whole sensible resistance ($R$) of the circuit thus formed be constant for all positions of the conductor. Let us further imagine the rails so placed that when the conductor slips along them it moves perpendicularly to the magnetic lines of force and to its own length. By experiment we know that when the conductor is moved along the rails cutting these lines of force, a current will be developed in the circuit, and that the action of the magnetic force on this current will cause a resistance ($f$) to the motion (due to electro-magnetic causes only); and, by equation (7), we find that this resistance $f = SLC$.

Let the motion be uniform, and its velocity be called $V$; and let the work done in the unit of time in overcoming the resistance to motion due to electro-magnetic causes be called $W$; then $W = VSLC$. But this force produces no other effect than the current, and the work done by the current must therefore be $W$, or equivalent to that done in moving the conductor against the force $f$; but, by equation (3), $W = C^2R$, and hence

$$R = \frac{VSL}{C}. \quad \text{.........................(8)}$$

It has already been shown that $C$ and $S$ can be obtained in absolute measure; hence the second member of equation (8) contains no unknown quantities, and, by the experiment described,

(NOTE - In this page from Maxwell's time, the letter, C, is used for *current.* The letter, I, is now used for current, and the letter, C, is now used for *capacity.* The letter, V, is here used for *velocity.* V, is now used for *voltage.)*

$$C = \frac{E}{R}, \quad \text{.............................(1)}$$

$$Q = Ct, \quad \text{.............................(2)}$$

$$W = C^2Rt, \quad \text{.........................(3)}$$

$$F = \frac{Q^2}{d^2}; \quad \text{.............................(4)}$$

$$f = \frac{CLm}{k^2}. \quad \text{.........................(5)}$$

$$C = \frac{Hk^2}{L}\tan d. \quad \text{.......................(6)}$$

the absolute resistance ($R$) of a wire might be determined. One curious consequence of these considerations is, that the resistance of a conductor in absolute measure is really expressed by a velocity; for, by equation (8), when $SL = C$ we have $R = V$, that is to say, the resistance of a conductor may be expressed or defined as equal to the velocity with which it must move, if placed in the conditions described, in order to generate a current equal to the product of the length of the conductor into the intensity of the magnetic field; or more simply, the resistance of a circuit is the velocity with which a conductor of unit length must move across a magnetic field of unit intensity in order to generate a unit current in the circuit. Moreover it can be shown that this velocity is independent of the magnitude of the fundamental units on which the expression of the magnetic intensity of the field or strength of the current is based, and hence that electrical resistance really is measured by an absolute velocity in nature, quite independently of the units of time and space in which it is expressed material might possess what we have called the first, fourth and fifth requisite qualities, to a high degree, although entirely wanting in the second and third. Weber's system, on the contrary, is found to fulfil the second and third conditions, but is defective in the fourth and fifth, for if the absolute or Weber's unit were adopted without qualification, the material standard by which a decimal multiple of convenient magnitude might be practically represented would require continual correction as successive determinations were made with more and more skill to determine the real value of the absolute unit with greater and greater accuracy. Few defects could be more prejudicial than this continued shifting of the standard.

"It then became a matter for consideration whether the advantages of an arbitrary material standard and those of the absolute system could not be combined; and the following proposal was made and adopted as the most likely to meet every requirement.

"It was proposed that a material standard should be prepared in such form and material as should ensure the most absolute permanency; that this standard should approximate as nearly as possible in the present state of science, to ten millions of meter/seconds, and the true ten millions of meter/seconds should be ascertained with increased accuracy, in order that the error, resulting from use of the new standard unit in dynamical calculations instead of the true absolute unit, may be corrected by those who require these corrections, but that the material standard itself shall under no circumstances be altered in substance or definition.

"By this plan the first condition is fulfilled; for the absolute magnitude of this standard will differ by only 2 or 3 percent from the Siemens mercury standard.

"The second and third conditions will be fulfilled with such accuracy as science at the time will allow.

"The fourth condition, of permanency, will be ensured so far as our knowledge of the electrical qualities of matter will permit; and even the fifth condition, referring to the reproduction, is rendered comparatively easy of accomplishment."

**THE COMMITTEE**, in coming to the foregoing conclusions, had made three crucial decisions, (1) that the absolute system, with material standard, was the most rational and practical procedure; (2) that while the British pounds and feet system was usable locally, nevertheless for general scientific acceptance the absolute unit should be expressed in metric measurement; (3) that the value of the absolute unit of resistance should be in the range of what was commonly being used in telegraphic practice at the time. For this Weber's proposal of one millimeter/second was far too small. Hence, the Committee, for a practical unit, chose the meter/second with a multiplier of $10^7$, equal to the distance of a quadrant of the earth meridian per second (one-thirtieth the velocity of light). The value of the absolute resistance unit chosen proved to be very nearly equal to that of the Siemens meter mercury column.

The value of $10^7$ meter/second for unit resistance would also relate to those unit values chosen later for the volt and ampere, in order to conform to Ohms law, and to the unit of work.

**TO AUGMENT** the efforts of the Committee, comments on their conclusions were solicited from eminent scientists in the chief countries of Europe, and from Joseph Henry in America. The replies generally approved the plans for the absolute unit. High importance was attached to a communication from Dr. Esselbach, a foreign telegraph engineer. He, as a practicing engineer, had independently and without consulting any of the members, come to the same conclusions and values as did the scientists of the Committee.

During the years from 1863 to 1870, under the chairmanship of Maxwell, with Thomson very active, the Committee carried out the two projects relating to the establishment of the resistance unit, (1) determination of the absolute unit of resistance, and (2) production of the material standards.

Weber had originally devised a revolving-coil method of determining resistance in absolute measure. Independently of him Thomson had designed another revolving-coil apparatus which proved to be superior. It was used and refined by Maxwell and others in making many determinations of the absolute resistance unit.

**THOMSON'S APPARATUS**, Fig. 16.1, consisted of two circular coils of copper wire, about one foot in diameter, placed side by side and connected in

Fig. 16.1 — REVOLVING COIL apparatus for determining the resistance of the coil in absolute dynamical measure. It was designed by Thomson and used by Maxwell and other members of the B.A. Committee on Electrical Standards of Resistance.



series to form a closed circuit. The coils revolved on a vertical axis and were driven by a belt from a power wheel. A small magnet with mirror attached was hung in the center of the two coils, and was enclosed for protection against air currents. The deflection of the magnet was read by a telescope from the reflection of a scale on a mirror.

Revolving in the earth's magnetic field, a voltage was generated in the coils, producing a current which was measured by the deflection of the magnet. With a coil of given turns and radius, driven at a given speed, the resulting magnet deflection would enable calculation of the resistance of the coil in absolute dynamical units.

When the coil was revolved, a torque, $T_1$, was exerted on the magnetic needle, whose moment was $ml$. When the needle deflected to an angle, $d$, the torque was:

$$T_1 = \frac{L^2 VH}{4k^2 R}\, ml \cos d$$

where, $L$, is the length of wire in the coils, $k$, is the radius of the coils, $V$, is the velocity in radians per second, $H$, is the strength of the earth's magnetic field, and $R$, the resistance of the coils.

An opposite torque, $T_2$, equal to $Hml \sin d$, is exerted on the magnetic needle by the earth's magnetism, $H$. By equating the two torques, $H$ and $ml$ are eliminated and the equation becomes:

$$\tan d = \frac{L^2 V}{4k^2 R}$$

Fig. 16.2 - MICHAEL FARADAY purchased the first 1 ohm wire-wound resistance standard offered for sale by the Committee of the British Association for Electrical Resistance Standards. Lent to the Science Museum, London by the Royal Institution.

Then the resistance, $R$, can be calculated in terms of the length of the coil wire, the velocity in radians, the radius of the coil, and the deflection and is expressed in absolute dynamical terms.

$$R = \frac{L^2 V}{4k^2 \tan d}$$

**Dr. A. MATTHIESSEN**, a committee member, made special investigations on the electrical properties of pure metals and of alloys in the solid and liquid states for the purpose of constructing material standards. During a period of several years enough knowledge was gained on the suitability of metals and alloys, and on the change in resistance with hardening, annealing, bending and ageing, to proceed with construction of the B.A. type standards. These were to serve as permanent standards.

Two equal standards were prepared with each of the following wires: platinum, gold-silver alloy, platinum-silver alloy, and platinum-iridium alloy. Twenty copies of the platinum-iridium alloy standards were distributed gratis in Britain and internationally. Copies were then offered for sale, Fig. 16.2. Michael Faraday was the first customer.

The B.A. Standards were not called absolute units, but simply resistance units of the British Association, so that any later improvement in experimental absolute measurement would not entail a change in the standard but only a small correction where needed in calculation.

**"OHM"** - The Committee had developed a new basic unit. Originally it went under such designations as "ten million meter/seconds," "earth quadrant/second," and "unit of 1862." But every unit in popular use requires a simple name. Latimer Clark suggested the name "Ohmad" for the unit of resistance, which, abbreviated to *ohm* was finally accepted. Thus was established a precedent for naming the electric and magnetic units for the founders of the science.

Electrical science owes a great deal of its exactness and progress to the initiative of the B.A. Committee in setting up the ohm as a definite unit with a definite name, thus providing the standard and the terminology for other units that followed. Electricity, in this respect, was in advance of the other far older sciences of the time.

**"MERCURY-OHM"** - The B.A. wire-type units prepared by Dr. Matthiessen, and the Siemens "mercury-ohm" were concurrently available and in use as the working standards of resistance. The B.A. ohm units were shown to correspond to the resistance of a square millimeter column 104.86 cm long measured at zero C. Both standards had been intended to represent as nearly as possible the value of the original $10^7$ meter-second absolute unit. But Henry A. Rowland in America, in 1878, announced he had found the units to be too low by 1.5 per cent. This initiated renewed activity in precise determination of the unit of resistance. As a result, in 1884, it was decided to base the standard ohm on the square millimeter mercury column extended to 106 centimeters. In 1893, the length was increased to 106.3 centimeters. The amount of mercury was specified at 14.4521 grams, and the column section left unspecified. This standard was adopted generally and was used as the "international ohm" for many years. Mercury had been selected because it could be obtained in a very pure form, and because the mercury column seemed to offer easy and accurate reproducibility. In practice, however, the mercury standard had problems, mainly in the construction of the glass tubing to obtain a straight, uniform bore of accurate section. As a result, reliable mercury ohm determinations were found difficult to make.

The first B.A. Committee was active from 1862 to 1867. To meet the requirements of land and submarine telegraphy, the ohm had been established in 1862. By 1867 a unit of electromotive force had been named the *volt* and a unit of electrical capacity the *farad,* with tentative values respectively of $10^6$ and $10^{-13}$ in the meter, gram, second system of units.

It was apparent to the Committee that these values must be a part of a general system of electrical measurements. In this system the units should be coherent and bear a definite relation to other units and the unit of work. The natural relations between these units, as a direct consequence of Ohms law, was that a unit of electromotive force maintained between two points of a conductor separated by a unit of resistance should produce a unit of current, and this current in a unit of time should convey a unit quantity of electricity. With two of the above units chosen and fixed the other electrical units would follow from the relations between them.

**A NEW B.A. COMMITTEE** was appointed in 1867 "for the Selection and Nomenclature of Dynamical and Electrical Units." One of the first matters to come to its attention was achievement of a unitary relationship between the dynamical and the electrical units. Gauss and Weber, in their pioneering work, had expressed measurements in millimeters, milligrams and seconds. The first B.A. Committee had used meters, grams and seonds for the absolute units.

The new committee, in its first report in 1873, recommended changing the basis of the absolute units to centimeters, grams, seconds. A main reason for the shift to a CGS system from the MGS system was the unit relationship between the gram and the mass of a cubic centimeter of pure water at its greatest density, whereas the gram per cubic meter was only a millionth of that mass of water.

**THE CGS SYSTEM** proposed by the Committee applied both to the dynamical and the electrical units. For the unit of *force* - that which gives a mass of one gram an acceleration of one centimeter per second per second - the Committee advocated the name *dyne.* For the quantity of *work* performed by a force of one dyne acting through one centimeter, they advoated the name *erg.* The unit of power thus became "erg per second." Hence, as one horsepower equals 746 watts, one watt equals $10^7$ ergs/second.

Using the CGS dimensions, two separate unit systems were developed, one for *electrostatics* and the other for *electromagnetics.*

The *electrostatic unit (esu)* was based on the interaction between electric charges at rest as described by Coulomb's force law. The unit of charge in the CGS electrostatic system of units was defined as that occurring when two charges, each of one esu, and one centimeter apart, repel each other with a force of one dyne.

The *electromagnetic unit (emu)* was based on the magnetic action of a current as described by the Biot-Savart law. A unit current in the CGS electromagnetic system of units was defined as that which, flowing in a circular conductor of one centimeter radius, acts on a unit magnetic pole at its center with a force on one dyne for every centimeter of conductor length.

**The CGS PRACTICAL UNITS** - The International Electrical Congress (IEC), the first international advisory body on electrical standards, assembled in Paris in 1881. This Congress, and others to follow in the next quarter century, served as forums for discussion and proposals on units and nomenclature, and as approval and promulgation bodies for the standards that were adopted.

The 1881 Congress sanctioned the CGS system as the basis for absolute units. These units were far too small and inconvenient for practical engineering, however. For example, the emu unit for the electromotive force induced when the interlinking magnetic flux was changing at unit rate, turned out to be only about one hundred millionth of the voltage of the batteries then in common use. Therefore, in addition to the esu and emu units, it was necessary to have parallel, convenient *practical* units that would be in everyday use.

The 1881 Congress accepted a B.A. Committee report recommending $10^8$ CGS emu units for the practical unit of the *volt.* This approximated the $1.07 \times 10^8$ emu units of the Daniell cell widely used in telegraphy. The Congress also accepted the $10^9$ CGS emu units which had been established for the *ohm.* With these two units determined, three other practical units were evaluated in decimal multiples of CGS emu units. The five practical recommended units were then as follows:

| | | | | | |
|---|---|---|---|---|---|
| 1 Volt | = | B.A. Adopted Standard | = | $10^8$ | emu |
| 1 Ohm | = | B.A. Adopted Standard | = | $10^9$ | emu |
| 1 Ampere | = | a ratio, $I = V/R = 10^8/10^9$ | = | $10^{-1}$ | emu |
| 1 Coulomb | = | 1 ampere·second | = | $10^{-1}$ | emu |
| 1 Farad | = | a ratio, $C = q/V = 10^{-1}/10^8$ | = | $10^{-9}$ | emu |

The second International Electrical Congress, meeting in Paris in 1889, adopted three more units in the practical system: the *joule,* the *watt* and the unit of inductance, the *quadrant,* now the *henry:*

| | | | | | |
|---|---|---|---|---|---|
| 1 Joule | = | unit of energy | = | $10^7$ | ergs |
| 1 Watt | = | unit of power | = | $10^7$ | $\frac{erg}{sec}$ |
| 1 Henry | = | a ratio, $V/I/\text{sec} = 10^8/10^{-1}$ | = | $10^9$ | emu |

To distinguish the CGS units from their practical counterparts, they were given prefixes. On the basis that the emu units were absolute units, they used the prefix "ab," thus: abvolt, abampere, etc. The same units from the CGS electrostatic system received the prefix "stat," giving statvolt, etc.

**THE MATERIAL STANDARDS** - Although the volt, ohm, ampere and other practical units had been defined in the absolute CGS electromagnetic system, their usage could be realized only through material standards. These would employ elements for reproducibility while achieving as accurately as possible the absolute values they represented.

At the fourth International Electrical Congress in Chicago in 1893, the volt, ampere and ohm were defined in terms of prototype material standards, and the various governments represented were urged to adopt them legally. The prototype for the ohm was the well-known Siemens mercury column. The prototype for the volt was a mercury-zinc reference cell developed by Latimer Clark in 1873. The output of the Clark cell was established as 1.456 volts at 62 degrees F. The prototype for the ampere was a Faraday silver nitrate electrolytic cell with which the current would be measured by the amount of deposition of silver.

**"INTERNATIONAL ELECTRIC UNITS"** - The International Conference of Electrical Units and Standards, the first great assembly of its kind, was held in London in 1908 at the invitation of the British Government. It was opened by Winston Churchill, as President of the Board of Trade. This Conference adopted as "fundamental units" the absolute system, including the ohm, ampere, volt, watt and joule, as defined in CGS terms.

The Conference also reactivated the sanction given at the Chicago Congress for reproducible material standards, even though they were contrary to the coherence principle adopted by the original B.A. Committee. Recognizing however, that the reproducible standards were not exactly equal to the absolute standards, the London Conference recommended for purposes of legislation, the inclusion of the material standards in a separate system called "International Electric Units." These units were to be calibrated in absolute units and then maintained as working standards. Four International Units established by their specifications were defined as follows:

**The International Ohm** is the resistance offered to the passage of an unvarying electric current by a column of mercury at the temperature of melting ice, of mass 14.4521 grams, of uniform cross-sectional area and of length 106.300 cm.

**The International Ampere** is the unvarying electric current which, when passed through a solution of silver nitrate in water deposits silver at the rate of 0.00111800 grams per second.

**The International Volt** is the steady electric pressure which, applied to a conductor of resistance 1 International Ohm, produces a current of 1 International Ampere.

**The International Watt** is the electrical energy per second expended when an unvarying electric current of 1 International Ampere flows under a pressure of 1 International Volt.

In 1910 delegates from the National Physical Laboratory in England, the Physikalisch Technische Reichsanstalt in Germany, the Central Electrical Laboratory in France and from the National Bureau of Standards in the U.S. met at the Bureau in Washington for a comparison of national electrical material standards. From the results of these inter-comparisons, the values were set for the national units, to be maintained by national laboratories and distributed where they were required. These International Units were to remain in use for nearly forty years. The 1910 meeting also adopted a new standard for voltage reference measurements - the "Weston Normal Cell," a cadmium-mercury cell, with a value of 1.01830 volts at 20 degrees C. The Weston cell replaced the earlier Clark cell.

**THE MAGNETIC UNITS** - In the early use of electricity for communications, the principal needs for measurement were the electric units. But with the development of dynamo-electric machines in the 1880s, whose operation was based on the interaction of electromagnetic fields, it became necessary to analyze and design magnetic circuits as well as electric circuits. This led to the need for appropriate standards and nomenclature for the magnetic quantities.

The units involved, and the relations between them, were generally those for magnetomotive force, magnetic field strength, magnetic flux and magnetic flux density. Among the factors to be resolved in setting up the magnetic circuit units were (1) determining the dimensions of the units, (2) establishing the connection between magnetic force and magnetic flux density, and (3) consideration of rationalization, or elimination of the factor $4\pi$ in certain formulas.

The impetus given by use of the CGS system for electric units carried over into considerations for the magnetic circuit units. But in the application of the CGS units, there was for years, lack of agreement as to the names and sizes of the units. Proposed names for magnetic circuit units, such as *gauss, weber, oersted* and *maxwell* brought objections, and the units represented different quantities due to misunderstandings.

**AS A FIRST STEP** toward standardization of the magnetic units, a committee of members of the American Institute of Electrical Engineers, meeting in Chicago in 1894, reported in favor of provisionally adopting working units in the CGS magnetic system with the following names:

(1) *gilbert* for magnetomotive force (the mmf around a straight conductor of infinite length carrying a current of one ampere is $4\pi/10$ gilberts).
(2) *weber* for magnetic flux (the flux resulting from one gilbert acting through a distance of one centimeter in air).
(3) *oersted* for magnetic reluctance (one unit of reluctance in a magnetic circuit will require one gilbert to establish one weber).
(4) *gauss* for magnetic flux density (one weber per normal square centimeter).

A year later the B.A. Committee of Electrical Standards recommended the tentative adoption of:

(1) *weber* as the unit of magnetic flux equal to $10^8$ emu.
(2) *gauss* as the CGS unit of magnetomotive force.

At the 1900 International Congress in Paris it was generally agreed that the working units of the magnetic circuit should be retained in the CGS system, but there was opposition to giving them names. The Congress finally voted on the adoption of two names:

(1) *gauss* for the CGS unit of magnetic field intensity, with the signature, $H$.
(2) *maxwell* for the CGS unit of magnetic flux.

The American delegates, however, mistakenly assumed that the name "gauss" had been assigned by the Congress to the unit of magnetic field, $B$, and so reported on their return. The discrepancy was discovered after publication of the Paris report.

No further official action on magnetic units was taken by the International Congress for the next thirty years. In the meanwhile in America it was the custom to use:

(1) *gilbert* for the working unit of magnetomotive force.

(2) *gilbert per centimeter* for the unit of magnetizing force, or field intensity, *H.*

(3) *maxwell* for the unit of magnetic flux produced by a gilbert-centimeter acting uniformly over one square centimeter.

(4) *gauss* for the unit of flux density, *B*, or one maxwell per square centimeter.

(5) *oersted* for the unit of reluctance, that is the reluctance of one cubic centimeter of free space, or one gilbert per centimeter to establish one maxwell of magnetic flux.

Various other units in common use were the "ampere-turn" for magnetomotive force, the "line" or "kiloline" for the CGS unit of magnetic flux, and the "maxwell per square inch" for the flux density.

In Europe "gauss" was used interchangeably for *H* and *B*; "maxwell" was replaced sometimes by the "line" or "kiloline." The "gilbert" and "oersted" were rarely used.

**THE "GAUSSIAN" SYSTEM** - Orignally, electrostatic and electromagnetic phenomena were thought to be two entirely different aspects or properties of electricity, leading to a dual view as regards the units. The electrical quantities, as employed in problems involving electrostatic phenomena were identical in magnitude with those of the CGS esu system. The magnetic quantities, as employed in working with magnetics, current, induced electromotive force, etc., were identical in magnitude with those of the CGS emu system. In situations involving independently either electrostatics or magnetostatics there was no connection between the systems of units.

However, whenever the electrical quantities, expressed in esu units, and the magnetic units in emu units, occurred in the same equation, there was a necessary connection between the two. That is, the electromagnetic forces and the electrostatic forces were necessary attributes of each other. This is recognized in the Gaussian system, originated in 1894, as a system of mixed CGS units.

The factor in the Gaussian system uniting the two CGS systems was first revealed in an experiment made some decades before the CGS units were developed. In 1856 Weber and Kohlrausch, in Germany, took a quantity of electricity, first in electrostatic measure, then in electromagnetic measure, and found a velocity relationship between the two measures of approximately $3 \times 10^{10}$ cm/sec. This was known as the speed of light, and Maxwell, having theorized that electromagnetism could have a wave character, concluded that light was an electromagnetic wave, and that its speed, $c = 3 \times 10^{10}$ cm/sec, was a property of space, and hence related dimensionally to the interaction of electric and magnetic fields in space.

In the development of the CGS units, the electric and magnetic force relationships in free space were left without dimension. That is, the electric permittivity constant, $\epsilon$, and the magnetic permeability constant, $\mu$, in the force formulas from which the CGS esu and emu units were derived, were fixed at unity. However, it had been shown by Maxwell that the two constants, $\epsilon$ and $\mu$, were related by the speed of light. Thus there was a necessary dimensional connection between the two because the speed of light must be equal to $1/\sqrt{\epsilon\mu}$ for the media in which its traveling, and this in free space equals $1/\sqrt{\epsilon_0 \mu_0}$.

The individual dimensions of the constants, $\epsilon_0$, and, $\mu_0$, in the Gaussian system were undetermined, but they had to be complementary, since the combination was related to $c$. Thus, with $\epsilon_0 = \mu_0 = 1$ in the two CGS systems, it was necessary, depending on whether $\epsilon_0$ or $\mu_0$ was made basic, that the quantities in the two systems, when appearing in the same equation, had to be related explicitly by using, $c$, or $c^2$, as a multiplier or divisor in the equation. That is, although $\epsilon_0$ and $\mu_0$ do not appear in the Gaussian equations, their hidden presence is taken care of by the use of $c$ and $c^2$.

For example, in the Maxwell equations for electrical and magnetic fields in motion in free space, the two quantities are related by $c$:

$$\text{curl } \mathbf{E} = -\frac{1}{c}\frac{\partial \mathbf{B}}{\partial t}, \text{ and, curl } \mathbf{B} = \frac{1}{c}\frac{\partial \mathbf{E}}{\partial t}$$

Thus, the Gaussian system reflected the fundamental symmetry of electromagnetics in free space. The system had the practical objection, however, that its units required in circuit analysis did not coincide with the accepted practical units based on the volt, ampere, ohm etc.

The Gaussian system had the convenience of unity for the *permittivity* and *permeability* constants. This meant, however, that in the CGS-esu system the electric flux density, **D**, and the electric field strength, **E**, were related in free space by a dimensionless constant, thus implying that the two related quantities were physically identical. Considered as having the same nature, they could logically have the same name. The same unity relationship held between magnetic field strength, **H**, and field density, **B**, and the term *gauss* was used indifferently by some for both **H** and **B**, ignoring that they could be dimensionally different.

The question of dimensions, nomenclature and the underlying definitions of quantities involved in the magnetic circuit brought on a great deal of animated debate and lack of international agreement. Some proposed that the free space permeability factor between $H$ and $B$ was zero, with the need of only a single unit and name for the magnetic force and the magnetic flux. Others insisted that permeability was not a mere numeric, but had a physical dimension, and that magnetizing force and flux, being separate entities, should have separate unit names. It appeared unlikely that any comprehensive system of units could make headway in the magnetic circuit for as long as the magnetic circuit units were wholly in the CGS system.

**AS LATE AS 1930** it came as a distinct surprise to many engineers to learn that no international agreement had yet been reached in the definition of such fundamental scientific quantities as the magnetic units. Sparking this reaction was a report following an International Electrotechnical Commission meeting at Bellagio, Italy in 1927 in which a change in the units of the magnetic circuit was introduced. The delegates from Italy submitted a proposal to establish a new international unit of magnetic flux in the "practical system," with the name maxwell. Their proposal would change the maxwell from its long-established value of 1 CGS unit to the new value of $10^8$ CGS units. The advantages of the proposal were elimination of the $10^8$ which entered into so many engineering calculations, and the establishment of a unit relation between a change of unit magnetic flux in one second to produce one volt. An alternate suggested name for the unit was "volt-second."

The Italian proposal and its discussion alerted the Commission delegates to the difference in usage of magnetic units in the various countries, both as to names and definitions. There were differences of opinion as to the dimensions of the magnetic units, and especially as to the dimension of the permeability constant $\mu_0$. The Commission decided that the whole question of the magnetic units and their names should be considered afresh, with a view to arriving at a sound, simple and satisfactory agreement.

For a resolution to the situation the Commission turned to consideration of another Italian proposal that had been made at the turn of the 20th century. In this proposal the mechanical, electrical and magnetic units were to be given a new start in a system in which the constant, $\mu_0$, was given a new dimension.

**THE MKSA SYSTEM** – On 13 October 1901, at a meeting of the Italian Electrical Engineering Association in Rome, a proposal was made for a new system of absolute measurement to be simultaneously applicable to all electrical, magnetic and mechanical units. Author of the proposal was Giovanni Giorgi (1871-1950), a noted professor of physics and mathematics at the University of Rome. Giorgi, as had others before him, took issue with the CGS system as ill-adapted to the connection between electrical and magnetic phenomena, and its relation to the practical units.

In a paper entitled: *Rational Units of Electromagnetism,* which was the cornerstone of his subsequent work, Giorgi described a system that would deal with the mechanical quantities, and also the primary electrical and magnetic quantities, such that they would have a one to one correspondence with the practical units in which they were measured: volts, amperes, ohm, farads, etc.

Commenting on his proposal, Giorgi wrote:

"It occurred to the writer that the solution of all difficulties . . . could at once be obtained if the whole of the electrotechnical practical units were taken together with the *metre* as a unit of length, the *kilogram* as unit of mass and the *second* as a unit of time, so that by adding one other arbitrary unit, a complete system of units of

Government Delegates, International Electrical Congress.

ST. LOUIS, 1904.

1. Herr W. Lützrodt, *Germany*.
2. Prof. Dr. S. Arrhenius, *Denmark and Sweden*.
3. Dr. R. T. Glazebrook, *Great Britain*.
4. Prof. Elihu Thomson, *United States*.
5. Prof. Moise Ascoli, *Italy*.
6. M. Guillebot de Nerville, *France*.
7. Señor Antonio Gonzalez, *Spain*.
8. Prof. H. J. Ryan, *United States*.
9. Ing. A. Maffezzini, *Italy*.
10. Dr. F. A. Wolff, Jr., *U. S. Bureau of Standards*.
11. Herr Bela Gati, *Hungary*.
12. M. Dennery, *France*.
13. Ormond Higman, Esq., *Canada*.
14. Dr. A. E. Kennelly, *United States*.
15. Señor M. Otamendi, *Spain*.
16. John Hesketh, Esq., *Australia*.
17. Capt. Ferrié, *France*.
18. Col. R. E. B. Crompton, *Great Britain*.
19. Prof. Jorge Newbery, *Argentine Republic*
20. Prof. L. Lombardi, *Italy*.
21. Marquis Luigi Solari, *Italy*.
22. Dr. S. W. Stratton, *United States*.
23. Prof. H. S. Carhart, *United States*.
24. Prof. John Perry, *Great Britain*.
25. J. C. Shields, Esq., *India*.

## GIORGI PRESENTED HIS SYSTEM TO THIS CONGRESS

25 world-famous electrical engineers, from 13 nations, as members of the 6th International Electrical Congress, met in St. Louis in 1904, where Giorgi read a paper on his rational system of units. This Congress also took the first steps in proposing an organization whose principle function would be to "facilitate the coordination and unification of national electrotechnical standards." As a result of this St. Louis proposal, the International Electrotechnical Commission was founded in 1906, with headquarters at Geneva, Switzerland. Originating with 13 member nations, the Commission now represents 43 member countries encompassing practically all the world's production and use of electrical energy.

The International Electrotechnical Commission was an outgrowth of the work in promoting standardization over the years by the various International Electrical Congresses. These pioneering assemblies of outstanding engineers and scientists first met in Paris in 1881, and in later years in Frankfort, Turin, Chicago and again in Paris. The St. Louis Congress, in 1904, was the sixth. By that time a critical need had developed for an organization permanently located that would coordinate the many national standards that by their diversity were hampering international cooperation and commerce.

The International Electrotechnical Commission grew with the enormous expansion of the electrical industry. It is now the world's foremost body for the production and coordination of international electrical standards.

absolute character was built up from four fundamentals.

"The whole system constitutes thus a useful extension of the set of practical units, of such character that it is equally fit for electrotechnical, for scientific and for common use of everyday life.

"It is not to be forgotten that the standards of measurement in all laboratories (even in the laboratories of pure physics) are all rated in ohms, volts, farads, etc., so that all units of every system require to be referred to the standards of practical units.

"It is not proposed to discard the existing systems of units, CGS electrostatic, CGS electromagnetic and other systems in scientific use. Each one will be employed according to the requirements of the subject and the preference of the user.

"The suggestion is to extend the group of practical electrotechnical units already in use in order to have them fitted into a coherent MKS system as described, capable of standing as an absolute system by itself.

"This will result in a great simplification of all practical calculations and of the learning of electrical theory in the schools. A great deal of waste of time and intellectual fatigue will be saved.

"Future practice will show which units are the most convenient for every particular purpose and the law of the 'survival of the fittest' will receive application."

In addition to various articles published in Italy in 1901 and 1902, Giorgi presented a paper on rationalizing the units of electricity to the Physical Society of London in 1902, and submitted another paper on his system to the International Electrical Congress meeting in St. Louis in 1904, before the group shown on the opposite page.

**GIORGI** maintained that a system using as its basic unit of energy, the *erg,* was poorly adapted to the current trends in science involving the connection between electricity and magnetism. The *joule,* the practical unit of work or energy was, Giorgi proposed, a better unit for this purpose. The joule, measured in the CGS system, is equal to $10^7$ grams · centimeters$^2$/seconds$^2$, or $10^7$ ergs.

Giorgi showed that the CGS joule was in fact equal to one kilogram · meter$^2$/second$^2$, or one MKS unit of work. Thus the joule could be directly related to its electrical equivalent of work, the

GIOVANNI GIORGI was born in Lucca, Italy, on 27 November 1871. After graduation in civil engineering from the Institute of Technology, Rome, he specialized in municipal engineering, involving steam and hydro generated electric power, in which he pioneered several concepts for more effective utilization and efficiency. From 1906 to 1923 he was director of the Technology Office for the City of Rome. In 1913 Giorgi joined the faculty of the University of Rome, and in 1939 became professor of electrical communications at the Royal Institute of Higher Mathematics.

Giorgi took an early interest in the systematization of electric units, and his chief fame rests on his proposal of a new absolute system of measurement applicable to electrical, magnetic and mechanical units. First presented in 1901, the Giorgi system attracted the support of eminent engineers in Europe and in America. But it was not until 1935 that the International Electrotechnical Commission meeting in Scheveningen, Netherlands, recommended adoption of the Giorgi system to supersede the CGS system. Giorgi's work is a part of the "International System of Units" (SI), which was confirmed in 1960 by the General Conference of Weights and Measures, and is now in general use, based on the meter, kilogram and second as well as the ampere, kelvin, mole and candela. The ampere was selected as the fourth unit of Giorgi's system in 1950, the year of his death.

volt-ampere-second, and to the unit of power, the watt, equal to doing work at the rate of one joule/second. The joule, composed of the basic mechanical units: $kg \cdot m^2/s^2$, was the basic energy unit of Giorgi's system. It was, without conversion factors, of the right size for adaptation to both mechanical and electrical systems.

However, simply replacing the centimeter and the gram by the meter and the kilogram in the absolute definition of the electric units did not give practical units in electricity and magnetism.

Giorgi showed that this difficulty could be solved by increasing the three mechanical units of the MKS system, the kilogram, meter and second, with a fourth unit of an electrical nature.

Giorgi wrote:

" . . . in the new system, a fourth fundamental dimension is introduced; this may be any one of the electric or magnetic magnitudes, for instance, the quantity of electricity ($Q$); then to write down the dimension of all the derived units becomes quite an easy matter.

"The fact that one unit force operating through one metre does an amount of work equal to one joule (= $10^7$ ergs) makes a very convenient link with the established electrical units and enables them to be incorporated as they stand, into an "absolute system." We thus get a system which is absolute in its scientific basis and practical in the size of the units employed.

"When we have one coulomb of electricity at a pressure of one volt, we have a store of energy equal to one joule. Whether the energy has been obtained by electrical work (a quantity of $Q$ coulombs being displaced through a difference of potential of $V$ volts) or has been obtained by doing mechanical work, it possesses the same dimensions."

Giorgi later suggested that the ohm or the ampere, rather than charge, might be used for the fourth unit.

**GIORGI'S PROPOSAL** was a major one, and would take several decades of consideration and discussion to gain acceptance. The system had the advantage of providing freedom from the use of a large number of decimal exponents found in the CGS system, and offered a single system applicable to all branches of science while retaining the practical units which were becoming firmly entrenched. During the years following Giorgi's proposal it received increasingly favorable response and support from eminent scientists and engineers, particularly in the electrical field.

Finally, in 1935, in its plenary session in Scheveningen, Netherlands, and in Brussels, the International Electrotechnical Commission voted unanimously "that the system with four fundamental units, comprising the three units: metre, kilogramme and second, and a fourth fundamental unit, to be chosen later, be adopted under the name of the Giorgi system." The MKS system was intended to remain in use simultaneously with the CGS system, in accordance with the suggestion advanced by Giorgi in 1901.

The 1935 Congress also confirmed three new units: *hertz* for the unit of frequency, *siemens* for the unit of electrical conductance, and the *weber* as the unit for magnetic flux, equal to $10^8$ maxwells.

Discussion of the Giorgi system continued at the 1938 Torquay, England, meeting of the Technical Committee on Electric and Magnetic Magnitudes and Units. The MKS unit of force, for which Giorgi had suggested the term "vis," was named the *newton,* equal to $10^5$ dynes.

**THE FOURTH BASE UNIT** - The 1935 adoption of the Giorgi system by the IEC meeting at Scheveningen, had left two important issues over for decision at some future date: (1) the selection of the fourth base unit of an electrical nature, (2) the question of "rationalization" of the units.

The question of rationalization was a legacy of the Gaussian system of units in which the factor $4\pi$ was omitted from the Coulomb force equation for electric charges and the Ampère force equation for current elements. As a result, in the traditional unrationalized form, $4\pi$ appeared in many MKS units as an added manipulative factor. By rationalization, $4\pi$ disappears from the MKS units, and some basic formulas. Eliminating $4\pi$ provided certain manipulative advantages that led, in future IEC meetings, to a continued lively interest in the eventual use of the rationalized system.

The 1935 Committee, in considering the fourth unit, recognized that any of the practical units:

ohm, ampere, volt, henry, farad, weber or coulomb which were already in use, could serve equally well as the base unit, because it was possible to derive each unit and its dimensions from any four others mutually independent. Giorgi had suggested that the logical fourth unit should be the coulomb. The Committee did not want to make an arbitrary choice, so the matter of the fourth unit was put into the hands of the International Committee of Weights and Measures for consultation and a future decision.

**THE DEFINING CONSTANT** - To provide a connecting link between mechanical and electrical units in the Giorgi system, the dimensional value of the permeability of free space could no longer be ignored, by being made unity, as was the case in the Gaussian system. A unit-defining constant of proportionality had to be introduced into the Giorgi system, based on the force law for electromagnetic interaction. An Advisory Committee of the International Electrotechnical Commission, meeting in Torquay, England, in June 1938, recommended that the MKS system should be complemented with a permeability space constant, $\mu_0$, having a value of $10^{-7}$ newtons per ampere squared, the newton being the new MKS unit of force.

The Committee stated that: "The MKS system can be made absolute by the assumption of any convenient value for $\mu_0$ (the measure of the permeability of free space); but, if the units of that system are to be the practical units of the CGS system, the value of $\mu_0$ must be $10^7$ times the CGS constant value (unity) in the CGS system.

The constant $\mu_0 = 10^{-7}$ used in defining the MKS system of units was based on consideration of the mechanical-electrical relationship in the force law between parallel current-carrying conductors. In the CGS electromagnetic system this force relationship can be written:

$$F = 2\frac{I_1 I_2 l}{d} K$$

where $F$ is the force in dynes, $I_1$ and $I_2$ are the parallel currents in abamperes (= 10 amperes) and $l$ and $d$ are the length and separation of the conductors in centimeters. $K$, the constant of permeability for free space, is unity in the CGS system.

In the MKS system, however, if $F$ is to be in newtons (= $10^5$ dynes) and $I_1$ and $I_2$ in amperes (= 0.1 abamperes), the expression above becomes:

$$F = 2\frac{I_1 I_2 l}{10^5\ 10^2\ d}, \text{ or } F = 2\frac{I_1 I_2 l}{d} \times 10^{-7}$$

When $I_1$ and $I_2$, $l$ and $d$, are all made equal to unity, and with $\mu_0 = 10^{-7}$, the force equation then becomes:

$$F = 2 \times 10^{-7}, \text{ or, } 2 \times \mu_0$$

The factor, $\mu_0$ , expresses the ratio of the magnetic flux to the magnetic field intensity per unit length, in free space, or:

$$\mu_0 = 10^{-7} \frac{\text{weber/meter}^2}{\text{ampere turn/meter}}$$

**A FINAL DECISION** on the fourth basic unit was reached at the 1950 Paris meeting of the Committee of Electric and Magnetic Magnitudes and Units. Their selection was the *ampere,* and the permeability constant value, $10^{-7}$, entered into its definition:

**Ampere** - The constant current which, when maintained in two parallel conductors of infinite length, of negligible cross-section, and separated by a distance of one meter in vacuo, would produce between those conductors a force equal to $2 \times 10^{-7}$ MKS units of force (newtons) per meter of length.

Other principal electric and magnetic units of the MKS, or Giorgi system, as a result of the 1950 Paris Committee decisions, became:

**Coulomb** - The quantity of electricity of one ampere-second.

**Volt** - The electromotive force or difference of potential equal to the joule/second or the watt/ampere.

**Volt per Meter** - The dimensions of the electric field strength.

**Ohm** - The electrical resistance defined as the relationship between voltage and current.

**Siemens** - The electrical conductance defined as the relationship between current and voltage.

**Henry** - The inductance of an electric circuit defined as the ratio between an electromotive impulsion and a current.

**Farad** - The capacitance of a circuit defined as the ratio between the quantity of electricity and voltage.

**Ampere per Square Meter** - Defined as the ratio between current and area.

**Weber** - The magnetic field product of a voltage and a time.

**Weber per Square Meter** - Defined as the magnetic flux per unit area. It was proposed in 1954 and confirmed in 1960 that this measure of magnetic flux density be named the **Tesla.**

**RATIONALIZATION** - In the formation of the classical system of CGS units in 1873, the simplest possible equations for the inverse square laws were used in establishing the basic esu and emu units. This was achieved by making the proportionality constant between the force and the point charges or between the force and the point current elements in the fundamental defining equations equal to 1. Thus, in the Coulomb equation:

$$F = K(Q_1 Q_2 / r^2)$$

where $K$ equals 1. The convenience of using a proportionality of 1, however, ignored the essential spherical geometry of the point sources whose emission area for unit radius is $4\pi$. Equating this to 1 was in effect like using a unit of length such as to make the surface area of a sphere one unit of length squared.

As a result of this suppression, the factor $4\pi$, or other multiple of $\pi$, then had to make its appearance in the frequently used equations that involved plane geometry, such as the capacitance of a parallel plate capacitor, while it did not appear in the expression for a cylindrical capacitor. Also, in the much used circuital law of magnetic windings, the magnetomotive force appeared in the form of $0.4\pi$ ampere-turns. In many ways $4\pi$ was looked upon as an unnecessary intruder.

**OLIVER HEAVISIDE** (1850-1925), the noted British engineer who made many scientific contributions to transmission line and electromagnetic theory, was one of the first to use the term. Rationalization was the name he gave to the process of transferring the factor $4\pi$ from one position to another in the basic equations of electrostatics and electromagnetism, for definite reasons.

Heaviside considered the CGS system of units as too irrational and complicated for practical use. He proposed, in 1891, two radical changes in the CGS system.

(1) adoption of a rational system which would cause $4\pi$ to appear in expressions relating to circular and spherical systems, and to be absent from expressions relating to rectilinear systems.

(2) use of a single multiplier, that is one power of 10 only, such as $10^8$, in passing from the CGS units to the practical units, instead of the various powers of 10 currently required for conversion.

Heaviside campaigned vigorously for his proposals in the journal *The Electrician,* exhibiting a foresight at a time when the advantages of rationalization were by no means apparent to those in electrical practice.

The principle of rationalization was received with favor by some far-seeing scientists and engineers. However, Heaviside's system retained unity for the value of the permeability constant, $\mu_0$, and for the permittivity constant, $\varepsilon_0$. As a result, Heaviside's system required changing existing equations by factors using $\sqrt{4\pi}$ and $1/\sqrt{4\pi}$. The units also made use of the single 10s multiplier. His system would therefore have forced changes in the value of the volt and ampere, and require recalibration of meters. These changes were considered unacceptable to practicing engineers.

For theoretical work, however, Hendrik A. Lorentz (1853-1928), professor at the University of Leyden, and famous for his theories on the electron and on relativity, adopted Heaviside's rationalizing principle in his publications. This led to a system called the Heaviside-Lorentz units which afforded the first use of the rationalization principle. Their systems did not, however, use the MKS units, and did not participate in the ultimate change brought about by rationalization.

**GIORGI**, when he made his original presentation for a new system of units, stated that: "it (the MKS system) permits the use of either the "rationalized" or "unrationalized" form for the desired units and formulae."

When the IEC Committee of 1935 adopted the Giorgi system, there was so much difference of opinion between those who had carried over from an unrationalized CGS form to the unrationalized MKSA units, and those who had logical reasons for preferring the rationalized form, that a Committee decision on which form should prevail was deferred.

With the interference of the World War, it was not until 1950 that further action was taken. At a meeting of the Advisory Committee of Electric and Magnetic Magnitudes and Units, in Paris, a major step was made in settling the question by a recommendation for total rationalization of the MKSA system. A committee of experts was then appointed to interpret "rationalization." The work of the Committee resulted, in effect, in a reassessment of the case for adoption in the rationalized form.

The Giorgi system of units was the single consistent absolute system that recognized the interdependence of the electric and the magnetic constants. Further, these constants were given values, based on the relationship $1/\varepsilon_0\mu_0 = c^2$, that made the system equivalent over a wide range, to the existing system or practical units.

In the unrationalized MKSA form:

Permeability Constant, $\mu_0$, was assigned the value: $10^{-7}$ henry/meter.

Permittivity Constant, $\varepsilon_0$, then had the value: $1/(9 \times 10^9)$ farad/meter.

In the rationalized MKSA form:

Permeability Constant, $\mu_0$, was assigned the value: $4\pi \times 10^{-7}$ henry/meter.

Permittivity Constant, $\varepsilon_0$, then had the value: $1/(36\pi \times 10^9)$ farad/meter.

In assigning the rationalized alternative, Giorgi had followed Heaviside in substance, but the values in the Giorgi system were such as to permit the ampere and the coulomb, and hence the other units, to remain undisturbed. Thus, the units of potential difference, $V$, current, $I$, electric field strength, $E$, magnetic flux, $B$, resistance, $R$, capacitance, $C$, self inductance, $L$, and almost all the other commonly used units were unchanged by rationalization. Only the MKSA provided this.

To accomplish rationalization the factor $4\pi$ had to be introduced into the fundamental Coulomb force equation for electric charges $Q_1$ and $Q_2$, and into the Ampère force equation for parallel currents, $I_1$ and $I_2$, of elemental lengths, $dl_1$ and $dl_2$, as shown below:

| | Unrationalized | Rationalized |
|---|---|---|
| Coulomb's Law | $F = \frac{1}{\varepsilon_0}\frac{Q_1 Q_2}{r^2}$ | $= \frac{1}{\varepsilon_0}\frac{Q_1 Q_2}{4\pi r^2}$ |
| Ampère's Law | $F = \mu_0 \frac{I_1 I_2\, dl_{1\,2}}{r^2}$ | $= \mu_0 \frac{I_1 I_2\, dl_{1\,2}}{4\pi r^2}$ |

Thus the basic force laws became more complicated. However, the new expression derived from the new forms of the inverse square laws became simpler. A large number of the quantities used in practical engineering were relieved of the $4\pi$, as shown in the Table on the next page. The numerical values calculated by using either unrationalized or rationalized expresseions were unchanged, since in changing the form of expression by rationalization the values of $\varepsilon_0$ and $\mu_0$ were also changed.

**THE CHOICE** for the MKSA system in either the unrationalized or rationalized form was that form which was most practical for the intended use. The unrationalized quantities had no particular advantages, but were an historical outgrowth of the form in which Maxwell had put his original equations of electrodynamics. The rationalized quantities and defining equations, on the other hand, provided simplicity in rectilinear geometry, and an appropriateness to the field-concept aspect of electromagnetic theory which was becoming more important as electrical energy was being increasingly directed by waves as well as by wires. Rationalization was in step with progress in the utilization of electric waves in free space.

In the rationalized system, the magnetic field strength, $H$, was measured in amperes/meter. By dividing the electric field strength measured in volts/meter, by the magnetic field strength, impedance measured in ohms was obtained. Meas-

COMPARISON OF QUANTITY VALUES IN UNRATIONALIZED AND RATIONALIZED MKSA UNITS

| QUANTITY | QUANTITY VALUE | |
|---|---|---|
| | In Unrationalized MKSA Units | In Rationalized MKSA Units |
| Permeability Constant | $10^{-7}$ henry/meter | $4\pi$ x $10^{-7}$ henry/meter |
| Permittivity Constant | 1/(9 x $10^{9}$) farad/meter | 1/($36\pi$ x $10^{9}$) farad/meter |
| Magnetomotive Force | $4\pi$ ampere-turn | ampere-turn |
| Magnetizing Force | $4\pi$ ampere-turn/meter | ampere-turn/meter |
| Magnetic Flux | weber | weber |
| Magnetic Flux Density | tesla (weber/meter$^{2}$) | tesla (weber/meter$^{2}$) |
| Reluctance | $4\pi$ ampere-turn/weber | ampere-turn/weber |
| Electric Charge | coulomb | coulomb |
| Electric Flux | $4\pi$ coulomb | coulomb |
| Electric Field Strength | volt/meter | volt/meter |
| Electric Flux Density | $4\pi$ coulomb/meter$^{2}$ | coulomb/meter$^{2}$ |

ured in the unrationalized system the answers were not in ohms.

With the development of microwaves and wave guides, beginning in the World War period, the concept of impedances had the same importance in the theory of wave propagation that ordinary circuit impedances had in the theory of transmission lines. It was necessary that the electromagnetic formulas for energy and field density have the same form to the corresponding circuit formulas. This was provided in the rationalizaed MKSA system.

After the World War, action on the adoption of the rationalized MKSA system picked up. The following editorial by S.A. Schelkunoff, and engineer with the Bell Telephone Laboratories, is reprinted, with permission, from Proceedings of the Institute of Radio Engineers, July 1948, copyright 1948 IRE (Now IEEE):

"**The end is in sight** for the age of diverse scientific units - or cgs electromagnetic units, of mixed electrostatic and electromagnetic units, of rationalized and unrationalized varieties of each. At a meeting of the I.R.E. Technical Committee on Wave Propagation held on March 4, 1947, a resolution was adopted to the effect that the rationalized mks system of units be recommended by The Institute of Radio Engineers as the preferred system of units. This action was prompted by the rapid and unmistakable trend toward universal adoption of this system, both in experimental and theoretical investigations, and by a desire to shorten the transition period.

"Although the rationalized mks system of units was first suggested in the middle of the last century, it remained almost unknown until about 15 years ago. In the last 15 years, however, it had made astonishingly rapid conquests. The reasons for this are many. In the mks system the electrical units are those already in common use in laboratory measurements: the volt, the ampere, the ohm, etc. If the system is of the rationalized variety, Maxwell's equations assume a form which is merely a generalization of the one-dimensional transmission-line equations.

"In recent years, the gap between circuit and transmission-line theories on the one hand and field theory on the other hand has been closed by waveguides and microwave circuits in general. This made it essential that there be no clash between the ideas, terminology, and units employed in these theories. Since the rationalized mks system fulfills this requirement, it is only natural that it should enjoy rapidly increasingly popularity. An added factor in this popularity is that the rationalized mks system is equally well adapted to electromechanical theories. It has begun to appeal to many physicists as well as to engineers - which is particularly fortunate, since the engineer of today must be somewhat of a physicist, and the physicist somewhat of an engineer."

**THREE MILESTONES** in the progress of the science of electrical measurement were marked by the year 1948.

On January 1, 1948 the absolute electrical units were made officially effective. This was the outcome of the establishment of national laboratories in Europe, America and Australia in which the development of improved apparatus and of new methods made precise measurements of the absolute type easier to perform. The availability of precision measurements brought on a movement for conversion of the ohm and other practical units to the absolute system. Action was taken in 1946 by the International Committee of Weights and Measures to set values for the new units. Adoption of the absolute units in 1948 marked the abandonment of the "International Electrical Units" which had been set up and remained in use since 1908.

On January 8, 1948 the Standards Committee of the Institute of Radio Engineers ratified a resolution that the Institute would henceforth use only the rationalized MKSA units. By this decision the trend toward universal use of the rationalized MKSA units became compelling.

Final approval of the MKSA rationalized system was made at the 1954 Philadephia meeting of the International Electrotechnical Commission. This concluded a more than half century interval of electrical measurement history since Giorgi's proposal in 1901. The use of MKSA units has now become widespread, with gradual abandonment of the CGS esu and emu unit systems.

The same post World War II momentum that resulted in approval of the absolute units and of rationalization, carried over into an international effort to develop a generally acceptable "international practical system of units" for international communication. The aim was to promote an international collaboration and agreement that would advance better understanding and effectiveness in science and technology, and in industrial trade relations, by a single, practical worldwide system of units, and be an essential link between peoples in a world of growing complexity.

In 1948 the 9th General Conference on Weights and Measures proceeded: "to study the establishment of a complete set of rules for units of measurement . . . to find out for this purpose, by official inquiry, the opinion prevailing in scientific technical and in educational circles in all countries . . . and to make recommendations on the establishment of a practical system of units of measurement suitable for adoption by all the signatories to the Meter Convention.

**THE INTERNATIONAL SYSTEM OF UNITS (SI)**

As a result of the information obtained from the official inquiry, the 11th General Conference, in 1960, gave the name "International System of Units" to a choice of seven well-defined units which by convention are regarded as dimensionally independent: the **meter**, the **kilogram**, the **second**, the **ampere**, the **kelvin**, the **mole** and the **candela.** These are the *base* units of "The International System of Units," for which the generally adopted abbreviation is "SI."

In addition to the base units, a second class of SI units contains *derived* units, formed by combining base units according to the algebraic relations linking the corresponding quantities, such as **volt, farad, ohm, weber,** etc.

A third class of SI units are called *supplementary* units, which may be regarded as either base units or derived units, such as **radian** and **steradian.**

The three classes form a coherent system - not requiring useless coefficients in going from one to the other.

The seven base SI units are as follows:

**Unit of Length** - The meter is the length equal to 1 650 763.73 wavelengths in vacuum of the radiation corresponding to the transition between the levels $2p_{10}$ and $5d_5$ of the krypton-86 atom.

**Unit of Mass** - The kilogram is the unit of mass (and not of weight or force); it is equal to the mass of the international prototype of the kilogram.

**Unit of Time** - The second is the duration of 9 192 631 770 periods of the radiation corresponding to the transition between the two hyperfine levels of the ground state of the cesium-133 atom.

**Unit of Electric Current** - The ampere is that

constant current which, if maintained in two straight parallel conductors of infinite length, of negligible circular cross-section, and placed 1 meter apart in vacuum, would produce between these conductors a force equal to $2 \times 10^{-7}$ newtons per meter of length.

**Unit of Thermodynamic Temperature** - The kelvin, unit of thermodynamic temperature, is the fraction 1/273.16 of the thermodynamic temperature of the triple point of water.

**Unit of Amount of Substance** - The mole is the amount of substance of a system which contains as many elementary entities as there are atoms in 0.012 kilogram of carbon 12.

**Unit of Luminous Intensity** - The candela is the luminous intensity, in the perpendicular direction, of a surface of 1/600 000 square meter of a blackbody at the temperature of freezing platinum wire under a pressure of 101 325 newtons per square meter.

## TWO MATERIAL ELECTRICAL STANDARDS

*"Although in the International System of Units (SI) the electric base unit is the ampere, the two units whose material representations have always served jointly as the departure point of electrical measurements in all laboratories are the unit of resistance and electromotive force; the ohm and the volt are in fact the units which we know how to realize and maintain with utmost accuracy."*
- from C.H. Page and P. Vigoureux, *The International Bureau of Weights and Measures 1875-1975.* National Bureau of Standards Special Publication 420, Washington, D.C.

**Voltage Reference Standard** - Cells of the Weston type, opposite page, are the internationally accepted voltage reference standard. They contain an acidified, saturated solution of cadmium sulfate, and are made of materials purified with extreme care. The electromotive force of the Weston cell at 20° C is of the order of 1.0184 volts.

Each national laboratory maintains a group of cells which are periodically checked for accuracy and stability, and renewed when necessary. Since Weston cells are extremely temperature sensitive, they are stored, for measurement purposes, in an oil bath or other enclosure in which the critical temperature of 20° C can be maintained with high accuracy.

**Resistance Reference Standards** -The best quality resistance standards are those of nominal one ohm value maintained in the national laboratories, where they are peridocally checked with modern apparatus for determining resistance in absolute units.

These resistors are wire-wound type, using an alloy such as manganin which has the least possible sensitivity to variation of the ambient temperature, and to the heating produced by the measuring current. The wire, of 1.2 to 1.5 mm in diameter, is wound in the form of a double helix to minimize self-inductance, and is annealed to assure the stability of the resistance value. The wire assembly is mounted in a hermetically sealed container filled with pure mineral oil or inert gas.

**TO ASSIGN VALUES** to the resistance standards and standard cells which constitute their reference groups, the principal national laboratories carry out absolute determinations. These are electromechanical experiments which relate the electric units to the units of length, mass and time in conformity with the definition of the ampere. These are difficult measurements, long and costly, which must, it is well understood, be repeated periodically; their precision is, for most of the electric units, inferior to the stability of physical standards which represent these units, and to the comparison measurements between these standards. For these reasons, and for reasons of continuity, the national laboratories do not modify the values of their reference standards each time they carry out absolute determinations. They do this only exceptionally, practically in the three following cases:

-When the difference between the result of the new determination and the value previously assigned exceeds the uncertainty of the measurements, and when some particular reasons can explain an abnormal change in the material standards (bad storage conditions, for example).
-When the progress of metrology permits improving the precision of the absolute determinations
-To improve international uniformity.

# THE TWO ELECTRICAL MATERIAL STANDARDS

## VOLTAGE REFERENCE STANDARD

The Cadmium Sulfate Standard Cell, developed by Dr. Edward Weston (1850-1936) in 1893. The electromotive force at 20° C is of the order of 1.01860 V

NEGATIVE LIMB

A- Cadmium Amalgam in 2 Phases

POSITIVE LIMB

B - Pure Mercury

C - Mercurous Sulfate $Hg_2SO_4$

BOTH LIMBS

D - Crystals of $CdSO_4$ + 8/3 $H_2O$

E - Saturated Solution of Cadmium Sulfate

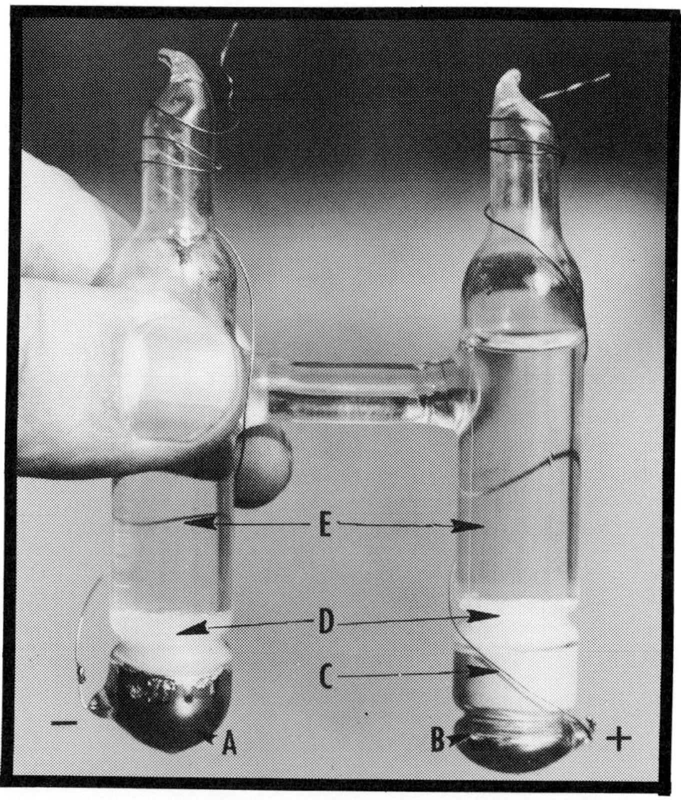


## RESISTANCE REFERENCE STANDARDS

Precision 1 ohm standards developed by various national laboratories.

The resistance standards are constituted of a metallic wire of an alloy having a low temperature coefficient, for example, *manganin,* (85% copper, 12% manganese, 3% nickel)

Illustrations courtesy International Bureau of Weights and Measures, Sevres, France.

## DETERMINATION OF THE ABSOLUTE AMPERE

### By the Current Balance and Electrodynamometer Methods

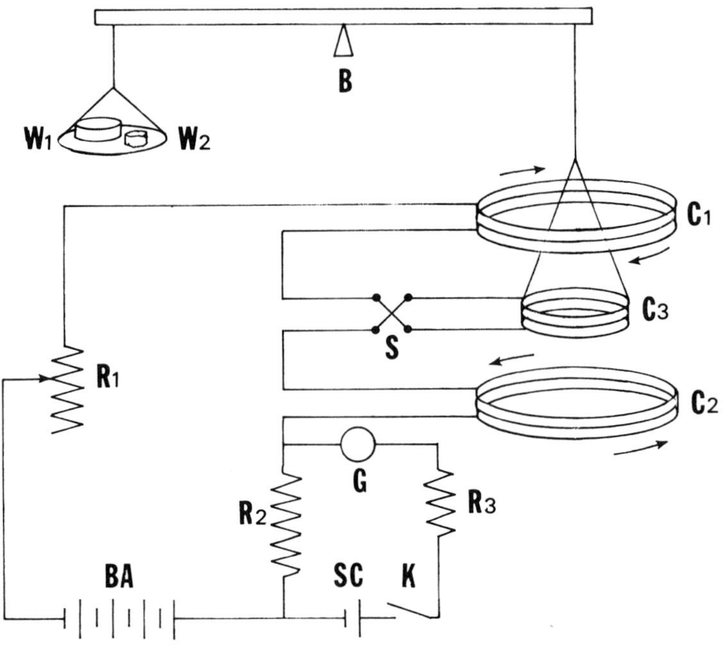

Illustrations courtesy the National Bureau of Standards

*The AMPERE is the constant current which if maintained in two parallel conductors of infinite length of negligible circular section, and placed 1 meter apart in a vacuum, will produce between these conductors a force equal to 2 x $10^{-7}$ mks units of force per meter of length.*

This definition, while simple and precise, is not appropriate for accurate realization of the ampere, because the mechanical force involved is too small. The ampere is therefore measured by other, more complicated arrangements.

The CURRENT BALANCE METHOD, one form of which is shown schematically, is equivalent to the principle of two parallel wires.

The current from the battery, BA, adjusted by rheostat, R1, flows through fixed coils C1 and C2 in opposite directions, so that their magnetic fields are in opposite direction in the region of the movable coil, C3. By reversing switch, S, the magnetic force on C3 can be either upward or downward. Force in the upward direction is balanced by weight, W1. Force in the downward direction is balanced by addition of weight, W2, and current, *i*, through the coil is:

$$2\,Ci^2 = mg$$

where, *C*, is a physical constant of the coil assembly, *m* the mass of the balancing weight, and *g* is the acceleration of gravity.

The current circulating in the coils is passed through a precision one ohm resistor, R2, and balanced at galvanometer, G, by a standard cell, Sc. On balance, the voltage drop, *ir*, through this circuit is equal to the voltage, *v*, of the standard cell. A current of 1 ampere may be established by a current of such strength that through a 1 ohm resistance the voltage drop is exactly 1 volt.

ELECTRODYNAMOMETER METHOD - This form of current balance is used for the absolute determination and corroboration of the ampere at the National Bureau of Standards, Washington, D.C. The principle is similar to that of the method above. In this apparatus a relatively small rotatable coil, with its axis vertical, is placed concentric with a long horizontal coil. A lever arm attached to the movable coil can be loaded to achieve a balance of the magnetic forces when the current in the fixed coil is reversed.

**COMPARISONS OF THE STANDARDS** - The two basic physical standards, the *volt* and the *ohm,* are maintained by the International Bureau of Weights and Measures, which was established by the Convention of the Meter in 1875, and is located at the Pavillon de Breteuil, near Paris, France, the birthplace of the metric system.

At periodic intervals the volt and ohm standards kept and calibrated at the various world national laboratories, are sent to France for comparison. These comparisons serve to evaluate the effective precision obtained by measurements of the absolute type carried out by the principal national laboratories.

*The study and maintenance of the volt and ohm standards is one of the basic activities of the International Bureau of Weights and Measures. The Bureau, by keeping its standards permanently to represent the volt and ohm as accurately as possible, has assured the uniformity of electrical measurement around the world.*

HEADQUARTERS of the International Bureau of Weights and Measures is located near Paris on the grounds of the Pavillon de Breteuil in the Parc de Saint-Cloud. The site of the headquarters was given by the French Government in 1876, and its upkeep is financed by Member States of the Metre Convention.

# INDEX OF NAMES

# REFERENCES


AMPÈRE, A.M. - *Sur l'état magnetique des corps qui transmettant un courant d'électricité*, Ann. Chem. Phys. XVI, Paris, 1821

BELL, E.T. -*Men of Mathematics,* Simon & Schuster, New York, 1965

BERKSON, W. - *Fields of Force,* Wiley, New York, 1974

BERMAN, M. - *Social Change and Scientific Organization - The Royal Institution 1799-1844* Cornell University Press, 1978

BERNAL, J.D. - *Science in History,* MIT Press, Cambridge, Mass. 1971

BOAS, M. - *The Scientific Renaissance,* Harper, New York, 1965

BRITISH ASSOCIATION for the ADVANCEMENT of SCIENCE - *Reports of the Committee on Electrical Standards,* Cambridge University Press, 1913

BRITISH STANDARDS INSTITUTION - *The M.K.S. System of Electrical and Magnetic Units,* British Standards 1637, 1950

BUCKLEY, H. - *A Short History of Physics,* Metheun, London, 1927

CAJORI, F. - *A History of Physics,* Dover, New York, 1962

CAMPBELL, L. and GARNETT, W. - *The Life of James Clerk Maxwell,* Macmillan, London, 1882

CAVENDISH, H. - *The Electrical Researches of the Honourable Henry Cavendish,* edited by J. Clerk Maxwell, 1879, Reprint, Frank Cass & Son, Ltd., London, 1967

CHIPMAN, R.A. - *The Earliest Electromagnetic Instruments,* U.S. National Museum Bulletin 240, pp. 121-136, Smithsonian Institution, Washington, D.C., 1964

COHEN, I.B. - *Franklin and Newton,* Harvard University Press, Cambridge, Mass., 1966

COULSON, T. - *Joseph Henry, His Life and Work,* Princeton University Press, Princeton, N.J. 1950

CROMBIE, A.C. - *Helmholtz,* Scientific American, Vol. 198, No. 3, Mar. 1958

CROWTHER, J.G. - *Fifty Years with Science,* Barrie & Jenkins, London, 1970

---- *British Scientists of the 19th Century,* Kegan Paul, London, 1935

DAMPIER, W.C. - *A History of Science,* Cambridge University Press, 1966

DIBNER, B. - *Galvani-Volta,* Burndy Library, Norwalk, Conn., 1952

---- *Oersted and the Discovery of Electromagnetism,* Burndy Library, Norwalk, Conn., 1969

---- *Benjamin Franklin, Electrician,* Burndy Library, Norwalk, Conn., 1961

---- *Early Electrical Machines,* Burndy Library, Norwalk, Conn., 1957

---- *Heralds of Science,* MIT Press, Cambridge, Mass., 1969

DICKINSON, H.W. - *James Watt,* Cambridge University Press, 1936

DICTIONARY of SCIENTIFIC BIOGRAPHY - Vols. I-XVI, Charles Scribner's Sons, New York

DUNNINGTON, H.W. - *Carl F. Gauss - Titan of Science,* Exposition Press, New York, 1955

DUNSHEATH, P - *A History of Electrical Power Engineering,* MIT Press, Cambridge, Mass., 1962

---- *Giants in Electricity,* Crowell, New York, 1967

ELECTRIC CENTURY, The, 1874-1974 - Reprints from *Electrical World,* 1973-4, McGraw-Hill, New York

ENCYCLOPEDIA BRITANNICA - 9th, 11th, 14th editions

ENCYCLOPEDIC DISTIONARY of PHYSICS - Pergamon Press, New York, 1961

EVERITT, C.W.F. - *James Clerk Maxwell, Physicist and Natural Philosopher,* Scribners, New York, 1975

FARADAY. M. - *Experimental Researches in Electricity,* Taylor, London, 1839-1855
---- *Diary,* Vols. I-VII, Basic Books, New York, 1964
FELICI, N.J. - *Electrostatics - Recent Developments and Some Modern Applications,* Journal of the British Association for the Advancement of Science, June 1966
FEYNMAN, R.P. - *The Feynman Lectures on Physics,* Addison-Wesley, Reading, Mass., 1963
FORGAN, S. - *Studies on Humphry Davy,* Science Reviews, London, 1980

GAUSS, C.F. - *Disquistiones arithmeticae,* Translated by A.A. Clarke, Yale University Press, New Haven, Conn., 1966
---- *Carl Friedrich Gauss Werke,* Vols. I-XII, Royal Scientific Society, Gottingen, Springer, Berlin, 1933
---- *Intensitas vis magneticae terrestris ad measuram absolutan revocata,* Gottingen, 1833
GEDDES, L.A., HUFF, H.E. - *The Discovery of Bioelectricity and Current Electricity,* IEEE Spectrum, Dec. 1971
GILBERD, E. D., Lord PENNEY - *William Gilberd,* Benhamand Co., Colchester, 1968
GILBERT, W. - *De Magnete,* Translated by P. Fleury Mottelay, 1893, Dover, New York, 1958
GILLMORE, C.S. -*Coulomb and the Evolution of Physics and Engineering in the 18th Century,* Princeton University Press, Princeton, N.J., 1971
GLAZEBROOK, R. T. - *The Giorgi (M.K.S.) System of Units,* Nature, Apr. 21, 1934
GRAY, A. - *Absolute Measurements in Electriciy and Magnetism,* Macmillan, London, Dover, N.Y., 1967
GRIEG, J. - *John Hopkinson - Electrical Engineer,* Science Museum, London, 1970
GULLWICK, E.G. - *The Fundamentals of Electromagnetism,* Cambridge University Press, 1966
GUPTA, M.S. - *George Simon Ohm and Ohm's Law,* Trans. IEEE on Education, Vol. E-22, Aug. 1980

HARRIS, F.K. - *Electrical Measurements,* Wiley, New York, 1952
HART, I. - *Makers of Science,* Oxford University Press, 1923
HEAVISIDE, O. - *Electromagnetic Theory,* The Electrician Publishing Co., London, 1893
HERTZ, H. - *Electric Waves,* Translated by D. F. Jones, London, 1893, Reprint Dover, New York, 1962
HULL, T. - *Gauss,* MIT Press, Cambridge, Mass., 1920
HUNT, I., DRAPER, W. - *Lightning in His Hand, the Story of Nikola Tesla,* Sage Books, Denver, Colo., 1964

JAFFE, B. - *Men of Science in America,* Simon & Schuster, New York, 1958
JEFFIMENKO, O.J. - *Electrostatic Motors,* Electret Scientific Company, Star City, W. Va., 1973
JONES, F. - *Life and Letters of Faraday,* London 1870
JOSEPHSON, M. - *Edison,* Mc Graw-Hill, New York, 1959

KENNELLEY, A. E. - *The Modern Age in Relation to Faraday's Discovery of Electromagnetism.* Nature, London, Aug. 29, 1931
---- *Magnetic Circuit Units,* Trans. AIEE, Apr. 1930
KING, R. - *Michael Faraday of the Royal Institution,* Royal Institution, London, 1973
KING, W.J. - *The Electrochemical Cell and the Electromagnet,* U.S. National Museum, Bulletin 228, pp. 231-271, Smithsonian Institution, Washington, D.C. 1962
---- *The Telegraph and the Telephone,* U.S. National Museum Bulletin 228, pp. 273-332, Smithsonian Institution, Washington, D.C. 1962
---- *The Early Arc Light and Generator,* U.S. National Museum Bulletin 228, pp. 333-407, Smithsonian Institution, Washington, D.C. 1962
KONDO, H. - *Michael Faraday* - A Centennial, IEEE Spectrum Aug. 1967
KUHLMANN, J.H. - *Design of Electrical Machinery,* Wiley, New York, 1950

LAMME, B.G. - *Autobiography,* G.P. Putnam's Sons, New York, 1926
LANGENBERG, D.N.,TAYLOR, B.N. - *Precision Measurements and Fundamental Constants,* National Bureau of Standards, Special Publication 343, U.S. Government Printing Office, 1971
LAW, R.J. - *The Steam Engine,* Science Museum, London, 1965
---- *James Watt and the Separate Condenser,* Science Museum, London, 1969
LENARD, P. - *Great Men of Science,* Macmillan, New York, 1934
LINCOLN, E.S. - *A Chronological History of Electrical Developments,* National Electrical Manufacturers Association, New York, 1946
LIPSON, H. - *Great Experiments in Physics,* Oliver & Boyd, Edinburgh, 1965

MacDONALD, D.K. - *Faraday, Maxwell and Kelvin,* Doubleday, New York, 1964
MacKENZIE,A.E.E. - *The Major Advancements in Science,* Cambridge University Press, 1960
MAGIE, W.F. - *A Source Book in Physics,* McGraw-Hill, New York, 1935
MARION, J.V. - *A Universe of Physics,* Wiley, New York, 1970
MASON, J. - *Why parts 'fit'; the role of IEC,* IEEE Spectrum, June 1980
MAXWELL, J.C. - *The Scientific Papers of . . .* Edited by W.A. Niven, Clarendon Press, London Reprint Dover, New York, 1952
---- *A Treatise on Electricity and Magnetism,* 3rd ed., Vols. I and II Edited by W.A. Niven, Clarendon Press, London, 1891, Reprint Dover, New York, 1954
---- *Elementary Treatise on Electricity,* Edited by W. Garnett, Oxford Press, 1888
MEYER, K. - *Prominent Danish Scientists Through the Ages, Hans Christian Oersted,* V. Meisen, Copenhagen
MEYER, H.W. - *A History of Electricity and Magnetism,* MIT Press, Cambridge, Mass.,1972
MOORE, A.D. - *Electrostatics and its Applications,* Wiley, New York, 1973
MORGAN, A.P. - *The Pageant of Electricity,* Appleton-Century, New York, 1939
MORRISON, P. and E. - *Heinrich Hertz,* Scientific American, Vol. 197, No. 6, Dec. 1957
MOTTELAY, P.F. - *Biographical History of Electricity and Magnetism,* McGraw-Hill, New York
MUIRHEAD,J.P. - *The Life of James Watt,* J. Murray, London, 1859

NEWMAN, J.R. - *James Clerk Maxwell,* Scientific American, Vol. 192, No. 6, June 1955

OESCHER, P.H. - *The Smithsonian Institution,* Praeger, New York, 1970
OHM, G.S. - *The Galvanic Circuit Investigated Mathematically,* Translated by William Francis, D. Van Nostrand, New York, 1891
O'NEILL, J. - *Prodigal Genius, the Life of Nikola Tesla,* Ives Washburn, New York, 1970

PAGE, C.H., VIGOUREUX, P. - *The International Bureau of Weights and Measures 1875-1975* National Bureau of Standards Special Publication 420, U.S. Government Printing Office, 1975
PANOFSKY and PHILLIPS - *Classical Electricity and Magnetism,* Addison-Wesley, Reading, Mass., 1962
PRIESTLEY, J. - *History and the Present State of Electricity, with Original Experiments* London, 1767, Johnson Reprint, New York, 1966
PLEDGE, H.T. - *Science Since 1500,* Science Museum, London, 1966
PARASINIS, D.S. - *Magnetism,* Science Today Series, Harper, New York, 1961

REINGOLD, N. - *The Papers of Joseph Henry,* Vols. I-III, Smithsonian Institution Press, Washington, D.C., 1972
RIDDING, A. - *S.Z. de Ferranti,* Science Museum, London, 1964

ROLLER, D.H.D. - *Francis Hauksbee,* Scientific American, Vol. 189, No. 2, Aug. 1953
RUNES, D.D. - *A Treasury of World Science,* Philosophical Library, New York, 1961

SANTILLANA, G. de - *Alessandro Volta,* Scientific American, Vol. 212, No. 1, Jan. 1965
SCHAFF, W. C. - *Carl Friedrich Gauss, Prince of Mathematicians,* Franklin Watts, New York, 1964
SCHWARTZ, G., BISHOP, P.W. -*Moments of Discovery*, Basic Books, New York, 1962
SHAMOS, M.H. - *Great Experiments in Physics,* Henry Holt, New York, 1959
SHEIRS, G. - *On the Origins of Electronic Devices,* IEEE Spectrum, Nov. 1972
SIEMENS, W. von - *Inventor and Entrepreneur,* Lund Humphries, London, 1966
SILSBEE, F.B. - *Systems of Electrical Units*
National Bureau of Standards, Monograph 56, U.S. Government Printing Office, 1962
STEWART, I. - *Gauss,* Scientific American, Vol. 237, No. 1, July 1977

TATON, R. - *Science in the 19th Century,* Basic Books, New York, 1965
---- *History of Science,* Vols. I-IV, Basic Books, New York, 1964
TESLA, N. - *The Problem of Increasing Human Energy,* Century Magazine, June 1900
---- *Tribute to Nikola Tesla,* Nikola Tesla Museum, Belgrade, 1961
---- *Lectures, Patents, Articles,* Nikola Tesla Museum, Belgrade, 1961
---- *Tesla at 75,* Time, July 20, 1931, pp. 29, 30
---- *Preceedings of the Tesla Symposium,*
Sponsored by the IEEE Power Engineering Society, New York, Jan. 30, 1976
THOMPSON, S.P. - *The Life of William Thomson,* Macmillan, London, 1872
TRICKER, R.A.R. - *The Contributions of Faraday and Maxwell to Electrical Science,*
Pergamon Press, New York, 1966
---- *Early Electrodynamics and the First Law of Circulation,* Pergamon Press, New York, 1965

WELLING, J.C. - *Life and Character of Joseph Henry,* Riverside Press, New York, 1904
WHITTAKER, Sir Edmund - *A History of the Theories of Aether and Electricity*
Thomas Nelson, London, 1954
WHEWELL, W. - *History of the Inductive Sciences,* London, 1857
WIGHTMAN, W.P.D. - *The Growth of Scientific Ideas,* Yale University Press, New Haven, Conn., 1969
WILLIAMS, L.P. - *Humphry Davy,* Scientific American, Vol. 202, No. 6, June 1960
---- *Michael Faraday,* Chapman & Hall, London, 1965
WOLFF, A. - *A History of Science, Technology and Philosophy in the 18th Century*
Ruskin House, London, 1952

YOUNG, O.B. - *The Real Beginnings of Radio, the Neglected Story of Mahlon Loomis,*
Saturday Review, Mar. 7, 1964